Compost and Liquid Organic Fertilizers

Abu Zahrim Yaser · Mariani Rajin · Sariah Saalah
Editors

Compost and Liquid Organic Fertilizers

Production Techniques and Applications

Springer

Editors
Abu Zahrim Yaser
Faculty of Engineering
Universiti Malaysia Sabah
Kota Kinabalu, Sabah, Malaysia

Mariani Rajin
Faculty of Engineering
Universiti Malaysia Sabah
Kota Kinabalu, Sabah, Malaysia

Sariah Saalah
Faculty of Engineering
Universiti Malaysia Sabah
Kota Kinabalu, Sabah, Malaysia

ISBN 978-981-95-8560-1 ISBN 978-981-95-8561-8 (eBook)
https://doi.org/10.1007/978-981-95-8561-8

This Springer imprint is published by the registered company Springer Nature Singapore Pte Ltd.
The registered company address is: 152 Beach Road, #21-01/04 Gateway East, Singapore 189721, Singapore

To Naimmaton Sa'adiah Mustain.
These past 20 years have been the most incredible journey of my life. Our first honeymoon wasn't in a resort, but in a humble compost lab at UKM—and the adventure continued at Sabah, where we collected kitchen waste and held our breath when our backyard home composting failed, not to mention when we turned leachate into liquid fertilizer. You stood beside me through it all—supporting every step, even when I built compost bins for the community or when my students made a mess. In all the chaos, the smell, and the hard work, you were my calm, my strength, and my love. And until today, you still are. Thank you, and Alhamdulillah.

—Abu Zahrim Yaser

My heartfelt appreciation goes to my parents, Mailin Duwat and Rajin Alip; my husband, Mohd Norjefri Mat Midin; my children,

Khaulah Amsyar and Adam Amsyar; and my family. Your love, support and encouragement continue to lift me. Your presence makes every milestone meaningful.

—Mariani Rajin

To my beloved parents, Samira Lante and Saallah Ganing; my husband, Dahlan Pajakkai; my children; and my family. Thank you for your enduring love, unwavering support, and constant encouragement throughout this journey.

—Sariah Saalah

MyKompos team members engaging the local community during the MyKompos@Pantai Dalit knowledge transfer programme, held on 18 October 2025 at Kg. Berungis, Tuaran

Preface

The global landscape of agriculture, waste generation, and sustainability is undergoing rapid change. As nations work to increase food production while reducing environmental pressures, the demand for organic fertilizers, particularly compost and liquid organic fertilizers (LOF), has risen significantly. According to market projections by Future Market Insights (2025), the global organic fertilizer market, valued at USD 9.3 billion in 2025, is projected to reach USD 17.0 billion by 2035, driven by expanding agricultural activities in Asia/Pacific, Central and South America, and Eastern Europe. Compost continues to be widely applied to enhance soil structure, increase organic matter content, and stimulate biological activity, while LOF, derived from nutrient-rich organic feedstocks, has gained prominence for its readily available nitrogen, phosphorus, and potassium. Reflecting this rising interest, the global LOF market, valued at USD 2.5 billion in 2023, is forecasted to reach USD 5.8 billion by 2032 (DataIntelo, 2023). As an example, in the Junín region of Peru, demand projections show LOF consumption potentially reaching 48,754–69,819 litres between 2025 and 2029 (Bonilla et al., 2025). These growth trends are reinforced by green-transition agendas and agricultural policies in regions such as China and the European Union, which continue to drive innovation and investment in organic fertilizer technologies.

Parallel to the demand for organic fertilizers, global waste generation poses an urgent and escalating challenge. Recent analyses estimate that total waste reached approximately 20 billion tonnes in 2017, equivalent to 2.63 tonnes per person annually, and may rise to 46 billion tonnes by 2050 under business-as-usual scenarios, based on global assessments by Maalouf and Mavropoulos (2023). Municipal solid waste (MSW) accounted for about 2.7 billion tonnes in 2019, marking a 30–50% increase over the past 15 years, with projections ranging from 2.89 to 4.54 billion tonnes by 2050. Alarmingly, nearly one-third of MSW remains uncollected, and roughly 42% of collected waste is still disposed of through open dumping or uncontrolled burning (Maalouf and Mavropoulos, 2023). These realities underscore the urgency of adopting circular bioeconomy approaches that convert organic waste into valuable products such as compost and liquid organic fertilizer.

Sustainable waste management has therefore become central to global environmental and agricultural policy, requiring integrated systems that combine regulation, infrastructure, technological innovation, and community participation. Composting remains one of the most effective and accessible methods for converting biodegradable waste into nutrient-rich soil amendments, supporting soil health while reducing reliance on synthetic fertilizers. Likewise, liquid organic fertilizers offer an efficient and environmentally responsible nutrient source that enhances crop productivity and reduces emissions associated with chemical fertilizer manufacture. Both compost and LOF provide practical, scalable, and economically viable solutions aligned with international sustainability frameworks, particularly the Sustainable Development Goals (SDGs) related to responsible consumption and production, climate action, food security, and life on land.

The scientific contributions presented in this book were also informed by the shared momentum fostered during the 2nd Symposium on Anaerobic Digestion and Composting 2025 (ADIC 2025). This symposium convened researchers, practitioners, industry partners, and policymakers from Malaysia, Indonesia, India, and China, enabling rich exchanges of knowledge and strengthening regional networks. The discussions reinforced the importance of advancing organic waste valorisation and contributed insights, case studies, and collaborations reflected in several chapters.

This book, *Compost and Liquid Fertilizers: Production Techniques and Applications*, is both timely and foundational. It integrates current scientific knowledge, experimental findings, field applications, and community-driven practices to guide researchers, students, practitioners, and policymakers toward effective strategies for organic waste transformation. By positioning compost and LOF as essential components of sustainable agriculture and resilient food systems, this volume contributes to global efforts toward environmental stewardship, climate mitigation, and a food-secure bioeconomy. It is our hope that the insights presented herein will inspire continued innovation and support practical solutions aligned with SDG 2 (Zero Hunger), SDG 12 (Responsible Consumption and Production), and SDG 13 (Climate Action).

The Editors belong to the Advanced Composting and Waste Processing (MyKompos) Research Group at the Faculty of Engineering, Universiti Malaysia Sabah (UMS). MyKompos serves as a dedicated platform for interdisciplinary research on composting technologies, liquid organic fertilizer development, and resource recovery from waste. The group bridges laboratory research with field applications through strong engagement with communities, local authorities, and industry partners. This mission is exemplified by the establishment of Kundasang Compost Living Lab, which functions as a real-world testbed for participatory research, technology demonstration, and knowledge transfer, enabling scalable and locally grounded solutions for circular bioeconomy practices. Finally, we would like to thank Universiti Malaysia Sabah for awarding the Kundasang Compost Living Lab the status of Recognised Living Lab (RLL) on 15 April 2026.

Kota Kinabalu, Sabah, Malaysia

Abu Zahrim Yaser
Mariani Rajin
Sariah Saalah

References

Bonilla, D., Martinez, A. and Chávez, R., 2025. Transforming Trout Waste into Liquid Fertilizer Product to Reduce Environmental Pollution. In *E3S Web of Conferences* (Vol. 624, p. 03002). EDP Sciences.

DataIntelo (2023) *Global Liquid Organic Fertilizer Market—Industry Analysis, Size, Share, Growth and Forecast 2023–2032.* Available at: https://dataintelo.com/report/global-liquid-organic-fertilizer-market (Accessed on 3 December 2025).

Future Market Insights (2025) *Organic Fertilizer Market Report: Global Industry Analysis 2020–2035.* Available at: https://www.futuremarketinsights.com/reports/organic-fertilizer-market (Accessed on 3 December 2025).

Maalouf, A. and Mavropoulos, A., 2023. Re-assessing global municipal solid waste generation. *Waste Management and Research, 41*(4), pp. 936–947.

Acknowledgements

The editors gratefully acknowledge the following individuals for their time and efforts in assisting the editors with the reviewing of manuscript. This book would not have been possible without the commitment of the reviewers.

- Asiah Nusaibah Masri.
- Azreen Ibrahim.
- Chinnapong Wangnai.
- Intan Sorfina Zainal Abidin.
- Januarius Gobilik.
- Khalisanni Khalid.
- Mailin Misson.
- Muhammad Hazwan Hamzah.
- Rafida Razali.
- Raimi Mohamed Redwan.
- Sandeep Kumar Mishra.
- Sunarti Abdul Rahman.

Contents

Introduction 1
Mariani Rajin and Sariah Saalah

Sustainable Solutions for Urban Garden Waste: The Role of Inoculum-Based Composting 5
Chandrashekhar Parab, Kunwar D. Yadav, and Vimalkumar Prajapati

Sustainable Agriculture: Transforming Rumen Waste into Compost 35
Rizkiy Amaliyah Barakwan, Danar Arifka Rahman, and Yulinah Trihadiningrum

Composting Processes and Techniques in Urban Forestry Nursery Application 49
Andy Russel Mojiol

The Usage of Compost from Leucaena Leucocephala Leaves and Moringa Oleifera Leaves for the Growth of Kappaphycus Alvarezii Seedling 71
Arofah Lyla Nurhayati and Indrian Rizka Amalia

Compost-biochar: Increasing Productivity of Strategic Food Crops and Remediator of Endosulfan Residue 83
Elisabeth Srihayu Harsanti, Anicetus Wihardjaka, Asep Nugraha Ardiwinata, Asep Kurnia, and Turnad Lenggo Ginta

Vermicompost in Agriculture: Role, Benefits, and Techniques for Sustainable Farming 99
Nurul Mayzaitul Azwa Jamaludin, Borhan Abdul Haya, and Mohamad Asrol Kalam

Smart Farming: AI-Powered Mechanization for Sustainable Composting Applications 115
Turnad Lenggo Ginta, Tri Martini, Elisabeth Srihayu Harsanti, Muizuddin Azka, Hens Saputra, Agung Setyo Darmawan, and Nur Islahudin

Sustainable Production of Compost and Liquid Fertilizers: From Waste to Resource 137
Intan Suhada Azmi and Mohd Jumain Jalil

Sustainable Liquid Fertilizers: A Review of Production from Food Waste and Fish Waste 149
Junidah Lamaming and Dayang Norafizan Awang Chee

Application of Liquid Organic Fertilizer in Organic Farming 165
Suhana Mohamed Sultan and Norahim Ibrahim

Effect of Preserved Compost Tea Microbial Isolates on the Growth Performance of Mustard Plants 185
Husnul Azan Tajarudin, Ng Chi Yan, Charles Ng Wai Chun, Yusuf Wibisono, and Rimi Akter

Compost Leachate from Food and Vegetable Wastes as Alternative Fertilizers for Brazilian Spinach (*Alternanthera sissoo*) 199
Abu Zahrim Yaser, Nursyafieqah Basri, Nur Amirah Sakirah Barli, Nur Fadilah Ramli, Sariah Saalah, Ahmad Safwan Ismail, Azreen Ibrahim, Nurhazwani Yusoff Azudin, and Mariani Rajin

Global Research Trends and Technical Insights in Biogas Production from Organic Waste 213
Nadia Isa and Pramila Tamunaidu

Optimizing Food Waste Anaerobic Digestion: Substrate Composition, Total Solids, Bulking Agents and Digestate as a Source of Fertilizer 231
Boanita Susanto Putri, I. D. A. A. Warmadewanthi, and Talent Nia Pramestyawati

Enhancing Phosphorus Recovery from Septic Tank Sludge Through Anaerobic Digestion as Biofertilizer- Effect of pH and Temperature 247
Rafidah Selaman, Ajimi Jawan, and Dewi Farhana Nazra

Decoding Veterinary Antibiotic Behavior in Aerobic Composting and Anaerobic Digestion of Manure 255
Kiki Gustinasari, Ahamd Rizqi Alima Fabri, Rizkiy Amaliyah Barakwan, Inggit Kresna Maharsih, Nimas Mayang Sabrina Sunyoto, and Jaisy Aghniarahim Putritamara

A Review on the Isolation of Facultative Anaerobes from Gut Bacteria with Bioplastic Degradation Potential 279
Lilian Jose, Ikki Matsuda, Cahyo Budiman, and Vijay Kumar

Controlled-Release Nanofertilizers: A Paradigm Shift in Agriculture 293
Nur Syafiqah Farhanah Dzulkharnien, Rosiah Rohani, and Rasul Sauket Ali

Bioproduct-Derived Nanofertilizers for Soil Health and Sustainable Agriculture 311
Effa Aida Nordin, M. N. A. Uda, Z. A. Arsat, M. K. R. Hashim, Uda Hashim, M. N. Afnan Uda, Maisara A. M. Akhir, Yuri Pamungkas, Widyastuti, Subash C. B. Gopinath, M. Isa, and N. H. Ibrahim

Bioproducts as a Renewable Source for Nanofertilizer Development 337
Ahmad Syahmi Faqihin, M. N. A. Uda, Khalisanni Khalid, Jasni Ismail, Meor Ahmad Faris, Z. A. Arsat, M. K. R. Hashim, Uda Hashim, M. N. Afnan Uda, Maisara A. M. Akhir, Yuri Pamungkas, Widyastuti, F. Abdullah, Muhammad Firdaus Abdul Muttalib, and N. H. Ibrahim

Fertilizer Runoff Management—Strategies and Practices to Minimise Excess Nutrients in Water Reservoirs 363
Qaisara Izzaty Mohd Azni, Fatin Shafiqa Shazili, and Noor Aziah Serri

Metagenomic Assessment of Microbial Diversity at Five Sewage Treatment Plants (STP) Across Sabah, Malaysia and Impact on Water Quality Parameters 385
M. R. Abd Rahman, A. Raidin, P. C. Lee, K. N. Balakrishnan, H. Roslie, R. Razali, I. M. Maliki, and K. F. Rodrigues

Conclusion and Future Perspectives 397
Abu Zahrim Yaser

Editors and Contributors

About the Editors

Abu Zahrim Yaser is an associate professor in the Chemical Engineering Programme at Universiti Malaysia Sabah (UMS), where he also serves as the dean of the Faculty of Engineering. He is the founder of the Kundasang Compost Living Lab and obtained his Ph.D. from Swansea University, UK. Dr. Zahrim's research primarily focuses on waste processing technology, and he has been leading multiple projects since 2006. To date, he has published 18 books, 12 of which are indexed by Scopus, along with over 130 research publications. He has served as a lead guest editor for three special issues in high-impact journals. His creative publication, Kompos, featured in the Festival of Ideas (Ministry of Higher Education, Malaysia), has garnered more than 3000 views, and he holds one patent and five copyrights related to environmental protection systems. He has delivered keynote speeches and invited talks on waste management and academic publishing while also chairing several international and national conferences. As an active reviewer for several high-impact journals, Dr. Zahrim has reviewed over 280 articles (Publons). His innovations in waste management have been adopted by various communities in Sabah, Malaysia, including Nabawan and Kundasang. He has also served as a visiting scientist at the University of Hull (UK) and is a professional engineer (Ir.) registered with the Board of Engineers Malaysia. Dr. Zahrim is a recipient of the

National Book Award (2023). His Scopus citations stand at 1800, with an h-index of 21.

Mariani Rajin is a senior lecturer at Universiti Malaysia Sabah (UMS). She holds a Ph.D. in Chemical Engineering from Universiti Sains Malaysia, and her research interests focus on composting technologies, enzymatic reactions, and enzyme-based applications, particularly in sustainable waste management and resource recovery. She has authored two books, contributed to ten book chapters, and published over 30 journal articles. To date, her research output has achieved a Scopus h-index of 10 with 268 citations, and she is an active reviewer for several well-regarded international journals. She previously served as the head of the Chemical Engineering Programme at UMS and remains actively engaged in community-based composting and knowledge transfer initiatives. She has chaired multiple community-based composting and technology transfer programmes across Sabah and regularly delivers outreach programmes on composting science and enzyme-assisted composting. In addition, she participates in international conferences primarily as a presenter, and has also contributed as an invited speaker.

Sariah Saalah is a senior lecturer in the Chemical Engineering Program at Universiti Malaysia Sabah (UMS). She obtained her Ph.D. in Chemical Engineering from Universiti Putra Malaysia, specializing in biopolymers. Her research interests include composting and sustainable waste management, biopolymers, nanomaterials, and polyurethane technology. To date, Dr. Sariah has authored one book, ten book chapters, and 34 journal articles, with 38 publications indexed in Scopus. Her scholarly work has received 637 citations, with an h-index of 14. She has been a speaker and invited speaker at international conferences, particularly in the areas of sustainable polymeric materials and waste valorization, and has also chaired an international conference. In addition to her academic activities, she is actively engaged in local community outreach on sustainable waste management, with an emphasis on composting-based initiatives. Dr. Sariah currently serves as an editorial board member of the Journal of Sustainable Engineering (SUSTEN), handling the Environmental and Process Control section, and has also acted as a reviewer for reputable international journals.

Contributors

F. Abdullah Faculty of Mechanical Engineering & Technology, Universiti Malaysia Perlis, Arau, Perlis, Malaysia

M. N. Afnan Uda Faculty of Engineering, Universiti Malaysia Sabah, Sabah, Malaysia

Maisara A. M. Akhir Faculty of Chemical Engineering & Technology, Universiti Malaysia Perlis, Arau, Perlis, Malaysia

Rimi Akter Division of Bioprocess Engineering Technology, School of Industrial Technology, Universiti Sains Malaysia, USM, George Town, Pulau Pinang, Malaysia

Rasul Sauket Ali Department of Chemical and Process Engineering, Faculty of Engineering and Built Environment, Universiti Kebangsaan Malaysia, UKM Bangi, Selangor, Malaysia

Indrian Rizka Amalia Achmad Baiquni Science, Technology and Education Park, National Research and Innovation Agency (BRIN), Yogyakarta, Indonesia

Asep Nugraha Ardiwinata Research Center for Estate Crops, Agricultural and Food Research Organization, National Research and Innovation Agency, Cibinong, West Java, Indonesia

Z. A. Arsat Faculty of Mechanical Engineering & Technology, Universiti Malaysia Perlis, Kampus Tetap Pauh Putra, Arau, Perlis, Malaysia;
Institute of Nano Electronic Engineering, Universiti Malaysia Perlis, Kangar, Perlis, Malaysia

Muizuddin Azka Research Center for Process and Manufacturing Industry Technology, National Research and Innovation Agency, Tangerang Selatan, Indonesia

Intan Suhada Azmi Faculty of Chemical Engineering, Universiti Teknologi MARA, Masai, Johor, Malaysia

Qaisara Izzaty Mohd Azni Bioprocess Technology Division, School of Industrial Technology, Universiti Sains Malaysia, Pulau Pinang, Malaysia

Nurhazwani Yusoff Azudin Faculty of Engineering, Universiti Malaysia Sabah, Sabah, Malaysia

K. N. Balakrishnan Biotechnology Research Institute, Universiti Malaysia Sabah, Jalan UMS, Kota Kinabalu, Sabah, Malaysia

Rizkiy Amaliyah Barakwan Environmental Engineering Study Program, Department of Biology, Faculty of Science and Technology, Universitas Airlangga, Surabaya, Indonesia;
Sustainable Environmental Systems and Infrastructure (Sustain) Research Group, Universitas Airlangga, Surabaya, Indonesia

Nur Amirah Sakirah Barli Faculty of Engineering, Universiti Malaysia Sabah, Sabah, Malaysia

Nursyafieqah Basri Faculty of Engineering, Universiti Malaysia Sabah, Sabah, Malaysia

Cahyo Budiman Biotechnology Research Institute, Universiti Malaysia Sabah, Kota Kinabalu, Sabah, Malaysia

Dayang Norafizan Awang Chee Faculty of Resource Science and Technology, Universiti Malaysia Sarawak, Kota Samarahan, Sarawak, Malaysia

Charles Ng Wai Chun Division of Bioprocess Engineering Technology, School of Industrial Technology, Universiti Sains Malaysia, USM, George Town, Pulau Pinang, Malaysia

Agung Setyo Darmawan Department of Mechanical Engineering, Faculty of Engineering, Universitas Muhammadiyah Surakarta, Surakarta, Indonesia

Nur Syafiqah Farhanah Dzulkharnien Department of Chemical and Process Engineering, Faculty of Engineering and Built Environment, Universiti Kebangsaan Malaysia, UKM Bangi, Selangor, Malaysia

Ahamd Rizqi Alima Fabri Faculty of Mathematics and Natural Sciences, Universitas Brawijaya, Malang, Indonesia;
Universitas Islam Malang, Malang, Indonesia

Ahmad Syahmi Faqihin Faculty of Mechanical Engineering & Technology, Universiti Malaysia Perlis, Arau, Perlis, Malaysia

Meor Ahmad Faris Institute of Nano Electronic Engineering, Universiti Malaysia Perlis, Kangar, Perlis, Malaysia

Turnad Lenggo Ginta Research Center for Process and Manufacturing Industry Technology, National Research and Innovation Agency, Tangerang Selatan, Indonesia

Subash C. B. Gopinath Institute of Nano Electronic Engineering, Universiti Malaysia Perlis, Kangar, Perlis, Malaysia

Kiki Gustinasari Study Program of Environmental Engineering, Department of Biosystem Engineering, Faculty of Agricultural Engineering, Universitas Brawijaya, Malang, Indonesia

Elisabeth Srihayu Harsanti Cibinong Science Center, Research Center for Horticulture, National Research and Innovation Agency, West Java, Indonesia; Research Center for Horticulture, Agricultural and Food Research Organization, National Research and Innovation Agency, Cibinong, West Java, Indonesia

M. K. R. Hashim Faculty of Mechanical Engineering & Technology, Universiti Malaysia Perlis, Kampus Tetap Pauh Putra, Arau, Perlis, Malaysia

Uda Hashim Faculty of Engineering, Universiti Malaysia Sabah, Kota Kinabalu, Sabah, Malaysia

Borhan Abdul Haya Faculty of Sustainable Agriculture, Universiti Malaysia Sabah, Sandakan, Sabah, Malaysia

Azreen Ibrahim Faculty of Engineering, Universiti Malaysia Sabah, Sabah, Malaysia

N. H. Ibrahim Faculty of Chemical Engineering & Technology, Universiti Malaysia Perlis, Arau, Perlis, Malaysia;
Faculty of Engineering, Universiti Malaysia Sabah, Kota Kinabalu, Sabah, Malaysia

Norahim Ibrahim Department of Biosciences, Faculty of Science, Universiti Teknologi Malaysia, Johor Bahru, Johor, Malaysia

M. Isa Institute of Nano Electronic Engineering, Universiti Malaysia Perlis, Kangar, Perlis, Malaysia

Nadia Isa Chemical Process Engineering Department, Malaysia-Japan International Institute of Technology (MJIIT), Universiti Teknologi Malaysia (UTM), Kuala Lumpur, Malaysia;
Process Engineering Technology Department, Universiti Kuala Lumpur Malaysian Institute of Chemical and Bioengineering Technology (UniKL MICET), Melaka, Malaysia

Nur Islahudin Industrial Engineering Program at Dian Nuswantoro University, Semarang, Indonesia

Ahmad Safwan Ismail Faculty of Engineering, Universiti Malaysia Sabah, Sabah, Malaysia

Jasni Ismail Institute of Nano Electronic Engineering, Universiti Malaysia Perlis, Kangar, Perlis, Malaysia

Mohd Jumain Jalil Faculty of Chemical Engineering, Universiti Teknologi MARA, Masai, Johor, Malaysia

Nurul Mayzaitul Azwa Jamaludin Faculty of Sustainable Agriculture, Universiti Malaysia Sabah, Sandakan, Sabah, Malaysia

Ajimi Jawan Faculty of Applied Sciences, Universiti Teknologi MARA Sabah Branch, Kota Kinabalu, Sabah, Malaysia

Lilian Jose Biotechnology Research Institute, Universiti Malaysia Sabah, Kota Kinabalu, Sabah, Malaysia

Mohamad Asrol Kalam Faculty of Sustainable Agriculture, Universiti Malaysia Sabah, Sandakan, Sabah, Malaysia

Khalisanni Khalid Agri-Nanotechnologi Program, Bioteknologi and Nanotechnology Research Centre, MARDI, Serdang, Selangor, Malaysia

Vijay Kumar Biotechnology Research Institute, Universiti Malaysia Sabah, Kota Kinabalu, Sabah, Malaysia

Asep Kurnia Research Center for Horticulture, Agricultural and Food Research Organization, National Research and Innovation Agency, Cibinong, West Java, Indonesia

Junidah Lamaming Faculty of Resource Science and Technology, Universiti Malaysia Sarawak, Kota Samarahan, Sarawak, Malaysia

P. C. Lee Biotechnology Research Institute, Universiti Malaysia Sabah, Jalan UMS, Kota Kinabalu, Sabah, Malaysia;
Faculty of Science and Technology, Universiti Malaysia Sabah Jalan UMS, Sabah, Malaysia

Inggit Kresna Maharsih Department of Biosystems Engineering, Faculty of Agricultural Technology, Study Program of Bioprocess Engineering, Universitas Brawijaya, Malang, Indonesia

I. M. Maliki Biotechnology Research Institute, Universiti Malaysia Sabah, Jalan UMS, Kota Kinabalu, Sabah, Malaysia

Tri Martini Research Center for Sustainable Production System and Life Cycle Assessment, National Research and Innovation Agency, Tangerang Selatan, Indonesia

Ikki Matsuda Wildlife Research Center, Kyoto University, Kyoto, Japan;
Chubu Institute for Advanced Studies, Chubu University, Kasugai, Aichi, Japan;
Institute for Tropical Biology and Conservation, Universiti Malaysia Sabah, Kota Kinabalu, Sabah, Malaysia

Andy Russel Mojiol Faculty of Tropical Forestry, Universiti Malaysia Sabah, Kota Kinabalu, Malaysia

Muhammad Firdaus Abdul Muttalib Faculty of Mechanical Engineering & Technology, Universiti Malaysia Perlis, Arau, Perlis, Malaysia

Dewi Farhana Nazra Faculty of Applied Sciences, Universiti Teknologi MARA Sabah Branch, Kota Kinabalu, Sabah, Malaysia

Effa Aida Nordin Faculty of Mechanical Engineering & Technology, Universiti Malaysia Perlis, Kampus Tetap Pauh Putra, Arau, Perlis, Malaysia

Arofah Lyla Nurhayati Fisheries Research Center, NationalResearchandInnovationAgency (BRIN), KST Soekarno, Bogor, West Java, Indonesia

Yuri Pamungkas Faculty of Medicine & Health, Institut Teknologi Sepuluh Nopember, Surabaya, Indonesia

Yuri Pamungkas Faculty of Medicine and Health, Institut Teknologi Sepuluh Nopember, Surabaya, Indonesia

Chandrashekhar Parab Department of Civil Engineering, Sardar Vallabhbhai National Institute of Technology, Surat, Gujarat, India

Vimalkumar Prajapati Division of Microbial and Environmental Biotechnology, Aspee Shakilam Biotechnology Institute, Navsari Agricultural University, Surat, Gujarat, India

Talent Nia Pramestyawati Department of Environmental Engineering, Institut Teknologi Sepuluh Nopember (ITS), Surabaya, Indonesia

Boanita Susanto Putri Department of Environmental Engineering, Institut Teknologi Sepuluh Nopember (ITS), Surabaya, Indonesia

Jaisy Aghniarahim Putritamara Department of Livestock Socio-Economics, Faculty of Animal Science, Universitas Brawijaya, Malang, Indonesia

Danar Arifka Rahman Environmental Engineering Study Program, Department of Biology, Faculty of Science and Technology, Universitas Airlangga, Surabaya, Indonesia;
Sustainable Environmental Systems and Infrastructure (Sustain) Research Group, Universitas Airlangga, Surabaya, Indonesia

M. R. Abd Rahman Jabatan Perkhidmatan Pembetungan Sabah, Kota Kinabalu, Sabah, Malaysia

A. Raidin Jabatan Perkhidmatan Pembetungan Sabah, Kota Kinabalu, Sabah, Malaysia

Mariani Rajin Faculty of Engineering, Universiti Malaysia Sabah, Kota Kinabalu, Sabah, Malaysia

Nur Fadilah Ramli Faculty of Engineering, Universiti Malaysia Sabah, Sabah, Malaysia

R. Razali Biotechnology Research Institute, Universiti Malaysia Sabah, Jalan UMS, Kota Kinabalu, Sabah, Malaysia

K. F. Rodrigues Biotechnology Research Institute, Universiti Malaysia Sabah, Jalan UMS, Kota Kinabalu, Sabah, Malaysia

Rosiah Rohani Department of Chemical and Process Engineering, Faculty of Engineering and Built Environment, Universiti Kebangsaan Malaysia, UKM Bangi, Selangor, Malaysia;
Research Centre for Sustainable Process Technology, Faculty of Engineering and Built Environment, Universiti Kebangsaan Malaysia, UKM Bangi, Selangor, Malaysia

H. Roslie Biotechnology Research Institute, Universiti Malaysia Sabah, Jalan UMS, Kota Kinabalu, Sabah, Malaysia

Sariah Saalah Faculty of Engineering, Universiti Malaysia Sabah, Kota Kinabalu, Sabah, Malaysia

Hens Saputra Research Center for Process and Manufacturing Industry Technology, National Research and Innovation Agency, Tangerang Selatan, Indonesia

Rafidah Selaman Faculty of Applied Sciences, Universiti Teknologi MARA Sabah Branch, Kota Kinabalu, Sabah, Malaysia

Noor Aziah Serri Bioprocess Technology Division, School of Industrial Technology, Universiti Sains Malaysia, Pulau Pinang, Malaysia

Fatin Shafiqa Shazili Bioprocess Technology Division, School of Industrial Technology, Universiti Sains Malaysia, Pulau Pinang, Malaysia

Suhana Mohamed Sultan Department of Electronic and Computer Engineering, Faculty of Electrical Engineering, Universiti Teknologi Malaysia, Johor Bahru, Johor, Malaysia

Nimas Mayang Sabrina Sunyoto Department of Agro-Industrial Technology, Faculty of Agricultural Technology, Universitas Brawijaya, Malang, Indonesia

Husnul Azan Tajarudin Division of Bioprocess Engineering Technology, School of Industrial Technology, Universiti Sains Malaysia, USM, George Town, Pulau Pinang, Malaysia

Pramila Tamunaidu Chemical Process Engineering Department, Malaysia-Japan International Institute of Technology (MJIIT), Universiti Teknologi Malaysia (UTM), Kuala Lumpur, Malaysia

Yulinah Trihadiningrum Department of Environmental Engineering, Faculty of Civil Planning and Geo Engineering, Institut Teknologi Sepuluh Nopember, Kampus ITS Sukolilo, Surabaya, Indonesia

M. N. A. Uda Faculty of Mechanical Engineering & Technology, Universiti Malaysia Perlis, Kampus Tetap Pauh Putra, Arau, Perlis, Malaysia;
Institute of Nano Electronic Engineering, Universiti Malaysia Perlis, Kangar, Perlis, Malaysia;
Centre of Excellence for Biomass Utilization, Universiti Malaysia Perlis, Arau, Perlis, Malaysia

I. D. A. A. Warmadewanthi Department of Environmental Engineering, Institut Teknologi Sepuluh Nopember (ITS), Surabaya, Indonesia

Yusuf Wibisono Program Studi Teknologi Bioproses, Fakultas Teknologi Pertanian, Universitas Brawijaya, Malang, Indonesia

Widyastuti Department of Material and Metallurgical Engineering, Institut Teknologi Sepuluh Nopember, Surabaya, Indonesia

Anicetus Wihardjaka Research Center for Food Crops, Agricultural and Food Research Organization, National Research and Innovation Agency, Cibinong, West Java, Indonesia

Kunwar D. Yadav Department of Civil Engineering, Sardar Vallabhbhai National Institute of Technology, Surat, Gujarat, India

Ng Chi Yan Division of Bioprocess Engineering Technology, School of Industrial Technology, Universiti Sains Malaysia, USM, George Town, Pulau Pinang, Malaysia

Abu Zahrim Yaser Faculty of Engineering, Universiti Malaysia Sabah, Jalan UMS, Sabah, Malaysia

Introduction

Mariani Rajin and Sariah Saalah

The global transition toward more sustainable agricultural systems has intensified interest in organic fertilizers and waste-derived nutrient sources. Evidence of soil degradation, nutrient imbalance and increased greenhouse gas emissions associated with synthetic fertilizer use has raised growing concern, signalling the urgent need for alternatives that are both productive and ecologically balanced. In response, researchers, practitioners and policymakers are exploring pathways that restore soil function while supporting reliable food production.

This book, *Compost and Liquid Fertilizers: Production Techniques and Applications*, brings together recent scientific advances, innovative application strategies, and emerging technologies that illustrate how organic residues can be transformed into high value agricultural inputs. The chapters collectively reflect the evolution of organic fertilizer research, from established composting systems to contemporary biotechnological, microbiome driven, and nanomaterial assisted approaches, providing an integrated view of how compost and liquid fertilizers are being reengineered to meet modern sustainability, productivity, and circular economy demands.

Composting continues to serve as one of the most practical and accessible means of converting organic residues into stable, nutrient-rich amendments that improve soil health and plant productivity. Its performance depends on feedstock characteristics, microbial processes and optimized management practices, all of which influence the quality and agronomic value of the end product. Chapters 2–8 advance current understanding of composting through studies on feedstock diversity, biological transformation pathways, application outcomes, and emerging technological approaches. Chapters 2 and 3 explore substrates such as urban garden waste and rumen waste, highlighting their nutrient contributions and the microbial dynamics that drive decomposition. These studies show how the combination of lignocellulosic

M. Rajin (✉) · S. Saalah
Faculty of Engineering, Universiti Malaysia Sabah, Kota Kinabalu, Sabah, Malaysia
e-mail: mariani@ums.edu.my

A. Z. Yaser et al. (eds.), *Compost and Liquid Organic Fertilizers*,
https://doi.org/10.1007/978-981-95-8561-8_1

materials, indigenous microorganisms and controlled composting phases contributes to high-quality compost output. Chapters 4 and 5 then illustrate compost use in diverse agricultural applications, including urban forestry nurseries and seaweed seedling development, demonstrating compost's versatility across both land-based and marine systems. Subsequently, Chapters 6–8 introduce innovation-oriented approaches, such as compost–biochar integration for pesticide remediation, vermicomposting for improved nutrient retention and AI-enabled systems that refine process control. Taken collectively, these contributions show how composting has transformed from a purely biological practice into a more integrated, technology-supported environmental solution.

As interest grows in extracting greater value from organic wastes, Chapters 9 and 10 introduce integrated concepts that convert a range of waste streams into both compost and liquid nutrient formulations. Chapter 9 outlines the biochemical principles underlying the transformation of agricultural and municipal organic waste into compost and liquid fertilizers, emphasizing microbial interactions, decomposition pathways and stabilization mechanisms. Chapter 10 examines the potential of food and fish waste as feedstocks for producing sustainable liquid fertilizers, focusing on their nutrient composition and conversion processes. It also discusses the environmental and economic benefits of waste-derived fertilizers, along with key challenges that influence their practical adoption. These chapters together illustrate how composting and liquid fertilizer production share core principles of microbial transformation and resource recovery, reflecting a broader shift toward circular management of organic materials.

Building on these foundations, Chapters 11–13 turn to the preparation, stabilization and use of liquid organic fertilizers (LOF) in agricultural contexts. Chapter 11 outlines the preparation of fermented plant juices, fermented fruit juices and fish amino acids, noting advantages related to nutrient availability and farmer adoption. Chapters 12 and 13 present experimental findings on preserved compost-tea microorganisms and compost leachates from food and vegetable wastes, demonstrating how microbial stabilization and dilution strategies expand the utility of liquid fertilizers across different crop systems. Collectively, these chapters position LOF as an adaptable extension of composting science that supports practical, field-level nutrient management.

Following the discussion on liquid fertilizers, the book extends its scope to biological processes capable of generating both renewable energy and fertilizer products. Anaerobic digestion, discussed in Chapters 14–17, serves a dual function by converting organic waste into biogas while simultaneously generating nutrient-rich digestate. Chapter 14 provides a comprehensive overview of current developments, challenges and global perspectives in anaerobic digestion, reflecting its growing relevance within integrated waste management and renewable energy initiatives. Chapter 15 evaluates how substrate composition, total solids and bulking agents affect methane yield in food waste anaerobic digestion, showing that optimizing these variables can also enhance the nutrient quality of digestate for agricultural use. Chapter 16 explores phosphorus recovery from septic tank sludge, demonstrating how operational conditions such as temperature and pH shape recovery efficiency and influence the availability of phosphorus as a fertilizer input. Chapter 17 evaluates microbial innovations that enhance anaerobic degradation, offering pathways to improve digestate stability and fertilizer performance. Chapter 18 complements this focus by reviewing gut-derived facultative anaerobes as potential agents for bioplastic degradation under low-oxygen conditions, highlighting their metabolic flexibility while emphasizing the need for biosafety evaluation prior to integration into engineered anaerobic waste management systems. Considered together, these chapters present anaerobic digestion as both a renewable energy solution and a practical pathway for sustainable nutrient recovery through the productive return of digestate to soils.

Recognizing nutrient recovery and efficiency as continuing priorities, the book also addresses technological innovations designed to improve fertilizer precision and performance. Chapter 19 introduces controlled-release nanofertilizers (CRNFs) as a more sustainable alternative to conventional fertilizers, using nanoparticles and polymer-based hydrogels to deliver nutrients slowly and precisely, improving plant uptake while reducing environmental losses. Chapters 20 and 21 continue this trajectory by exploring broader nanotechnology applications within organic fertilizer systems. Chapter 20 evaluates bioproduct-derived nanofertilizers, noting their potential to improve nutrient use efficiency, plant growth and soil health while acknowledging risks such as phytotoxicity and disruption to beneficial soil microorganisms. Chapter 21 explores renewable feedstocks such as agricultural residues, microbial biomass and plant-based polymers for producing bio-based nanofertilizers, demonstrating that green synthesis can enable controlled nutrient release and reduced environmental impact, while acknowledging challenges in scalability and regulation.

The final section addresses governance frameworks and emerging nutrient sources that support sustainable fertilizer use. Chapter 22 examines fertilizer runoff management by explaining nutrient dynamics, the causes of excess discharge, and the role of compost and liquid organic fertilizers in protecting soil and water quality, while emphasizing best management practices, regulatory frameworks, monitoring systems, and predictive technologies that enable environmentally responsible nutrient application. Chapter 23 demonstrates how sewage and treated effluents serve as underutilized nutrient resources that can contribute to fertilizer production when supported by appropriate safety standards and monitoring, highlighting the importance of microbial assessment in understanding effluent characteristics and ensuring effective treatment. Together, these chapters underscore the importance of aligning governance, technology and resource recovery, reinforcing the book's overarching theme of closing nutrient loops within a circular bioeconomy.

In sum, this volume presents an integrated view of how organic wastes may be transformed into high-value agricultural inputs through composting, fermentation, anaerobic digestion and nanotechnology-enabled processes. The chapters collectively demonstrate that sustainable nutrient management is both a technical capacity and an ecological requirement for resilient food systems. As agriculture responds to the combined pressures of climate change, resource scarcity and population growth, the knowledge presented here supports informed decision-making and long-term strategic planning for sustainable fertilizer development. This volume serves as both a reference and a call to action, inviting researchers, practitioners and communities to rethink waste, reclaim nutrients and redesign agriculture for generations to come.

Sustainable Solutions for Urban Garden Waste: The Role of Inoculum-Based Composting

Chandrashekhar Parab, Kunwar D. Yadav, and Vimalkumar Prajapati

Abstract The rapid expansion of urban populations and green spaces has intensified challenges in managing municipal solid waste (MSW), particularly garden waste (GW), which is difficult to process due to its fibrous lignocellulosic structure and low nutrient content. Conventional disposal methods like landfilling and incineration are resource-intensive and contribute to environmental degradation through greenhouse gas emissions. Composting, especially when supplemented with microbial inoculants, offers a sustainable and efficient alternative. These inoculants—comprising bacteria, fungi, and actinomycetes—accelerate the breakdown of organic matter, enhance humus formation, and produce compost enriched with plant-available nutrients. The resulting product not only improves soil health but also helps manage increasing GW volumes in space-constrained urban settings. While practical implementation faces challenges such as limited public awareness, technological constraints, and policy gaps, the growing emphasis on circular economy principles and advances in microbial technologies offer significant potential for scaling. Inoculum-based composting thus represents a promising and resilient approach to urban GW management.

Keywords Cellulase · CMC · Consortia · Effective inoculum · Lignocellulose

1 Introduction

As per the United Nations Department of Economic and Social Affairs (2022), the global population has surpassed 7.8 billion and is projected to climb to approximately 9.7 billion by 2050. This population surge has driven rapid urbanization, with people

C. Parab (✉) · K. D. Yadav
Department of Civil Engineering, Sardar Vallabhbhai National Institute of Technology, Surat, Gujarat, India
e-mail: chandrashekharparab94@gmail.com

V. Prajapati
Division of Microbial and Environmental Biotechnology, Aspee Shakilam Biotechnology Institute, Navsari Agricultural University, Surat, Gujarat, India

A. Z. Yaser et al. (eds.), *Compost and Liquid Organic Fertilizers*,
https://doi.org/10.1007/978-981-95-8561-8_2

flocking to cities for better opportunities and quality of life (Humbal et al., 2023). Developing countries, especially India, have felt the brunt of this trend. India now stands as the world's most populous nation, with a population exceeding 1.4 billion (Hussain et al., 2024). One major consequence of this population growth and urbanization is the escalating volume of municipal solid waste (MSW). In 2016, global MSW generation stood at 2.01 billion tonnes, and it is anticipated to soar to around 3.4 billion tonnes by 2050 (Kaza et al., 2018). Organic waste constitutes the largest portion of MSW globally, making up roughly 44–50% of the total waste generated (Chen et al., 2020; Liu et al., 2018).

As urban expansion continues and environmental concerns intensify, the role of green spaces—such as parks, gardens, and public recreational areas—has become increasingly significant in mitigating the adverse effects of urbanization. Often described as the "lungs of urban landscapes," these areas contribute to improved air quality by sequestering carbon dioxide and releasing oxygen, thus improving urban air quality. Additionally, they help alleviate the urban heat island effect, thereby enhancing the livability of cities (Amani-Beni & Khalilinezhad, 2023). However, the growth of such green infrastructure also brings with it the challenge of increased GW generation. This type of waste encompasses a diverse range of organic matter, including tree trimmings, bark, twigs, fallen leaves, grass cuttings, and soil. A substantial portion originates from municipal gardens, public parks, institutional campuses, office landscapes, and research facilities. Studies indicate that GW accounts for approximately 10–20% of MSW in many developed countries, with global annual generation exceeding 100 million tonnes (Inghels et al., 2016; Reyes-Torres et al., 2018). The effective handling of GW has become increasingly important for sustainable urban governance, especially in developing areas where the rate of urban growth often surpasses the capacity of current waste management infrastructure.

A significant challenge in the effective management of GW stems from its heterogeneous and structurally resistant composition. Up to 75% of the dry matter in GW comprises lignocellulosic compounds—including cellulose, hemicellulose, and lignin—which are biopolymers characterized by their recalcitrant structure and resistance to microbial degradation (Parab & Yadav, 2024). This biochemical complexity often results in prolonged composting durations and suboptimal nutrient recovery during conventional treatment processes (Nair et al., 2022). The low levels of macronutrients in GW—typically 0.5–1.5% nitrogen, 0.1–0.2% phosphorus, and 0.4–0.8% potassium—further limit its effectiveness as a direct soil amendment (Reyes-Torres et al., 2018). Consequently, without further processing or nutrient supplementation, the agronomic value of untreated GW remains limited.

The composting process involves the biological degradation of organic matter, resulting in the formation of a stable and humified material suitable for soil application. Despite the high proportion of biodegradable material in municipal solid waste and the growing strain on waste management systems, efficient composting practices remain underutilized in many regions. Alarmingly, approximately 33% of global waste is still disposed of in open dumps, whereas only about 5.5% undergoes composting (Fig. 1). In the context of GW, composting faces several operational challenges. The process demands substantial land area, significant labour input, and extended timeframes. Owing to the recalcitrant lignocellulosic nature of GW, the composting cycle can span between 3 and 6 months. Such extended durations in composting not only reduce operational efficiency but also increase the risk of nutrient degradation, leading to lower compost quality (Waqas et al., 2023; Khalil et al., 2008; Neugebauer & Solowiej, 2017).

To address the limitations of conventional GW composting, the application of microbial inoculants has gained significant attention as an innovative and effective strategy. These inoculants, comprising selected bacterial strains or microbial consortia, facilitate more efficient composting by accelerating the degradation of resistant lignocellulosic materials. Their activity accelerates the composting process while simultaneously improving the nutrient profile and overall quality of the final compost (Oviedo-Ocaña et al., 2022a). Studies have demonstrated that introducing lignocellulolytic microbes—such as *Bacillus*, *Streptomyces*, and *Pseudomonas*—can increase the breakdown rate of cellulose and hemicellulose by 15–45%. Their ability

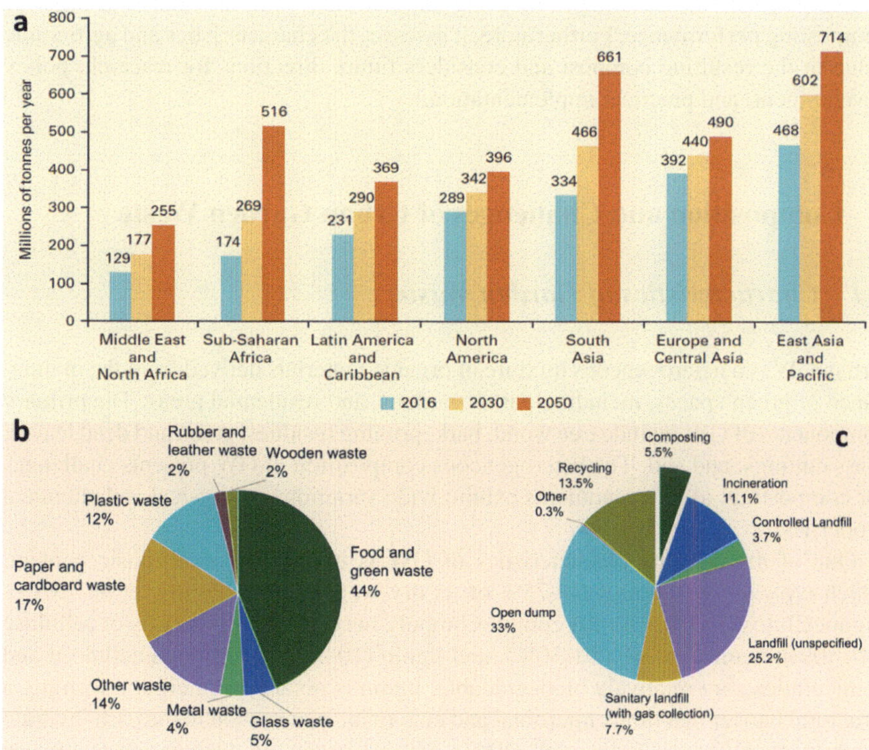

Fig. 1 Solid waste generation (**a**), composition of solid waste (**b**), and disposal options used globally for solid waste (**c**) (*Source* World Bank. What a waste 2.0)

to decompose fibrous plant residues is complemented by their role in promoting humic substance formation, thereby supporting soil fertility and ecological function (Chukwuma et al., 2021; Jurado et al., 2014; Shinde et al., 2022). Furthermore, the use of microbial consortia supports a more robust thermophilic phase—a critical period during which elevated temperatures aid in pathogen reduction and improve the safety of the composted product (Liu et al., 2018).

Although microbial inoculants offer significant potential for enhancing the composting of GW, their large-scale application remains relatively nascent. Several critical research questions must be resolved to support their broader adoption. These include determining the appropriate dosages of inoculants, elucidating microbial interactions within diverse composting environments, and developing cost-effective, scalable solutions tailored to the needs of urban waste management systems. Furthermore, the integration of microbial inoculants with complementary strategies—such as the addition of bulking agents or nutrient supplements—can significantly enhance composting efficiency and improve the quality of the resulting compost. This chapter examines inoculum-assisted composting as an effective and sustainable approach for the management of urban GW. It examines the inherent challenges in composting GW, investigates the microbial processes that underpin compost transformation, and emphasizes the advantages of using targeted microbial inoculants to improve composting performance. Furthermore, it assesses the characteristics and agronomic value of the resulting compost and considers future directions for research, policy development, and practical implementation.

2 Composition and Challenges of Urban Garden Waste

2.1 Characteristics of Garden Waste

Urban GW is a heterogeneous mixture of organic materials derived from the maintenance of green spaces, including parks, gardens, and residential areas. The primary components of GW include tree wood, bark, pruning residues, green and dead leaves, grass cuttings, and soil. The heterogeneous composition of GW presents challenges for composting, as its constituents exhibit wide variations in physical and chemical properties.

One of the defining characteristics of GW is its high lignocellulosic content, which typically constitutes 60–75% of its dry weight (Muniyasamy et al., 2013). Lignocellulose is a structurally complex biopolymer consisting primarily of cellulose (30–40%), hemicellulose (20–30%), and lignin (10–30%). Although cellulose and hemicellulose are relatively biodegradable, lignin is notably recalcitrant, forming a structural barrier that limits microbial and enzymatic penetration to the carbohydrate components (Vargas-García et al., 2007). This resistance to degradation is a major factor contributing to the slow composting rates of GW.

In addition to its recalcitrant lignocellulosic structure, GW typically contains low concentrations of essential nutrients such as nitrogen, phosphorus, and potassium. The typical nutrient composition of GW includes nitrogen ranging from 0.5 to 1.5% of its dry weight, phosphorus between 0.1 and 0.2%, and potassium varying from 0.4 to 0.8%. These low nutrient levels can limit its direct use as a sole nutrient source in biological processes, often necessitating supplementation to support microbial activity and decomposition. Such nutrient deficiencies can hinder microbial dynamics during composting, as the biodegradation process relies on maintaining a favorable carbon-to-nitrogen ratio, generally around 25–30:1. Nitrogen immobilization is a common consequence of GW's high C/N ratio (typically >50:1), resulting in delayed decomposition and inferior compost characteristics. (Reyes-Torres et al., 2018). To better illustrate the composition of GW, Table 1 provides a summary of its typical components, lignocellulosic content, and nutrient levels.

The heterogeneous nature of GW, combined with its high lignocellulosic content and low nutrient levels, presents significant challenges for composting. These challenges include prolonged processing times, reduced microbial activity, and the need for additional nutrient supplementation to achieve optimal composting conditions. In the following sections, we will explore these challenges in greater detail and discuss strategies to overcome them, with a particular focus on the role of microbial inoculants.

Table 1 Typical composition of urban garden waste

Component	Percentage of dry weight (%)	Lignocellulosic content	Nutrient content
Tree wood	20–30	High (Cellulose, Lignin)	Low N, P, K
Bark	10–20	High (Lignin)	Low N, P, K
Pruning residues	15–25	High (Cellulose, Hemicellulose)	Moderate N, Low P, K
Dead leaves	10–20	Moderate (Cellulose, Lignin)	Low N, P, K
Green leaves	5–15	Moderate (Cellulose)	Moderate N, Low P, K
Grass clippings	10–20	Low (Cellulose)	High N, Moderate P, K
Soil	5–10	Negligible	Variable (Depends on soil type)

Source Adapted from (Reyes-Torres et al., 2018) and (Muniyasamy et al., 2013)

2.2 Environmental and Social Impacts

Urban GW, when improperly managed, poses significant environmental and social challenges. Landfilling of organic waste contributes significantly to greenhouse gas emissions, primarily through the release of methane—a highly potent gas produced during anaerobic decomposition (Lou & Nair, 2009; Kiehbadroudinezhad et al., 2024). This contributes climate change and wastes valuable organic material that could otherwise enrich soil health. Additionally, the transportation of GW to landfills increases carbon footprints and strains municipal waste management systems. On a social level, inefficient waste management can lead to community dissatisfaction, health hazards from pests and odors, and missed opportunities for community engagement in sustainable practices. Conversely, composting GW offers environmental benefits, such as reducing landfill dependency, sequestering carbon in soil, and improving soil fertility (Ansar et al., 2025; Ayilara et al., 2020). Socially, community composting initiatives can foster environmental awareness, create green jobs, and strengthen community bonds by involving residents in shared sustainability goals (McSweeney, 2019; Roseland, 2000). Thus, addressing urban GW through sustainable methods like composting remains crucial for mitigating environmental impacts and enhancing social well-being.

2.3 Current Waste Management Practices

Current urban GW management systems rely on a mix of traditional methods, including landfilling, incineration, and conventional composting, each with its own set of advantages and limitations (Hutton et al., 2009). Despite its environmental drawbacks, landfilling remains the most prevalent waste disposal method due to its cost-effectiveness and ease of implementation. However, it is associated with significant environmental issues, such as greenhouse gas emissions, leachate production, and inefficient land use (EPA, 2023; Ghosh et al., 2023). Incineration offers advantages in waste volume reduction and energy generation; however, the release of toxic byproducts, including dioxins and particulate pollutants, raises significant environmental and health concerns (World Bank, 2022). Traditional composting, though more sustainable, often faces challenges in urban settings, such as slow decomposition rates, odor issues, and the need for significant space and labor. Additionally, many urban areas lack the infrastructure and public awareness to implement composting effectively. These limitations highlight the need for innovative solutions, such as inoculum-based composting, to address the inefficiencies of current practices and align with sustainability goals.

3 Inoculum-Based Composting: An Overview

Inoculum-based composting represents a sophisticated approach to composting that entails the deliberate introduction of specialized microbial inoculants into organic waste. This technique is designed to expedite the decomposition process and elevate the quality of the resulting compost. Unlike conventional composting methods, which depend on the spontaneous activity of naturally present microorganisms, inoculum-based composting incorporates a concentrated blend of advantageous microbes, including bacteria, fungi, and actinomycetes. These microbes are specifically chosen or engineered to flourish under composting conditions, thereby facilitating the efficient breakdown of intricate organic compounds such as cellulose, lignin, and proteins (Greff et al., 2022). The key difference between inoculum-based composting and traditional methods lies in the speed and efficiency of the process. Traditional composting can take several months to produce mature compost, whereas inoculum-based systems can significantly reduce this time by optimizing microbial activity (Karnchanawong & Nissaikla, 2014). Additionally, microbial inoculants help maintain optimal conditions for decomposition, such as temperature, moisture, and aeration, while suppressing pathogens and reducing foul odors. This technique yields a superior compost that is abundant in nutrients and beneficial microorganisms, thereby promoting soil health and plant growth. By harnessing the capabilities of microbial inoculants, inoculum-based composting presents a more sustainable and efficient strategy for dealing with urban GW. The microbial inoculants employed in composting are varied, with each type performing a distinct function in the decomposition process. These inoculants can generally be classified into bacteria, fungi, and actinomycetes, each of which plays a role in different phases of composting and the degradation of particular organic substances.

3.1 Bacteria

Bacteria are the primary decomposers in composting, making up 80–90% of microorganisms in the pile. The process begins with mesophilic bacteria working at moderate temperatures (68–113°F), breaking down simple compounds like sugars and proteins. They generate heat through metabolic activity, raising the temperature of the compost pile and facilitating thermophilic decomposition. (Finore et al., 2023; Soong et al., 2025). Certain bacteria, including *Bacillus subtilis, Pseudomonas fluorescens,* and various Streptomyces species, are renowned for their capacity to break down cellulose through cellulolytic activities. *Bacillus subtilis,* in particular, has garnered significant research attention due to its ability to produce and secrete cellulase enzymes. A notable study uncovered a cold-adapted GH5-CBM3 endo-cellulase from *Bacillus subtilis 1AJ3,* which demonstrated remarkable efficiency in converting switchgrass and coffee grounds into simpler sugars (Zeng et al., 2024). While *Pseudomonas fluorescens* also exhibits cellulolytic capabilities, detailed investigations into its cellulase secretion are relatively sparse in scientific literature. In contrast, Streptomyces species, such as *Streptomyces sp. MS2A*, have been well-documented for their robust cellulolytic enzyme systems. These systems encompass endo-β-1, 4 glucanase, exoglucanase, and β glucosidases, all of which play essential roles in the degradation of complex lignocellulosic materials (Ashwini John & Selvarajan, 2023). These bacterial strains are not only effective in degrading cellulose but also have potential applications in biofuel production and waste management.

3.2 Fungi

Fungi excel at breaking down intricate organic compounds like lignin, cellulose, and hemicellulose, which are prevalent in green lignocellulosic waste. They release a range of extracellular enzymes, such as cellulases, hemicellulases, and ligninases, that dismantle these resilient materials into simpler substances that can be more easily decomposed by other microbes (Feng et al., 2024). For example, the fungus *Clonostachys rogersoniana* has been found to markedly boost the degradation rates of cellulose, hemicellulose, and lignin when used in conjunction with other microorganisms in a three-strain microbial (Costa et al., 2023; Gomes et al., 2020). Saprophytic fungi are primarily responsible for decomposing dead organic matter. They obtain nutrients from plant debris and contribute significantly in recycling nutrients back into ecosystems. Examples include *Trichoderma spp.*, known for their aggressive colonization abilities and disease suppression characteristics, and *Pleurotus ostreatus* (Oyster Mushroom), which thrives on agricultural waste and contributes significantly to decomposition (Kredics et al., 2022). Molds such as *Aspergillus* and *Penicillium* are commonly found in compost piles. They help decompose organic matter by secreting enzymes that break down complex carbohydrates (Wright et al., 2016). This collaborative method not only speeds up the composting process but also enhances the quality of the resulting compost. Moreover, fungi play a vital role in forming humic acid and stabilizing organic matter during the thermophilic phase of composting, which is essential for producing high-quality compost (Zhu et al., 2024).

3.3 Actinomycetes

Actinomycetes play a vital role in the degradation and composting of green lignocellulosic waste due to the production of extracellular enzymes like cellulases, chitinases, and proteases (Savitha et al., 2022). A well-known genus, *Streptomyces*, is particularly effective in degrading lignocellulosic materials (Wei et al., 2019). Beyond decomposition, actinomycetes are crucial for nutrient cycling, as they convert complex organic compounds into simpler, plant-available forms and participate in nitrogen fixation and phosphorus solubilization (Bhatti et al., 2017). Actinomycetes also play a significant role in humus formation during the compost curing phase, contributing to the development of stable organic matter that enhances soil structure, aeration, and water retention (Zhao et al., 2017). Typically appearing as white, filamentous structures in compost, actinomycetes thrive in aerobic conditions, especially in environments with a neutral to slightly alkaline pH. Their thermophilic nature enables them to function effectively during the high-temperature phase of composting. In practical applications, the addition of specific actinomycetes strains to compost has been shown to accelerate the breakdown of cellulose, hemicellulose, and lignin, thereby improving compost quality (Wei et al., 2019). Environmentally, these organisms support sustainable agriculture by enhancing soil health and nutrient availability while suppressing plant pathogens. Overall, actinomycetes are essential to the composting process due to their multifunctional roles in organic matter degradation, nutrient transformation, and soil enrichment.

4 Scientific Basis of Inoculum-Based Composting

Inoculum-based composting relies on the synergistic activity of selected microorganisms to decompose organic waste, resulting in the formation of stable and nutrient-rich compost. This process encompasses several crucial stages driven by microbial activity: decomposition, mineralization, and humification. Each stage plays a vital role in breaking down organic matter and forming high-quality compost.

4.1 Decomposition

Decomposition serves as the foundation of the composting process, during which microorganisms break down complex organic materials into simpler compounds. This stage is essential for converting raw GW, such as leaves, grass clippings, and plant trimmings, into a stable, nutrient-rich organic material. A diverse community of microorganisms, including bacteria, fungi, and actinomycetes, drives the decomposition process, with each group specializing in breaking down specific components of organic matter.

During the initial phase of composting, mesophilic microorganisms, which thrive at moderate temperatures (20–45 °C), take the lead. These microbes, mostly bacteria, start decomposing readily degradable compounds like sugars, starches, and proteins. They secrete enzymes such as amylases, proteases, and lipases, which break down complex molecules into simpler forms, including glucose, amino acids, and fatty acids. The heat generated from microbial metabolism during this phase causes the compost pile's temperature to increase.

As temperatures increase (45–70 °C), thermophilic bacteria and fungi take over. These heat-loving microbes are highly effective at breaking down tougher materials like cellulose, hemicellulose, and lignin. Cellulose-degrading bacteria (e.g., *Cellulomonas*) and fungi (e.g., *Trichoderma*) produce cellulase enzymes that break down cellulose into glucose. Likewise, fungi that degrade lignin, such as *Phanerochaete chrysosporium,* release enzymes like lignin peroxidase to break down lignin, a complex polymer that provides rigidity to plant cell walls. The thermophilic phase, characterized by high temperatures, is vital for reducing pathogens, as it kills weed seeds and harmful microorganisms.

As the easily degradable organic matter is depleted, the compost pile cools down, allowing mesophilic microorganisms to regain dominance. Actinomycetes, which flourish in moderate temperatures, take center stage in breaking down the remaining complex compounds, such as chitin and cellulose. During this phase, the compost continues to stabilize, and the formation of humic substances commences, signaling the onset of the humification stage (Nozhevnikova et al., 2019; Rawat & Johri, 2013; Chinakwe et al., 2019; Finore et al., 2023; Insam & Bertoldi, 2007).

Microbial inoculants play a crucial role in enhancing the composting process by introducing high concentrations of targeted microorganisms that accelerate the decomposition of various types of organic matter. These inoculants are designed to enrich the microbial diversity and activity within the compost, thereby speeding up the breakdown of complex compounds. For instance, cellulose degraders such as *Trichoderma* and *Bacillus* species effectively decompose cellulose, a primary component of plant residues. Lignin, a more recalcitrant material resistant to bacterial attack, can be broken down with the help of lignin-degrading fungi like *Phanerochaete chrysosporium.* In addition to decomposers, nitrogen-fixing bacteria such as *Azotobacter* contribute to the compost's nutrient profile by converting atmospheric nitrogen into bioavailable forms, enhancing nitrogen content in the final product. By optimizing microbial activity through the use of specific inoculants, this approach not only accelerates the composting timeline but also improves the overall quality of the compost, making it richer in nutrients and more beneficial for soil application (Oviedo-Ocaña et al., 2022a).

The rate and efficiency of decomposition in composting are influenced by several key factors. Maintaining a balanced carbon-to-nitrogen (C:N) ratio of 25–30:1 is crucial for optimal microbial activity. An excess of carbon can slow down the decomposition process, while too much nitrogen can lead to ammonia loss and the production of foul odors (Gaudel et al., 2024). Adequate moisture levels, typically between 50 and 60%, are essential for the survival of microorganisms and the activity of their enzymes. Oxygen is another critical component, as aerobic decomposition is generally faster and produces fewer odors compared to anaerobic processes. Additionally, smaller particle sizes enhance the surface area available for microbial action, thereby accelerating the breakdown process (Mishra & Yadav, 2022). The end result of decomposition is the conversion of raw organic waste into simpler compounds, including humic acids, fulvic acids, and inorganic nutrients such as ammonium, nitrates, and phosphates (Yu et al., 2019). These compounds collectively form nutrient-rich compost, which improves soil structure, enhances water retention, and supports healthy plant growth.

4.2 Mineralization

Mineralization is a crucial phase in composting, where organic matter is converted into inorganic nutrients that plants can easily absorb (Cambardella et al., 2003). Microorganisms drive this process by breaking down complex organic compounds, such as proteins, nucleic acids, and lipids, into simpler inorganic forms like ammonium (NH_4^+), nitrates (NO_3^-), phosphates (PO_43^-), and sulfates (SO_42^-) (Xie et al., 2023). These nutrients are vital for plant growth and soil health, making mineralization an essential step in creating high-quality compost.

Nitrogen mineralization is a vital process in composting, where organic nitrogen compounds, such as proteins and amino acids, are converted into inorganic forms like ammonium (NH_4^+) and nitrate (NO_3^-) that plants can readily use. During the ammonification stage, proteolytic bacteria like *Bacillus* and *Pseudomonas* break down proteins and amino acids into ammonium through the action of enzymes like proteases and deaminases. In the subsequent nitrification stage, nitrifying bacteria such as *Nitrosomonas* and *Nitrobacter* oxidize ammonium into nitrites (NO_2^-) and then into nitrates (NO_3^-). These nitrates are highly soluble and can be easily absorbed by plants (Hoang et al., 2022). Similarly, phosphorus mineralization involves the conversion of organic phosphorus (e.g., phospholipids, nucleic acids) into inorganic phosphate (PO_43^-) by phosphate-solubilizing microbes like *Pseudomonas* and *Bacillus* (Billah et al., 2019) These microbes secrete organic acids and phosphatases that mobilize phosphate for plant use (Bohórquez-Sandoval et al., 2024). In sulfur mineralization, sulfur-containing compounds such as proteins and sulfolipids are decomposed into sulfate (SO_42^-) by sulfur-oxidizing bacteria like *Thiobacillus*. Sulfatase enzymes are critical in liberating sulfur from organic material, thereby making it available to plants (Rana et al., 2020).

Microbial inoculants boost the mineralization process by introducing concentrated populations of specialized microorganisms that expedite the decomposition of organic matter. Nitrogen-fixing bacteria, such as *Azotobacter* and *Rhizobium*, incorporated in inoculants, can fix atmospheric nitrogen, thereby enriching the compost's nitrogen content (Harindintwali et al., 2020). Similarly, phosphate-solubilizing microbes like *Pseudomonas* and *Bacillus* enhance phosphorus availability by converting both organic and inorganic phosphates into forms that plants can absorb. In addition, sulfur-oxidizing bacteria such as *Thiobacillus* play a crucial role in promoting sulfur mineralization, contributing to a well-balanced nutrient composition in the compost.

Several factors influence the rate and efficiency of mineralization during composting. The carbon-to-nitrogen (C:N) ratio plays a crucial role, with an optimal range of 25–30:1 supporting active microbial growth; imbalances can hinder the process—excess carbon slows decomposition, while excess nitrogen may result in ammonia volatilization. Temperature is another key factor, with mineralization peaking during the thermophilic phase (45–70 °C), when microbial activity is at its highest. pH also affects microbial efficiency, with a neutral to slightly alkaline range

(6.5–8.0) being most favorable, as extreme acidity or alkalinity can suppress microbial processes. Additionally, proper moisture content (50–60%) and sufficient aeration are vital, as aerobic conditions promote faster and more efficient mineralization compared to anaerobic environments (Nahm, 2005).

Mineralization in compost offers several significant benefits. It enhances nutrient availability by converting organic nutrients into inorganic forms that plants can readily absorb, thereby increasing the compost's fertilizing potential. This process also improves soil health, as the release of essential nutrients like nitrogen, phosphorus, and sulfur boosts soil fertility and supports a diverse microbial ecosystem. Moreover, mineralization contributes to sustainability by recycling nutrients from organic waste, reducing dependence on synthetic fertilizers, and promoting environmentally friendly agricultural practices. However, there are challenges associated with mineralization. Nutrient loss can occur when ammonium volatilizes as ammonia gas under overly alkaline or poorly aerated conditions. Additionally, rapid nitrogen mineralization without simultaneous phosphorus or sulfur release may result in nutrient imbalances. Incomplete mineralization, particularly during the thermophilic phase, can also lead to the survival of pathogens, posing a risk to compost quality and plant health.

4.3 Humification

Humification is the final and most critical phase of composting, where organic matter is transformed into stable, complex molecules known as humic substances. These substances include humic acids, fulvic acids, and humin, which are essential components of soil organic matter. They significantly improve soil structure, nutrient retention, and water-holding capacity (Chen et al., 2023). Humification, a gradual and microbially driven process, mainly takes place during the cooling and maturation stages of composting. It signifies the shift from breaking down organic waste to producing nutrient-rich, stable compost that supports long-term soil health.

Humic substances are dark-colored, complex organic compounds formed during composting that are highly resistant to further decomposition. They originate from the polymerization and condensation of simpler organic molecules such as phenols, quinones, and amino acids. These substances are crucial for soil health and are classified into three main categories (Guo et al., 2019). Humic acids are large molecules that dissolve in alkaline conditions but remain insoluble in acidic environments; they enhance soil structure and improve nutrient retention. Fulvic acids are smaller, more soluble molecules known for their ability to chelate nutrients, thereby increasing their availability to plants (Plaza et al., 2005). Humin is the most stable and insoluble component, playing a vital role in long-term carbon storage in the soil.

Humification is driven by a series of intricate biochemical reactions influenced by microbial activity and environmental conditions. One key process is polymerization, where simple organic molecules such as phenols and quinones are chemically linked to form larger, more stable polymers. This transformation is facilitated by oxidative enzymes secreted by fungi and actinomycetes, including species like *Phanerochaete chrysosporium* and *Streptomyces* (Lu et al., 2023). Another critical step involves condensation reactions, such as Maillard reactions, where amino acids and sugars interact to form dark-colored, nitrogen-rich polymers that enhance the stability of humic substances (Mu et al., 2024). Additionally, the microbial degradation of complex plant materials—particularly lignin and cellulose—by lignin-degrading fungi and actinomycetes releases essential precursors for humic compound synthesis (Zhao et al., 2017).

Microbial inoculants can significantly enhance the humification process by introducing targeted microorganisms that accelerate the decomposition of complex organic matter and stimulate the formation of humic substances. Lignin-degrading fungi, such as *Phanerochaete chrysosporium* and *Trametes versicolor*, are commonly used in inoculants to break down lignin, releasing phenolic compounds that serve as precursors for humic acid synthesis (Yang et al., 2020). Inoculants containing *Streptomyces* species, a group of actinomycetes, aid in the polymerization of organic molecules, contributing to the development of stable humic substances (Javed et al., 2021). Additionally, microbial strains that produce oxidative enzymes like laccases and peroxidases help drive the polymerization and condensation reactions essential to humification (Bui et al., 2023).

Several factors influence the rate and quality of humification during composting. A balanced carbon-to-nitrogen (C:N) ratio of 25–30:1 is crucial, as it provides sufficient carbon for humic substance synthesis while supporting microbial growth with adequate nitrogen. Humification primarily takes place during the cooling and maturation phases, where moderate temperatures (20–40 °C) favor the activity of microbes involved in this process; excessive heat can suppress these organisms. The pH level also plays a vital role, with a neutral to slightly alkaline range (6.5–8.0) being optimal, as extreme pH conditions can impair microbial function and enzyme activity. Proper moisture content (50–60%) and adequate aeration are necessary to maintain aerobic conditions that facilitate microbial activity essential for humification. Importantly, humification is a gradual process that unfolds over weeks to months, depending on the composting environment and the composition of the organic materials (Li et al., 2021).

Humification offers numerous benefits that contribute to soil health and sustainable agriculture. One key advantage is the improvement of soil structure, as humic substances act as natural binding agents that enhance soil aggregation, porosity, aeration, and water infiltration. Nutrient retention is also significantly improved, particularly through the high cation exchange capacity (CEC) of humic acids, which enables them to hold and gradually release essential nutrients like calcium, magnesium, and potassium. Additionally, humic substances increase the soil's water-holding capacity, reducing irrigation needs. In terms of environmental impact, humin plays a vital role in long-term carbon sequestration, helping to mitigate climate change by lowering greenhouse gas emissions. Moreover, fulvic acids promote plant growth by chelating micronutrients, making them more accessible to plants and supporting improved root development and nutrient uptake.

5 Sourcing and Preparing Indigenous Inoculum

5.1 Isolation from Existing Compost:

Many researchers focus on isolating microorganisms from mature, high-quality compost samples, as these native microbes are well-adapted to the composting environment. The isolation process typically targets microorganisms with strong hydrolytic activity (such as cellulase, amylase, and protease enzymes) and nitrogen-fixing capabilities. For example, Yu et al. (2018) used a systematic method to prepare microbial inoculants for composting. A 10 g sample of fresh compost was added to 90 mL of sterile water (1:9, w/w) in a plastic bottle. The bottle was shaken at 200 rpm for 30 min, then left to stand for 15 min to allow the microorganisms to separate into the supernatant. The supernatant was serially diluted from 10 to 105 times, and 0.1 mL aliquots were plated on beef extract peptone agar for bacterial isolation and rose bengal agar for fungal isolation. All plates were incubated at 28 °C. After clear colonies appeared, distinct colonies were streaked onto fresh plates to obtain pure cultures. Single colonies were transferred to liquid media—bacteria to beef extract peptone medium and fungi to potato dextrose medium—and cultured at 28 °C and 200 rpm. Once the liquid cultures became turbid, 10 μL of each suspension was inoculated onto filter paper discs (6 mm in diameter) placed on selective media. Selective Medium A was designed to identify cellulase-producing microorganisms and consisted of CMC-Na (10.0 g), K_2HPO_4 (1.0 g), $MgSO_4 \cdot 7H_2O$ (0.1 g), $FeSO_4 \cdot 7H_2O$ (0.1 g), $MnSO_4$ ($5 \times 10^{-}4$ g), peptone (10.0 g), yeast extract (10.0 g), and 1.0 L of water with a natural pH. Medium B contained 0.1 g of Azure B per liter of Potato Dextrose Agar (PDA) and was used to detect manganese peroxidase (MnP) activity. Medium C was formulated with 0.4 mL of guaiacol per liter of PDA to screen for lignin peroxidase (LiP) and laccase activity. After inoculation, all plates were incubated at 28 °C to allow for microbial growth and enzyme expression. In parallel, 0.7 mL of each microbial suspension was mixed with 0.7 mL of 50% (v/v) glycerol in sterile 2 mL tubes and stored at −20 °C for long-term preservation of the isolates.

The media were monitored daily for microbial growth, with enzyme activity indicated by specific color changes. On Selective Medium B, microorganisms that caused a fading of the blue color at the colony edges were identified as positive for manganese peroxidase (MnP) activity. On Selective Medium C, colonies that turned rufous indicated the presence of lignin peroxidase (LiP) or laccase activity. For Selective Medium A, after two days of incubation, 1 g/L Congo red solution was poured over the medium to stain it, followed by washing with 1 mol/L sodium chloride after 30 min. Microorganisms that caused a fading of the red color at colony edges were selected as cellulase producers. The selected isolates were then tested for antagonistic interactions using an antagonism assay. Microorganisms that showed no antagonism toward one another were chosen for further use. Finally, three compatible and functionally distinct microorganisms were identified and combined into a composite inoculum using the quadratic regression orthogonal design (QROD) method.

5.2 Specific Microbial Strains:

Researchers often utilize pure or mixed cultures of well-characterized microbial strains recognized for their efficiency in composting. These strains can be sourced from established culture collections or commercial microbial suppliers. In a recent study, Feng et al. (2024) employed three non-antagonistic lignocellulose-degrading strains: *Bacillus amyloliquefaciens* (J1), *Clonostachys rogersoniana* (B2), and *Streptomyces thermoviolaceus* (J3). Strains J1 and J3 were maintained on lysogeny agar (LA) plates, while B2 was stored on potato dextrose agar (PDA) plates. For inoculum preparation, J1 and J3 were subcultured on LA plates, and B2 on PDA plates, each incubated for 10 days at 28 °C to promote spore formation. These strains exhibited different optimal survival temperatures, with J1 thriving at approximately 45 °C, B2 at 30 °C, and J3 at 55 °C, making them suitable for various composting conditions.

5.3 Solid-State Fermentation:

In certain approaches, inoculum preparation involves using solid substrates such as rice bran, wheat bran, or other agricultural by-products inoculated with targeted microorganisms and incubated under controlled conditions of moisture and aeration. Khalib et al. (2018) introduced a comprehensive and cost-effective method for developing indigenous microbial inoculum (IMO) using locally sourced materials to enhance composting. Their method includes five sequential phases aimed at cultivating and enriching beneficial microbes. During the first phase, termed IMO (I), 200 g of rice was placed in a small plastic container, covered with white paper, and positioned under bamboo clumps to avoid direct sunlight. After 72 h, the rice was observed to be covered with white mold (mycelium) and emitted a slightly sour odor, indicating the successful growth and collection of indigenous microorganisms from the environment.

In the second phase, IMO (II), 100 g of the previously prepared IMO I was blended with 100 g of granulated brown sugar in a 1:1 ratio and fermented in a cool, shaded location for five days. This was followed by the third phase, IMO (III), where 3 g of IMO II was mixed with 3.0 kg of rice bran and 0.75 L of fermented rice wash water until a semi-moist consistency (approximately 65% moisture) was achieved. The mixture was then shaped into a mound, covered with dried leaves, and left to ferment for five days. In the fourth phase, IMO (IV), 1.0 kg of IMO III was combined with 1.0 kg of soil and 0.25 L of fermented rice wash water, maintaining 65% moisture while ensuring the temperature remained below 70 °C during the five-day incubation. Finally, in the fifth phase, IMO (V), 8.0 kg of dried goat dung was mixed with 0.8 kg of IMO IV in a 10:1 ratio, and 2.0 L of fermented rice wash water was added to maintain adequate moisture. The mixture was monitored daily to prevent temperatures from exceeding 70 °C. After 14 days, the final inoculum, IMO V, was obtained. This stepwise enrichment method cultivates a diverse and resilient microbial population, making IMO V a potent and practical inoculant for improving composting efficiency in both urban and rural contexts.

5.4 Enrichment

Researchers may use enrichment techniques to selectively cultivate specific groups of microorganisms by providing targeted substrates—such as cellulose for cellulolytic microbes—as the primary carbon source in the growth medium. For instance, Mishra and Yadav (2021) formulated five specialized microbial inoculants to optimize composting, each designed for distinct organic inputs and microbial communities. Inoculant-1 (C1) was developed by dissolving 50 g of jaggery in water to serve as an energy-rich substrate for microbial proliferation (Fig. 3). To this base, 1 kg of fresh cow dung and 1 L of cow urine were added, followed by 10 L of water to create a nutrient-rich fermentation broth. The mixture was loosely covered with cloth to allow gas exchange and was stirred daily with a wooden stick to maintain aerobic conditions. After three days of fermentation, during which the microbial population reached its peak, the solution was diluted to 50 L and uniformly sprayed over 500 kg of GW. This approach enhances microbial colonization and accelerates the decomposition of organic matter.

Inoculant-2 (C2) was developed using processed food waste collected from a hostel, which was manually sorted and shredded to a particle size of 0.1–0.5 mm. The material was then covered and left undisturbed for seven days to promote fungal colonization. After visible fungal growth was established, the inoculated waste was mixed with water to create 50 L of slurry, which was then applied in layers over 500 kg of GW within an aerated in-vessel composting system (Fig. 2). Inoculant-3 (C3) involved creating a microbial-rich slurry from 25 kg of dry cow dung mixed with water. This slurry was directly sprayed over 500 kg of GW to enhance microbial activity. Inoculant-4 (C4) utilized mature compost derived from flower waste; 25 kg of this compost was blended with water to produce 50 L of liquid inoculum, which was also applied to GW in a similar manner. Lastly, Inoculant-5 (C5) was formulated by grinding fresh vegetable waste to 0.1–0.5 cm, covering it for seven days to induce fungal development, and then mixing 25 kg of the partially decomposed material with water to make 50 L of inoculant. This mixture was sprayed over 500 kg of GW and processed in an aerated in-vessel system to ensure efficient decomposition.

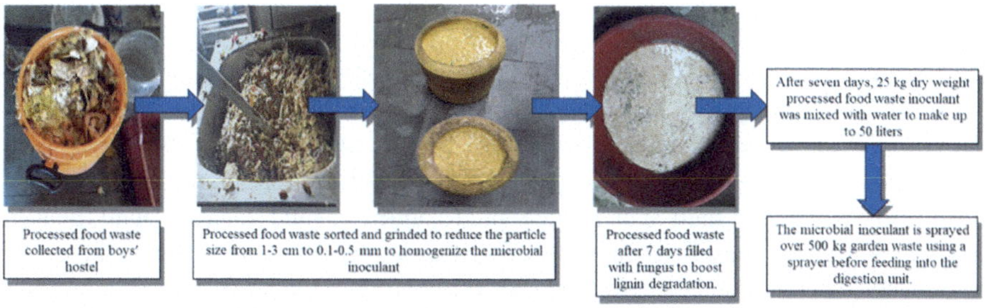

Fig. 2 Preparation of enrichment inoculum from food waste as described by Mishra and Yadav (2021)

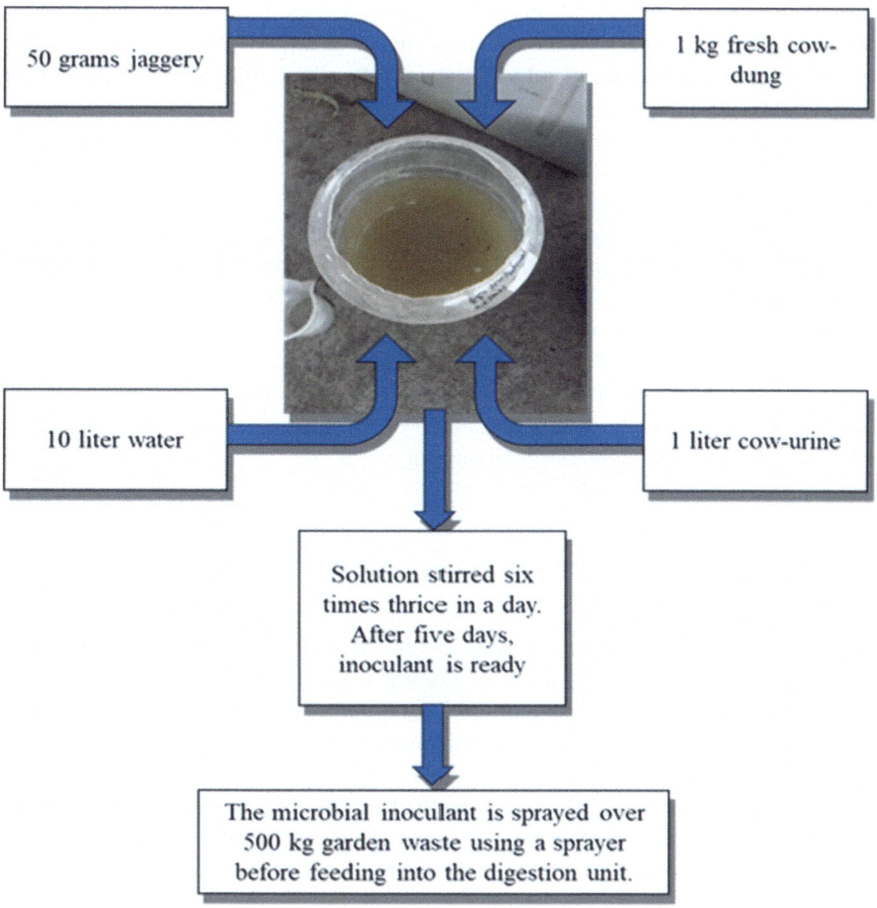

Fig. 3 Preparation of enrichment inoculum from jaggery cow dung and cow urine as described by Mishra and Yadav (2021)

5.5 From Defoliating Insects

Moringa Hairy Caterpillar, *Spodoptera litura*, *Achaea Janata*, *Euproctis fraternal*, *Diacrisia obliqua*, leaf webber are some examples of insects that primarily feed on the leaves (foliage) of plants, often leading to defoliation (partial or complete removal of leaves), which can reduce photosynthesis and overall plant health. Various researchers attempted to isolate cellulose degrading microbes from the gut of these insects and inoculated with the GW composting. isolated cellulose-degrading bacteria from the gut of *Moringa* hairy caterpillars (*Spodoptera litura*) actively feeding on *Moringa oleifera* leaves. To reduce the presence of transient microbes, the caterpillars were starved for 24 h prior to dissection. They were surface-sterilized using ethanol and immobilized on ice for 10 min. Dissection was performed under aseptic conditions using sterilized instruments in a laminar airflow cabinet. The excised gut was suspended in 1 mL of sterile 0.9% saline solution and homogenized using a micro pestle.

The resulting homogenate was inoculated into a basal salt medium containing $NaNO_3$ (0.25 g), KH_2PO_4 (0.2 g), $MgSO_4$ (0.02 g), NaCl (0.02 g), and $CaCl_2 \cdot 6H_2O$ (0.01 g) per 100 mL, to enrich cellulose-degrading bacteria. Cultures were incubated for 7 days at 37 °C and 100 rpm in a shaker incubator. At two-day intervals, 1 mL of the culture was serially diluted up to 10^{-5}, and aliquots were plated on LB agar supplemented with 1% carboxymethyl cellulose (CMC) to isolate and purify strains. Positive isolates were screened for cellulolytic activity by determining their hydrolytic capacity (HC), calculated as the ratio of the diameter of the clear zone to that of the colony. The most efficient cellulose-degrading strains, based on HC values, were subsequently inoculated into GW composting systems to assess their ability to enhance organic matter degradation.

5.6 Commercially Available Inoculants

Several commercial inoculants are available to optimize composting processes by enhancing microbial activity and accelerating organic matter breakdown. EM-1 (Effective Microorganisms) is one of the most commonly used products; it contains a synergistic blend of lactic acid bacteria, yeasts, and photosynthetic bacteria that help accelerate decomposition, suppress odors, and improve compost quality (Jusoh et al., 2013). Bokashi Bran, a fermented product enriched with beneficial microbes, is typically used in small-scale composting systems such as Bokashi bins and is known for promoting anaerobic fermentation and nutrient preservation (Kruker et al., 2023; Lew et al., 2021).

Commercially available Compost Accelerators consist of diverse microbial consortia, including bacteria, fungi, and enzyme-producing microbes, formulated to speed up the composting process while improving the nutritional profile of the final compost. These products are usually applied either during the initial layering of compost or as liquid sprays during turning or maintenance, depending on system design and manufacturer instructions.

The application of these inoculants not only improves the efficiency and rate of composting but also enhances the microbial diversity and nutrient value of the resulting compost, making it more beneficial for soil structure, fertility, and plant health. Some of the studies which used microbial inoculum for the composting of GW are summarized in Table 2.

6 Benefits of Inoculum-Based Composting for Urban Sustainability

6.1 Environmental Benefits

Inoculum-based composting offers significant environmental advantages, making it a cornerstone of sustainable urban waste management. By accelerating the decomposition of organic waste, this method reduces the release of greenhouse gases, particularly methane, which is emitted when organic matter decomposes anaerobically in landfills. According to the EPA (2023), composting can divert up to 30% of municipal solid waste from landfills, significantly lowering the carbon footprint of urban areas. Additionally, inoculum-based composting minimizes the need for chemical fertilizers by producing nutrient-rich compost that enhances soil health. The introduction of beneficial microorganisms improves soil structure, increases water retention, and promotes the growth of healthy plants, thereby supporting urban green spaces and reducing the urban heat island effect. By closing the loop on organic waste, this method contributes to a circular economy, reducing reliance on synthetic inputs and fostering long-term environmental sustainability.

6.2 Economic Benefits

The economic benefits of inoculum-based composting are substantial, particularly for urban areas grappling with high waste management costs. By diverting organic waste from landfills, municipalities can significantly reduce waste disposal fees, which are often calculated based on volume or weight. For example, cities like San Francisco have reported savings of millions of dollars annually through composting initiatives (Farhidi et al., 2022). Furthermore, the compost produced can be sold to local farmers, landscapers, or community gardens, generating additional revenue. Small-scale urban composting initiatives can also create green jobs, from waste collection and processing to compost marketing and distribution. These economic benefits make inoculum-based composting a cost-effective solution for urban waste

Table 2 Microbial inoculum used in various studies in composting of GW

Feedstock	Inoculant	Concentration	Reference
GW	Mature compost	15 20, 25, and 30% (w/w, dw)	(Yang & Zhang, 2022)
	Commercial microbial inoculum, which consisted of *Trichoderma spp.* (60%, v/v) and *Phanerochaete chrysosporium Burdsall* (40%, v/v)	0.5% (v/w, dw)	
GW	Composite inoculum *Bacillus sp.* (B1 and B2 and *Aspergillus sp.* (A)	B1: 2.97 × 10^4 CFU/g (dw) B2: 6.69 × 10^6 CFU/g (dw) A: 1.83 × 10^5 CFU/g (dw)	(Yu et al., 2018)
	Phanerochaete chrysosporium		
	EM (Effective Microorganisms)	0.1% (w/dw)	
GW, textile waste, paper and cardboard waste	*Streptomyces cellulosae, Achromobacter xylosoxidans*, and *Serratia liquefaciens*	4%, 6%, and 8%	(Biyada et al., 2022)
GW	*Bacillus amyloliquefaciens* (J1) *Clonostachys rogersoniana* (B2) *Streptomyces thermoviolaceus* (J3)	J1:B2 = 1:1, J1:J3 = 1:1, B2:J3 = 3:2, and J1: B2: J3 = 3:3:2	(Feng et al., 2024)
GW, household wet biodegradable waste	Commercial inoculum containing lactic acid bacteria, yeast and phototrophic bacteria	6.25 mL/kg	(Manu et al., 2019) (Manu et al., 2017)
GW	Cellulolitic (*Aspergillus fumigatus*) and Ligninolitic (*Geotrichum sp.*) Fungi	10–20%	(Irawan et al., 2019)
Switchgrass and eucalyptus	Green waste compost (GWC) and grape pomace compost (GPC)	0.1 g compost/g mixture (dw)	(Simmons et al., 2014)
GW, unprocessed food waste, processed food waste	*Bacillus sp.* F3 × 3 and *Paenibacillus sp* F1A5	4.85 × 10^5 UFC/mL 1.44 × 10^5 UFC/mL	(Oviedo-Ocaña et al., 2022b) (Soto-Paz et al., 2022)
GW, olive mill solid sludge	Thermophilic Inoculum	2% (10^8 UFC/g)	(Bouhia et al., 2021)

(continued)

Table 2 (continued)

Feedstock	Inoculant	Concentration	Reference
GW, silage and flue gas desulfurized gypsum	*Trichoderma spp.* (60%, v/v) and *Phanerochaete chrysosporium Burdsall* (40%, v/v),	5 ml per kg of GW (dw)	(Yang et al., 2023)
GW and sheep manure	Commercial inoculum and cellulase	4% and 2%	(Liu et al., 2018)

management while simultaneously creating opportunities for local entrepreneurship and economic growth.

6.3 Social and Community Benefits

Inoculum-based composting fosters social and community benefits by encouraging collective action and environmental stewardship. Community-based composting initiatives, such as neighborhood composting hubs or school composting programs, provide opportunities for residents to engage in sustainable practices and learn about the importance of waste reduction. These initiatives can strengthen social cohesion by bringing people together around a shared goal of environmental sustainability. For instance, community gardens that use locally produced compost often become hubs for education, recreation, and social interaction, enhancing the quality of urban life. Additionally, composting programs can raise awareness about the environmental impact of waste and empower individuals to adopt more sustainable lifestyles. By integrating composting into urban life, cities can build more resilient, connected, and environmentally conscious communities.

7 Challenges and Limitations

7.1 Technical Challenges

Inoculum-based composting, while highly effective, faces several technical challenges that can hinder its widespread adoption. Maintaining optimal composting conditions, such as temperature, moisture, and aeration, is critical for microbial activity and efficient decomposition. Fluctuations in these parameters can slow down the process or lead to incomplete composting. Ensuring the efficacy of microbial inoculants is another challenge, as the performance of introduced microorganisms can vary depending on the type of organic waste, environmental conditions, and competition with indigenous microbes. For example, lignin-degrading fungi may struggle to thrive in environments dominated by cellulose-rich materials. Additionally, urban settings often lack the space and infrastructure needed for large-scale composting, requiring innovative solutions like in-vessel systems or modular composting units. Addressing these technical challenges requires ongoing research, monitoring, and adaptation to local conditions.

7.2 Social and Behavioral Barriers

The success of inoculum-based composting in urban areas is heavily influenced by social and behavioral factors. Public acceptance and participation are often limited by a lack of awareness about the benefits of composting or misconceptions about its complexity and odor. Many urban residents may also lack the time, space, or motivation to participate in composting initiatives. Educational campaigns and hands-on workshops can help overcome these barriers by demonstrating the simplicity and benefits of composting. However, changing long-standing behaviors and attitudes toward waste management requires sustained effort and community engagement. Social equity is another concern, as low-income communities may have limited access to composting facilities or resources, exacerbating existing disparities in environmental sustainability efforts.

7.3 Policy and Regulatory Constraints

Policy and regulatory frameworks play a crucial role in promoting or hindering the adoption of inoculum-based composting. In some regions, outdated regulations or lack of supportive policies may restrict the implementation of composting programs. For instance, zoning laws may prohibit composting facilities in urban areas, while health and safety regulations may impose stringent requirements that increase costs and complexity. On the other hand, proactive policies, such as mandates for organic waste diversion, subsidies for composting infrastructure, or tax incentives for compost producers, can significantly boost adoption. Cities like San Francisco and Seattle have successfully implemented such policies, achieving high rates of organic waste diversion. However, inconsistent enforcement or lack of political will can undermine these efforts. To fully realize the potential of inoculum-based composting, policymakers must prioritize waste management reforms and create an enabling environment for sustainable practices.

8 Future Prospects and Innovations

8.1 Emerging Technologies

The future of inoculum-based composting is being shaped by cutting-edge technologies that enhance efficiency, scalability, and sustainability. Advances in microbial inoculants, such as genetically engineered strains and synthetic microbial consortia, are enabling faster decomposition and improved compost quality. For example, researchers are developing lignin-degrading microbes with enhanced enzymatic activity to break down tough plant materials more effectively. Smart composting systems, equipped with IoT-based sensors, are revolutionizing urban composting by providing real-time monitoring of temperature, moisture, and microbial activity. These systems can optimize composting conditions remotely, reducing labor costs and ensuring consistent results. Additionally, blockchain technology is being explored to track compost production and distribution, ensuring transparency and quality control. These innovations are making inoculum-based composting more accessible and efficient, particularly in space-constrained urban environments.

8.2 Integration with Circular Economy Models

Inoculum-based composting is a natural fit for circular economy models, which aim to minimize waste and maximize resource efficiency. By converting organic waste into nutrient-rich compost, this method closes the loop on organic matter, reducing the need for synthetic fertilizers and promoting sustainable agriculture. Urban composting initiatives can be integrated with other circular economy practices, such as urban farming, green infrastructure, and renewable energy production. For instance, compost can be used to enrich urban gardens or green roofs, while biogas generated from composting can power local facilities. Collaborative efforts between municipalities, businesses, and communities can create closed-loop systems where waste is seen as a resource rather than a burden. This integration not only enhances environmental sustainability but also creates economic opportunities and strengthens community resilience.

8.3 Policy Recommendations

To promote widespread adoption, policymakers should implement supportive measures, including mandatory organic waste diversion programs, financial incentives for composting infrastructure, and public awareness campaigns on the benefits of composting. Investments in research and development are essential to advancing microbial inoculants and innovative composting techniques. Furthermore, updating zoning laws to facilitate urban composting facilities and streamlining regulatory frameworks can help overcome logistical barriers. By fostering collaboration between governments, research institutions, and community stakeholders, inoculum-based composting can be effectively integrated into urban waste management strategies, contributing to a more circular and sustainable economy.

9 Conclusion

Inoculum-based composting emerges as a promising and sustainable solution for managing urban GW, addressing the environmental, economic, and social challenges posed by traditional waste management practices. The increasing volume of GW due to rapid urbanization and population growth necessitates innovative approaches to reduce landfill dependency, minimize greenhouse gas emissions, and enhance soil health.

The chapter has highlighted the critical role of microbial inoculants in accelerating the composting process and improving the quality of the final compost product. By introducing specific bacteria, fungi, and actinomycetes, inoculum-based composting can effectively degrade lignocellulosic compounds, reduce processing time, and enhance nutrient availability. The synergistic activity of these microorganisms drives the decomposition, mineralization, and humification processes, resulting in a stable, nutrient-rich compost that benefits urban agriculture and green spaces. In conclusion, inoculum-based composting offers a multifaceted solution to urban GW management, aligning with environmental sustainability goals, economic benefits, and social well-being. By addressing the challenges and leveraging technological advancements, this method can play a crucial role in creating resilient, sustainable, and environmentally conscious urban communities.

References

Amani-Beni, M., & Khalilnezhad, M. R. (2023). Book Review: Mertens, E. Resilient city landscape architecture for climate change; Birkhäuser: Basel, Switzerland, 2022; ISBN: 978–3035622348. *Land*, *12*(5), 1082. https://doi.org/10.3390/land12051082

Ansar, A., Du, J., Javed, Q., Adnan, M., & Javaid, I. (2025). Biodegradable waste in compost production: A review of its economic potential. *Nitrogen, 6*(2), 24. https://doi.org/10.3390/nitrogen6020024

Ashwini John, J., & Selvarajan, E. (2023). Genomic analysis of lignocellulolytic enzyme producing novel Streptomyces sp.MS2A for the bioethanol applications. *International Journal of Biological Macromolecules, 250*, 126138. https://doi.org/10.1016/j.ijbiomac.2023.126138

Ayilara, M. S., Olanrewaju, O. S., Babalola, O. O., & Odeyemi, O. (2020). Waste management through composting: Challenges and potentials. *Sustainability (Switzerland), 12*(11), 1–23. https://doi.org/10.3390/su12114456

Bhatti, A. A., Haq, S., & Bhat, R. A. (2017). Actinomycetes benefaction role in soil and plant health. *Microbial Pathogenesis, 111*, 458–467. https://doi.org/10.1016/j.micpath.2017.09.036

Billah, M., et al. (2019). Phosphorus and phosphate solubilizing bacteria: Keys for sustainable agriculture. *Geomicrobiology Journal, 36*(10), 904–916. https://doi.org/10.1080/01490451.2019.1654043

Biyada, S., et al. (2022). Bio-augmentation as an emerging strategy to improve the textile compost quality using identified autochthonous strains. *Applied Sciences (Switzerland), 12*(6), 3160. https://doi.org/10.3390/app12063160

Bohórquez-Sandoval, L. J., García-Molano, J. F., Pascual-Valero, J. A., & Ros-Muñoz, M. (2024). A comprehensive review on organic waste compost as an effective phosphorus source for sustainable agriculture. *International Journal of Recycling of Organic Waste in Agriculture, 13*(4), 132441. https://doi.org/10.57647/ijrowa-ybbx-kp03

Bouhia, Y., et al. (2021). Effect of microbial inoculation on lipid and phenols removal during the co-composting of olive mill solid sludge with green waste in bioreactor. *Waste and Biomass Valorization, 12*(3), 1417–1429. https://doi.org/10.1007/s12649-020-01077-3

Bui, V. K. H., Truong, H. B., Hong, S., Li, X., & Hur, J. (2023). Biotic and abiotic catalysts for enhanced humification in composting: A comprehensive review. *Journal of Cleaner Production, 402*, Article 136832. https://doi.org/10.1016/j.jclepro.2023.136832

Cambardella, C. A., Richard, T. L., & Russell, A. (2003). Compost mineralization in soil as a function of composting process conditions. *European Journal of Soil Biology, 39*(3), 117–127. https://doi.org/10.1016/S1164-5563(03)00027-X

Chen, L., et al. (2023). Improving the humification by additives during composting: A review. *Waste Management, 158*, 93–106. https://doi.org/10.1016/j.wasman.2022.12.040

Chen, D. M.-C., Bodirsky, B. L., Krueger, T., Mishra, A., & Popp, A. (2020). The world's growing municipal solid waste: trends and impacts. *Environmental Research Letters 15*(7), 74021. https://doi.org/10.1088/1748-9326/ab8659.

Chinakwe, E. C., et al. (2019). Effect of temperature changes on the bacterial and fungal succession patterns during composting of some organic wastes in greenhouse. *Journal of Advances in Microbiology, 15*(1), 1–10. https://doi.org/10.9734/jamb/2019/v15i130075

Chukwuma, O. B., Rafatullah, M., Tajarudin, H. A., & Ismail, N. (2021). A review on bacterial contribution to lignocellulose breakdown into useful bio-products. *International Journal of Environmental Research and Public Health, 18*(11). https://doi.org/10.3390/ijerph18116001

Costa, D. A. L., Williams, T. C. R., do Vale, L. H. F., & Filho, E. X. F. (2023). Characterization of mannanases from Clonostachys byssicola involved in the breakdown of lignocellulosic substrates. *Biocatalysis and Agricultural Biotechnology, 50*, 102680. https://doi.org/10.1016/j.bcab.2023.102680

EPA. (2023). *Composting at home*. United States Environmental Protection Agency. Retrieved February 24, 2025, from https://www.epa.gov/recycle/composting-home

Farhidi, F., Madani, K., & Crichton, R. (2022). How the US economy and environment can both benefit from composting management. *Environmental health insights, 16*. https://doi.org/10.1177/11786302221128454

Feng, X., Qiu, M., & Zhang, L. (2024). Construction of lignocellulose-degrading compound microbial inoculum and its effects on green waste composting. *Journal of Environmental Management, 370*, Article 122502. https://doi.org/10.1016/j.jenvman.2024.122502

Finore, I., et al. (2023). Thermophilic bacteria and their thermozymes in composting processes: A review. *Chemical and Biological Technologies in Agriculture, 10*(1), 1–22. https://doi.org/10.1186/s40538-023-00381-z

Gaudel, G., et al. (2024). Soil microbes, carbon, nitrogen, and the carbon to nitrogen ratio indicate priming effects across terrestrial ecosystems. *Journal of Soils and Sediments, 24*(1), 307–322. https://doi.org/10.1007/s11368-023-03609-5

Ghosh, A., Kumar, S., & Das, J. (2023). Impact of leachate and landfill gas on the ecosystem and health: Research trends and the way forward towards sustainability. *Journal of Environmental Management, 336*, Article 117708. https://doi.org/10.1016/j.jenvman.2023.117708

Gomes, H. A. R., et al. (2020). Evaluation of different secretomes produced by Clonostachys byssicola as tools to holocellulose breakdown. *International Biodeterioration & Biodegradation, 148*, Article 104880. https://doi.org/10.1016/j.ibiod.2019.104880

Greff, B., Szigeti, J., Nagy, Á., Lakatos, E., & Varga, L. (2022). Influence of microbial inoculants on co-composting of lignocellulosic crop residues with farm animal manure: A review. *Journal of Environmental Management, 302*, Article 114088. https://doi.org/10.1016/j.jenvman.2021.114088

Guo, X., Liu, H., & Wu, S. (2019). Humic substances developed during organic waste composting: Formation mechanisms, structural properties, and agronomic functions. *Science of the Total Environment, 662*, 501–510. https://doi.org/10.1016/j.scitotenv.2019.01.137

Harindintwali, J. D., Zhou, J., & Yu, X. (2020). Lignocellulosic crop residue composting by cellulolytic nitrogen-fixing bacteria: A novel tool for environmental sustainability. *Science of the Total Environment, 715*, Article 136912. https://doi.org/10.1016/j.scitotenv.2020.136912

Hoang, H. G., et al. (2022). The nitrogen cycle and mitigation strategies for nitrogen loss during organic waste composting: A review. *Chemosphere, 300*, Article 134514. https://doi.org/10.1016/j.chemosphere.2022.134514

Humbal, A., Chaudhary, N., & Pathak, B. 2023. Urbanization trends, climate change, and environmental sustainability BT—Climate change and urban environment sustainability. In B. Pathak & R. S. Dubey (eds.), *Climate change and urban environment sustainability* (pp. 151–166). Springer Nature Singapore. https://doi.org/10.1007/978-981-19-7618-6_9

Hussain, S., et al. (2024). Navigating the impact of climate change in India: A perspective on climate action (SDG13) and sustainable cities and communities (SDG11). *Frontiers in Sustainable Cities, 5*, 1308684. https://doi.org/10.3389/frsc.2023.1308684

Hutton, B., Horan, E., & Norrish, M. (2009). Waste management options to control greenhouse gas emissions–landfill, compost or incineration. In *Paper for the international solid waste association (ISWA) conference, Portugal* (pp. 12–15).

Inghels, D., Dullaert, W., & Bloemhof, J. (2016). A model for improving sustainable green waste recovery. *Resources, Conservation and Recycling, 110*, 61–73. https://doi.org/10.1016/j.resconrec.2016.03.013

Insam, H., & de Bertoldi, M. (2007). Chapter 3 Microbiology of the composting process. In L. F. Diaz, M. de Bertoldi, W. Bidlingmaier, & E. B. T.-W. M. S. Stentiford (eds.), *Compost science and technology* (pp. 25–48). Elsevier. https://doi.org/10.1016/S1478-7482(07)80006-6

Irawan, B., et al. (2019). Effect of induced compost by cellulolitic (*Aspergillus fumigatus*) and ligninolitic (*Geotrichum sp.*) fungi inoculum application on vegetative growth of red chili (*Capsicum annuum l.*). *Journal of Pure and Applied Microbiology, 13*(2), 815–821. https://doi.org/10.22207/JPAM.13.2.16

Javed, Z., Tripathi, G. D., Mishra, M., & Dashora, K. (2021). Actinomycetes—The microbial machinery for the organic-cycling, plant growth, and sustainable soil health. *Biocatalysis and Agricultural Biotechnology, 31*, Article 101893. https://doi.org/10.1016/j.bcab.2020.101893

Jurado, M., et al. (2014). Exploiting composting biodiversity: Study of the persistent and biotechnologically relevant microorganisms from lignocellulose-based composting. *Bioresource Technology, 162*, 283–293. https://doi.org/10.1016/j.biortech.2014.03.145

Jusoh, M. L. C., Manaf, L. A., & Latiff, P. A. (2013). Composting of rice straw with effective microorganisms (EM) and its influence on compost quality. *Iranian Journal of Environmental Health Science & Engineering, 10*(1), 17. https://doi.org/10.1186/1735-2746-10-17

Karnchanawong, S., & Nissaikla, S. (2014). Effects of microbial inoculation on composting of household organic waste using passive aeration bin. *International Journal of Recycling of Organic Waste in Agriculture, 3*(4), 113–119. https://doi.org/10.1007/s40093-014-0072-0

Kaza, S., Yao, L., Bhada-Tata, P., & Van Woerden, F. (2018). *What a waste 2.0 a global snapshot of solid waste management to 2050*. World Bank. Retrieved February 24, 2025, from https://www.epa.gov/recycle/composting-home

Khalib, S. N. B., Zakarya, I. A., & Izhar, T. N. T. (2018). Composting of garden waste using indigenous microorganisms (IMO) as organic additive. *International Journal of Integrated Engineering, 10*(9), 53–59. https://doi.org/10.30880/ijie.2018.10.09.026

Khalil, A., Domeizel, M., & Prudent, P. (2008). Monitoring of green waste composting process based on redox potential. *Bioresource Technology, 99*(14), 6037–6045. https://doi.org/10.1016/j.biortech.2007.11.043

Kiehbadroudinezhad, M., Merabet, A., & Hosseinzadeh-Bandbafha, H. (2024). Chapter Six—Landfill source of greenhouse gas emission. In M. R. Rahimpour, M. A. Makarem, & M. B. T.-A. and T. D. in G. G. E. Meshksar (eds.), *Capture and conversion* (pp. 123–145). Elsevier. https://doi.org/10.1016/B978-0-443-19231-9.00023-5

Kredics, L., et al. (2022). Trichoderma green mould disease of cultivated mushrooms. In N. Amaresan, A. Sankaranarayanan, M. K. Dwivedi, & I. S. Druzhinina (eds) *Advances in trichoderma biology for agricultural applications* (pp. 559–606). Springer International Publishing. https://doi.org/10.1007/978-3-030-91650-3_21

Kruker, G., Guidi, E. S., Santos, J. M., Mafra, Á. L., & Almeida, J. A. (2023). Quality of Bokashi-type biofertilizer formulations and its application in the production of vegetables in an ecological system. *Horticulturae* 9(12). https://doi.org/10.3390/horticulturae9121314

Lew, P. S., Nik Ibrahim, N. N., Kamarudin, S., Thamrin, N. M., & Misnan, M. F. (2021). Optimization of Bokashi-composting process using effective microorganisms-1 in smart composting bin. *Sensors, 21*(8). https://doi.org/10.3390/s21082847.

Li, L., Guo, X. P., Zhao, T. N., Liu, L., & Li, T. Y. (2021). Identifying the key environmental factors and bacterial communities in humification and their relationships during green waste composting. *Applied Ecology and Environmental Research, 19*(1), 45–62. https://doi.org/10.15666/aeer/1901_045062

Liu, L., Wang, S., Guo, X., Zhao, T., & Zhang, B. (2018). Succession and diversity of microorganisms and their association with physicochemical properties during green waste thermophilic composting. *Waste Management, 73*, 101–112. https://doi.org/10.1016/j.wasman.2017.12.026

Lou, X. F., & Nair, J. (2009). The impact of landfilling and composting on greenhouse gas emissions—A review. *Bioresource Technology, 100*(16), 3792–3798. https://doi.org/10.1016/j.biortech.2008.12.006

Lu, X.-L., Ding, K., Dong, X.-X., Li, G., & Ma, J. (2023). A bioaugmentation strategy for promoting the humification process during composting by microbial inoculants: A review. In K. R. Hakeem (ed.). IntechOpen. https://doi.org/10.5772/intechopen.1001316

Manu, M. K., Kumar, R., & Garg, A. (2019). Decentralized composting of household wet biodegradable waste in plastic drums: Effect of waste turning, microbial inoculum and bulking agent on product quality. *Journal of Cleaner Production, 226*, 233–241. https://doi.org/10.1016/j.jclepro.2019.03.350

Manu, M. K., Kumar, R., & Garg, A. (2017). Performance assessment of improved composting system for food waste with varying aeration and use of microbial inoculum. *Bioresource Technology, 234*, 167–177. https://doi.org/10.1016/j.biortech.2017.03.023

McSweeney, J. (2019). *Community-scale composting systems a comprehensive practical guide for closing the food system loop and solving our waste crisis.* Chelsea Green Publishing.

Mishra, S. K., & Yadav, K. D. (2021). Application of locally available microbial inoculant to accelerate green waste composting at a community level. *Bioresource Technology Reports, 16*, Article 100859. https://doi.org/10.1016/j.biortech.2008.12.006

Mishra, S. K., & Yadav, K. D. (2022). Assessment of the effect of particle size and selected physico-chemical and biological parameters on the efficiency and quality of composting of garden waste. *Journal of Environmental Chemical Engineering, 10*(3), Article 107925. https://doi.org/10.1016/j.jece.2022.107925

Mu, D., et al. (2024). Effect of Maillard reaction based on catechol polymerization on the conversion of food waste to humus. *Chemosphere, 353*, Article 141560. https://doi.org/10.1016/j.chemosphere.2024.141560

Muniyasamy, S., Anstey, A., Reddy, M. M., Misra, M., & Mohanty, A. (2013). Biodegradability and compostability of lignocellulosic based composite materials. *Journal of Renewable Materials, 1*(4), 253–272. https://doi.org/10.7569/JRM.2013.634117

Nahm, K. H. (2005). Factors influencing nitrogen mineralization during poultry litter composting and calculations for available nitrogen. *World's Poultry Science Journal, 61*(2), 238–255. https://doi.org/10.1079/WPS200455

Nair, L. G., Komal, A., & Verma, P. (2022). An insight into the principles of lignocellulosic biomass-based zero-waste biorefineries: A green leap towards imperishable energy-based future. *Biotechnology and Genetic Engineering Reviews, 38*(2), 288–338. https://doi.org/10.1080/02648725.2022.2082223

Neugebauer, M., & Sołowiej, P. (2017). The use of green waste to overcome the difficulty in small-scale composting of organic household waste. *Journal of Cleaner Production, 156*, 865–875. https://doi.org/10.1016/j.jclepro.2017.04.095

Nozhevnikova, A. N., Mironov, V. V., Botchkova, E. A., Litti, Y. V., & Russkova, Y. I. (2019). Composition of a microbial community at different stages of composting and the prospects for compost production from municipal organic waste (review). *Applied Biochemistry and Microbiology, 55*(3), 199–208. https://doi.org/10.1134/S0003683819030104

Oviedo-Ocaña, E. R., Soto-Paz, J., Domínguez, I., Sanchez-Torres, V., & Komilis, D. (2022a). A systematic review on the application of bacterial inoculants and microbial consortia during green waste composting. *Waste and Biomass Valorization, 13*(8), 3423–3444. https://doi.org/10.1007/s12649-022-01687-z

Oviedo-Ocaña, E. R., Soto-Paz, J., Torres, V. S., Castellanos-Suarez, L. J., & Komilis, D. (2022b). Effect of the addition of the Bacillus sp., Paenibacillus sp. bacterial strains on the co-composting of green and food waste. *Journal of Environmental Chemical Engineering, 10*(3), 107816. https://doi.org/10.1016/j.jece.2022.107816

Parab, C., & Yadav, K. D. (2024). A review on green waste composting, role of additives and composting methods for process acceleration. *Environmental Science and Pollution Research*, 0123456789. https://doi.org/10.1007/s11356-024-35429-6

Plaza, C., Senesi, N., Polo, A., & Brunetti, G. (2005). Acid−base properties of humic and fulvic acids formed during composting. *Environmental Science & Technology, 39*(18), 7141–7146. https://doi.org/10.1021/es050613h

Rana, K., Rana, N., & Singh, B. (2020). Chapter 10—Applications of sulfur oxidizing bacteria. In R. Salwan, V. B. T.-P. and B. A. of E. Sharma (eds.), *Physiological and biotechnological aspects of extremophiles* (pp. 131–136). Academic Press https://doi.org/10.1016/B978-0-12-818322-9.00010-1

Rawat, S., & Johri, B. N. (2013). Role of thermophilic microflora in composting. In T. Satyanarayana, J. Littlechild, & Y. Kawarabayasi (eds) *Thermophilic microbes in environmental and industrial biotechnology: biotechnology of thermophiles* (pp. 137–169). Springer Netherlands. https://doi.org/10.1007/978-94-007-5899-5_5

Reyes-Torres, M., Oviedo-Ocaña, E. R., Dominguez, I., Komilis, D., & Sánchez, A. (2018). A systematic review on the composting of green waste: Feedstock quality and optimization strategies. *Waste Management, 77*, 486–499. https://doi.org/10.1016/j.wasman.2018.04.037

Roseland, M. (2000). Sustainable community development: Integrating environmental, economic, and social objectives. *Progress in Planning, 54*(2), 73–132. https://doi.org/10.1016/S0305-9006(00)00003-9

Savitha, T., Khalifa, A., & Sankaranarayanan, A. (2022). Isolation of actinobacteria from compost samples. In D. Dharumadurai (ed.), *Methods in actinobacteriology* (pp. 39–43). Springer US. https://doi.org/10.1007/978-1-0716-1728-1_9

Shinde, R., et al. (2022). Potential of lignocellulose degrading microorganisms for agricultural residue decomposition in soil: A review. *Journal of Environmental Management, 320*, Article 115843. https://doi.org/10.1016/j.jenvman.2022.115843

Simmons, C. W., Reddy, A. P., Simmons, B. A., Singer, S. W., & Vandergheynst, J. S. (2014). Effect of inoculum source on the enrichment of microbial communities on two lignocellulosic bioenergy crops under thermophilic and high-solids conditions. *Journal of Applied Microbiology, 117*(4), 1025–1034. https://doi.org/10.1111/jam.12609

Soong, W. S., Chew, J., & Gew, L. T. (2025). Insights of bacterial communities in kitchen and agricultural waste composts for environmental sustainability: A systematic review. *International Journal of Environmental Science and Technology, 22*(3), 2073–2094. https://doi.org/10.1007/s13762-024-05898-w

Soto-Paz, J., et al. (2022). Optimization of lignocellulolytic bacterial inoculum and substrate mix for lignocellulose degradation and product quality on co-composting of green waste with food waste. *Bioresource Technology, 359*(April), Article 127452. https://doi.org/10.1016/j.biortech.2022.127452

United Nations Department of Economic and Social Affairs, P.D. (2022). *World population prospects 2022: Summary of results*. p. UN DESA/POP/2022/TR/NO. 3.

Vargas-García, M. C., Suárez-Estrella, F., López, M. J., & Moreno, J. (2007). In vitro Studies on lignocellulose degradation by microbial strains isolated from composting processes. *International Biodeterioration & Biodegradation, 59*(4), 322–328. https://doi.org/10.1016/j.biortech.2022.127452

Waqas, M., et al. (2023). Composting processes for agricultural waste management: A comprehensive review. *Processes, 11*(3), 731. https://doi.org/10.3390/pr11030731

Wei, Y., et al. (2019). Improved lignocellulose-degrading performance during straw composting from diverse sources with actinomycetes inoculation by regulating the key enzyme activities. *Bioresource Technology, 271*, 66–74. https://doi.org/10.1016/j.biortech.2018.09.081

World Bank. (2022). *Solid waste management*. Retrieved February 24, 2025, from https://www.worldbank.org/en/topic/urbandevelopment/brief/solid-waste-management

Wright, C., Gryganskyi, A. P., & Bonito, G. (2016). Fungi in composting. In D. Purchase (ed.), *Fungal applications in sustainable environmental biotechnology* (pp. 3–28). Springer International Publishing. https://doi.org/10.1007/978-3-319-42852-9_1

Xie, S., Tran, H.-T., Pu, M., & Zhang, T. (2023). Transformation characteristics of organic matter and phosphorus in composting processes of agricultural organic waste: Research trends. *Materials Science for Energy Technologies, 6*, 331–342. https://doi.org/10.1016/j.mset.2023.02.006

Yang, L., et al. (2020). Enhanced lignin degradation in tobacco stalk composting with inoculation of white-rot fungi Trametes hirsuta and Pleurotus ostreatus. *Waste and Biomass Valorization, 11*(7), 3525–3535. https://doi.org/10.1007/s12649-019-00692-z

Yang, W., Yin, Z., & Zhang, L. (2023). Silage and flue gas desulfurized gypsum as additives improve green waste composting. *Journal of Environmental Chemical Engineering, 11*(2), Article 109385. https://doi.org/10.1016/j.jece.2023.109385

Yang, W., & Zhang, L. (2022). Addition of mature compost improves the composting of green waste. *Bioresource Technology, 350*, 126927. https://doi.org/10.1016/j.biortech.2022.126927

Yu, K., et al. (2018). Application of quadratic regression orthogonal design to develop a composite inoculum for promoting lignocellulose degradation during green waste composting. *Waste Management, 79*, 443–453. https://doi.org/10.1016/j.wasman.2018.08.018

Yu, H., Xie, B., Khan, R., & Shen, G. (2019). The changes in carbon, nitrogen components and humic substances during organic-inorganic aerobic co-composting. *Bioresource Technology, 271*, 228–235. https://doi.org/10.1016/j.biortech.2018.09.088

Zeng, X., et al. (2024). Screening, gene cloning and expression of cellulase-producing strain Bacillus subtilis Xh-16. *Current Microbiology, 81*(12), 452. https://doi.org/10.1007/s00284-024-03961-w

Zhao, Y., et al. (2017). Effect of thermo-tolerant actinomycetes inoculation on cellulose degradation and the formation of humic substances during composting. *Waste Management, 68*, 64–73. https://doi.org/10.1016/j.wasman.2017.06.022

Zhu, L., et al. (2024). Fungi play a crucial role in sustaining microbial networks and accelerating organic matter mineralization and humification during thermophilic phase of composting. *Environmental Research, 254*, Article 119155. https://doi.org/10.1016/j.envres.2024.119155

Sustainable Agriculture: Transforming Rumen Waste into Compost

Rizkiy Amaliyah Barakwan, Danar Arifka Rahman, and Yulinah Trihadiningrum

Abstract Rumen waste, a by-product of livestock digestion, presents significant potential for transformation into sustainable agricultural inputs. Traditionally regarded as an environmental liability, this waste is rich in organic matter, nutrients, and microbial biomass, making it a viable resource for compost and liquid fertilizer production. This article explores innovative strategies for valorizing rumen waste through technological advancement, particularly the incorporation of nutrient enrichment and local inoculants. Rumen waste has the potential to be recycled into organic fertilizer with various composting technology developments. The rumen composting process that is rich in nutrients can be combined, such as an anaerobic-anoxic-oxic system. In addition, the rumen can be mixed with other organic materials to adjust and enrich the nutrient content in composting. Rumen waste can also be a source of organic material degrading microorganism cultures that can be added to the composting process of other organic materials. In addition, rumen can also be used as a bioactivator to accelerate the composting process. Composting involving rumen waste can improve the nutrient quality of the compost fertilizer produced.

Keywords Composting · Nutrient · Organic fertilizer · Rumen · Waste

R. A. Barakwan (✉) · D. A. Rahman
Environmental Engineering Study Program, Department of Biology, Faculty of Science and Technology, Universitas Airlangga, Kampus C Mulyorejo, 60115 Surabaya, Indonesia
e-mail: rizkiy.amaliyah@fst.unair.ac.id

Sustainable Environmental Systems and Infrastructure (Sustain) Research Group, Universitas Airlangga, Kampus C Mulyorejo, 60115 Surabaya, Indonesia

D. A. Rahman
e-mail: danar.arifka@fst.unair.ac.id

Y. Trihadiningrum
Department of Environmental Engineering, Faculty of Civil Planning and Geo Engineering, Institut Teknologi Sepuluh Nopember, Kampus ITS Sukolilo, Surabaya, Indonesia
e-mail: trihadiningrum@gmail.com

A. Z. Yaser et al. (eds.), *Compost and Liquid Organic Fertilizers*,
https://doi.org/10.1007/978-981-95-8561-8_3

1 Introduction

Sustainable agriculture is increasingly recognized as a critical strategy for mitigating environmental challenges and enhancing resource efficiency in modern farming systems (Alfadhly et al., 2024). Livestock waste management, particularly that associated with ruminants, has emerged as a major concern due to its substantial contribution to environmental pollution and soil degradation (Ramdinthara & Shanthi Bala, 2019; Saxena & Sandhwar, 2023; Scanes, 2017). As of 2023, the total number of ruminant livestock in Indonesia is approximately 63 million, leading to an average waste of 15 kg/day and 946 million kg of livestock waste per day in total (Sahara, 2024). Considering the increasing pressure on natural resources, converting waste by-products into value-added agricultural inputs is aligned with integrated waste management and circular economy principles (Khoshnevisan et al., 2021; Ramirez et al., 2021).

Rumen waste, composed primarily of undigested feed residues and complex microbial biomass, has traditionally been viewed as a disposal liability (Bohórquez-Sandoval et al., 2020). Conventional approaches to its management, which often involve landfilling or direct application as raw manure, are increasingly investigated due to associated risks of nutrient imbalance and potential contamination (Pancapalaga et al., 2021; Al Zahra et al., 2024). Recent advances in waste valorization emphasize that targeted processing of rumen waste can improve nutrient availability and reduce environmental impacts (Pancapalaga et al., 2021).

The transformation of rumen waste into compost and liquid fertilizers offers a promising avenue for sustainable agriculture by recycling nutrients back into the soil (Sharma et al., 2024). Processes such as aerobic composting and controlled fermentation enhance the stabilization of organic matter and promote the development of bioactive compounds that can stimulate plant growth (Martins Oliveira et al., 2019; Zouitina et al., 2019). The integration of advanced biotechnological interventions, including nutrient enrichment and incorporation of local inoculant, not only accelerates the decomposition process but also improves the quality and efficacy of the produced compost and fertilizers.

Despite the recognized potential of this approach, the specific utilization of rumen waste as a primary input for high-grade compost and liquid fertilizers remains underexplored in the literature. Existing studies have mainly focused on managing livestock wastes into compost (Cruz et al., 2024; Giagnoni et al., 2020) or fertilizer (Lesik et al., 2019; Pancapalaga et al., 2021) only. This chapter aims to address the identified research gaps by thoroughly examining the conversion pathways of rumen waste into compost. Emphasis is placed on technological innovation by composting system combination, nutrient enrichment, rumen as bioactivator, and the addition of local inoculants. By presenting an in-depth analysis supported by recent advances, this report aspires to make a significant contribution to sustainable agricultural practices and waste management solutions.

2 Characteristic of Rumen Waste

Understanding ruminant characteristics is essential for optimizing its transformation into valuable agricultural inputs. Chemically, rumen waste is rich in organic constituents. Studies have shown that dried cattle rumen digestion contains approximately 96.34% dry matter, with carbohydrates constituting about 70.64% of this content. Additionally, it contains around 14.40% ash and 40.79% nitrogen-free extract, indicating a substantial presence of digestible nutrients (Agolisi et al., 2020). In recent studies, researchers have shown that rumen waste typically exhibits a neutral to slightly acidic pH (Lesik et al., 2019), high moisture (often exceeding 60%) (Colorado et al., 2022), and a substantial proportion of nutrient on a dry matter (Ihoeghian et al., 2022) as shown on Table 1. These conditions, such as high nutrients, moist, temperature, and pH, favor the growth of diverse microorganisms (Xing et al., 2020). The microbial ecosystem within the rumen is not only diverse but also functionally significant. Bacteria dominate this ecosystem, accounting for a substantial portion of microbial biomass. These microbes are adept at degrading structural carbohydrates, leading to the production of VFAs and gases such as methane and carbon dioxide (Zhang et al., 2017). The presence of such a robust microbial community in rumen waste suggests its utility in bioconversion processes, where these microbes can facilitate the transformation of organic waste into valuable end-products like compost and liquid fertilizers.

Furthermore, the high moisture content in rumen waste supports rapid microbial growth. The organic matter content is especially crucial because it represents the carbon-rich fraction necessary for the composting process. Nutrient contents in terms of total nitrogen and phosphorus are important for evaluating the agronomic potential of the resulting compost. Research indicates that total nitrogen levels in these substrates typically range from 0.7 to 2.5% (dry weight), with phosphorus

Table 1 Rumen waste characteristics

Characteristic	Value	References
Moisture contents (%)	56–89	(Colorado et al., 2022; Ihoeghian et al., 2022; Sugito & Ratnawati, 2020)
C/N ratio	23.7	(Ihoeghian et al., 2022)
Nitrogen (%)	0.7–2.05	(Ihoeghian et al., 2022; Lesik et al., 2019)
Phosphorus (%)	1.9–2.5	(Lesik et al., 2019)
Organic C (%)	48.6	(Ihoeghian et al., 2022)
Potassium (%)	0.7	(Pancapalaga et al., 2021)
pH	5.6–6.8	(Darwin et al., 2022; Ihoeghian et al., 2022)
Total solids (%)	10.85	(Ihoeghian et al., 2022)
Volatile solids (%)	89	(Ihoeghian et al., 2022)

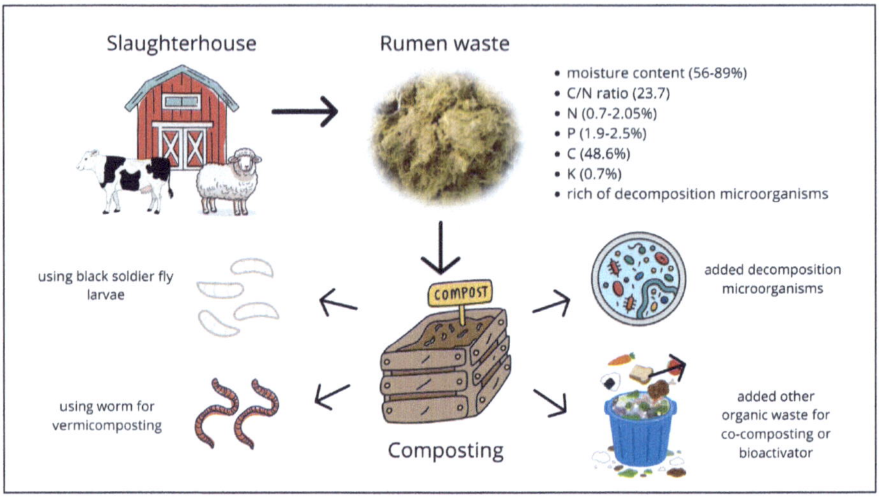

Fig. 1 Source, characteristics, and composting processes of rumen waste

around 2% (Ihoeghian et al., 2022; Lesik et al., 2019). These values influence both the compost quality and the nutrient cycling when the compost or liquid fertilizer is applied to soil. The carbon-to-nitrogen (C/N) ratio, often around 20 for rumen waste analogues (Ihoeghian et al., 2022), is favorable to ensure the balance between microbial proliferation and the stabilization of organic matter. This balance plays a key role in reducing nitrogen losses during compost maturation and optimizing nutrient release once the compost is used as a soil amendment (Waqas et al., 2023). Source, characteristics, and potential composting systems of rumen waste can be seen in Fig. 1.

3 Composting Process of Rumen Waste

The abundance of organic matter and nutrients in rumen waste makes it a valuable resource for composting, as a sustainable practice in agriculture. Composting process transforms complex organic residues, which are rich in undigested fibers and microbial biomass from the rumen, into stabilized organic matter that can enhance soil fertility. Incorporating composting into waste management strategies not only minimizes environmental pollution but also recycles nutrients for crop production (Sayara et al., 2020). During the composting process, rumen waste passes through several key stages: mesophilic, thermophilic, and finally, maturation phase (Waqas et al., 2023). The essence of the composting process converts this waste into stable, nutrient-rich compost through microbial activity. Various composting methods can be used, each with different operational conditions and results. The choice of method

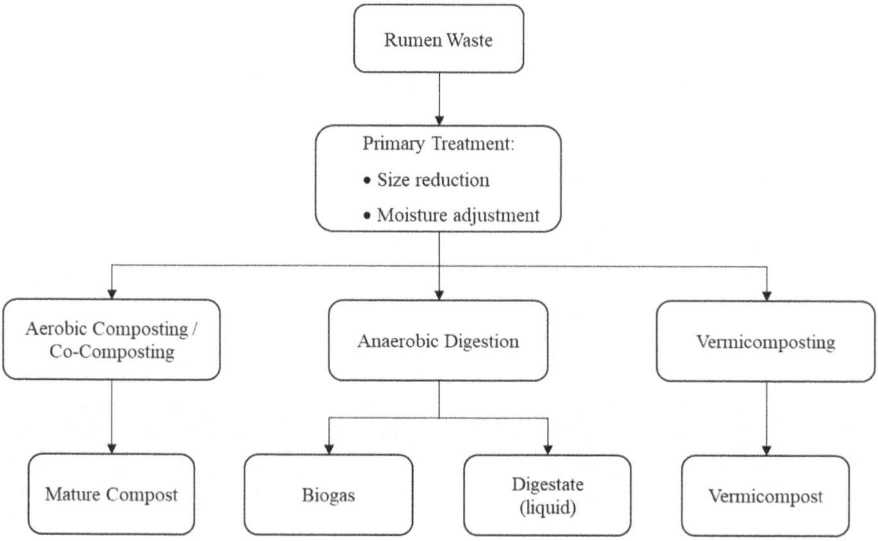

Fig. 2 Schematic overview of rumen waste-to-compost pathways

depends on factors such as available resources, desired compost quality, and environmental considerations. The schematic overview presented in Fig. 2 illustrates key transformation routes of rumen waste into valuable agricultural inputs via aerobic composting, anaerobic digestion, and vermicomposting.

3.1 Aerobic Co-composting Process

Aerobic composting involves the decomposition of organic matter by aerobic microorganisms in the presence of oxygen. This method is characterized by higher temperatures that facilitate the breakdown of complex organic compounds and the elimination of pathogens (Cai et al., 2019). Studies have shown that aerobic composting of rumen waste can lead to significant reductions in organic matter and the production of stable compost suitable for agricultural use (Afifah et al., 2022; Guo et al., 2019). However, maintaining optimal oxygen levels and moisture content is crucial to ensure efficient decomposition and prevent the formation of anaerobic zones that could lead to odor issues and emissions (Jiang et al., 2022), since it can produce ammonia and greenhouse gases (Chen et al., 2021; Jiao et al., 2021). Composting techniques included in aerobic composting are windrow, aerated static pile, and in-vessel composting.

Co-composting is an advanced composting technique that integrates multiple organic waste streams to enhance the degradation process and improve the quality of the resulting compost. Co-composting involves combining this nutrient-rich byproduct with other organic materials, such as leaf litter or contaminated soils, to optimize microbial activity and nutrient balance. Modifications to co-composting, such as the addition of bioactivators will further stimulate microbial activity, leading to increased degradation rates and compost quality (Budiharto et al., 2023). This method not only accelerates the composting process but also contributes to the bioremediation of polluted environments (Giagnoni et al., 2020).

3.2 Anaerobic Digestion

In contrast to aerobic composting, anaerobic digestion occurs in the absence of oxygen, relying on anaerobic microorganisms to break down organic matter (Lin et al., 2019). This process is slower than aerobic composting, but has the advantage of producing biogas (Enitan-Folami & Swalaha, 2023), which can be utilized as a renewable energy source. In the context of rumen waste, anaerobic composting can effectively reduce waste volume and produce a nutrient-rich digestate that can be used to improve soil quality (Lesik et al., 2019; Agolisi et al., 2020). However, this process requires careful management to control odor and methane emissions, which is a potent greenhouse gas.

3.3 Vermicomposting

Vermicomposting utilizes earthworms to decompose organic waste, including rumen content. The worms consume the waste and excrete castings that are rich in nutrients and beneficial microorganisms (Mohite et al., 2024). This method not only accelerates the composting process but also enhances the quality of the final product. Research indicates that vermicomposting can significantly reduce the carbon-to-nitrogen ratio and increase the availability of essential nutrients like nitrogen and phosphorus in the compost (Cruz et al., 2024; Mohite et al., 2024). Moreover, the process is environmentally friendly and low-tech, making it accessible for various scales of operation (Pattnaik & Mishra, 2019).

3.4 Mixed Methods

Combining different composting methods can leverage the advantages of each to optimize the composting process. For instance, initiating composting with an aerobic phase to rapidly decompose easily degradable materials, followed by a vermicomposting phase to further stabilize the compost and enhance its nutrient content, has been shown to produce high-quality compost (Iqbal et al., 2010). Such integrated approaches can enhanced efficiency of the composting process (Lim et al., 2017), improve product quality (Kasinski et al., 2016), and be more cost-effective (Lim et al., 2017).

4 Innovation of Rumen Waste in the Composting Process

4.1 Anaerobic-Anoxic-Oxic (A^2/O) System

The A^2/O composting method is adopted from the biological high nutrient removal in wastewater treatment systems. This treatment involves three different stages, namely: anaerobic, anoxic, and oxic phases. Organic matter is converted into methane and ammonia in the anaerobic phase. Ammonia, organic matter residues, and sulfide compounds are oxidized through the nitrification process during the oxic phase, and the denitrification process occurs in the anoxic phase (Vo et al., 2022).

Research by Ratnawati et al. (2016) showed that the optimum anaerobic-anoxic–oxic period for rumen waste composting was 25–10–15 days. The water content decreased from 83 to 72% in the end of composting. The concentration of organic and inorganic nitrogen decreased from 2.04 to 1.19% and 0.29 to 0.07% at the end of the composting process, respectively. The water content of the compost exceeds the technical requirements for organic fertilizer based on SNI 19-7030-2004, which is a maximum of 50%. Meanwhile, the concentration of organic nitrogen meets the minimum limit, which is more than 0.4%. This study proves that the composting method with the A^2/O system is appropriate for processing solid waste with high nutrient content such as cow rumen.

The concentration of organic nitrogen decreases due to the transformation and transfer of organic nitrogen concentration during composting process. Biological transformation occurs through the processes of ammonification, nitrification, denitrification, and immobilization. Transfer of organic nitrogen concentration from rumen waste to leachate and atmosphere. Conversion of nitrogen to NH_3, N_2O, and N_2 (Ratnawati et al., 2016; Vo et al., 2022).

4.2 Nutrient Enrichment in Rumen Waste Composting

Adjustment of nutrient concentration of compost can be done by mixing or adding different organic materials during composting. Composting with two or more organic materials is called co-composting. In rumen waste composting, in addition to balancing nutrient concentration, the addition of other organic materials is useful for improving manure handling, reducing odor, flies, and other vector problems, and reducing weed seeds and pathogens.

Research by Roy et al. (2013) showed that the addition of biogas sludge (R/S = 40/1) and cow blood (R/B = 10/1) in rumen (R) composting can increase organic matter removal, protein concentration, and adjust the C/N ratio. The increase in organic matter removal is caused by decomposition by anaerobic microorganisms in rumen waste, cow blood, and biogas sludge. The increase in protein concentration can be caused by the high nitrogen content in cow blood and biogas sludge can increase the growth rate of nitrifying microorganisms during the anaerobic period of the composting. The C/N ratio at the end of composting: 100% rumen, a mixture of rumen with biogas sludge, and a mixture of rumen with cow blood were 31.42, 23.88, and 22.83, respectively. The C/N ratio of compost that approaches the C/N ratio requirements for organic fertilizers in Indonesia (10–20) is the result of composting a mixture of rumen with cow blood and biogas sludge (National Standardization Institute, 2004). However, the pH values at the end of composting: 100% rumen, rumen with biogas sludge mixture, and rumen with cow blood mixture were 7.58, 7.93, and 8.15, respectively. The pH value in the composting process should not be more than 8.5 because pH more than 8.5 encourages the conversion of nitrogen compounds into ammonia, which can increase alkalinity and slow down the composting process due to the loss of nitrogen compounds that are volatilized into ammonia and cause odor problem (Roy et al., 2013).

Research by Sugito and Ratnawati (2020) showed that composting with a mixture of rumen waste and rice straw obtained optimum results at a composition of 40% rumen and 60% rice straw. It was caused the rice straw having a high carbon content compared to rumen. Meanwhile, the nitrogen content of rice straw is lower than rumen. Therefore, the addition of organic material rich in carbon can balance nutrients for microorganisms during composting of rumen waste. The C/N ratio decreased from 15.91 to 13.25 at the end of composting. The concentration of phosphorus and potassium increased, respectively from 4.27 to 6.12% and 5.46 to 7.55%. The pH value fluctuated from 6.71 to 6.93 during composting. Changes in the C/N ratio during the composting process indicate the decomposition process of organic materials. The C/N ratio value at the end of composting met the requirements for organic fertilizer based on SNI 19-7030-2004, namely 10–20 (National Standardization Institute, 2004). The highest phosphorus value of 6.12% met the standard value for organic fertilizer, which is >0.1%. The highest potassium value of 7.55% met the requirements for organic fertilizers, which is >0.2%. Phosphorus content is needed by plants for root growth, assimilation, and increasing the number of flowers, fruits, and seeds. Meanwhile, potassium content is useful for plant growth and development (opening and closing stomata; the process of transporting nutrients from roots to leaves; accumulation and translocation of sucrose; filling seeds, tubers, and root growth; cellulose synthesis; strengthening cell walls, stems, and enzymatic systems; and protein synthesis). The concentration of phosphorus and potassium increases due to the breakdown of the C chain due to the increasing number of microorganisms in the process of degrading organic compounds in composting.

4.3 Addition of Local Inoculant

The rumen waste is rich in microorganisms. Microorganisms identified in the rumen waste include *Bacillus sp.*, *cellulolytic* microorganisms, and *Azospirillum sp.* (Indasah & Fitriani, 2021). *Bacillus sp.* and *cellulolytic* microorganisms are microorganisms that have a role in the decomposition of organic matter in the composting process, while *Azospirillum sp.* has a role in binding nitrogen in compost and increasing plant growth (Arifin et al., 2019; Ariyanti et al., 2023; Rosmalia, 2019). In addition, anaerobic fungi is one of the microbes that play an important role in rumen fiber degradation because they can produce cellulase enzymes and penetrate feed particles. *Caecomyces, Neocallimastix,* and *Piromyces* are rumen anaerobic fungi isolated from buffalo rumen with different morphological characteristics (Agustina et al., 2022).

Indasah and Fitriani (2021) stated that the addition of microorganism cultures from cow rumen to fruit waste composting increased the total bacterial population; concentration of nitrogen, phosphorus, and potassium; and the C/N ratio that met organic fertilizer standards based on SNI 19-7030-2004 (National Standardization Institute, 2004).

4.4 Rumen Waste as Composting Bioactivator

The decomposition of organic waste in composting process can be accelerated by adding bioactivators. One alternative bioactivator that can be used is the rumen waste (Dewilda et al., 2021). The rumen waste contains many enzymes, bacteria, protozoa, and fungi that have the potential to degrade organic compounds in the composting process (Pancapalaga et al., 2021; Suhardjadinata et al., 2018).

Dewilda et al. (2021) stated that in aerobic composting of food waste, the most optimal composting process is with the addition of rumen bioactivator. Composting with the addition of rumen as a bioactivator runs the fastest, which was 6–7 days compared to other bioactivators. The average C/N ratio and water content at the end of composting reached 19.55 and 29.29%, respectively. Composting of food waste with rumen bioactivator also has the highest potassium and phosphorus content compared to the addition of other bioactivators, namely 2.2 and 0.36%, respectively. The values of these parameters meet the organic fertilizer standards according to SNI 19-7030-2004 (National Standardization Institute, 2004).

Research by Pancapalaga et al. (2021) stated that the addition of 15% rumen bioactivator to goat manure composting provided optimum compost results compared to the addition of 5 and 10% of rumen. All concentrations of N (1.59%), P (0.15%), K (0.72%), C-organic (30.6%), C/N ratio (19.24), and water content (45.86%) met the organic fertilizer standards in Indonesia based on SNI 19-7030-2004 (National Standardization Institute, 2004). The increase in potassium due to the composting process occurs due to the release of K + ions from cation exchange and decomposition of dissolved organic matter in the rumen contents (Yu & Zhang, 2017).

5 Conclusions

Rumen is organic waste that is rich in nutrients needed by plants. Application of rumen waste in the composting process to be recycled into organic compost fertilizer can be used as: raw material for composting; can be composted together with other organic materials; as a bioactivator; and the addition of rumen microorganism culture results in composting process other organic materials. The combination of anaerobic-anoxic–oxic composting systems can also be an alternative in transforming rumen into organic fertilizer. Development of rumen waste processing with composting produces compost with nutrient content that meets the requirements for organic fertilizer in Indonesia.

References

Afifah, N., Auvaria, S. W., Nengse, S., Utama, T. T., & Yusrianti. (2022). Comparative study of composting methods by windrow, hollow brick, and vermicomposting. *Jurnal Kesehatan Lingkungan: Jurnal dan Aplikasi Teknik Kesehatan Lingkungan, 19*(1), 121–128. https://doi.org/10.31964/JKL.V19I1.468

Agolisi, H. M., Abonuusum, A., & Jacob, A. (2020). Nutritional value of dried rumen digesta from cattle, sheep and goat: A case of Bolgatanga Abattoir, Ghana. *Animal Health Journal, 2*(1), 7–15. https://doi.org/10.47941/ahj.445

Agustina, S., Wiryawan, K. G., Suharti, S., & Meryandini, A. (2022). The enrichment process and morphological identification of anaerobic fungi isolated from buffalo rumen. *Biodiversitas, 23*(1), 469–477. https://doi.org/10.13057/biodiv/d230150

Al Zahra, W., Midelaar, C. E., Oosting, S. J., & Boer, I. J. M. (2024). Nutrient imbalances of smallholder dairy farming systems in Indonesia: The relevance of manure management. *Agricultural Systems, 218*, Article 103961. https://doi.org/10.1016/j.agsy.2024.103961

Alfadhly, N. K. Z., Al-Temimi, A. A., Alkanan, Z. T., Altemimi, A. B., Younis, M. I., Giuffre, A. M., & Abedelmaksoud, T. G. (2024). Sustainable agriculture development for food safety and nutrition. *Food Systems, 7*(3), 491–504. https://doi.org/10.21323/2618-9771-2024-7-3-491-504

Arifin, Z., Gunam, I. B. W., Antara, N. S., & Setiyo, Y. (2019). Isolation of cellulose degrading bacteria from compost. *Jurnal Rekayasa Dan Manajemen Agroindustri, 7*(1), 30–37.

Ariyanti, M., Rosniawaty, S., & Nadiyah, F. (2023). Pengaruh Aplikasi Bacillus sp. dan Kompos Tandan Kosong Kelapa Sawit terhadap Pertumbuhan Bibit Kelapa Sawit. *Agrikultura, 34*(2), 306. https://doi.org/10.24198/agrikultura.v34i2.43170

Budiharto, I. W., Wijaya, S. K., Budiono, D., & Sari, C. B. (2023). Utilisation of ruminant waste microorganisms in an implementation co-composting method in bioremediation of used oil contaminated soil. *Environment and Ecology Research, 11*(6), 973–978. https://doi.org/10.13189/EER.2023.110608

Cai, X., Huang, X., Ji, H., Xie, Q., & Ge, Y. (2019). A study on aerobic composting of organic waste. *AIP Conference Proceedings, 2106*, 20025. https://doi.org/10.1063/1.5109348

Chen, Y. X., Lu, Y. M., Mo, J. T., Pan, J. J., & Zhang, J. (2021). Mechanism and control strategy of ammonia and nitrous oxide during composting of municipal solid wastes. *E3S Web of Conferences, 271*, 02003. https://doi.org/10.1051/e3sconf/202127102003

Colorado, A., Morales, O., Ossa, D., Amell, A., & Chica, E. (2022). Modeling the optimal condition for drying rumen contents using a solar greenhouse dryer. *Case Studies in Thermal Engineering, 30*, Article 101678. https://doi.org/10.1016/j.csite.2021.101678

Cruz, G. S. J., Hermes, P. H., Miriam, S. V., & Lopez, A. M. (2024). Benefits of vermicompost in agriculture and factors affecting its nutrient content. *Journal of Soil Science and Plant Nutrition, 24*(3), 4898–4917. https://doi.org/10.1007/s42729-024-01880-0

Darwin, Humaira, T., & Muliawati, A. (2022). Identification and characterization of acidosis on in vitro rumen fermentation with feeds based on grass, rice bran, concentrate, and tofu pulp. *Journal of Applied Biology and Biotechnology, 10*, 53–58. https://doi.org/10.7324/JABB.2022.10s108

Dewilda, Y., Silvia, S., Riantika, M., & Zulkarnaini. (2021). Food waste composting with the addition of cow rumen using the takakura method and identification of bacteria that role in composting. *IOP Conference Series: Materials Science and Engineering, 1041*(1), 012028. https://doi.org/10.1088/1757-899x/1041/1/012028

Enitan-Folami, A. M., & Swalaha, F. M. (2023). Anaerobic treatment system: a sustainable clean environment and future hope of renewable energy production. In *Handbook of environmental chemistry* (pp. 169–198). Springer Science and Business Media Deutschland GmbH. https://doi.org/10.1007/698_2021_792

Giagnoni, L., Martellini, T., Scodellini, R., Cincinelli, A., & Renella, G. (2020). Co-composting: An opportunity to produce compost with designated tailor-made properties. In *Organic waste composting through nexus thinking: Practices, policies, and trends* (pp. 185–211). https://doi.org/10.1007/978-3-030-36283-6_9

Guo, X. X., Liu, H. T., & Wu, S. B. (2019). Humic substances developed during organic waste composting: Formation mechanisms, structural properties, and agronomic functions. *Science of the Total Environment, 662*, 501–510. https://doi.org/10.1016/j.scitotenv.2019.01.137

Ihoeghian, N. A., Amenaghawon, A. N., Ajieh, M. U., Oshoma, C. E., Ogofure, A., Erhunmwunse, N. O., Edosa, V. I. O., Tongo, I., Obuekwe, I. S., Isagba, E. S., Emokaro, C., Ezemonye, L. I. N., Lag-Brotons, A. J., Semple, K. T., & Martin, A. D. (2022). Anaerobic co-digestion of cattle rumen content and food waste for biogas production: Establishment of co-digestion ratios and kinetic studies. *Bioresource Technology Reports, 18*, Article 101033. https://doi.org/10.1016/J.BITEB.2022.101033

Indasah, I., & Fitriani, N. (2021). Rotten fruit and cow rumen as local microorganisms for producing high-quality compost. *International Journal of Integrated Engineering, 13*(3), 9–19. https://doi.org/10.30880/ijie.2021.13.03.002

Iqbal, M. K., Shafiq, T., & Ahmed, K. (2010). Effect of different techniques of composting on stability and maturity of municipal solid waste compost. *Environmental Technology, 31*(2), 205–214. https://doi.org/10.1080/09593330903426703

Jiang, J., Wang, Y., Yu, D., & Hou, R. (2022). Enabling environment sanitation and financing by composting technologies. In *Current developments in biotechnology and bioengineering: Advances in composting and vermicomposting technology*, pp. 345–366. Elsevier. https://doi.org/10.1016/B978-0-323-91874-9.00013-9

Jiao, M., et al. (2021). Opportunities and challenges in a livestock manure cleaner composting process [畜禽粪污清洁堆肥——机遇与挑战]. *Journal of Agro-Environment Science, 40*(11), 2361–2371. https://doi.org/10.11654/jaes.2021-1138

Kasinski, S., Slota, M., Markowski, M., & Kaminska, A. (2016). Municipal waste stabilization in a reactor with an integrated active and passive aeration system. *Waste Management, 50*, 31–38. https://doi.org/10.1016/j.wasman.2016.02.012

Khoshnevisan, B., Duan, N., Tsapekos, P., Awasthi, M. K., Liu, Z., Mohammadi, A., Angelidaki, I., Tsang, D. C. W., Zhang, Z., Pan, J., Ma, L., Aghbashlo, M., Tabatabei, M., & Liu, H. (2021). A critical review on livestock manure biorefinery technologies: Sustainability, challenges, and future perspectives. *Renewable and Sustainable Energy Reviews, 135*, Article 110033. https://doi.org/10.1016/j.rser.2020.110033

Lesik, M. M. N. N., Dadi, O., Wahida, Andira, G., & Laban, S. (2019). Nutrient analysis of liquid organic fertilizer from agricultural waste and rumen liquid. *IOP Conference Series: Earth and Environmental Science, 343*(1), 012178. https://doi.org/10.1088/1755-1315/343/1/012178

Lim, L. Y., Bong, C. P. C., Lee, C. T., & Klemes, J. J. (2017). Review on the current composting practices and the potential of improvement using two-stage composting. *Chemical Engineering Transactions, 61*, 1051–1056. https://doi.org/10.3303/CET1761173

Lin, L., Xu, F., Ge, X., & Li, Y. (2019). Biological treatment of organic materials for energy and nutrients production—Anaerobic digestion and composting. In *Advances in bioenergy* (pp. 121–181). Elsevier Inc. https://doi.org/10.1016/bs.aibe.2019.04.002

Martins Oliveira, O. A., Amaral, A. G., Pereira, K. A., & Campos, J. C. D. (2019). Use of modifying additives of ruminal fermentation in beef cattle[Utilização de aditivos modificadores da fermentação ruminal em bovinos de corte]. *Revista em Agronegocio e Meio Ambiente, 12*(1), 287–311. https://doi.org/10.17765/2176-9168.2019v12n1p287-311

Mohite, D. D., Chavan, S. S., Jadhav, V. S., Kanase, T., Kadam, M. A., & Singh, A. S. (2024). Vermicomposting: A holistic approach for sustainable crop production, nutrient-rich bio fertilizer, and environmental restoration. *Discover Sustainability, 5*(1), 60. https://doi.org/10.1007/s43621-024-00245-y

National Standardization Institute. (2004). *Compost spesification from domestic of organic waste* (p. 12). National Standardization Institute.

Pancapalaga, W., Suyatno, S., & Sedlacek, D. (2021). The use of rumen contents as bio-activators for fermentation in goat manure fertilizer production. *E3S Web of Conferences, 226*, 5–9. https://doi.org/10.1051/e3sconf/202122600048

Pattnaik, A. K., & Mishra, G. (2019). Sustainable agriculture through vermicomposting technology. *International Journal of Psychosocial Rehabilitation, 23*(5), 735–739. https://doi.org/10.37200/IJPR/V23I5/PR190682

Ramdinthara, I. Z., & Shanthi Bala, P. (2019). Issues and challenges in smart farming for sustainable agriculture. In *Modern techniques for agricultural disease management and crop yield prediction* (pp. 1–22). IGI Global. https://doi.org/10.4018/978-1-5225-9632-5.ch001

Ramirez, J., McCabe, B. K., Jensen, P. D., & Speight, R. (2021). Wastes to profit: A circular economy approach to value-Addition in livestock industries. *Animal Production Science, 61*(6), 541–550. https://doi.org/10.1071/AN20400

Ratnawati, R., Trihadiningrum, Y., & Juliastuti, S. R. (2016). Composting of rumen content waste using anaerobic-anoxic-oxic (A22/O) system. *Journal of Solid Waste Technology and Management, 42*(2), 98–106. https://doi.org/10.5276/JSWTM.2016.98

Rosmalia, A. (2019) 'Peranan Bakteri Azospirillum sp. dan Kaitannya dengan Peningkatan Produksi Hijauan Pakan. *ReaerchGate*, 1–10.

Roy, B. C., Khan, M. R. I., Rahman, M. M., Salleh, M. A. M., & Ahsan, M. A. (2013). Development of a convenient method of rumen content composting. *Journal of Animal and Veterinary Advances, 12*(December), 1439–1444.

Sahara, D. (2024). Livestock waste as alternative energy for rural households: A review. *IOP Conference Series: Earth and Environmental Science, 1292*(1). https://doi.org/10.1088/1755-1315/1292/1/012002

Sandoval, L. B., Molano, F. G., Arango, W. M., Bejarano, J. C., & Soler, N. P. (2020). Vermicomposting: A transformation alternative for rumen content generated in slaughterhouses[Vermicompostaje: Una alternativa de transformación del contenido ruminal generado en mataderos]. *Revista Facultad Nacional de Agronomia Medellin, 73*(2), 9201–9212. https://doi.org/10.15446/rfnam.v73n2.80104

Saxena, D., & Sandhwar, V. K. (2023). Cow farm wastes: A bioresource for sustainable development. In *Bio-based materials and waste for energy generation and resource management: Present and emerging waste management practices: Volume 5 of advanced zero waste tools* (pp. 411–429). Elsevier. https://doi.org/10.1016/B978-0-323-91149-8.00017-X

Sayara, T., Salimia, R. B., Hawamde, F., & Sanchez, A. (2020). Recycling of organic wastes through composting: Process performance and compost application in agriculture. *Agronomy, 10*(11). https://doi.org/10.3390/agronomy10111838

Scanes, C. G. (2017). Impact of agricultural animals on the environment. In *Animals and human society* (pp. 427–449). Elsevier. https://doi.org/10.1016/B978-0-12-805247-1.00025-3

Sharma, P., Sharma, P., & Thakur, N. (2024). Sustainable farming practices and soil health: A pathway to achieving SDGs and future prospects. *Discover Sustainability, 5*(1), 250. https://doi.org/10.1007/s43621-024-00447-4

Sugito, S., & Ratnawati, R. (2020). Aerobic composting of rumen content waste and rice straw at different C/N ratios. *Journal of Physics: Conference Series, 1469*(1). https://doi.org/10.1088/1742-6596/1469/1/012008

Suhardjadinata, S., Pangesti, D., & Tedjaningsih, T. (2018). Aplikasi pupuk organik limbah rumah potong hewan untuk meningkatkan kesuburan tanah dan produktivitas padi. *Jurnal AGRO, 5*(1), 39–47. https://doi.org/10.15575/1675

Vo, D.T., Phan, H.V.T., Hoang, L.T.T.T., Nguyen, V.K., Tran, T.N., & Dao, M.T. (2022). Application of an anaerobic-anoxic-oxic-oxic (AAO/O) model to the treatment of real domestic wastewater. *Journal of Chemistry, 2022*. https://doi.org/10.1155/2022/9456026

Waqas, M., Hashim, S., Humphries, U. W., Ahmad, S., Noor, R., Shoaib, M., Naseem, A., Hlaing, P. T., & Lin, H. A. (2023). Composting processes for agricultural waste management: A comprehensive review. *Processes, 11*, 731. https://doi.org/10.3390/PR11030731

Xing, B. S., Han, Y., Wang, X. C., Wen, J., Cao, S., Zhang, K., Li, Q., & Yuan, H. (2020). Persistent action of cow rumen microorganisms in enhancing biodegradation of wheat straw by rumen fermentation. *Science of the Total Environment, 715*, Article 136529. https://doi.org/10.1016/j.scitotenv.2020.136529

Yu, X. Z., & Zhang, F. F. (2017). Distribution of phenolic compounds in rice seedlings under cr exposure. *Applied Environmental Biotechnology*, 2(2), 29–36. https://doi.org/10.26789/AEB.2017.01.004

Zhang, L., Chung, J., Jiang, Q., Sun, R., Zhang, J., Zhong, Y., & Ren, N. (2017). Characteristics of rumen microorganisms involved in anaerobic degradation of cellulose at various pH values. *RSC Advances, 7*(64), 40303–40310. https://doi.org/10.1039/C7RA06588D

Zouitina, M., Echarrafi, K., El Hassani, I., & El Haji, M. (2019). Optimization of solid waste composting: A literature review and perspective for fast composting. *Advances in Intelligent Systems and Computing, 913*, 73–83. https://doi.org/10.1007/978-3-030-11881-5_7

Composting Processes and Techniques in Urban Forestry Nursery Application

Andy Russel Mojiol

Abstract Composting is a viable and environmentally sustainable method for converting organic waste into nutrient-dense soil, essential for urban forestry nurseries. These nurseries are essential for improving urban environments by growing healthy plants. It also offers a budget-friendly method for managing waste while increasing soil quality. This chapter explores composting practices and methods adapted for urban forestry nurseries, contrasting them with industrial composting in terms of scale (small vs. large), processes (manual vs. mechanized), and waste handling capacity (targeted organic inputs vs. diverse bulk waste). It starts with the basics of composting, involving important elements like carbon-to-nitrogen ratios (C:N), moisture levels, and aeration. Different methods such as aerobic composting, vermicomposting, static pile composting, and Bokashi composting are studied and described to show their efficacy across different scales and conditions. The chapter also profiles the stages of composting, from the breakdown of materials and removal of pathogens during the thermophilic phase to the development of compost. It addresses challenges unique to urban nurseries, including limited space, waste sorting, and odour management, planning practical solutions to overcome these issues. Additionally, the advantages of composting are emphasized, such as improved soil richness, decreased reliance on chemical fertilizers, and reduced environmental effect. The use of compost in nursery practices like potting mixes, soil amendments, and mulching explains its flexibility in helping healthy plant growth. In conclusion, the chapter highlights the significance of adopting sustainable composting methods to amend nursery operations and encourage urban forestry programs. This discussion serves as a thorough guide for employing composting practices that meet the specific needs of urban forestry nurseries.

Keywords Composting · Urban forest nursery · Soil amendment · Organic waste · Green waste management

---

A. R. Mojiol (✉)
Faculty of Tropical Forestry, Universiti Malaysia Sabah, Kota Kinabalu, Malaysia
e-mail: andy@ums.edu.my

A. Z. Yaser et al. (eds.), *Compost and Liquid Organic Fertilizers*,
https://doi.org/10.1007/978-981-95-8561-8_4

1 Introduction

Urban forestry nurseries play a crucial role in creating and maintaining green spaces within cities. They are responsible for growing trees, shrubs, and various plants that not only beautify urban environments but also enhance air quality and provide ecological benefits. However, these nurseries often face the significant challenge of managing large amounts of organic waste, including plant trimmings, leaves, and root balls (Kodoh et al., 2023). Composting emerges as a strategic solution to this issue, transforming this waste into valuable soil amendments. This process not only reduces waste volumes but also produces nutrient-rich compost that can diminish reliance on chemical fertilizers and promote plant health.

The pursuit of such sustainable substrate alternatives is a central focus in modern nursery management. Conventional peat-based media, while effective, are associated with ecological degradation, driving the search for renewable substitutes. In a comprehensive review, Pascual et al. (2018) titleholder compost as a viable peat alternative, demonstrating its ability to maintain essential soil properties such as structure, water retention, and nutrient supply. Aligning with this, the simple adoption of composting methods like windrows or bins allows nurseries to close the nutrient loop, enhance soil quality, and support wider urban greening initiatives.

However, while Pascual et al. (2018) provide a broad overview of organic substrates, their work does not specifically address the unique logistical constraints of urban nurseries, such as severe space limitations that may preclude large windrows, or the optimization of composting for urban waste streams. Furthermore, the success of urban trees depends not only on the substrate used during nursery production but also on how these practices influence long-term establishment. Here, the research of Allen et al. (2017) offers a critical perspective by establishing a clear link between nursery production systems and post-planting tree survival, underscoring the need for high-quality growing media. Yet, their work does not delve into the specific composting methods that could produce such media in an urban context.

Therefore, a clear niche exists at the intersection of waste management and advanced nursery practice: the development of urban-adapted composting techniques that are both practical and scalable. This chapter seeks to fill that gap by moving beyond general discussions to provide a focused analysis of composting principles and methods tailored for urban forestry nurseries. Building upon foundational knowledge, our unique contribution lies in evaluating techniques from simple bins to more innovative methods like vermicomposting and Bokashi explicitly designed for spatial constraints and diverse urban waste. By integrating waste recycling directly with nursery soil science, this discussion aims to equip nurseries with the knowledge to select appropriate methods, manage waste effectively, and ultimately support a more sustainable and resilient urban forest.

2 Composting Processes

2.1 Overview of Composting

Composting is a fascinating natural process where tiny organisms, fungi, and other decomposers work together to break down organic materials, turning them into rich, nutrient-packed humus. Not only does this process help recycle waste, but it also creates a fantastic product that boosts soil health, improves its structure, retains moisture better, and provides essential nutrients that plants need to thrive (Rynk et al., 2021).

In urban forestry nurseries, composting plays a crucial role by transforming organic waste, like fallen leaves, grass clippings, leftover plant materials, and even wood chips into a valuable resource. The compost produced can be used in potting mixes, as a soil enhancer, or as mulch. This not only reduces the reliance on synthetic fertilizers but also supports sustainable practices in the nursery industry. By embracing composting, we can contribute to a healthier environment while nurturing our plants (Mojiol & Lim, 2022). The Fig. 1 shows the finishing product of a clean compost and suitable to be use in any urban greening initiatives.

Fig. 1 The finishing outcome of a fresh clean compost and be ready to be use in urban greening

3 Types of Composting Processes

The practice of composting, while universally aimed at recycling organic matter, encompasses a diverse array of methods, each distinguished by the specific environmental conditions and microbial communities they foster. These approaches, ranging from the slow, cold decomposition in a simple pile to the meticulously managed, high-temperature processes of hot composting or vermicomposting, each present a unique set of benefits and inherent challenges. For instance, a passive cold compost system requires minimal effort but is slower and may not eradicate all pathogens or weed seeds, whereas a carefully turned hot compost pile achieves rapid, sanitized results at the cost of greater labor and management. This critical distinction makes the selection of an appropriate composting strategy absolutely paramount for an urban forestry nursery, where the specific requirements extend beyond mere waste reduction. The nursery's need to produce robust, healthy seedlings demands a final compost product that is not only rich in nutrients but also free from pests and diseases, all while operating within the spatial and logistical constraints of an urban environment. Therefore, the choice of method must be a deliberate one, carefully weighing factors like processing speed, space availability, and desired compost quality to ensure the final product effectively supports the rigorous health and growth objectives of the nursery's valuable stock.

- *Aerobic Composting:*

Aerobic composting is an efficient method of waste conversion that leverages the presence of oxygen to support a thriving population of aerobic microorganisms. These microbes are the engine of the process, as they actively break down organic matter, and their activity generates significant metabolic heat. This leads to a core advantage of the system: the production of high temperatures that accelerate decomposition, effectively reduce pathogens and weed seeds, and, when the system is well-managed, result in minimal unpleasant odors. This combination of speed and odor control makes it particularly suitable for commercial nurseries, which often require a rapid and tidy conversion of their organic waste, such as plant trimmings and spent soil, into high-quality, usable compost. However, the success of this method is not automatic; it hinges on carefully maintaining three critical conditions. First, adequate aeration—through turning or forced air—is essential to sustain the oxygen-dependent microbes. Second, moisture levels must be balanced to support microbial life without creating waterlogged, anaerobic conditions. Finally, a proper carbon-to-nitrogen (C:N) ratio must be achieved, as this provides the ideal nutritional balance for the microorganisms, ensuring efficient and complete decomposition of the material, a principle highlighted by Manrique & Alvarado (2023).

- *Anaerobic Composting:*

While aerobic composting is a common practice, its anaerobic counterpart offers a distinct, albeit slower, alternative. Anaerobic composting occurs in an oxygen-deprived environment, a condition that significantly decelerates the decomposition process compared to aerobic methods. This enclosed process does have a unique by-product: biogas, which can be captured as a potential energy source. However, this same lack of oxygen leads to the production of other gases, namely methane and hydrogen sulfide, which are responsible for the characteristically unpleasant odors associated with this technique. Due to these odor issues and the slower processing time, anaerobic systems are less frequently adopted in nursery settings where aesthetics and space turnover are priorities. Nevertheless, as noted by López et al. (2008), this method retains its utility for specific types of organic waste or in situations where space and resources for managing an aerobic pile are constrained, presenting a viable, specialized solution for circumstances.

- *Vermicomposting:*

Vermicomposting offers a highly efficient and sustainable method for organic waste management by employing earthworms to decompose matter into nutrient-rich castings, a form of compost. This process is particularly advantageous for nurseries, as the resulting vermicompost is a superior growth medium; its fine texture and high concentration of plant-available nutrients, beneficial microbes, and growth hormones significantly enhance seed germination and promote vigorous seedling development. Furthermore, the system's remarkable versatility is a key benefit, as it can be successfully adapted to small, contained spaces like stacked bins, making it an ideal solution for urban nurseries that often operate with limited square footage. Beyond its spatial efficiency and high-quality output, vermicomposting is also an exceptionally clean and low-maintenance process; once the system is properly established with a healthy worm population, it operates odorlessly and requires only periodic feeding and moisture management, ensuring it is not a labor-intensive burden (Manrique & Alvarado, 2023).

4 Phases of Composting

Composting is an exciting natural process that transforms everyday organic waste into a rich, nutrient-dense soil amendment. This transformation is not a single event but a dynamic journey that unfolds in several distinct, sequential phases, each driven by unique microbial communities and marked by specific environmental conditions. The process begins with the mesophilic phase, where mesophilic, or moderate-temperature, microbes rapidly break down the most easily degradable compounds, causing the pile's temperature to quickly rise. As the heat intensifies, the process smoothly transitions into the thermophilic phase, the most critical stage for efficient composting. Here, thermophilic bacteria and fungi take over, thriving in the high temperatures that effectively break down complex proteins, fats, and cellulose while also neutralizing pathogens and weed seeds. After this period of intense microbial activity, the readily available food sources are depleted, and the pile begins to cool, entering the cooling and maturation phase. During this extended period, mesophilic organisms return, along with larger decomposers like earthworms and insects, to finish the work by breaking down the remaining tough compounds like lignin. This final stage, often called "curing," is essential for stabilizing the compost, allowing it to mature into a humus-rich, non-phytotoxic product that is safe and beneficial for plants. Grasping the progression of these phases from initial breakdown through thermophilic treatment to final maturation is key to managing a compost pile effectively, as it allows one to monitor temperature, moisture, and aeration to ensure the process runs efficiently, ultimately leading to a high-quality final result.

1. ***Initial Decomposition (Mesophilic Phase)***

The captivating process of composting begins with what is known as the mesophilic phase, a crucial initial stage driven by microorganisms that thrive in moderate temperatures, typically between 20 °C and 50 °C. As soon as organic waste is combined, these mesophilic microbes, which include a variety of bacteria and fungi, eagerly jump into action. Their primary role is to break down the most readily available components of the pile, such as simple sugars, starches, and proteins. This microbial activity is not a silent process; it is exothermic, meaning it releases energy in the form of heat. Consequently, as the microorganisms consume and digest these easy meals, their metabolic processes cause the compost pile's internal temperature to steadily rise. This heating effect is a clear sign that the decomposition process is underway. While highly active, this initial mesophilic phase is relatively short-lived, typically lasting only from a few days to a couple of weeks. Its duration is not fixed and depends heavily on factors such as the specific composition of the materials added to the pile and the surrounding environmental conditions, which can either accelerate or slow down the microbial workforce (Papale et al., 2021).

2. ***Thermophilic Phase***

Following the initial mesophilic stage, the compost pile enters the more intense thermophilic phase, where the temperature steadily climbs into a range of 50–70 °C. This heat is generated by a new team of heat-loving, or thermophilic, microbes that take charge. These microorganisms are the powerhouses of the composting process, specializing in breaking down the tougher, more resilient organic materials that the initial microbes couldn't handle, such as the complex structures of cellulose, lignin, and fats. This vigorous activity is crucial for several reasons. Firstly, the sustained high temperatures act as a natural pasteurizer, effectively eliminating harmful pathogens, parasites, and the seeds of weeds, which ensures the final compost is not only rich but also safe and healthy for use in the garden. Secondly, the thermophiles dramatically accelerate the decomposition of these stubborn materials, which helps to shorten the overall composting timeline. Furthermore, when managed correctly by ensuring adequate aeration, this hot, aerobic environment suppresses the odor-causing bacteria that thrive in low-oxygen conditions, thereby minimizing unpleasant smells. The duration of this critical thermophilic phase is not fixed; it can last for several weeks, with its length being directly influenced by factors such as the size of the compost pile, the specific mix of materials used, and, most importantly, how much air it receives to sustain the hard-working microbial community (Fig. 2).

3. ***Maturation Phase***

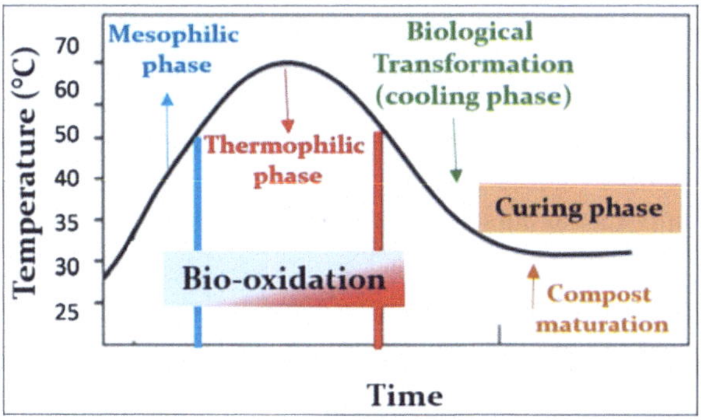

Fig. 2 Temperature changes in the composting process (*Source* Papale et al., 2021)

Following the intense heat of the thermophilic phase, the composting process enters a crucial period of refinement known as the maturation phase. As the pile's temperature gradually decreases, it creates a welcoming environment for mesophilic microorganisms to return and recolonize the material. These microbes meticulously break down the more complex and resilient organic compounds that survived the initial hot phase, such as lignin and complex carbohydrates. It is during this extended period of stabilization that the compost truly transforms, developing into a mature humus a dark, crumbly, and earthy-smelling substance that is densely packed with nutrients. This maturation is not merely a waiting game; it is vital for producing high-quality compost, as it ensures the breakdown of potential phytotoxins and pathogens, making the final product safe for use in gardens and nurseries. The duration of this phase is flexible, typically spanning from several weeks to several months, and is ultimately determined by the desired level of stability and the compost's intended application. By understanding and respecting this complete cycle, one can effectively transform waste into a rich resource, closing the loop in an environmentally friendly way and creating a powerful ally for any garden.

5 Techniques for Composting in Urban Forest Nurseries

Urban forestry nurseries must strategically align their composting operations with their unique operational profile, as the choice of method is not one-size-fits-all but is instead dictated by a careful balance of scale, available resources, and specific production needs. For a small-scale nursery with limited space and manpower, simpler techniques like passive pile composting or vermicomposting may be ideal; these methods require minimal infrastructure and technical expertise, allowing the nursery to effectively recycle its green waste into a valuable soil amendment without a significant investment. Conversely, a larger nursery managing substantial volumes of woody debris and leaf litter might find that the efficiency of windrow composting justifies the initial investment in equipment like a front-end loader or a specialized turner, as this method accelerates decomposition and produces a more uniform product on a commercial scale. Ultimately, each composting technique presents a distinct set of trade-offs between cost, labor intensity, processing speed, and space requirements. Therefore, the most successful approach involves a deliberate assessment of these factors to select a system that not only manages waste sustainably but also actively supports the nursery's core mission of cultivating healthy, robust trees.

6 Windrow Composting

Windrow composting is a method where organic materials are piled into long, narrow rows, known as windrows. These piles are regularly turned to promote good airflow and ensure everything breaks down evenly. This approach works particularly well for nurseries that have enough space for larger composting setups (Basheer et al., 2019). Windrow composting is viable for small nurseries with adjustments to size, turning frequency, and feedstock management. It offers a middle ground between passive piles and high-tech systems, balancing efficiency and simplicity.

Windrow composting stands out as a highly effective method for managing organic waste, particularly for operations like nurseries that generate significant volumes of material. Its primary advantage lies in its capacity to handle large quantities, making it a scalable solution that can grow in tandem with the business needs. The process itself is designed for efficiency, as the regular turning of the long, piled rows actively promotes fast decomposition. This mechanical agitation introduces essential oxygen, fuels the aerobic microbes, and prevents the pile from becoming compacted and stagnant.

However, to harness these benefits, certain challenges must be acknowledged and planned for. The very nature of windrow composting demands a substantial commitment of both space and equipment; it will need a sufficient land area for the rows themselves and machinery, such as a front-end loader or a specialized windrow turner, to aerate them effectively. Consequently, this can become a labor-intensive process, especially as the scale of the operation increases (Fig. 3).

Success in windrow composting is therefore guided by a few critical best practices. It begins with creating a balanced recipe for a pile, specifically by maintaining a carbon-to-nitrogen (C:N) ratio of between 25:1 and 30:1. This balance provides the ideal diet for the decomposing microorganisms. Simultaneously, moisture levels must be carefully monitored and kept within the 40% to 60% range, as this is the sweet spot for microbial activity without creating waterlogged, anaerobic conditions. The cornerstone of the entire process, however, is the consistent and regular turning of the windrows. This crucial step ensures even decomposition throughout the pile, manages temperature, and reintroduces the oxygen necessary to avoid unpleasant odors and produce a uniform, high-quality compost. By diligently adhering to these guidelines, it can transform the nursery's organic waste into a valuable, nutrient-rich resource that will nourish plants and support a more sustainable operation.

Fig. 3 Picture **a** A small urban nursery in Berlin uses 1.5 m-wide windrows turned biweekly with a handheld turner, producing compost in 10–12 weeks for potting mixes. Picture **b** The practice of scalable industrial Windrow composting in Germany using Backhus Windrow Turner to uncovering, turning and adding water in a single operation

7 Static Pile Composting

Static pile composting is a method where it creates piles of organic waste and let them breathe through perforated pipes or vents. This means it don't have to turn the piles as often, making it a great option for nurseries that may not have a lot of staff available for intensive upkeep. It's an efficient way to manage composting while saving time and effort (Íñiguez-Covarrubias et al., 2018).

While windrow composting is effective for large volumes, the static pile method offers a compelling alternative, particularly noted for its low-maintenance and less labour-intensive nature. This is because the piles are not turned, significantly reducing the physical effort and machinery required. This inherent advantage makes the system especially suitable for smaller-scale operations or nurseries with limited space, where dedicating extensive land and manpower to composting is not feasible. Furthermore, static piles are highly effective for processing materials that decompose slowly, such as wood chips and yard waste, as the method allows for a long, stable curing period.

However, this approach is not without its challenges. The most significant trade-off for reduced labour is slower decomposition compared to the aerated turning of windrow composting. This decelerated process necessitates careful monitoring of aeration and moisture levels within the pile. Without proper management, the compost can quickly become compacted and develop anaerobic conditions, leading to unpleasant odors and inefficient decomposition.

To mitigate these challenges and ensure successful composting, several best practices are essential. First, using perforated pipes or vents integrated into the base of the pile is critical to ensure adequate airflow from the bottom up, actively preventing anaerobic pockets. Second, strategically layering materials with a mix of carbon-rich "browns" and nitrogen-rich "greens" during pile construction promotes even decomposition and a balanced carbon-to-nitrogen ratio. Finally, covering the pile with a breathable material, such as a compost cloth or layer of straw, serves the dual purpose of retaining essential moisture and heat while still allowing gases to exchange, creating an optimized environment for microbial activity (Fig. 4).

Fig. 4 The static pile composting uses a holed pipes or vents to ensure adequate airflow

8 Vermicomposting

Vermicomposting is an amazing way to transform the organic waste into rich, nutrient-packed compost using earthworms. This method is particularly beneficial for small nurseries or anyone nurturing seedlings, as the compost produced can significantly boost the health and growth of young plants (Edwards, 1998).

Vermicomposting offers a uniquely efficient and rewarding method of recycling organic waste into a powerful resource for a small garden. One of its most significant advantages is the production of exceptionally high-quality compost. Unlike traditional composting, the process involves worms digesting organic matter, which creates a castings-rich final product teeming with beneficial microbes and nutrients, making it a superior, all-natural fertilizer for a plant. Furthermore, this system is remarkably space efficient. Its compact nature, often managed in a contained bin, makes it an ideal solution for urban dwellers or anyone with limited outdoor space, such as an apartment balcony or small patio. A well-managed vermicomposting system is surprisingly low maintenance, operating with minimal upkeep and remaining virtually odorless, seamlessly practicability in a home or small-scale nursery (Fig. 5).

However, to ensure this smooth operation, a degree of attentive management is required. It needs to periodically monitor the health of the worm population and their living conditions, ensuring it have the right balance of food, moisture, and temperature to thrive. This leads to another consideration: capacity. Vermicomposting is best suited for handling smaller, consistent quantities of kitchen scraps, meaning it might not be the sole solution for large families or those with substantial garden waste (Fig. 6).

Appreciatively, achieving success is straightforward by following a few key practices. It begins with choosing the right worms; species like red wigglers (*Eisenia fetida*) are the titleholders of vermicomposting due to their prolific appetite and adaptability to life in a bin. Feeding it a balanced diet of fruit and vegetable scraps, while avoiding oily or acidic foods, is crucial for keeping the workforce healthy and productive. Finally, maintaining the proper moisture level is essential the bedding should feel like a wrung-out sponge. By embracing these simple guidelines, it can effortlessly maintain a thriving vermicomposting system, creating a sustainable cycle that enriches the surrounding garden while effectively reducing household waste.

9 Bokashi Composting

Bokashi composting is a unique way to recycle organic waste using a special bran that's been treated with helpful microorganisms. This process happens in sealed containers, which means it doesn't require air. It's a fantastic option for urban gardens or nurseries where space is tight. One of the best things about Bokashi composting is that it can handle a variety of organic materials, including meat and dairy, which

Fig. 5 Vermicomposting process using dry leaf mix with wood waste in a green house

typically can't compost using traditional methods. This makes a versatile and effective choice to reduce waste and enrich their soil. EM Bran Bokashi is crucial for enhancing the soil's fungal and microbial profile. Without these beneficial microbes, soil nutrients cannot be utilized by plants. By incorporating EM Bran Bokashi into compost, it becomes enriched with vital microbial and fungal life, supporting the growth of healthy plants (Zaman et al., 2016).

Bokashi composting offers a unique and efficient method for recycling organic waste, primarily due to its compact and space-efficient design, making it an ideal solution for urban settings like apartments or homes with small yards. Unlike traditional composting, the process is remarkably fast, typically completing the initial fermentation stage in just two to four weeks. A significant advantage is its ability to process a wider variety of organic waste, including items that are typically problematic for other systems, such as meat, dairy, and cooked foods, by fermenting them in

Fig. 6 Using *Eisenia fetida* is an effective worm in vermicomposting process

an anaerobic environment. However, this method also presents specific challenges; for instance, it produces a pre-compost material that is pickled and acidic, requiring a subsequent phase of decomposition in soil to become finished compost. Furthermore, the process depends on specialized materials, namely airtight containers and inoculated Bokashi bran, which can represent an additional cost compared to more passive compost piles (Fig. 7).

To ensure success, several best practices should be followed. The process involves layering kitchen scraps with Bokashi bran inside an airtight container to create the oxygen-free environment essential for fermentation. As the materials break down, it's crucial to drain the excess liquid, known as "Bokashi tea," regularly to prevent odor and potential mess; this potent liquid can be diluted and used as a nutrient-rich fertilizer. Finally, once the fermentation period is complete, the pre-compost must be buried in a garden trench or added to a traditional compost bin, where it will quickly break down further, enriching the soil with beneficial microbes and nutrients.

Fig. 7 Using EM Bran Bokashi as a treatment for composting organic waste

10 The Differences Between Nursery Composting and Industrial Composting

Composting practices can vary significantly depending on the scale and purpose, with urban forestry nurseries and industrial facilities adopting distinct approaches. Below is a comparison of nursery composting and industrial composting, highlighting key differences in scale, process, and waste handling capacity (Table 1).

The Nursery composting is ideal for small-scale, sustainable waste management in urban settings, prioritizing soil health and plant growth. While the industrial composting is designed for large-scale waste diversion, requiring advanced infrastructure and adhering to stringent environmental regulations. Industrial systems

Table 1 Differences between nursery composting and industrial composting

Aspect	Nursery composting	Industrial composting
Scale	Small to medium-scale operations, tailored to the needs of urban nurseries	Large-scale operations designed to handle massive volumes of organic waste
Process	Focuses on methods like windrow, static pile, vermicomposting, and Bokashi	Primarily uses windrow or in-vessel systems with mechanized aeration and turning
Waste handling capacity	Processes limited quantities of organic waste, such as plant trimmings and leaves	Handles diverse waste streams, including municipal green waste, food waste, and manure
Space requirements	Requires minimal space; adaptable to urban settings with limited area	Needs extensive land for large windrows or specialized facilities for in-vessel systems
Labor and equipment	Relies on manual labor or simple tools for turning and monitoring	Uses heavy machinery (e.g., turners, shredders) and automated systems for efficiency
Decomposition time	Slower due to smaller piles and less intensive management	Faster due to optimized conditions (e.g., controlled aeration, temperature monitoring)
End-product use	High-quality compost used for potting mixes, soil amendments, and mulching in nurseries	Bulk compost often sold for agricultural, landscaping, or municipal applications
Environmental controls	Minimal odor and pathogen control, suited for low-impact urban environments	Strict regulations for odor, leachate, and emissions management to meet environmental standards
Cost	Low-cost, budget-friendly solutions with minimal infrastructure	High initial investment but cost-effective per unit of waste processed

handle more diverse waste and require advanced technology, whereas nurseries prioritize sustainability and soil enrichment. Both methods contribute to waste reduction but differ significantly in scale, complexity, and end-use applications. By understanding these differences, urban nurseries can select composting methods that align with their operational capabilities and sustainability goals, while industrial facilities can optimize processes for efficiency and scalability.

11 Benefits of Composting in Urban Forest Nurseries

Integrating composting into urban forest nursery operations yields significant advantages, fostering both sustainable management and enhanced environmental health. These benefits are multifaceted, ranging from improved soil quality to economic efficiency.

- *Improved Soil Health*

Composting serves as a foundational practice for revitalizing soil health, fundamentally transforming its physical properties to create a more hospitable environment for plant growth. By introducing organic matter, compost acts as a natural soil conditioner, binding minute soil particles together to form stable aggregates. This enhancement in soil structure is pivotal, as it directly improves aeration, allowing plant roots to breathe more easily, and simultaneously boosts the soil's water retention capacity, ensuring moisture is available for longer periods rather than quickly draining away. The culmination of these improvements—better structure, oxygen flow, and moisture availability—creates an exceptionally favorable growing environment, which is particularly beneficial for developing seedlings, leading to healthier and more robust plants. Furthermore, this process doesn't just benefit plants mechanically; it also stimulates the soil's biological engine. The organic matter in compost provides a rich food source, thereby fostering a surge in beneficial microbial activity. This vibrant ecosystem of bacteria and fungi is essential for the natural decomposition of organic materials and the critical cycling of nutrients, making them bioavailable to plants and completing a cycle of sustainable growth, as supported by the research of Landis and Khadduri (2008).

- *Cost Savings*

One of the most compelling advantages for nurseries adopting composting is the significant reduction in operational costs. By transforming their own plant waste into a rich, organic soil amendment, nurseries can drastically curtail their reliance on expensive external inputs, particularly chemical fertilizers and other commercial soil conditioners. This shift from purchasing to producing creates a direct and sustainable cycle of resource efficiency. The benefits extend beyond just fertilizer replacement; this homemade compost serves as an excellent, cost-effective substitute for peat moss within growing media. Peat moss, while effective, is a purchased material whose extraction raises environmental concerns, making its replacement a dual victory for both a nursery's budget and its ecological footprint. As supported by research from López Núñez and colleagues (2008), this practice of internal resource cycling not only lowers immediate expenses but also firmly anchors the nursery's operations in a more sustainable and self-sufficient model of plant cultivation.

- *Environmental Benefits*

Composting offers profound environmental benefits, primarily by playing a crucial role in waste diversion and climate change mitigation. When organic waste, such as food scraps and yard trimmings, is sent to landfills, it decomposes anaerobically (without oxygen), a process that generates significant amounts of methane—a potent greenhouse gas with a global warming potential many times greater than carbon dioxide. By diverting this organic stream into composting facilities, we not only reduce the volume of waste in landfills but also prevent the release of these harmful emissions, directly aligning with urban sustainability initiatives focused on waste reduction and climate impact mitigation. Furthermore, the process of composting transforms what would have been a source of pollution into a valuable resource. The end product is a nutrient-rich humus that, when added to soil, completes a natural cycle by returning essential organic matter and nutrients back to the earth. This enhances soil structure, water retention, and fertility, which in turn reduces the reliance on synthetic fertilizers. By supporting healthier plant growth and creating more resilient ecosystems, composting demonstrates a circular economy in action, turning waste into a cornerstone of environmental health (Arborist Now, 2024).

- *Support for Urban Agriculture and Green Spaces*

Building on the direct benefits of producing compost within urban forest nurseries, this nutrient-rich material serves as a vital catalyst for broader urban sustainability initiatives, particularly in supporting urban agriculture and enhancing green spaces. By providing a local, high-quality source of organic matter, the compost becomes the foundation for fertile soil in community gardens and productive rooftop farms, directly contributing to more robust and healthy plant growth. This cycle of local production and use not only bolsters food security and community engagement but also creates a cascade of environmental benefits. The resulting lush, healthy vegetation plays a critical role in mitigating the urban heat island effect a phenomenon where city areas experience significantly higher temperatures than their rural surroundings. Through a process called evapotranspiration, thriving plants release moisture into the air, while their shade reduces heat absorption by surfaces, collectively cooling the urban environment. Therefore, the simple act of recycling local green waste into compost within city nurseries ultimately fosters a greener, more resilient, and more livable urban ecosystem (Arborist Now, 2024).

Integrating composting into urban forest nurseries extends its benefits far beyond simply enriching the soil, creating a powerful ripple effect that enhances both operational efficiency and the broader urban environment. For instance, certain composts act as a natural line of defence by suppressing soil-borne pathogens and pests, which can significantly reduce a nursery's reliance on chemical pesticides. This not only creates a healthier growing environment for seedlings but also aligns with growing public demand for more sustainable horticultural practices. This connection to the community is another key advantage, as composting initiatives themselves serve as a powerful tool for public engagement. By involving residents in programs that divert organic waste, nurseries can foster environmental education, promote sustainable practices, and strengthen social cohesion around a shared green goal. This community involvement directly supports the logistical success of urban composting, which thrives on the diversification of feedstocks. By actively utilizing a wide variety of urban organic materials from household food scraps to municipal yard trimmings these programs simultaneously address two critical issues: they dramatically reduce the volume of waste sent to landfills while transforming it into a valuable, locally sourced resource for the nursery. Ultimately, this integrated approach demonstrates how urban forest nurseries can evolve from mere plant producers into active hubs of sustainability, where enhanced plant growth and streamlined operations contribute directly to the achievement of broader urban environmental goals.

12 Challenges and Solutions

While urban nurseries are increasingly drawn to composting for its clear environmental and educational benefits, they often encounter a unique set of logistical obstacles that can make implementation challenging.

- *Space Limitations*

Several efficient methods are available to compost in environments with limited space. *Static Pile Composting* offers a straightforward approach by creating a stationary mound of organic materials that decomposes gradually over time. This method is particularly advantageous as it requires less physical space and can be effectively managed in small yards or garden corners. However, for an even more compact solution that can be brought indoors, *Vermicomposting* is an excellent alternative. This process utilizes specific worms, such as red wigglers, to actively break down food scraps into a nutrient-dense, soil-like amendment. It's contained nature makes it ideal for small apartments, balconies, or even under a sink. Taking indoor convenience a step further, *Bokashi Composting* provides a unique anaerobic system. Unlike the other methods, Bokashi uses a specialized bran inoculated with microorganisms to ferment all food waste including items traditionally difficult to compost like meat and dairy in a sealed container. This final method is exceptionally well-suited for apartment dwellers and can be easily adapted for use in nurseries or classrooms, as it is odorless and compact, completing a suite of options for turning waste into a resource in virtually any setting.

- *Waste Segregation*

Effective sorting of compostable materials is the critical first step in producing high-quality compost, as it prevents contamination from non-organic items that can disrupt the decomposition process or introduce toxins. For nurseries, where organic waste like plant trimmings and soiled pots is abundant, enhancing this waste segregation requires a dedicated and informed approach. The cornerstone of this effort is comprehensive staff training, which moves beyond simply listing compostable items to actively educating team members on the "why" and "how." This involves clarifying what materials are suitable such as untreated plant debris and certain biodegradable pots while explicitly identifying common contaminants like plastic plant tags or synthetic fertilizers. To cement this knowledge and foster a culture of collective responsibility, nurseries can implement ongoing educational programs. These initiatives, which might take the form of initial onboarding workshops, refresher sessions, and clear visual guides posted at sorting stations, ensure that every staff member, regardless of their role, understands both the environmental importance and the practical process of correct waste segregation. By investing in this continuous education, nurseries empower their teams, transforming waste management from a routine task into a conscious contribution to the nursery's sustainability goals and the production of nutrient-rich compost that, in turn, nourishes future plants.

- *Odour Control*

In densely populated urban and suburban settings, the process of composting organic waste, while environmentally beneficial, can be problematic due to the generation of unpleasant odors. These odors are not an inevitable byproduct of decomposition but are typically a sign that the compost pile has entered an anaerobic state, a condition where a lack of oxygen causes putrefying bacteria to produce foul-smelling gases like hydrogen sulfide and ammonia. The primary technique to prevent this and maintain an efficient, odor-free process is to ensure aerobic composting. This is achieved by actively managing the pile to provide adequate oxygen, which supports the preferred oxygen-loving microbes that break down waste without creating significant smells. The most direct method for maintaining these ideal conditions is through proper aeration, which involves regularly turning the compost pile. This simple action introduces fresh air, redistributes moisture and microbes, and prevents the material from becoming compacted and oxygen-deprived. For an added layer of odor control, especially in sensitive areas, incorporating odor-neutralizing materials such as dry wood chips, shredded leaves, or even a thin layer of finished compost can be highly effective. These "brown" materials act as a bulking agent to improve airflow structurally and simultaneously absorb and filter out any potential odors before they can become a nuisance, making the practice of composting more manageable and community friendly.

13 Applications of Compost in Nursery Practices

Compost serves as a cornerstone of modern, sustainable nursery practices, offering a multitude of benefits that enhance plant health from the ground up. Its primary application begins with the very foundation of plant growth: the potting mix. By blending compost with other components like peat moss and perlite, nurseries can create a nutrient-rich, well-aerated medium that is ideal for seedlings. This carefully balanced environment is crucial for fostering strong root development and vigorous early growth, giving young plants the best possible start. Beyond individual containers, compost's value extends to the broader planting beds as a powerful soil amendment. When incorporated into the soil, it significantly improves fertility by releasing a steady supply of essential nutrients, while simultaneously enhancing soil structure and water retention. This dual action creates a more resilient growing environment, reducing the need for frequent watering and synthetic fertilizers. Furthermore, compost demonstrates its versatility by also functioning as an effective, natural mulch. A layer of compost spread over the soil surface works to suppress competing weeds, retain vital soil moisture by reducing evaporation, and help regulate soil temperature, protecting plant roots from extreme heat or cold. By integrating compost in these diverse ways from potting mixes to soil conditioning and mulching, urban nurseries can drastically reduce their reliance on synthetic products. This holistic approach not only streamlines operations and improves plant quality but also actively contributes to building a more sustainable and resource-efficient urban environment.

14 Novelty Contribution and Comparison and of This Work with Others

From the article published by Pascual et al. (2018) entitled "Organic substrate for transplant production in organic nurseries. A review" emphasizing sustainable alternatives to peat-based substrates. It synthesizes global research on compost-derived substrates, their physicochemical properties, and their impact on plant growth, aligning with urban forestry nurseries' needs for eco-friendly practices. Their key contributions is a compost as peat alternative, and highlights compost's role in replacing peat (reducing ecological degradation) while maintaining soil structure, water retention, and nutrient supply. It discusses compost types (e.g., green waste, vermicompost) suitable for nursery substrates.

In term with similarities with their study, both highlight compost's role in improving soil health and reducing reliance on synthetic fertilizers. However, the differences is that Pascual et al. (2018), focus broadly on organic substrates (including peat alternatives), while this paper specifically targets urban forestry nurseries, addressing unique challenges like space constraints and urban waste recycling. In terms of novelty, this study provides practical, scalable composting techniques (e.g., Bokashi, vermicomposting) tailored to urban settings, whereas Pascual et al. (2018) discussed generally on organic media without urban-specific adaptations.

Additionally, other article published by Allen et al. (2017) entitled "Nursery Production Systems and Urban Tree Survival" evaluates how nursery practices (e.g., container vs. bare-root production) affect urban tree survival post-planting. It stresses the need for high-quality growing media but does not deeply explore composting methods. Allen et al. (2017) offers a foundational understanding of how nursery practices affect tree survival in cities. So, this paper could extend its insights by explicitly linking compost techniques (e.g., vermicompost for root development) to the paper's recommendations for resilient urban trees. Both acknowledge the importance of nursery practices in urban forestry success. However, Allen et al. (2017) focussing on tree establishment and production systems, while this paper details work composting as a waste-to-resource strategy for nurseries, linking it to urban sustainability. In terms of novelty, this study provides introduce innovative composting techniques (e.g., static pile with aeration pipes, EM Bokashi) not covered by Allen et al. (2017) along with urban-specific solutions for odor and space limitations.

As unique contributions of this paper are urban-focused composting techniques, thus it provides practical, space-efficient methods (e.g., vermicomposting for small nurseries, Bokashi for meat/dairy waste) tailored to urban constraints. In term of integration of waste management and plant health: it bridges the gap between organic waste recycling and nursery soil enhancement; a niche less explored in the cited works. Long story short, its novel adaptations on the discussion of aerobic vs. anaerobic urban composting and cost–benefit analysis offers actionable insights absent in Pascual et al. (2018) and Allen et al. (2017). Whereas both authors have provided foundational knowledge on organic substrates and nursery systems, and this paper advances the field by tailoring composting solutions to urban forestry nurseries, emphasizing scalability, sustainability, and real-world applicability.

15 Conclusion

In summary, composting is essential for urban forest nurseries as it facilitates the sustainable recycling of organic waste while enhancing plant health and minimizing environmental impact. By employing suitable composting methods such as windrow composting, vermicomposting, and Bokashi composting, nurseries can improve their operational efficiency and contribute positively to urban greening efforts. This chapter serves as a detailed guide for implementing effective composting processes tailored to the specific needs of urban forestry nurseries, promoting sustainability and resilience within urban ecosystems. While prior research covers composting science, agricultural applications, or urban forestry broadly, this paper bridges the gap by providing a technical, nursery-focused composting guide with urban adaptability. Its novelty lies in actionable strategies for small-scale, sustainable waste recycling in city-based nurseries, filling a niche underexplored in existing literature.

References

Allen, K. S., Harper, R. W., Bayer, A., & Brazee, N. J. (2017). A review of nursery production systems and their influence on urban tree survival. *Urban Forestry & Urban Greening, 21*, 183–191.

Arborist Now (2024) *Composting in Urban Environments: Solutions for City Dwellers*. Arborist Now, Inc. https://www.arboristnow.com/news/composting-in-urban-environments-solutions-for-city-dwellers. Retrieved 13 Mar 2025.

Basheer, S., Nazir, M., Rashid, H., Nasir, A., & Hussain, E. M. (2019). Development of efficient windrow composting technique for food waste and its optimization. *Earth Science Pakistan, 3*(1), 18–26.

Edwards, C. A. (1998). The use of earthworms in the breakdown and management of organic wastes. In C. A. Edwards (Ed.), *Earthworm ecology* (pp. 327–354). CRC Press.

Íñiguez-Covarrubias, G., Gómez-Rizo, R., Ramírez-Meda, W., & de Jesús Bernal-Casillas, J. (2018). Composting of food and yard wastes under the static aerated pile method. *Advances in Chemical Engineering and Science, 8*(04), 271–284.

Kodoh, J., Jaunis, O., Maid, M., Lintangah, W., & Mojiol, A. R. (2023). Urban forestry in Kota Kinabalu: Home garden practices by urban villages for biodiversity, socio-economic and environmental services. *IOP Conference Series: Earth and Environmental Science, 1145*(1), Article 012012.

Landis, T. D., & Khadduri, N. (2008). Composting applications in forest and conservation nurseries. *Forest Nursery Notes, 28*(2), 9–18. https://www.researchgate.net/publication/272790673_Using_organic_fertilizers_in_forest_and_native_plant_nurseries. Retrieved 13 Mar 2025.

López, R., Cabrera, F., Madejón, E., Sancho, F., & Álvarez, J. M. (2008). Urban composts as an alternative for peat in forestry nursery growing media. *Dynamic Soil, Dynamic Plant, 2*(Special Issue 1), 60–66.

Manrique-Vega, S. M., & Alvarado-Sanabria, O. (2023). Compost increases soil fertility and promotes the growth of five tropical species used in urban forestry. *Floresta e Ambiente, 30*(4), Article e20230018.

Mojiol, A. R., & Lim, W. S. (2022). The status and future of urban forestry in Sabah, Malaysia. In: Maruthaveeran, S., Chen, W. Y., & Morgenroth, J. (Eds.), *Urban forestry and arboriculture in Malaysia: An interdisciplinary research perspective* (pp. 19–41). Singapore: Springer Nature Singapore. https://doi.org/10.1007/978-981-19-5418-4_2

Papale, M., Romano, I., Finore, I., Lo Giudice, A., Piccolo, A., Cangemi, S., Di Meo, V., Nicolaus, B., & Poli, A. (2021). Prokaryotic diversity of the composting thermophilic phase: The case of ground coffee compost. *Microorganisms, 9*(2), 218.

Pascual, J. A., Ceglie, F., Tuzel, Y., Koller, M., Koren, A., Hitchings, R., & Tittarelli, F. (2018). Organic substrate for transplant production in organic nurseries. A review. *Agronomy for Sustainable Development, 38*(3), 35.

Rynk, R., Black, G., Gilbert, J., Biala, J., Bonhotal, J., Schwarz, M., & Cooperband, L. (Eds.). (2021). *The composting handbook: A how-to and why manual for farm, municipal, institutional and commercial composters*. Academic Press.

Zaman, M., Ahmed, M., & Gogoi, P. (2016). Effect of bokashi on plant growth, yield and essential oil quantity and quality in patchouli (*Pogostemon Cablin* Benth.). *Biosciences Biotechnology Research Asia, 7*(1), 383–387.

The Usage of Compost from Leucaena Leucocephala Leaves and Moringa Oleifera Leaves for the Growth of Kappaphycus Alvarezii Seedling

Arofah Lyla Nurhayati and Indrian Rizka Amalia

Abstract Seaweed is one of the leading export commodities from Indonesia's fisheries sector. One of them is cottoni (*Kappaphycus alvarezii*). High quality of the carrageenan produced can be obtained by using seedlings from generative propagation. In contrast to vegetative propagation, generative propagation requires a comparatively slower growth period and seedlings that perish more quickly. On the other hand, there is compost, natural fertilizer, which is commonly used to promote plant growth. The similarity of properties between plant and seaweed led to this research being carried out with the aim to looking for methods that can accelerate seaweed seedling growth but remains environmentally friendly. The treatment were A. plantlets soaked in compost made from fermented lamtoro (*Leucaena leucochepala*) leaves solution, B. plantlets soaked in compost made from fermented moringa (*Moringa oleifera*) leaves solution, and C. control without soaked in compost solution. The soaking process done in 15 min, 1 time a week, for 5 weeks maintenance in the sea. The result showed that compost able to promote the growth of cottoni seedlings, as evidenced by an increase in weight, the emergence of the new shoots, and darker thallus color during observation, compared to the control.

Keywords Compost · Cottoni · Plantlet · Lamtoro · Moringa

A. L. Nurhayati (✉)
Fisheries Research Center, NationalResearchandInnovationAgency (BRIN), KST Soekarno, Bogor, West Java, Indonesia
e-mail: arof003@brin.go.id

I. R. Amalia
Achmad Baiquni Science, Technology and Education Park, National Research and Innovation Agency (BRIN), Yogyakarta, Indonesia

A. Z. Yaser et al. (eds.), *Compost and Liquid Organic Fertilizers*,
https://doi.org/10.1007/978-981-95-8561-8_5

1 Introduction

Indonesia is the second largest seaweed producer in the world based on Ministry of Maritime Affairs and Fisheries (MMA, 2021). However, the carrageenan quality will decrease along with the decrease in the quality of the seedlings used. Therefore, seedling renewal through generative propagation is very necessary. The weakness of seedlings produced by generative propagation is that the growth rate and survival ability of the plantlets are still low when transferred from the laboratory to the sea. Therefore, some methods are needed to help plantlets survive in new media while still paying attention to environmental safety. One of it is using compost.

Compost is a product of composting. Composting is the aerobic bioconversion of organic waste facilitated by microorganisms, including fungi and bacteria (Izydorczyk et al., 2024). Compost commonly used to promote plant growth because it is rich in essential plant nutrients and helps suppress plant pests and diseases. The similarity of properties between plant and seaweed led to this study. In this study, we soaked plantlets in to compost solution. Plantlets are young seaweed seeds that have developed body structure resembling adult seaweed (Sulistiani & Yani, 2014).

The main ingredients to make compost in this research are lamtoro (*Leucaena leucochepala*) leaves and moringa (*Moringa oleifera*) leaves. These two types of leaves are used because they are widely found in Jepara, Central Java, Indonesia and are not optimally utilized by the surrounding community even though both have high protein content. Lamtoro leaves contain 24.1% crude protein (Nafifa, 2018) while moringa leaves contain 28.25% protein (Sicca, 2025). In addition, waste from lamtoro and moringa leaves is included in the lignocellulose group. The selection of raw materials is based on the consideration that lignocellulose is a derivative of agricultural waste, forest waste, and the main component of biomass and it shows a lower carbon footprint compared to conventional fertilizers, thus complying with the principles of sustainability and the needs of a circular economy (Izydorczyk et al., 2024).

2 Materials and Methods

2.1 Time and Location of Research

The entire series of research activities were carried out from April to July 2020. The research located at Main Center for Brackish Water Aquaculture (BBPBAP) Jepara, Central Java, Indonesia.

2.2 Compost Preparation

The method of making compost using lamtoro and moringa leaves is relatively simple. Lamtoro leaves are taken from their branches and collected until a total weight of 1 kg is obtained. One kg of lamtoro leaves, 333 g of fruit waste; 33.3 g of ammonium sulphate fertilizer; 67 g of manure; 13.3 g of molasses; 67 mL of liquid probiotic starter; and 1.2 L of brackish water are mixed until all components are evenly mixed and then put into a jar (Fig. 1). The same composition is also used for moringa leaves where moringa leaves replace lamtoro leaves while the other components are the same as making compost using lamtoro leaves (Fig. 2).

The jar is closed, the part between the lid and the neck of the jar is covered with clear tape. The lid of the jar is weighed down and then labeled with the fertilizer identity and date of production. The jar is stored in a dark room for 4 weeks. In the fourth week, it is checked whether all the ingredients have fermented or not. If a fresh sour smell is smelled, the fermentation process is successful and the compost can be used immediately. Conversely, if an unpleasant odor is smelled, the fermentation

Fig. 1 Lamtoro leaves compost **a** before fermentation; **b** after fermentation (Private documentation: Nurhayati, 2020—unpublished)

Fig. 2 Moringa compost before fermentation (left); after fermentation (right) (Private documentation: Nurhayati, 2020—unpublished)

process is unsuccessful and the compost has not ready to be used. The use of immature compost can interfere with plant growth and development because there is a change in the N element to a form that is not available to plants (Sutriana & Baharuddin, 2019). Incase the compost has not ready yet, we can add extra probiotic and brackish water then keep the compost for another 1 week.

2.3 Procedure

In this research, plantlets of cottoni (*Kappaphycus alvarezii*) was used. This type of seaweed was chosen because it is type of seaweed for export purposes from Indonesia (Siregar, 2023; TEMPO, 2024). The treatment is applied by taking 50 g of compost then diluting it in brackish water until the pH in range 6–7. Treatment A using lamtoro leaves and treatment B using moringa leaves. The plantlets are weighed and measured in length for their data recorded before treatment, then put into a basin that has been filled with compost solution. The plantlets were soaked for 15 min (Fig. 3).

The plantlets that have been soaked with compost solution are then put into a round container and then covered with a tight netting (Fig. 4). This aims to prevent small crabs and shrimp from entering the container, which is feared will interfere with the process of recording plantlet growth data. The plantlets were then planted in a seaweed nursery in the sea for 5 weeks (Fig. 5). The seedlings in the maintenance container were shaken in seawater to reduce mud deposits on the surface of the thallus with an intensity of 3 times a week. Thallus is all parts of seaweed body (Sulistiani & Yani, 2014). The maintenance container is lifted from the nursery to clean the seedlings from mud, barnacles, moss, and to record the seedlings growth (Fig. 6) once a week. Observations of the physical quality of the maintenance water were carried out at every growth sampling (Fig. 7).

Fig. 3 Soaking plantlets of cottoni in to compost solution (Private documentation: Nurhayati, 2020—unpublished)

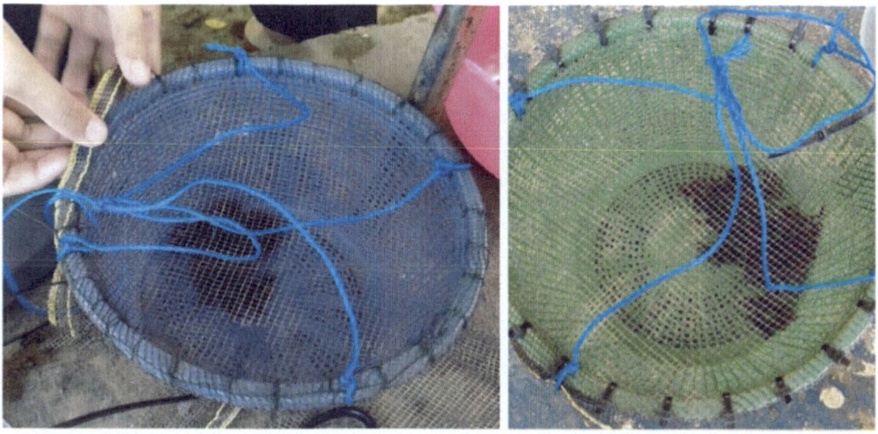

Fig. 4 Plantlet growing container (Private documentation: Amalia, 2020—unpublished)

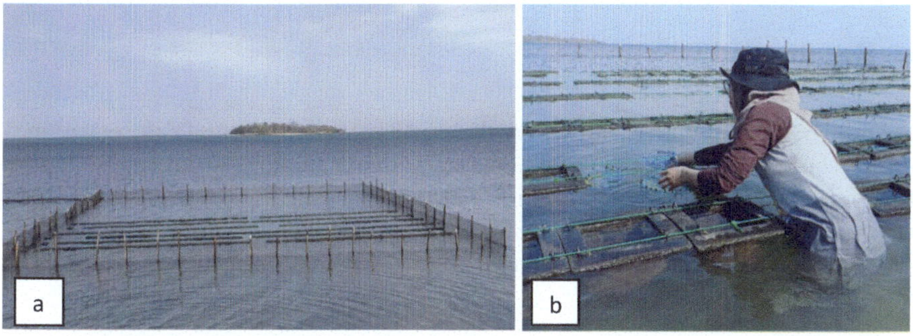

Fig. 5 **a** Location of seaweed nursery; **b** Plantlets planting process in the nursery (Private documentation: Nurhayati, 2020—unpublished)

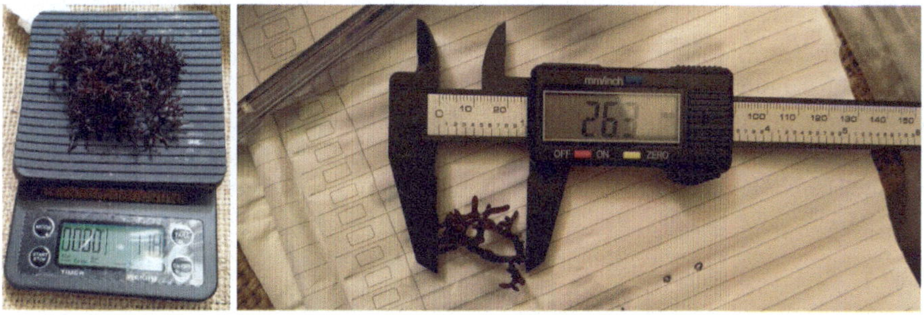

Fig. 6 Documentation of plantlets growth data (Private documentation: Amalia, 2020—unpublished)

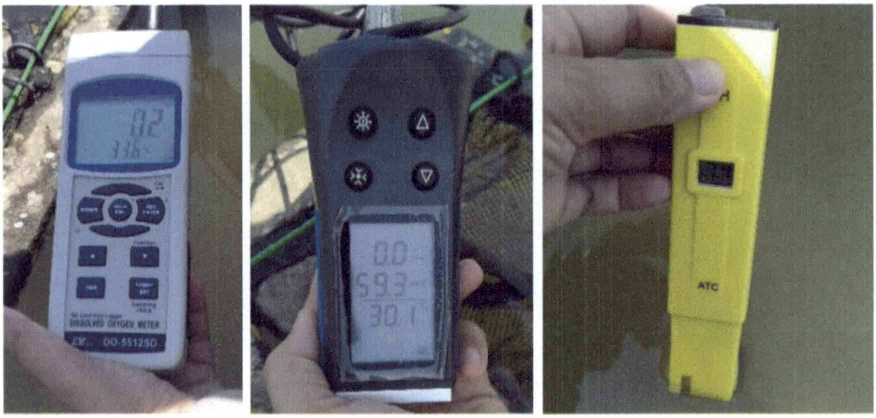

Fig. 7 Documentation of environmental data (Private documentation: Nurhayati, 2020—unpublished)

3 Results and Discussion

The primary data of this research were the seedling weight (Table 1), the emergence of new shoots/thallus (Fig. 8), and color changing of thallus (Figs. 9 and 10) after the compost application. Two composts were applied such as compost from lamtoro leaves and moringa leaves, compared with control without soaking in to compost. The secondary data of this research was environmental condition of seawater as growth media (Table 2).

3.1 Plantlet Growth and Weight Changes

The weight result during the study untill the second week of treatment (Table 1), it showed that both compost from fermented lamtoro leaves and fermented moringa leaves were able to support the growth of cottoni plantlets at the acclimatization stage compared with control plantlets. This result is in accordance with the results obtained by Nurhayati et al. (2023) which stated that compost from fermented lamtoro

Table 1 Plantlets growth data

Plantlets	Weight (g) per week					
	0	1	2	3	4	5
Lamtoro leaves	14.5	11.8	14.2	11.7	8.1	5.9
Moringa leaves	13.5	20.7	20.7	5.6	2.5	1
Control	14.37	12.1	0	0	0	0

Fig. 8 New shoots/thallus shown by the white circle (Private documentation: Amalia, 2020—unpublished)

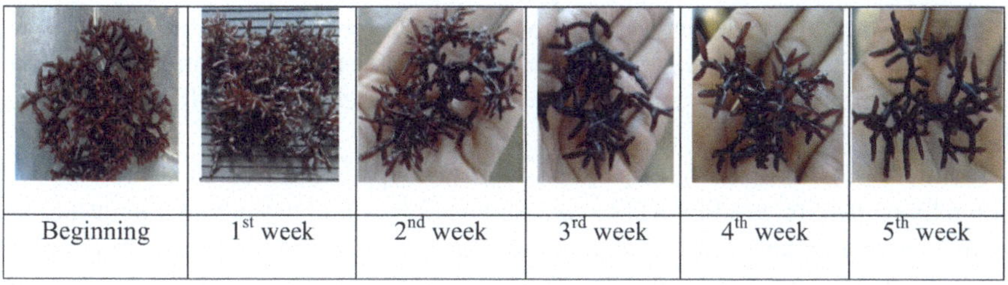

Fig. 9 Treatment plantlets growth during research (Private documentation: Amalia, 2020—unpublished)

Fig. 10 Control plantlets growth during research (Private documentation: Nurhayati, 2020—unpublished)

leaves and moringa leaves was able to increase the growth of *Gracilaria* sp. plantlets. The weight of plantlets using fermented moringa leaves is higher than those using lamtoro leaves. This may happen due to the components of compost from fermented moringa leaves are already in the form of short chains that can be directly utilized by

Table 2 Environmental data

Parameter	Week					
	0	1	2	3	4[1]	5[1]
pH	7	8.3	7.5	8	7.5	7.9
Temperature (°C)	28	28	29	29.3	29.3	29.3
DO (mg/L)	5	4.3	4.5	5.6	3.4	2.8
Current speed (km/h)	0	0	0	0	0	0.5

[1]*: it rained several times*

plantlets so that the growth effect is very striking since the first week. The decrease in treatment plantlets weight found in the third week which is 1 week longer than control plantlets. It was shown that the treatment plantlets more capable to adapted with the unstable environmental condition compared with the control plantlets. The decrease weight occurred concurrently with increasing plantlets fracture (Fig. 11). Therefor the recording of thallus length was not continued because the thallus began to break in the fourth week. However based on its physical condition, the treatment thallus was still alive thus the recording of weight was continued until the end of the observation period.

Plantlets growth also could be seen from the appearance of some new shoots shown by white circle in Fig. 8. The new shoots only found in treatment plantlets but not in control plantlets. This implied that the meristem area of the treatment plantlets was in the best condition to divide and grow into the new shoots/thallus. Meristem commonly found in plant and consist of actively dividing cells. Unlike plants in general, seaweed meristem tissue is a simpler meristem tissue, spread throughout the parts that actively grow, especially at the vegetative growth points (Sulistiani & Yani, 2014). Meanwhile, meristems in plants are divided into apical, lateral, and

Fig. 11 Fracture plantlets (Private documentation: Amalia, 2020—unpublished)

intercalary meristems which are clearly for the growth of roots, stems, and leaves (Ningsih et al., 2024). The result that showed the emergence of new shoots in this study were consistent with the research of Nurhayati et al. (2023), who also found that more new shoots emerged after *Gracilaria* sp. plantlets were soaked in to compost solution.

3.2 Color Variations of Thallus

There was a color changing happened during growing period. At the beginning of sampling, treatment thallus color tended to be bright red especially at the thallus tips as the effect of compost soaking. The thallus color became progressively darker as the research going (Fig. 9). The opposite phenomenon was found in the control thallus where the thallus which were originally dark brown gradually turns red, then orange, and finally died (Fig. 10).

In the second week after enlargement in the nursery, the color of the treatment thallus tips was still red but the color at the thallus base began to darker. In the third week, the color of the treatment thallus base became dark red, and new tissues were visible at the thallus branch tips which were indicated by the brighter brown color. These new tissues were the favourite feed for the *Crustacean* and *Portunidae.* In the fourth and fifth weeks, a decrease in the size of the thallus begins to be seen, especially in the area with marks of *Crustacean* and *Portunidae* bit or claws. The color of remaining treatment thallus that had not been bitten or pinched by *Crustacean* and *Portunidae* becomes darker (Fig. 11). The color change in treatment thallus was thought to be influenced by the provision of compost before enlargement in the nursery. The addition of organic fertilizer with a frequency of once a week could provide a darker thallus color. One of the things that influences the rate of photosynthesis was the color or pigment in the thallus. If the pigment color become darker, the rate of photosynthesis will become higher (Fadilah & Pratiwi, 2017). Higher photosynthesis will stimulate the growth of plantlets. This means darker color of treatment thallus indicating that it still alive even though had been broken. The opposite phenomenon found in control thallus where the broken thallus could not survive in environmental conditions due to unavailability of additional nutrients from compost, so that the thallus covered in mud or dirt became red, then orange in color which showed that it had rotted and died.

The time required to soak the plantlets could affect the color of the thallus according to plantlets age. The longer the young plantlets were soaked in the compost solution, the redder the thallus will become. Soaking time in this study was only 15 min due to the plantlet become brighter red and softer after soaking time more than 15 min. Softer plantlets make it more easily damaged. Based on consideration that soaking time did not affect the plantlet growth (Mukhlis et al., 2016), it was prefered to soak the plantlet for 15 min in this research. For the use of artificial fertilizer, the soaking time could be done for 12 h (Mukhlis et al., 2016). The soaking period in this research was once a week. This was accordance to Fadilah and Pratiwi (2017) which stated that the optimal frequency of giving organic fertilizer could be carried out once a week.

3.3 Environmental Factors and External Disturbances

The environmental data were almost the same from the beginning up till the fifth week, except for several times it rained at night which caused sudden change in environmental conditions and no data was recorded.

The growth pattern of the plantlets tended to decline from the third to the fifth week. This weight loss was influenced by the large number of plantlets lost and exited through the gaps in the maintenance container (Fig. 12. right), so that the calculation of the plantlet weight was biased. The fractures on the plantlets were likely caused by fluctuations in environmental conditions because the maintenance location was at the sea and it rained several times at night. In addition, the observation results showed that several members of *Crustacea* and *Portunidae* could still enter the maintenance container (Fig. 12. left). This was also suspected as a factor in the plantlets breaking during the maintenance period. In the maintenance of *K. alvarezii* plantlets in the nursery, external factors such as the entry of several types of invertebrates could be avoided by using tighter containers, with the consequence of having cleaning it more frequently because it was more prone to being clogged by mud or other dirt.

In addition, one of the causes of the plantlets breaking easily during maintenance at the sea was the presence of white spots that spread to the surface of the thallus or were known as symptoms of *ice-ice* disease (Fadilah & Pratiwi, 2017). The presence of these symptoms could reduce the growth during seedling/thallus maintenance. Ice-ice symptoms appear due to fluctuations in environmental conditions such as changes in salinity, temperature, high light intensity especially during the day (Santoso & Nugraha, 2008; Arisandi et al., 2011). The fluctuations of environmental conditions

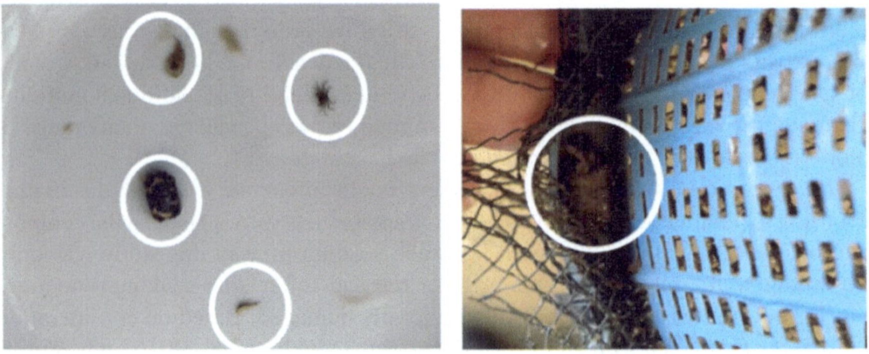

Fig. 12 Members of the class Crustacea and the order Portunidae (high lighted by white circle) were found in the maintenance container (Private documentation: Amalia, 2020—unpublished)

happened due to there were some rain during the fourth and fifth week which cause the plantlets to experience environmental shock. This environmental shock had affected to the growth of the plantlets.

4 Conclusions

Compost was not only able to promote the growth of plants but also the growth of seaweed *K. alvarezii* seedlings. It was proven by weight increase, new shoots/thallus occur during the cultivation, and darker color of plantlets.

Acknowledgements We would like to thank BBPBAP Jepara for providing the facilities and infrastructure to assist in this research. We also thank to Mrs. Ita Rizkiyanti for assisting us during composting process.

Funding This research was conducted using independent funds from the researcher and did not receive financial assistance from any party.

Conflict of Interest The authors declare that they have no conflicts of interest.

Author Contribution All authors contributed to the study conception and design. Material preparation, data collection and analysis were performed by Arofah Lyla Nurhayati and Indrian Rizka Amalia. The first draft of the manuscript was written by Arofah Lyla Nurhayati and all authors commented on previous versions of the manuscript. All authors read and approved the final version of the manuscript.

References

Arisandi, A., Marsoedi, Happy, N., & Sartimbul, A. (2011). Kecepatan dan Presentase Infeksi Penyakit *Ice-Ice* pada *Kappaphycus alvarezii* di perairan Bluto, Sumenep—Speed and Percentage of Ice-Ice Disease Infection in *Kappaphycus alvarezii* in Bluto Waters, Sumenep. *Jurnal Ilmiah Perikanan dan Kelautan, 3*(1), 47–51.

Fadilah, S., & Pratiwi, D. A. (2017). Pengaruh Frekuensi Penambahan Pupuk Terhadap Pertumbuhan Rumput Laut *Kappaphycus alvarezii*—The Effect of Fertilizer Addition Frequency on the Growth of Kappaphycus alvarezii Seaweed. *Prosiding Seminar Nasional Tahunan XIV Hasil Penelitian Perikanan dan Kelautan Universitas Gadjah Mada*, Yogyakarta, pp. 219–225.

Izydorczyk, G., Dawid, S., Małgorzata, M., Katarzyna, M., Mateusz, S., Filip, G., Rafał, T., Konstantinos, M., & Katarzyna, C. (2024). Lignocellulosic biomass fertilisers: Production, characterization, and agri-applications. *Science of the Total Environment* 1–16. https://doi.org/10.1016/j.scitotenv.2024.171343

Ministry of Maritime Affairs and Fisheries of the Republic of Indonesia. (2021). Ekspor Rumput Laut Kering dari Batam Melonjak Drastis - Dried Seaweed Exports from Batam Soar Dramatically. Siaran Pers Kementerian Kelautan Dan Perikanan No: SP.253/SJ.5/III/2021. https://www.kkp.go.id/news/news-detail/ekspor-rumput-laut-kering-dari-batam-melonjak-drastis65c1d1327e619.html#:~:text=Seperti%20diketahui%2C%20Indonesia%20merupakan%20produsen,memiliki%20nilai%20ekonomi%20yang%20tinggi

Mukhlis, Zakirah, R. Y. & Rusaini. (2016). Pertumbuhan Eksplan Rumput Laut *Kappaphycus alvarezii* pada Lama Perendaman Berbeda dalam Larutan Pupuk Urea dan TSP—Growth of *Kappaphycus alvarezii* Seaweed Explants at Different Immersion Times in Urea and TSP Fertilizer Solutions. *Prosiding Simposium Nasional III Kelautan dan Perikanan Universitas Hassanudin*, Makassar, pp. 434–446.

Murbandono, L. (1996). Membuat Kompos—Making Compost. *Penebar Swadaya*, Jakarta.

Nafifa, R. S. (2018). Kajian Nilai Nutrisi Tanaman Pada Program I-Jalapi Terhadap Pertumbuhan Sapi Di Labangka—Study of the Nutritional Value of Plants in the I-Jalapi Program on Cattle Growth in Labangka. *MA thesis*, Mataram University, Mataram.

Ningsih, M. S., Susilo, E., Rahmadina, Qolby, F. H., Tanjung, D. D., Anis, U., Eka, S. N., Panggabean, N. H., Priyadi, S., Nasution, J., Sari, N. Y., Baharuddin, R., & Wisnubroto, M. P. (2024). Dasar-dasar Fisiologi Tumbuhan—Fundamentals of Plant Physiology. CV Hei Publishing Indonesia, First Edition. Ed. Tri Putri Wahyuni, S.Pd. ISBN: 978-623-89104-4-1

Nurhayati, A. L., Permana, D., Aulia, R. & Nurhidayat. (2023). Enhancement growth of plantlet *Gracilaria* sp. with additional fermented fertilizer. In: *5th International Conference on Fisheries and Marine Science*, IOP Conf. Series: Earth and Environmental Science (pp. 1–6). https://doi.org/10.1088/1755-1315/1273/1/012036

Purwani, J., Achdiat & Dwiwanti, S. (2012). Lama Pengomposan dan Cara Aplikasi Pupuk Hijau Legum Terhadap Pertumbuhan dan Hasil Caisim (*Brassica juncea* l)—The effect of composting time and application method of legume green fertilizer on the growth and yield of Caisim (*Brassica juncea* l). *Instalasi Sayuran*, STPP Bogor.

Santoso, L., & Yudha, T. N. (2008). Pengendalian Penyakit *Ice-Ice* untuk Meningkatkan Produksi Rumput Laut Indonesia—ice-ice disease control to increase Indonesian seaweed production. *Jurnal Saintek Perikanan, 3*(2), 37–43.

Sicca, S. P. (2025). Berapa Banyak Kandungan Protein dalam Daun Kelor?—How much protein is in Moringa leaves? *Kompas.com.* (https://health.kompas.com/read/25A10120000268/berapa-banyak-kandungan-protein-dalam-daun-kelor-?page=all)

Siregar, N. A. (2023). PPI Lepas Ekspor Rumput Laut Ke China—PPI Releases Seaweed Exports to China. *PPINews*. https://ppinews.ptppi.co.id/blog/ppi-lepas-ekspor-rumput-laut-ke-china

Sulistiani, E. & Yani, S. A. (2014). Kultur Jaringan Rumput Laut Kotoni (*Kappaphycus alvarezii*)—Kotoni Seaweed Tissue Culture (*Kappaphycus alvarezii*). SEAMEO BIOTROP South East Asian Regional Centre for Tropical Biology. 2nd Printed: p. 141. ISBN: 978-979-8275-44-9

Sutejo, M. M. & Kartasapoetra, A. G. (1990). Pupuk dan Cara Pemupukan—fertilizers and fertilization methods. *Rineka Cipta*, Jakarta.

Sutriana, S., & Raisa, B. (2019). Uji Tingkat Kematangan Kompos Terhadap Produksi Tiga Varietas Bawang Merah (*Allium ascolanicum* L) Pada Tanah Gambut—test of compost maturity level on the production of three varieties of Shallots (*Allium ascolanicum* L) on Peat Soil. *Jurnal Ilmiah Pertanian, 16*(1), 25–35.

TEMPO. (2024). Mengenal 5 Ragam Rumput Laut di Perairan Indonesia: Jenis, Kegunaan, dan Manfaatnya—Get to Know 5 Types of Seaweed in Indonesian Waters: Types, Uses, and Benefits. *Tempo*. https://www.tempo.co/lingkungan/mengenal-5-ragam-rumput-laut-di-perairan-indonesia-jenis-kegunaan-dan-manfaatnya-4566

Compost-biochar: Increasing Productivity of Strategic Food Crops and Remediator of Endosulfan Residue

Elisabeth Srihayu Harsanti, Anicetus Wihardjaka, Asep Nugraha Ardiwinata, Asep Kurnia, and Turnad Lenggo Ginta

Abstract Agricultural waste such as rice hush, coconut shells, empty oil palm bunches, bamboo, corncob, and cattle manure were abundantly available in the agricultural environment, where it can be utilized to improve crop productivity, soil quality, and the environment quality. These wastes can be utilized as compost through composting or as biochar through pyrolysis. Both products can be utilized in remediating endosulfan residues from agricultural soils where endosulfan insecticides have been widely used since the green revolution. This paper was prepared to provide an overview of the role of the combination of compost and biochar in increasing the productivity of food crops while remediating endosulfan insecticide residues. Through a desk literature review, this article synthesizes evidence on how compost–biochar integration not only improves soil physical, chemical, and biological properties but also accelerates degradation and immobilization of endosulfan. This work is innovative because it uses locally available agricultural wastes to frame the compost–biochar synergy as a multipurpose, resource-efficient solution that simultaneously improves soil health, increases yields, and reduces pesticide pollution. This approach provides a pathway for sustainable, climate-resilient, and pollution-free agriculture, so its contribute to a more sustainable and productive agricultural future.

Keywords Biochar-compost · Endosulfan remediation · Food security · Sustainable agriculture

---

E. S. Harsanti (✉) · A. Kurnia
Research Center for Horticulture, Agricultural and Food Research Organization, National Research and Innovation Agency, Cibinong, West Java, Indonesia
e-mail: elis010@brin.go.id

A. Wihardjaka
Research Center for Food Crops, Agricultural and Food Research Organization, National Research and Innovation Agency, Cibinong, West Java, Indonesia

A. N. Ardiwinata
Research Center for Estate Crops, Agricultural and Food Research Organization, National Research and Innovation Agency, Cibinong, West Java, Indonesia

T. L. Ginta
Research Center for Process and Manufacturing Industry Technology, National Research and Innovation Agency, Tangerang Selatan, Indonesia

A. Z. Yaser et al. (eds.), *Compost and Liquid Organic Fertilizers*,
https://doi.org/10.1007/978-981-95-8561-8_6

1 Introduction

The availability of sufficient, equitable and save food is a fundamental indicator of the national welfare and food security. Increasing food sufficiency is in line with the rate of human growth and population. The availability of food is an indicator of the stability of sustainable national food security. The state must play an active role in ensuring the food availability for its population, including through efforts to increase and stabilize the production and productivity of food crops such as agricultural intensification, agricultural extensification, and food diversification.

The success of agricultural crop production and productivity cannot be separated from efforts to provide planting media with guaranteed soil fertility, sufficient essential nutrients, and input of production facilities, especially external nutrient sources. The need for natural nutrients and external inputs has been carried out before the green revolution era with natural materials such as litter and livestock manure. Plant litter and livestock manure are significantly agricultural resources. Organic manure has been known to have beneficial effects on soil fertility and many other soil properties, contributing to the overall soil health (Verma et al., 2024).

Since the green revolution era, the provision of production facilities, especially inorganic fertilizers and pesticide have been intensified to boost food crop production. Traditional agricultural intensification has relied heavily on chemical fertilizers and pesticides since the green revolution, yet this has contributed to declining soil organic matter and persistent chemical contamination. Gradually, the depletion of organic matter content occurs due to the absence of organic matter input into the soil, which is one of the causes of the decline in food crop production. Rice fields in Indonesia generally contain low organic matter, namely 65 percent of ±5 million ha of irrigated rice fields have an organic matter content of less than 2 percent. The decrease in organic C content in the soil is accelerated by continuous cultivation practices, intensive soil cultivation, and without returning plant residues/biomass after harvest (Wihardjaka & Harsanti, 2021). In addition to inorganic fertilizers, farmers also use pesticides intensively, such as organophosphate pesticides and organochlorine pesticides, including endosulfan.

Endosulfan is one of an organochlorine insecticide detected which is relatively slow degrade, toxic, and persistent in soil. These residues not only threaten human health but also degrade soil and environmental quality. At the same time, large volumes of agricultural wastes (e.g., rice husks, corn cobs, coconut shells, livestock manure) remain underutilized, often contributing to pollution if left unmanaged. While compost and biochar have been individually studied for their benefits to soil fertility and crop productivity, the novelty of this review is in highlighting their combined application as a synergistic solution to two interconnected challenges: low soil fertility and pesticide contamination. Specifically, this article synthesizes evidence on how compost–biochar integration: (1) Improves the physical, chemical, and biological properties of soil, (2) Enhances the productivity of strategic food crops, and (3) Functions as a natural remediator of endosulfan residues. By explicitly framing compost–biochar as a dual-function soil amendment-both a productivity enhancer and a bioremediation tool, this review contributes to rethinking waste-to-resource pathways for sustainable agriculture.

Sustainable agricultural systems pay attention to environmental, social, and economic pillars through the utilization of biological resources such as biomass and/or organic agricultural waste to obtain economic benefits, social benefits, and environmental conservation. The contribution to sustainable agriculture often involves: (a) the production of goods and services (economic function); (b) the management of natural resources (ecological function); and (c) the contribution to rural dynamics (social function). The harmonious combination of these three interconnected functions constitutes the backbone of sustainable agriculture (Jaroenkietkajorn et al., 2024; Shahpari et al., 2021; Voisin et al., 2025).

Sustainable agricultural systems refer to an ecological approach that is not exploitative, and is adapted to natural conditions (Robinson, 2024). One of the inputs of organic materials into agricultural soil is compost and biochar (biochar). Compost combined with biochar will have added value in providing a source of nutrients, improving the physical, chemical, and biological properties of the soil and functioning in reducing contaminants/pollutants in the soil. This paper reviews the role of the combination of compost and biochar in increasing the productivity of food crops while remediating endosulfan insecticide residues to support sustainability agriculture.

This review article follows a desk study approach, which primarily involves analysing and synthesizing existing literature on the subject matter. This review helps gather a comprehensive understanding of the theoretical frameworks, experimental results, and practical applications of these substances in agriculture (Voisin et al., 2025).

The literature review specifically focuses on the roles of compost and biochar in enhancing crop productivity, particularly for food crops. Compost is known for its potential to improve soil fertility, enhance moisture retention, and increase microbial activity, all of which contribute to higher agricultural productivity. Biochar, on the other hand, is recognized for its ability to improve soil structure, increase nutrient availability, and reduce greenhouse gas emissions. Both substances have been widely studied for their positive impacts on crop yields and soil health. By reviewing various studies, the article highlights the effectiveness of these organic amendments in boosting crop productivity in diverse agricultural contexts.

Additionally, this review article emphasizes the role of compost and biochar in remediating endosulfan residues in soil. Endosulfan, a persistent pesticide, poses significant risks to both human health and the environment. The article examines the scientific evidence surrounding the ability of compost and biochar to mitigate the negative impacts of endosulfan contamination in soils. Through literature studies, the review identifies how these materials can help in degrading or immobilizing the pesticide residues, thereby reducing its toxicity and improving the safety of agricultural lands. This comprehensive review contributes to a better understanding of sustainable agricultural practices that can both enhance productivity and remediate harmful chemical residues (Karam et al., 2022).

1.1 Endosulfan Insecticide

Endosulfan is one of the organochlorine insecticides which is a persistent organic pollutants (POPs) compound that is semi-volatile, persistent, bioaccumulative in living things, toxic to humans and other living things (Harsanti et al., 2018, 2019, Wihardjaka et al., 2022). Endosulfan (6,7,8,9,10,10-hexachloro-1,5,5a,6,9,9a-hexahydro-methano-2,4,3-benzodioxathiepine 3-oxide) is an ester compound containing chlorine and sulfur groups (ATSDR, 2024; Berdowska & Bandurska, 2025). Endosulfan consists of two isomers α-endosulfan and β-endosulfan, and its metabolites (endosulfan sulfate, endosulfan diol, endosulfan ether, endosulfan hydroxy ether, and endosulfan lactone). Endosulfan sulfate is the main metabolite of endosulfan because it is formed in the environment and its toxicity is almost the same as α-endosulfan. Endosulfan in the soil will be metabolized into endosulfan sulfate which is estimated to be immobile because it has a Koc value of 9,800, likewise α- and β-endosulfan each have Koc values of 10,600 mL g^{-1} and 13,500 mL g^{-1} so it is estimated that its movement tends to be limited in the soil (Harsanti et al., 2018). The chemical structure of endosulfan and its metabolite (endosulfan sulfate) is presented in Fig. 1.

Endosulfan is insoluble in water or non-polar with a solubility in water of around 0.53 mg L^{-1} (ATSDR, 2024). Due to its high toxicity with a toxicity level of class I (highly hazardous) or class II (moderately hazardous), endosulfan is preferred by farmers because it is considered'really great' or efficacious in the past. Endosulfan has moderate-persistent properties, stable reactions, and the adsorption capacity of α-endosulfan is lower than β-endosulfan on clay. Endosulfan could easily be leached into groundwater and tendency to be absorbed into the soil (Rasool et al., 2022; Vaikosen et al., 2022). The complete characteristics of endosulfan are shown in Table 1.

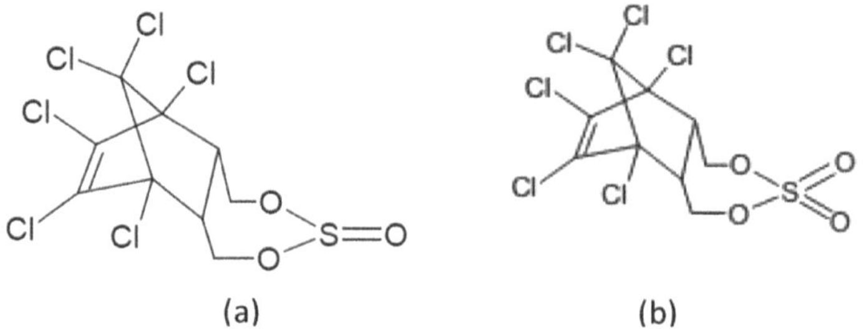

Fig. 1 Structure of endosulfan **a** and endosulfan sulfate **b** (ATSDR, 2024; Berdowska & Bandurska, 2025; Vaikosen et al., 2022)

Table 1 Characteristic of endosulfan

Characteristics	Remark
Ion strength	Stabile reaction
Soil adsorption coefficient (Kd)	12,400
Solubility (mg L^{-1} water)	α-isomer: 0.32, β-isomer: 0.33, sulfate endosulfan: 0.22 at t 22 °C α-isomer: 0.53, β-isomer: 0.28, sulfate endosulfant: 0.13 at t 25 °C
Vapor pressure (at t 25 °C)	α, β-endosulfan:1 × 10^{-5} mmHg; sulfate endosulfan:1 × 10^{-11} mmHg
Polarity	Non polar
Toxicity	LD50 = 18–160 mg kg^{-1} (*rats*), 7.36 mg kg^{-1} (*mice*), 77 mg kg^{-1} (*dogs*); (EPA-kls I; WHO-II)
Soil adsorption (K_d)	0.01–0.045 mg g^{-1} (α-endosulfan) 0.0942–0.2722 mg g^{-1} (β-endosulfan)
Persistency (1/2-ʎ)	50–135, 600 days (moderate-persistent)
Biological reaction	Potency of high bioaccumulation Potency of high biology
Weight of molecule (WM)	406.93 (α-endosulfan, β-endosulfan) 422.95 (endosulfan sulfate)
Soil partition coefficient (Log K_{ow}) Carbon organic coefficient (Log K_{oc})	α-endosulfan: 3.83 β-endosulfan:3.62; Sulphate endosulfan:3.66 α-endosulfan: 4.03 β-endosulfan: 4.13; Sulphate endosulfan:-

Source (ATSDR, 2024; Berdowska & Bandurska, 2025; Hwang & Kim, 2022; Vaikosen et al., 2022)—no data yet

Endosulfan degradation is moderate in sandy clay soil, has a half-life of about 60–800 days (Berdowska & Bandurska, 2025). The two endosulfan isomers have different degradation times in soil, α-isomer has a half-life of about 35 days, and β-isomer about 150 days. Both isomers have different degradation times in neutral soil conditions and both will persist longer in acidic soil conditions. Degradation of endosulfan in soil and water is slow to hydrolyze, slow to hydrolyze, α- and β-Endosulfan is rapidly oxidized to endosulfan sulfate but β-endosulfan is slower, and both are slowly oxidized to endosulfan sulfate when in air (Hwang & Kim, 2022).

Endosulfan is an insecticide belonging to the organochlorine group, has very high toxicity (included in class I), and its use is limited (Berdowska & Bandurska, 2025; Sharma et al., 2022). The United State Environmental Protection Agency (US EPA) and WHO classify endosulfan as Category I (Highly Acute Toxic) and Category II (moderately dangerous), respectively. Exposure to endosulfan is harmful to human health, including inhibiting ATPase which is an enzyme that transfers nerve impulses, and disrupting the performance of endocrine glands (Berdowska & Bandurska, 2025).

Liu et al. (2024) reported, endosulfan, recognized as an endocrine disruptor, has emerged as an important risk factor for human breast cancer that was enhanced cell proliferation, increased colony formation rate, and improved cell migration ability.

Endosulfan is highly toxic to fish and aquatic invertebrates. Bioaccumulation in shellfish (*Mytelus edulis*) can reach 600 times the concentration in water, and in fish for 2 years the residue in catfish can increase 7 times (Hussein et al., 2019; Fadilah et al., 2023). For other organisms, endosulfan is quite toxic to bees but not toxic to parasites and mites. According to Navarrete-Rodríguez et al. (2016), the dose for endosulfan exposure through intake is 0.015 mg kg^{-1} for adults and 0.0015 mg kg^{-1} for children. Chronic intake exposure is 0.006 mg kg^{-1} day^{-1} for adults and 0.0006 mg kg^{-1} day^{-1} for children.

1.2 Compost of Agricultural Wastes

Abundant agricultural waste can harm human health and reduce environmental quality. Biomass could be a source of toxic substances for the environment and is the habitat of pathogens. Moreover, it contributes to the increase of greenhouse gas emissions as a result of decaying processes (Gęca et al., 2022). Agricultural waste biomass such as crop residues and solid livestock manure must be utilized as an effort to reduce its impact on the environment, including by composting so that compost products can be effectively utilized and returned to agricultural soil.

Compost is a mixture of organic material residues (animal waste, litter, straw, other agricultural waste) processed through thermophilic decomposition at a temperature of 45– >100 °C. Composting can be done in situ and ex situ. Rice straw from the harvest that is not transported and composted in the field (Fig. 2). In the composting process, indigenous decomposing microorganisms are needed to accelerate the formation of compost. The ex situ composting process at the farmer level is carried out by piling and mixing the composted materials, adding nitrogen fertilizer and water.

Ex situ composting stages include (i) Collection and sorting to gather and separate organic waste from non-compostable materials, (ii) Shredding and chopping to reduce particle size to accelerate decomposition, (iii) Moisture and aeration Management to maintain moisture levels (50–60%) and ensure aeration for microbial activity, (iv) Carbon-to-nitrogen (C:N) ratio Balance to maintain an optimal C:N ratio of 25–30:1 for effective decomposition, (v) Pile formation and turning to stack waste in windrows or static piles, turning regularly for oxygen supply, (vi) Temperature monitoring to keep composting temperatures between 50–65 °C to kill pathogens and weed seeds, and (vii) Curing and maturation to allow compost to mature for several weeks to stabilize nutrients. Composting material's C/N ratio varies greatly (Table 2).

Compost functions as a soil conditioner that plays an important role in improving the physical, chemical, and biological properties of the soil. Compost increases and

Fig. 2 Straw left on the land is composted in situ using a simple technique (*Source* personal collection)

Table 2 C/N ratio of several composted materials

Material	C/N ratio	Material	C/N ratio
Cattle manure	19:1	Poultry carcass	4:1
Cattle carcass	10:1	Sawdust	442:1
Corn silage	40:1	Sheep manure	16:1
Corn stalk	68:1	Swine carcass	14:1
Dairy manure	20:1	Swine manure	12:1
Grass clippings	17:1	Turkey litter	16:1
Horse manure	30:1	Wheat straw	127:1
Leaves	54:1	Wood chips	600:1

Source Keena et al. (2022)

can reduce the need for inorganic fertilizers. The provision of compost from livestock manure improves the yield and yield components of food crops without application due to the supply of macronutrients, for example in the US, cow manure and chicken manure compost can increase sorghum and corn yields by 15.8% and 95.8% respectively (Rayne & Aula, 2020).

2 Biochar of Agricultural Wastes

Agricultural waste can be composted, especially materials containing low lignin, while materials with high lignin or high C/N ratio should be turned into biochar through the pyrolysis process (Wihardjaka & Harsanti, 2021). Agricultural waste is abundantly available in the environment, such as rice husks, corn cobs, sorghum biomass, coconut shells, empty oil palm bunches, bamboo, wood, etc. have the potential to be used as biochar. A total of 13.7 million tons of rice husks can be converted into 7.1 million tons of biochar or about 51.8% of raw material (Nurida, 2014), presented in Table 3. Meanwhile, other research showed that biochar production from rice husks is around 35% of the original raw material (Karam et al., 2022). Pyrolysis is a thermochemical procedure that decomposes organic materials like biomass or trash in the absence or with minimal oxygen (Zahoor et al., 2025). Incomplete combustion of agricultural waste at a temperature of 250–300 °C in an oxygen-deficient environment reduces the volume of biological waste and can also be a method of bioenergy production, which produces the final product in the form of biochar (Dorner et al., 2022).

Biochar is a carbon-rich end product resulting from the thermochemical conversion of biomass. Biochars are chemically heterogeneous, and characterized by negative surface charge, high aromaticity, high ion-exchange capacity, high porosity, and high surface area. These characteristics facilitate strong sorption and retention of metal cations as well as many organic compounds (Dorner et al., 2022). The functional group content, surface area, porosity and sorption capacity of biochar vary depending on several factors including the type of biomass raw material. The sorption capacity of biochar is indicated by the iodine (I_2) absorption capacity (Ardiwinata, 2020). Biochar from rice husks and coconut shells has an iodine absorption capacity of 426 and 457 mg g^{-1}, respectively, which is higher than compost (312 mg g^{-1}) and organic fertilizer (412 mg g^{-1}) (Ardiwinata, 2020). High iodine absorption capacity indicates better biochar quality in absorbing contaminant compounds. Biochar derived from wood shows more pores compared to chicken manure as raw material (Ratmini et al., 2018; Yang et al., 2024), as shown in Fig. 3.

The quality of biochar as a soil amendment is determined by pH, volatile content, ash content, water holding capacity, BD, pore volume and specific surface area (Nurida, 2014). Biochar has a very excellent nest structure and a large interior specific surface area, allowing it to hold certain molecules to the inner surface while adsorbing other substances in gas or liquid form, which contains 87–97% C and includes the components O, H, S, and N (Harsanti et al., 2024). Biochar can increase plant development by enhancing soil structure, water retention, nutrient and organic matter

Table 3 Biochar potential from several agricultural wastes

Agricultural biomass	Production (t yr^{-1})	Ratio of biochar: biomass	Potency of biochar (t yr^{-1})
Rice husks	13,612,243	0.26	1,769,605
Coconut shells	539,644	0.25	67,456
Empty oil palm bunches	6,400,000	0.50	960,000
Cacao rind	1,208,557	0.33	199,411
Corn cobs	3,652,372	0.13	142,443

Source Nurida (2014)

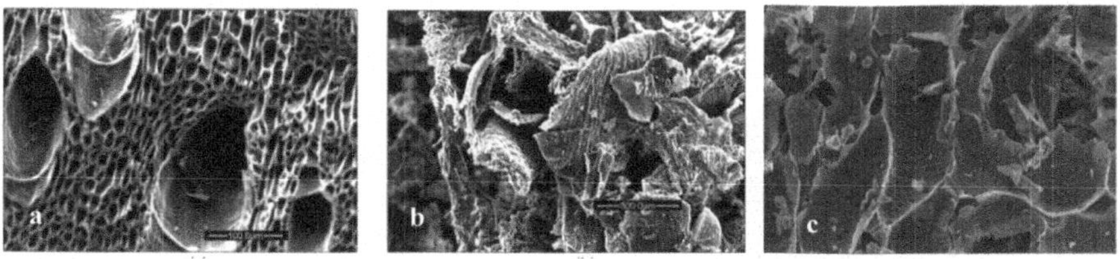

Fig. 3 SEM of biochar from wood **a** and chicken manure **b** in slow combustion (*Source* Ratmini et al., 2018), and **c** SEM of biochar from corn cob with 500 × magnification (*Source* Harsanti et al. 2018, 2019)

content, and increasing cation exchange capacity. Cation exchange capacity (CEC) can be increased by the presence of biochar through an electrostatic bonding mechanism on surfaces that contain many functional groups, such as carbonyl, carboxyl, and hydroxyl (Harsanti et al., 2024). The higher CEC value occurs since the carboxyl groups readily deprotonate in water, resulting in a more negative charge on the biochar/activated charcoal surfaces (Gęca et al., 2022).

Biochar is possible to detoxify toxic compounds from pesticide residues that are still left in the soil. Biochar can bind compounds from pesticides through several possible mechanisms such as surface charge change, complexity of adsorbed molecules, and coordination bonds with metal mediators (Gęca et al., 2022). Possible bonding mechanisms between biochar and organic compounds from pesticides are shown in Fig. 4. Biochar can enhance the degradation of a variety of organic contaminants. Dorner et al., (2022) reported that biochar was shown to enhance the abiotic reductive degradation of the explosive compound 1,3,5 trinitroperhydro-1,3,5-triazine by hydrogen sulphide, and biochar was shown to accelerate the abiotic reduction of nitroglycerine by hydrogen sulphide (Dorner et al., 2022). Dorner et al. (2022); Xu et al. (2015) further reported that the abiotic degradation of a variety of halogenated organics was sorbed to biochar via reactive sulfur species (sulfide, thiosulfate, sulfite), including the halogenated heterocyclic amines 2,6-dichloropyridine and 2,6-dibromopyridine, the nitrated and halogenated aromatics 3 bromonitrobenzene, 1,3,5-tribromobenzene, and the chlorinated aliphatic insecticide γ-hexachlorocyclohexane.

3 Biocompost Increase Productivity of Food Crops

The provision of rice straw compost can increase the yield of lowland rice in rainfed agroecology by 5.4–10.0 percent (Wihardjaka, 2011; Huang et al., 2023). The provision of 5 tons ha^{-1} of rice straw significantly increased the average yield of dry milled rice of the Ciherang variety by 0.7 tons ha^{-1}, potassium absorption by 0.07–0.28 g of K per plant, and the content of available K in the soil by 0.22–0.30 cmol kg^{-1} (Wihardjaka, 2016).

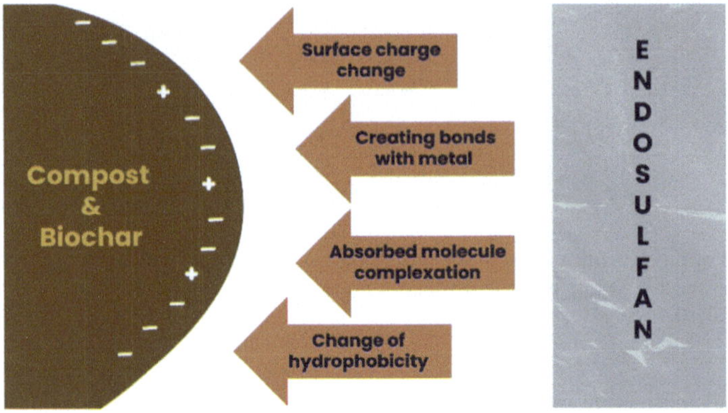

Fig. 4 Mechanisms that allow binding between biochar and pesticide residues (*Source* Gęca et al., 2022)

Straw composting contributes to efforts to mitigate greenhouse gas emissions, especially methane from rice fields. Straw compost emits lower methane than fresh straw in rice fields (Allen et al., 2020; Dash et al., 2023). Cow dung compost produced from a biodigester can reduce methane emissions from rice plants by 59.1 percent compared to fresh straw that is directly embedded in the rice field soil (Wihardjaka, 2018).

Application of compost from livestock manure improves soil organic C content and delays soil organic matter depletion. The application of 10, 25, 40, and 50 tons compost chicken manure per hectare (ha) increased SOM levels by 0.85, 1.50, 1.72, and 1.95 percentage points, respectively relative to the control treatment with no addition of manure (Rayne & Aula, 2020). The increase in SOM will indirectly improve soil and plant productivity.

The role of compost can be multifunctional, especially if the compost is combined with biochar. Application of soil ameliorant (combined corncob biochar and livestock manure) can increase caysim biomass 5.5–26.2% compared without soil ameliorant (Harsanti et al., 2022). The combination of compost and biochar, in addition to improving the physical, chemical, and biological properties of the soil, also increases carbon reserves/sequestration, reduces the impact of climate change through adaptation-mitigation efforts, and of course improves environmental quality. As a soil conditioner, compost + biochar is given to planting preferably on an economic scale, where farmers can make efforts.

4 Compost as Remediator of Endosulfan Residue

Soil amendments or ameliorants act as remediators of soil contaminants such as pesticide residues, and they can improve the physical, chemical, and biological properties of agricultural fields. Agricultural waste such as rice husks, oil palm bunches, corn cobs, cow manure, chicken manure and others can be used as remediation materials or referred to as remediators in agricultural soil contaminated with pesticides, including endosulfan which is one of the active ingredients of organochlorine. In Indonesia, the use of endosulfan has been banned since 2011, but in reality in the field pesticide products containing the active ingredient endosulfan are still found in several areas, including in the rice fields of Jombang Regency, East Java. Around 40.78% of the rice field area is contaminated with endosulfan with a concentration range of 0.0032–0.2196 mg kg^{-1} (maximum residu limits/MRL = 0.0085 mg kg^{-1}) and 18.1% of them exceed the MRL (Harsanti et al., 2018, 2019). However, endosulfan residue can be treated with remediation efforts.

According to Harsanti et al. (2018b), the application of ameliorant (a mixture of cow manure compost and biochar with a ratio of 4:1) can reduce the concentration of endosulfan in the soil by 68.1% during panicle initiation and 62.8% during the ripening phase compared to without ameliorant and the endosulfan residue carried in the rice grains is < MRL (0.1 mg kg^{-1}). In addition to manure and corn cob biochar, rice productivity significantly increases by 8–26%. The application of biochar-agricultural waste compost such as corn cobs and cow manure or chicken manure with a ratio of 1:4 can reduce the endosulfan residue carried by percolation water undetected.

Application of cow manure and chicken manure compost mixed with biochar (4:1) can also increase the microbial population (bacteria and fungi) in the soil of the rhizosphere area, however, the microbial population is higher in the combination of cow manure and biochar than chicken manure and biochar. Manure and biochar provide organic matter that uses microbes as a food source to support microbial activity in degrading pesticide residues. One of the bacteria that can degrade organochlorine is *Bacillus sp* which is detected in soil applied with biochar remediator + cow manure or chicken manure in a ratio of 1:4 with a population of 5×10^3–13×10^3 sel g^{-1} (Harsanti et al., 2018).

The application of biochar combined with livestock manure compost can significantly affect the absorption of endosulfan sulfate (endosulfan derivative) in caysim and the productivity of caisim. In contrast to rice plants, the application of a mixture of corn cob biochar and cow dung compost with a ratio of 1:1, 2:4, and 1:4 can reduce endosulfan sulfate residues by 29.2%, 74.4%, and 33.2%, respectively; while the combination of biochar + chicken manure compost with a ratio of 1:4 reduces endosulfan sulfate residues by 21.2% (Harsanti et al., 2022). Several research results show the ability of remediator to increase productivity (Table 4).

Endosulfan residue remediation efforts can be approached by physical, chemical and biological mechanisms. The remediation mechanism using remediator (a mixture of biochar and livestock manure compost) can be physically possible from the role

Table 4 Yield of several commodities given biochar ameliorants and livestock manure as remediator

Comodities	Kind of remediator/ameliorant	Yield ($t\ ha^{-1}$)	
		With remediator	Without remediator
Rice[1]	Biochar of corn cobs + compost of cow manure, 2 $t\ ha^{-1}$	10.80	9.50
Rice[1]	Biochar of corn cobs + compost of chicken manure, 2 $t\ ha^{-1}$	10.35	9.50
Rice[2]	Compost of cow manure, 10 $t\ ha^{-1}$	8.70	7.80
Corn[3]	Biochar of bamboo waste 10 $t\ ha^{-1}$	7.83	5.77
Corn[3]	Compost of cow manure 20 $t\ ha^{-1}$	7.42	6.37
Caysim[4]	Biochar of corn cobs + compost of cow manure, 2 $t\ ha^{-1}$	14.51	12.13
Caysim[4]	Biochar of corn cobs + compost of chicken manure, 2 $t\ ha^{-1}$	13.71	12.13
Shallot[5]	Biochar of coconut shells 10 $t\ ha^{-1}$ + compost of chicken 10 $t\ ha^{-1}$	19.02	17.84

Source [1]Harsanti et al. (2018b); [2]Apriyani et al. (2021); [3]Lelu et al. (2018); [4]Harsanti et al. (2022); [5]Vinolina et al. (2024)

of biochar in providing a home for microbes, the chemical mechanism is shown from the role of biochar functional groups that have the potential to bind pesticide contaminants such as endosulfan and biologically the role of microbes carried in ameliorants or in soil trapped in the biochar cavity and using it as a carbon source. Two factors that play an important role in remediation efforts are organic matter and microbial populations. Therefore, both factors need to be managed by increasing their availability. This has been shown by the statistical model $Y = 0.787–0.0274X_1–0.00817X_2–0.369X_3–0.002X_4–0.003X_5$ with a value of $R^2 = 0.675$ ($R = 0.821$). The five factors tested were fungi, soil temperature, bacteria population, C-soil organic, and soil pH, however, only 2 factors had a significant effect, namely the bacterial population and soil C-organic content (Harsanti et al., 2021a).

The results of research on endosulfan degradation technology in the environment (soil and water) through bioremediation, photocatalysis, adsorption, and membranes were effectively conducted in situ, utilizing bioremediation by plants, fungi, and bacteria (Mudhoo et al., 2019). In addition, for the purpose of filtering contaminants in water, it is still limited to endosulfan degradation. The rice husk biochar (RHB) can improve soil fertility and increase fertilizer effectiveness when mixed together due to its ability to retain nutrients and high silica content (Karam et al., 2022). Pesticide adsorption depends on organic matter, soil type, and environmental conditions in the soil (pH and water content) and microbial activity are important factors that determine pesticide degradation (Rasool et al., 2022).

The references explain how organic materials play a role in improving soil quality or degrading pesticides, demonstrating their respective capabilities. Based on productivity data for several food crops and vegetables, the combination of compost-biochar in agricultural fields demonstrates multiple roles: enhancing productivity and degrading endosulfan due to the role of biochar and the microorganisms residing in its pores, thereby constituting a sustainable, environmentally friendly technology. A synergistic solution to two interconnected challenges: low soil fertility and pesticide contamination. This finding is highlighted in this review. The presence of organic carbon and microorganisms (bacteria) plays a role in remediation functions.

5 Conclusion

The utilization of agricultural waste in the form of compost or biochar has promising prospects in efforts to maintain food security while improving the quality of the agricultural environment. The combination of compost and biochar plays a multifunctional role in ensuring environmental sustainability, meeting food needs, and preventing human health problems. The combination of compost and biochar is effective in increasing food crop productivity, reducing endosulfan pesticide residues in the soil, and improving the physical, chemical, and biological properties of the soil. Given the importance of this role, assistance and guidance to farmers should continue to be carried out so that the provision of location-specific resources for both materials is always sustainable to support the realization of a sustainable agriculture.

References

Allen, J., & et al. (2020). Rice straw management effects on greenhouse gas emissions and mitigation options. In *Sustainable Rice Straw Management*. Springer Open (pp. 145–160). https://doi.org/10.1007/978-3-030-32373-8_9

Apriyani, S., & et al. (2021). Effect of inorganic fertilizer and farmyard manure to available P, growth and rice yield in rainfed lowland Central Java. *IOP Conference Series: Earth and Environmental Science, 648*(1). https://doi.org/10.1088/1755-1315/648/1/012190

Ardiwinata, A. N. (2020). Pemanfaatan Arang Aktif dalam Pengendalian Residu Pestisida di Tanah: Prospek dan Masalahnya. *Jurnal Sumberdaya Lahan, 14*(1), 49. https://doi.org/10.21082/jsdl.v14n1.2020.49-62

ATSDR. (2024). *Toxicological Profile for Endosulfan [Update 2024], ATSDR's Toxicological Profiles*. https://doi.org/10.1201/9781420061888_ch19

Berdowska, A., & Bandurska, K. (2025). Health hazards associated with exposure to endosulfan: A mini-review. *Toxics*. Multidisciplinary Digital Publishing Institute (MDPI) (p. 455). https://doi.org/10.3390/toxics13060455

Dash, P. K., & et al. (2023). Mitigation of greenhouse gases emission through value-added straw amendments in rice–green gram system. *International Journal of Environmental Science and Technology, 20*, 1019–1036. https://doi.org/10.1007/s13762-021-03768-3

Dorner, M., & et al. (2022). Biochar-mediated abiotic and biotic degradation of halogenated organic contaminants—A review. *Science of the Total Environment, 852*(June), 158381. https://doi.org/10.1016/j.scitotenv.2022.158381

Fadilah, A., & Rahardja, B. S. (2023). Sublethal toxicity effects of the organochlorine insecticide endosulfan on oxygen consumption level and gill histopathology of Tilapia (Oreochromis Niloticus) fry. *Journal of Aquaculture Science,* 8(2), 99–109. https://doi.org/10.20473/joas.v8i2.48700

Gęca, M., Wiśniewska, M., & Nowicki, P. (2022). Biochars and activated carbons as adsorbents of inorganic and organic compounds from multicomponent systems—A review. *Advances in Colloid and Interface Science, 305*(April). https://doi.org/10.1016/j.cis.2022.102687

Harsanti, E. S., Ardiwinata, A. N., & Kurnia, A. (2022). Remediasi cemaran metabolit endosulfan menggunakan limbah pertanian pada pertanaman caisim. *Seminar Hasil Penelitian Pertanian XI: Penguatan Penelitian dalam rangka mendukung implementasi Smart Eco-Bioproduction, 20 November 2021* (pp. 98–103). Fakultas Pertanian Universitas Gadjah Mada.

Harsanti, E. S., Ardiwinata, A. N., & Wahyuni, S. (2018). Remediation of paddy soil contaminated by endosulfan residues by applying a combination of farmyard manures and corn cob biochar. In *Proceeding of International Workshop and Seminar Innovation of Environmental-Friendly Agricultural Technology Supporting Sustainable Food Self-Sufficiency*. IAARD Press (pp. 233–249).

Harsanti, E. S., & et al. (2018). Preliminary study of environmental risk from endosulfan usage during the Green Revolution: Case study in central paddy area of Jombang District, East Java, Indonesia. In *Competition and cooperation in social and political sciences* (1st ed., pp. 259–266). London: Routledge. https://doi.org/10.1201/9781315213620-32

Harsanti, E. S., & et al. (2019). The usage of corn cob waste to remediate paddy soil contaminated by endosulfan. In *AIP Conference Proceedings* (p. 040033). https://doi.org/10.1063/1.5115671

Harsanti, E. S., & et al. (2021). A statistical model for remediation plan of endosulfan-contaminated lowland rice fields with agricultural waste. *IOP Conference Series: Earth and Environmental Science, 648*(1). https://doi.org/10.1088/1755-1315/648/1/012207

Harsanti, E. S., & et al. (2024). Nitrogen losses (N2O and NO3–) from mustard (Brassica juncea L.) cropping applied urea coated bio-charcoal. *Plant, Soil and Environment, 70*(1), 1–10. https://doi.org/10.17221/282/2023-PSE

Huang, X., & et al. (2023). Rice straw composting improves the microbial diversity of paddy soils to stimulate the growth, yield, and grain quality of rice. *Sustainability (Switzerland), 15*(2). https://doi.org/10.3390/su15020932

Hussein, M. M. A. & et al. (2019). Endosulfan toxicity in Nile tilapia (Oreochromis niloticus) and the use of lycopene as an ameliorative agent. *Comparative Biochemistry and Physiology, Part C, 224*(108573). https://doi.org/10.1016/j.cbpc.2019.108573

Hwang, J. I., & Kim, J. E. (2022). Uptake of endosulfan isomers from soils by leafy vegetable lettuce: A comparative study between model-predicted and field-experimented results. *Science of the Total Environment, 844*, 157056. https://doi.org/10.1016/j.scitotenv.2022.157056

Jaroenkietkajorn, U., & et al. (2024). Challenges and opportunities of bio-circular-green economy for agriculture. *Circular Economy and Sustainability, 4*, 1729–1750. https://doi.org/10.1007/s43615-024-00355-9

Karam, D. S., & et al. (2022). An overview on the preparation of rice husk biochar, factors affecting its properties, and its agriculture application. *Journal of the Saudi Society of Agricultural Sciences, 21*(3), 149–159. https://doi.org/10.1016/j.jssas.2021.07.005

Keena, M. A., Augustin, C., & Rahman, S. (2022). *Composting Animal Manure: A guide to the process and management of animal manure compost*. https://www.ndsu.edu/agriculture/extension/publications/composting-animal-manures-guide-process-and-management-animal-manure-compost

Lelu, P. K., Situmeang, Y. P., & Suarta, M. (2018). Aplikasi biochar dan kompos terhadap peningkatan hasil tanaman jagung (Zea mays L.). *Gema Agro, 23*(1), 24–32. https://doi.org/10.22225/ga.23.1.655.24-32

Liu, Z., & et al. (2024). Endosulfan promotes cell growth, migration and invasion via CCL5/CCR5 axis in MCF-7 cells. *Ecotoxicology and Environmental Safety, 288*(June), 117344. https://doi.org/10.1016/j.ecoenv.2024.117344

Mudhoo, A., & et al. (2019). Endosulfan removal through bioremediation, photocatalytic degradation, adsorption and membrane separation processes: A review. *Chemical Engineering Journal, 360*, 912–928. https://doi.org/10.1016/j.cej.2018.12.055

Navarrete-Rodríguez, G., & et al. (2016). Endosulfan: its isomers and metabolites in commercially aquatic organisms from the Gulf of Mexico and the Caribbean. *Journal of Agricultural Science, 8*(1), 8. https://doi.org/10.5539/jas.v8n1p8

Nurida, N. L. (2014). Potensi Pemanfaatan Biochar untuk Rehabilitasi Lahan Kering di Indonesia. *Jurnal Sumberdaya Lahan Edisi Khusus* (pp. 57–68).

Rasool, S., Rasool, T., & Gani, K. M. (2022). A review of interactions of pesticides within various interfaces of intrinsic and organic residue amended soil environment. *Chemical Engineering Journal Advances, 11*, 100301. https://doi.org/10.1016/j.ceja.2022.100301

Ratmini, N.P. S., Juwita, Y., & Sasmita, P. (2018). Pemanfaatan Biochar Untuk Meningkatkan Produktivitas Lahan Sub Optimal. In *Prosiding Seminar Nasional Lahan Suboptimal: Tantangan dan Solusi Pengembangan PAJALE dan Kelapa Sawit Generasi Kedua (Replanting) di Lahan Suboptimal.* Unsri Press (pp. 502–509). http://conference.unsri.ac.id/index.php/lahansuboptimal/article/view/1253

Rayne, N., & Aula, L. (2020). Livestock manure and the impacts on soil health: A review. *Soil Systems, 4*(4), 1–26. https://doi.org/10.3390/soilsystems4040064

Robinson, G. M. (2024). Global sustainable agriculture and land management systems. *Geography and Sustainability* (pp. 637–646) Beijing Normal University Press. https://doi.org/10.1016/j.geosus.2024.09.001

Shahpari, S., & et al. (2021). An integrated economic, environmental and social approach to agricultural land-use planning. *Land, 10*(364). https://doi.org/10.3390/land10040364

Sharma, A., & et al. (2022). Exposure to endosulfan can cause long term effects on general biology, including the reproductive system of mice. *Frontiers in Genetics, 13*(November). https://doi.org/10.3389/fgene.2022.1047746

Vaikosen, E. N., & et al. (2022). Kinetic and isotherm studies on the adsorption-desorption of technical-grade endosulfan in loamy soils under Theobroma cacao L cultivation, Southwestern Nigeria. *Environmental Science: Advances,2*, 257–277. https://doi.org/10.1039/d2va00090c

Verma, S., & et al. (2024). Effect of organic manure on different soil properties: A review. *International Journal of Plant & Soil Science, 36*(5), 182–187. https://doi.org/10.9734/ijpss/2024/v36i54515

Vinolina, N. S., Saragih, D., & Marbun, P. (2024). The effects of applying poultry waste and coconut shell biochar on the growth and production of shallots (Allium ascalonicum L.). In *IOP Conference Series: Earth and Environmental Science.* Institute of Physics. https://doi.org/10.1088/1755-1315/1413/1/012047

Voisin, R., & et al. (2025). Comprehensive assessment instruments for place-based sustainable agriculture: a scoping review. *Sustainability Science* (pp. 1037–1054). Springer. https://doi.org/10.1007/s11625-025-01629-3

Wihardjaka, A. (2018). Application of environmental friendly agriculture models as guaranttee in improving quantity and quality of rice yields. *Pangan, 27*, 155–161.

Wihardjaka, A., & Harsanti, E. S. (2021). Dukungan Pupuk Organik untuk Memperbaiki Kualitas Tanah pada Pengelolaan Padi Sawah Ramah Lingkungan. *PANGAN, 30*(1), 53–64.

Wihardjaka, A., Harsanti, E. S., & Ardiwinata, A. N. (2022). Effect of fertilizer management on potassium dynamics and yield of rainfed lowland rice in Indonesia. *Chilean Journal of Agricultural Research, 82*(1), 33–43. https://doi.org/10.4067/S0718-58392022000100033

Wihardjaka, A. (2016). Peran Jerami Padi Dalam Memperbaiki Hasil Gabah Dan Serapan Kalium Di Lahan Sawah Tadah Hujan Di Kabupaten Pati, Jawa Tengah. *Agric, 27*(1), 15. https://doi.org/10.24246/agric.2015.v27.i1.p15-22

Xu, W., Pignatello, J. J., & Mitch, W. A. (2015). Reduction of nitroaromatics sorbed to black carbon by direct reaction with sorbed sulfides. *Environmental Science and Technology,49*(6), 3419–3426. https://doi.org/10.1021/es5045198

Yang, L., & et al. (2024). Exploring the relationship between biochar pore structure and microbial community composition in promoting tobacco growth. *Plants, 13*(2952). https://doi.org/10.3390/plants13212952

Zahoor, A., & et al. (2025). Agricultural lignocellulose biochar material in wastewater treatment: A critical review and sustainability assessment. *Environmental Functional Materials.* https://doi.org/10.1016/j.efmat.2024.12.005

Vermicompost in Agriculture: Role, Benefits, and Techniques for Sustainable Farming

Nurul Mayzaitul Azwa Jamaludin, Borhan Abdul Haya, and Mohamad Asrol Kalam

Abstract Vermicomposting has evolved as a sustainable method of managing organic waste and enhancing soil health in agricultural settings. This biological process, powered by earthworms and microbes, transforms agricultural, industrial, and household waste into nutrient-dense vermicompost. The final product improves soil structure, increases nutrient availability, boosts microbial activity, and promotes crop growth and resilience. Vermicomposting outperforms conventional composting in terms of decomposition speed, humification efficiency, and adaptability for small-scale and decentralised systems. This study covers the biological and ecological mechanisms of vermicomposting, the importance of feedstock variety, technical improvements, and the agronomic, environmental, and socioeconomic advantages. A focus is given on vermicomposting's incorporation into climate-smart and regenerative agriculture techniques, as well as its contributions to circular economy models. The study also addresses existing research limitations, such as microbial dynamics and economic feasibility, and proposes future avenues for further acceptance and innovation. Vermicomposting is a multidimensional technique for improving soil fertility, reducing environmental pollution, and promoting sustainable food systems worldwide.

Keywords Circular economy · Earthworms · Organic waste · Regenerative farming · Sustainable agriculture · Soil health

1 Introduction

The shift to sustainable agriculture is crucial for tackling the urgent global issues of soil degradation, climate change, and food poverty. Conventional agricultural methods, which depend significantly on synthetic inputs, frequently deteriorate soil health and exacerbate environmental contamination. Regenerative and circular

N. M. A. Jamaludin · B. A. Haya · M. A. Kalam (✉)
Faculty of Sustainable Agriculture, Universiti Malaysia Sabah, Sandakan, Sabah, Malaysia
e-mail: asrolkalam@ums.edu.my

A. Z. Yaser et al. (eds.), *Compost and Liquid Organic Fertilizers*,
https://doi.org/10.1007/978-981-95-8561-8_7

farming methods have gained prominence, with vermicomposting seen as a vital technique for organic waste recycling and soil rehabilitation (Biswas et al., 2024). Vermicomposting is a biological process in which earthworms and microbes transform organic waste into vermicompost, a nutrient-dense, physiologically active substance. This procedure diverts organic waste from landfills and generates a superior soil supplement that improves physical structure, nutritional composition, and microbial diversity in soils (Kumari et al., 2024). The resultant vermicompost enhances crop yield, mitigates pests and diseases, and diminishes reliance on chemical fertilisers, thereby fostering agronomic productivity and environmental sustainability (Yatoo et al., 2021).

In contrast to traditional composting, vermicomposting is more efficient, devoid of odour, and appropriate for decentralised applications, hence benefiting smallholder farmers and rural communities. Earthworms, especially epigeic species such as *Eisenia fetida*, are essential for waste fragmentation, substrate aeration, and the enhancement of microbial activity, leading to accelerated decomposition and humification (Suthar, 2008). This current review consolidates existing information on vermicomposting, emphasising its processes, advantages, technical advancements, and contribution to climate-smart agriculture. It also recognises current deficiencies and suggests methods to improve acceptance and scalability. This review highlights the potential of vermicomposting as a transformative tool for sustainable agricultural systems by analysing it through the perspectives of agroecology, soil science, and circular economy. Enhancing agricultural output and promoting environmental sustainability.

This review presents a more integrated and system-oriented viewpoint on organic waste bioconversion and vermicomposting, building on numerous recent studies that have examined various facets of these topics. Concentrated on vegetable waste bioprocessing in rural regions, with less attention to the vermicomposting process and its agricultural ramifications. Ratnasari et al. (2023) conducted a biochemical evaluation of the vermicomposting of organic and inorganic waste, focusing on additive effects and microbial processes, although omitted field-level applications and socio-economic integration. Raza et al. (2022) investigated the use of agricultural wastes, manure, and biochar in vermicomposting, assessing feedstock efficacy and compost performance. This study employs a comprehensive and multidisciplinary perspective, encompassing biological principles, technical advancements, agronomic benefits, climatic resilience, and rural development dimensions of vermicomposting. The innovation resides in connecting vermicomposting to regenerative and climate-smart agriculture, emphasising its multifunctionality in bolstering circular economy frameworks, and delineating actionable solutions for broader implementation, particularly among smallholder farmers.

2 The Living Engine: Earthworms and Microbiomes in Action

The effectiveness of vermicomposting lies in the ability of earthworms to fragment organic materials, thereby enhancing the surface area accessible for microbial colonization and enzymatic activity. This leads to a more efficient decomposition of organic matter compared to conventional composting methods, as demonstrated by studies showing significant reductions in total organic carbon, lignin, and cellulose during vermicomposting (Cai et al., 2018; Kusum et al., 2023). The overall flow and application of the vermicomposting process are illustrated in Fig. 1, which outlines the key biological stages driven by earthworms and microbes, as well as the downstream use of vermicompost in sustainable agriculture systems.

Earthworms, particularly species like *Eisenia fetida*, dramatically alter the composition of microbial communities present during vermicomposting, creating favorable conditions for beneficial bacteria and fungi. Such modifications minimize antibiotic resistance genes in treated bio-waste (Domínguez et al., 2019, 2021). Additionally, the enzymatic activities from earthworm excretion continue to facilitate organic matter breakdown post-excretion, enriching the microbial population in the resulting vermicompost (Abad & Shafiqi, 2024). The end product not only serves as a high-quality organic fertilizer but also contributes positively to soil health by improving its physical and chemical characteristics, including nutrient content and microbial diversity (Domínguez, 2018; Motcha, 2017; Vuković et al., 2021).

In essence, the vermicomposting process exemplifies an efficient recycling method, transforming organic waste into valuable resources while simultaneously promoting soil sustainability (Hrebeckova et al., 2023; Rosa et al., 2023; Saha et al., 2022).

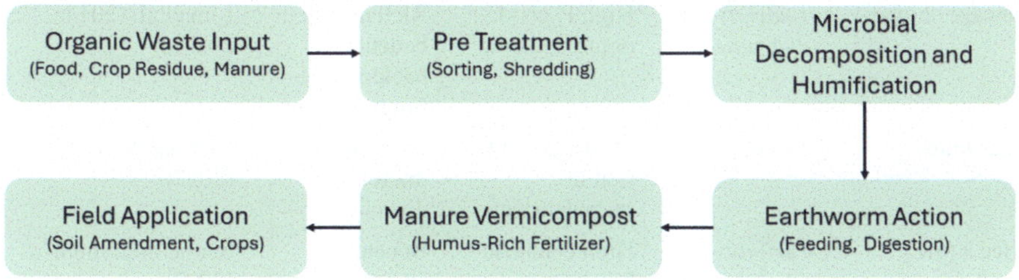

Fig. 1 Schematic flowchart vermicomposting process and application pathways

3 Feedstock Diversity and Optimization for Targeted Benefits

Diverse organic waste, including animal manure, sewage sludge, leaf litter, rice straw, and municipal solid waste, are appropriate for vermicomposting. When processed by earthworms, these chemicals improve nutrient dynamics and promote plant growth (Bai et al., 2025; Ritu & Sudesh, 2022). Earthworms play a vital role in the decomposition of organic waste, enhancing microbial activity and lowering the carbon-to-nitrogen (C:N) ratio, hence increasing compost fertility (Ali et al., 2015; Suthar & Singh, 2007). As summarised in Table 1, different types of organic waste and earthworm species significantly influence the nutrient composition and microbial activity of vermicompost, emphasizing the importance of feedstock selection in optimizing compost quality and agronomic outcomes.

The substrate utilised significantly affects the physicochemical and biological characteristics of vermicompost. Some feedstocks can more efficiently augment nutritional content and microbial enzymatic activity than others (Ose et al., 2021). Research indicates that earthworm activity enhances enzymatic activities and increases humic substance levels, correlating with enhanced soil health and plant

Table 1 Types of organic waste and their impact on vermicompost quality

Organic waste type	Earthworm species	Key nutrient improvements (NPK)	Microbial activity	References
Cattle manure	*Eisenia fetida*	High N, moderate P and K	Enhanced microbial biomass and enzymes	Bai et al. (2025)
Sewage sludge	*Eudrilus eugeniae*	High P, good organic matter	Rich in beneficial bacteria and fungi	Lim et al. (2014a, 2014b)
Leaf litter	*Eisenia fetida*	Moderate NPK, rich in micronutrients	Increased microbial diversity	Ritu and Sudesh (2022)
Rice straw	*Perionyx excavatus*	High cellulose breakdown, improved C:N ratio	Boosted enzymatic activity	Suthar and Singh (2007)
Municipal solid waste	*Lampito mauritii*	Variable NPK, high humus content	Moderate microbial activity	Lalander et al. (2015)
Palm oil mill waste	*Perionyx ceylanensis*	Improved K and micronutrients	High actinomycete and fungal content	Rupani et al. (2017)

responses (Chatterjee et al., 2014; Pramanik & Chung, 2010). In addition to enhancing soil quality, vermicomposting serves as an eco-friendly waste management approach. It converts organic waste into useful fertiliser while mitigating environmental contamination (Lalander et al., 2015). Its capacity to successfully treat a diverse array of waste kinds underscores its promise as a crucial element in circular and sustainable agricultural systems (Ali et al., 2015).

4 Technological Innovations in Vermicomposting Systems

Vermicomposting is crucial for sustainable solid waste management, converting various organic wastes from agricultural, residential, and industrial sources into nutrient-rich compost (Rupani et al., 2017). The utilisation of efficient earthworm species, such as Lampito mauritii and Perionyx ceylanensis, enhances substrate adaption and compost quality (Karmegam & Daniel, 2008), making the system adaptable and scalable for diverse waste types. Technological advancements like high-rate vermicomposting have improved efficiency, enabling rapid processing without the need for pre-composting or significant alterations (Hussain & Abbasi, 2016; Lim et al., 2014a). These methods are highly effective for addressing difficult waste materials, improving both environmental and economic benefits.

Beyond waste transformation, vermicomposting has shown potential in bioremediation. Earthworms aid in degrading harmful compounds such as glyphosate in contaminated soils, contributing to land restoration (Masín et al., 2018). Applications also include agro-industrial waste treatment, such as palm oil mill residues and wastewater, producing valuable organic fertilizers and mitigating pollution (Rupani et al., 2017). Emerging innovations involve the integration of nanotechnology for instance, Fe_3O_4 nanoparticle-enriched vermicompost which enhances soil remediation and nutrient availability (Ghafghazi, 2025). These developments suggest new directions for waste management systems. Finally, global interest in vermicomposting continues to rise, with patent trends and collaborative research indicating its growing relevance in sustainable agriculture and circular economy frameworks (Mhamdi, 2024). Mainstream adoption of this technology could significantly advance organic waste valorization and environmental resilience.

5 Agronomic Value of Vermicompost in Soil and Crop Systems

The agronomic value of vermicompost in soil and crop systems is increasingly recognized due to its positive effects on soil health, nutrient dynamics, and crop productivity. Vermicompost, produced by the interaction of earthworms with organic waste, enhances soil microbial diversity, which is crucial for nutrient cycling and disease suppression. Research demonstrates that vermicomposting substantially enhances populations of beneficial microbes, hence fostering plant growth and sustainability in agricultural activities (Domínguez et al., 2019; Pathma & Sakthivel, 2014).

The use of vermicompost has been associated with enhanced vegetative growth and increased fruit yield in a variety of crops. Research indicates that utilising vermicompost in conjunction with inorganic fertilisers improves the production and quality of tomatoes and other crops, demonstrating their viability as a sustainable alternative to traditional fertilisers (Kashem et al., 2015; Qasim et al., 2023). Moreover, the combination of vermicompost with chemical fertilisers has been shown to improve nutrient absorption, resulting in higher yields than when using chemical fertilisers alone (Aslam et al., 2019; Qasim et al., 2023).

Vermicompost improves the structure of the soil, its ability to hold water, and its ability to exchange cations, which makes the soil more fertile. It can help crops that are missing nutrients, especially because it has more of the critical plant nutrients that plants need, such phosphates, nitrates, and potassium (Nunes et al., 2018). Also, using it has been shown to make the soil less compact and more active, which is good for the health of the soil (Maulina et al., 2023; Moustafa et al., 2022).

Furthermore, researchers have demonstrated that vermicompost may inhibit specific plant diseases and promote the growth of beneficial crops, highlighting its significance in integrated pest control (Sun et al., 2024). Vermicompost is an important part of organic farming and sustainable agriculture since it may enhance the physical and chemical qualities of soil (Blouin et al., 2019; Lim et al., 2014b).

6 Environmental Sustainability and Climate Resilience

Vermicomposting makes the soil better by adding organic matter and nutrients. Water-soluble nutrients in vermicompost help plants grow and produce more (Nejatzadeh, 2024). Vermicomposting gives plants helpful bacteria and fungus that help plants stay healthy and help nutrients move around, which is important for keeping the ecosystem in balance (Pathma & Sakthivel, 2014; Pereira et al., 2022). Vermicompost also makes the soil healthier by storing more carbon and releasing less greenhouse gases (Amante, 2024; Atakli et al., 2022). Vermicomposting turns organic waste into soil amendments. Food waste and agricultural byproducts are transformed into vermicompost, hence diminishing landfill waste and its ecological consequences (Almaramah et al., 2024; Zaman & Yaacob, 2022). A closed-loop system that turns garbage into organic fertilisers helps the environment and the economy (Rose et al., 2022; Zaman & Yaacob, 2022). Vermicomposting is good for the environment and makes farming more resilient. Vermicompost treatment enhances soil water retention, especially in areas susceptible to drought (Karademir & Kibar, 2022). Enhanced water retention capacity boosts crop yield and fortifies agricultural systems against climate change (Karademir & Kibar, 2022; Pereira et al., 2022).

Vermicompost makes farming more nutritious, which helps keep food safe and healthy for everyone. Vermicompost's nutrients help crops grow better and make them more profitable (Iswahyudi et al., 2024; Mohite et al., 2024). Crops produced with vermicompost have more essential nutrients, which helps make food production more sustainable (Oyege & Sridhar, 2023).

7 Social and Economic Opportunities in Vermicomposting

Vermicomposting offers substantial socio-economic prospects by converting organic waste into valuable commodities while promoting sustainable agriculture methods. Vermicomposting is a good way to deal with garbage, and it may provide jobs, especially in rural areas where there is a lot of organic waste. This technique deals with problems with getting rid of trash and encourages local business by setting up community-based microenterprises that make and sell vermicompost (Fernando & Arunakumara, 2021; Moledor et al., 2016).

Turning organic waste into vermicompost may improve soil health and increase crop yields, which can help farmers make more money. Research indicates that crops cultivated using vermicompost have enhanced growth metrics relative to those treated with conventional fertilisers. This rise in production can improve food security and lessen reliance on synthetic inputs (Fernando & Arunakumara, 2021; Readyhough et al., 2021; Singh et al., 2019). Vermicomposting may also help with environmental problems that come from throwing away organic waste. It can make the soil better and the ecosystem healthier, which is good for the community (Atapattu et al., 2025; Gebrehana et al., 2025).

Additionally, educational programs focused on vermicomposting may give people in the community more power by teaching them about eco-friendly practices and getting them involved in activities that are good for the environment. Such training can enhance community cohesiveness and collaborative problem-solving for agricultural waste management (Gough & Accordino, 2013; Moledor et al., 2016).

8 Incorporating Vermicompost into Regenerative and Climate-Smart Farming

Much research has shown that vermicompost helps plants grow and produce more. For example, Masri et al. (2022) showed that adding Gliricidia sepium residue vermicompost to organic choy sum mustard significantly increased yields and dry matter biomass. Adhikari et al. (2023) also observed that vermicompost considerably improved growth metrics and yields of cabbage, underscoring its efficacy as a growth promoter attributable to its advantageous physical characteristics, including aeration, drainage, and water retention. Hafez et al. (2020) talk about how the nutrients in vermicompost help plants grow strong and increase the quality of crops by making nutrients more available in the soil.

Additionally, studies on continuous cropping systems have stressed the economic benefits of using vermicompost. Zhang et al. (2023) showed that using vermicompost increased the yield of continuous cropping pepper in Southwest China and saved money by lowering fertiliser expenditures and making the soil more fertile. This study is consistent with the conclusions of Mohite et al. (2024), who saw vermicomposting as a comprehensive approach to sustainable crop production that encompasses environmental restoration and the generation of nutrient-rich fertilisers.

The ecological benefits of vermicompost are also very important. The approach helps to improve the structure of the soil, the diversity of microbes, and the cycling of nutrients, all of which are necessary for long-term sustainability in agriculture (Zaman & Yaacob, 2022). For instance, vermicompost helps people use less artificial fertilisers, which is good for the environment since it helps good microbes grow (Yatoo et al., 2021). Abad and Shafiqi (2024) also say that using vermicompost can reduce the demand for watering and pest management, which is another way to assist climate-smart farming. The broader role of vermicomposting in circular agriculture and climate-smart systems is illustrated in Fig. 2.

Vermicompost is also very important for regenerative agriculture since it helps recycle organic waste, reduces methane emissions from landfills, and improves soil health and fertility (Ducasse et al., 2022; Muniya Naik et al., 2024). Using vermicompost and other organic materials together has also been found to make plants more resistant to pests and diseases, which lowers the negative effects on the environment that come with using agrochemicals (Zhang et al., 2023). This is an important part of creating a sustainable agricultural paradigm that can handle the effects of climate change.

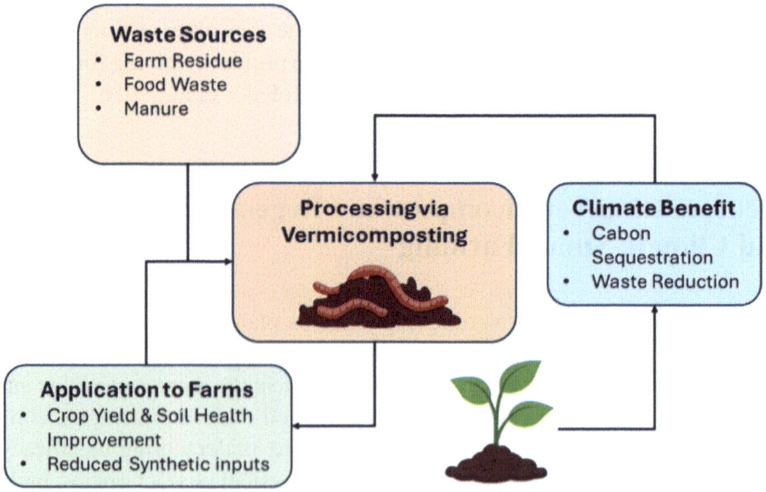

Fig. 2 Vermicomposting in circular agriculture and climate-smart system

Using vermicompost as a main part of soil management not only increases agricultural productivity, but it also fits with larger goals of sustainability and caring for the environment. This makes it an important tool for moving towards regenerative and climate-smart farming methods.

9 Current Gaps and Research Priorities

The nutritious content of vermicompost is one of its main functions. Vermicompost is a good organic fertiliser that can replace synthetic fertilisers, which helps cut down on pollution (Mishra et al., 2022; Ramnarain et al., 2018). Studies have shown that vermicompost has a lot more nitrogen than regular organic fertilisers, which leads to better crop development and yield (Abad & Shafiqi, 2024; Lim et al., 2014a, 2014b; Ramnarain et al., 2018). This phenomenon has been shown in several crops, such as watermelon and peas, where vermicomposting markedly enhanced growth metrics and fruit quality (Bayram et al., 2021; Hasan et al., 2024). Studies show that vermicompost also increases the number of microbes in the soil, which makes the soil more fertile and the plants healthier (Oyege & Sridhar, 2023; Vuković et al., 2021).

The use of vermicompost has had good impacts on the soil, making its physical, chemical, and biological qualities better (Oyege & Sridhar, 2023; Prakash et al., 2024). The amendment not only helps the soil hold water and get air, but it also helps the structure of the soil, which reduces the danger of erosion in places that are prone to drought (Abad & Shafiqi, 2024; Hasan et al., 2024). This skill is important since climate change is making things harder for farmers, which hurts their output. For instance, using vermicompost in farming in Bangladesh has been related to less reliance on chemical fertilisers, which is in line with the aims of sustainable agriculture (Rahman & Barmon, 2019).

Even if it has certain benefits, there are still significant limitations in our knowledge of the precise microbial dynamics that happen throughout the vermicomposting process and how they affect the stability and availability of nutrients in the compost that is made (Kolbe et al., 2019; Oyege & Sridhar, 2023). The impact of several feedstocks on the microbiological efficacy and characteristics of vermicompost has not been well investigated (Mishra et al., 2022; Zaman & Yaacob, 2022). Furthermore, the economic feasibility of vermicomposting across various agricultural systems, especially among small-scale farmers, requires additional investigation to ascertain how this technique may be broadly implemented while optimising farmers' revenue (Prakash et al., 2024; Singh et al., 2020).

Recent studies show that farmers need thorough training and education programs on vermicomposting to learn how to use these eco-friendly methods successfully (Dada, 2024; Prakash et al., 2024). Moreover, improving consumer knowledge and looking into how the vermicompost market works might greatly increase its use (Ose et al., 2021; Tiwari et al., 2023). In general, vermicompost has many benefits for farming, but it is important to fill up these research gaps to get the most out of it for sustainable farming.

10 Conclusion

Vermicomposting is an effective and sustainable technique for transforming organic waste into valuable compost using earthworms and microbes. It improves soil vitality, fosters plant growth, and reduces the need for synthetic fertilisers. This method assists farmers in waste management, environmental conservation, and the enhancement of more resilient agricultural systems. It successfully connects with climate-smart and regenerative agriculture by recycling nutrients and minimising pollution. Vermicomposting can create employment and entrepreneurship opportunities, especially in rural areas. Additional research, education, and support are required to enable successful adoption by farmers.

References

Abad, Q., & Shafiqi, S. (2024). Vermicompost: Significance and benefits for agriculture. *Journal for Research in Applied Sciences and Biotechnology,3*(2), 202–207. https://doi.org/10.55544/jrasb.3.2.36

Adhikari, R., Katel, S., Chhetri, P. K., Simkhada, P., Chaudhari, P., & Yadav, S. P. S. (2023). Effect of different sources of organic fertilizers on crop growth and yield of cabbage. *Journal of Agriculture and Applied Biology,4*(1), 83–94. https://doi.org/10.11594/jaab.04.01.09

Ali, U., Sajid, N., Khalid, A., Riaz, L., Rabbani, M. M., Syed, J. H., & Malik, R. N. (2015). A review on vermicomposting of organic wastes. *Environmental Progress & Sustainable Energy, 34*(4), 1050–1062. https://doi.org/10.1002/ep.12100

Almaramah, S. B., Abu-Elsaoud, A. M., Alteneiji, W. A., Albedwawi, S., El-Tarabily, K. A., & Raish, S. M. A. (2024). The impact of food waste compost, vermicompost, and chemical fertilizers on the growth measurement of red radish (Raphanus sativus): A sustainability perspective in the United Arab Emirates. *Foods, 13*(11), 1608. https://doi.org/10.3390/foods13111608

Amante, G. (2024). Advancing agricultural sustainability: Vermicomposting as a biochemical pathway for improved soil health and climate resilience. *Middle East Research Journal of Agriculture and Food Science,4*(02), 86–94. https://doi.org/10.36348/merjafs.2024.v04i02.006

Aslam, Z., Bashir, S., Hassan, W., Bellitürk, K., Ahmad, N., Niazi, N. K., Khan, A. U., Khan, M. I., Chen, Z., & Maitah, M. (2019). Unveiling the efficiency of vermicompost derived from different biowastes on wheat (Triticum aestivum L.) plant growth and soil health. *Agronomy,9*(12), 791. https://doi.org/10.3390/agronomy9120791

Atakli, S. B., ŞAhİN, S., Ceritoglu, M., & Çağatay, H. F. (2022). Vermicompost enhances the effectiveness of arbuscular mycorrhizal fungi, cowpea development and nutrient uptake. *Legume Research - An International Journal*. https://doi.org/10.18805/lrf-698

Atapattu, A. J., Nuwarapaksha, T. D., Udumann, S. S., & Dissanayaka, N. S. (2025). Integrating organic fertilizers in coconut farming: Best practices and application techniques. *Crops, 5*(2), 17. https://doi.org/10.3390/crops5020017

Bai, X., Wei, L., Xu, J., Li, Q., Xue, Z. X., & Wang, X. (2025). Effects of cattle manure and sludge vermicompost on nutrient dynamics and yield in strawberry cultivation with distinct continuous cropping histories in a greenhouse. *Frontiers in Plant Science, 15*. https://doi.org/10.3389/fpls.2024.1514675

Bayram, C. A., Büyük, G., & Kaya, A. (2021). Çiftlik Gübresi, Vermikompost Ve Bitki Büyüme Düzenleyicilerinin Karpuz Meyvesinde Verim Ve Meyve Kaliteye Etkileri. *Kahramanmaraş Sütçü İmam Üniversitesi Tarım Ve Doğa Dergisi, 24*(1), 64–69. https://doi.org/10.18016/ksutarimdoga.vi.701708

Biswas, S., Singh, S., & Bera, A. (2024). Regenerative nutrient management practices for enhancing plant nutrition and soil health. In *Regenerative agriculture for sustainable food systems* (pp. 303–339). Springer.

Blouin, M., Barrere, J., Meyer, N., Lartigue, S., Barot, S., & Mathieu, J. (2019). Vermicompost significantly affects plant growth. A meta-analysis. *Agronomy for Sustainable Development, 39*(4). https://doi.org/10.1007/s13593-019-0579-x

Cai, L., Gong, X., Sun, X., Li, S., & Yu, X. (2018). Comparison of chemical and microbiological changes during the aerobic composting and vermicomposting of green waste. *PLoS ONE, 13*(11), e0207494. https://doi.org/10.1371/journal.pone.0207494

Chatterjee, R., Bandyopadhyay, S., & Jana, J. C. (2014). Evaluation of vegetable wastes recycled for vermicomposting and its response on yield and quality of carrot (Daucus carota L.). *International Journal of Recycling of Organic Waste in Agriculture,3*(2). https://doi.org/10.1007/s40093-014-0060-4

Dada, E. O. (2024). Vermiculture and vermicomposting: Twin earthworm technologies, suitable for households and smallholder farmers to convert wastes to wealth—A narrative review. *Fudma Journal of Agriculture and Agricultural Technology, 10*(1), 134–143. https://doi.org/10.33003/jaat.2024.1001.17

Domínguez, J. (2018). *Earthworms and vermicomposting*. https://doi.org/10.5772/intechopen.76088

Domínguez, J., Aira, M., Crandall, K. A., & Pérez-Losada, M. (2021). Earthworms drastically change fungal and bacterial communities during vermicomposting of sewage sludge. *Scientific Reports, 11*(1). https://doi.org/10.1038/s41598-021-95099-z

Domínguez, J., Aira, M., Kolbe, A. R., Gómez-Brandón, M., & Pérez-Losada, M. (2019). Changes in the composition and function of bacterial communities during vermicomposting may explain beneficial properties of vermicompost. *Scientific Reports,9*(1). https://doi.org/10.1038/s41598-019-46018-w

Ducasse, V., Capowiez, Y., & Peigné, J. (2022). Vermicomposting of municipal solid waste as a possible lever for the development of sustainable agriculture. A review. *Agronomy for Sustainable Development, 42*(5). https://doi.org/10.1007/s13593-022-00819-y

Fernando, K. M. C., & Arunakumara, K. K. I. U. (2021). Sustainable organic waste management and nutrients replenishment in the soil by vermicompost: A review. *Agrieast Journal of Agricultural Sciences, 15*(2), 32. https://doi.org/10.4038/agrieast.v15i2.105

Gebrehana, Z. G., Gebremikael, M. T., Beyene, S., Wesemael, W., & Neve, S. D. (2025). Suppression potential of selected vermicomposts against root-knot nematode (Meloidogyne incognita) under in vitro, pot, and field conditions. *Frontiers in Plant Science, 16*. https://doi.org/10.3389/fpls.2025.1532800

Ghafghazi, L. (2025). Soil responses to nano-enhanced bioremediation techniques: Comparative analysis of bio-based vermicompost, Fe_3O_4 nanoparticles, and their novel synergistic effects against phthalate contamination. https://doi.org/10.21203/rs.3.rs-6787119/v1

Gough, M. Z., & Accordino, J. (2013). Public gardens as sustainable community development partners. *Urban Affairs Review, 49*(6), 851–887. https://doi.org/10.1177/1078087413477634

Hafez, E. M., Omara, A. E., Alhumaydhi, F. A., & El-Esawi, M. A. (2020). Minimizing hazard impacts of soil salinity and water stress on wheat plants by soil application of vermicompost and biochar. *Physiologia Plantarum, 172*(2), 587–602. https://doi.org/10.1111/ppl.13261

Hasan, A. K., Islam, S. S., Jahan, M., Kheya, S. A., Uddin, M. R., Islam, M. S., & Khomphet, T. (2024). Synergistic effects of vermicompost and biochar amendments on soil fertility and wheat productivity in Bangladesh floodplain soils. *Scientifica, 2024*(1). https://doi.org/10.1155/sci5/6624984

Hrebeckova, T., Kralikova, N., Hanc, A., & Wiesnerova, L. (2023). Problems associated with vermicomposting of dog excrement in practice using Eisenia andrei. *Waste Management and Research, 41*(2), 328–336. https://doi.org/10.1177/0734242X221123143

Hussain, N., & Abbasi, S. A. (2016). Vermicomposting transforms Allelopathic parthenium into a benign organic fertilizer. *Journal of Environmental Management, 180*, 180–189. https://doi.org/10.1016/j.jenvman.2016.05.013

Iswahyudi, I., Sutanto, A., Widodo, W., Warkoyo, W., Yuniwati, E. D., Garfansa, M. P., Ekalaturrahmah, Y. A. C., Sugiono, S., Ramadani, S. D., & Setyobudi, R. H. (2024). Impact of microplastics on the carbon and nitrogen of vermicompost. *Environmental Quality Management, 34*(2). https://doi.org/10.1002/tqem.22325

Karademir, S., & Kibar, B. (2022). Influence of different vermicompost doses on growth, quality and element contents in curly lettuce (Lactuca sativa L. Var. Crispa). *Kahramanmaraş Sütçü İmam Üniversitesi Tarım Ve Doğa Dergisi, 25*(Ek Sayı 2), 430–440. https://doi.org/10.18016/ksutarimdoga.vi.1047470

Karmegam, N., & Daniel, T. (2008). Investigating efficiency of Lampito mauritii (Kinberg) and Perionyx ceylanensis Michaelsen for vermicomposting of different types of organic substrates. *The Environmentalist, 29*(3), 287–300. https://doi.org/10.1007/s10669-008-9195-z

Kashem, M. A., Sarker, A., Hossain, I., & Islam, M. S. (2015). Comparison of the effect of vermicompost and inorganic fertilizers on vegetative growth and fruit production of tomato (*Solanum lycopersicum* L.). *Open Journal of Soil Science, 05*(02), 53–58. https://doi.org/10.4236/ojss.2015.52006

Kolbe, A. R., Aira, M., Gómez-Brandón, M., Pérez-Losada, M., & Domínguez, J. (2019). Bacterial succession and functional diversity during vermicomposting of the white grape marc Vitis vinifera v. Albariño. *Scientific Reports,9*(1). https://doi.org/10.1038/s41598-019-43907-y

Kumari, B., Prakash, O., Pal, P., Singh, P. K., Singh, M. P., Kumar, P., Sarangi, P. K., Prakash, A., Pandey, B., & Manohar, S. (2024). Advancement in sustainable management and valorization of solid waste through composting and vermitechnology. In *Recent trends in management and utilization of industrial sludge* (pp. 359–397). Springer.

Kusum, A., Kumari, N., Choudhari, B., Sahu, D., & Naya, N. (2023). Optimizing household waste and cow dung proportions for quality vermicompost production. *International Journal of Environment and Climate Change, 13*(9), 341–346. https://doi.org/10.9734/ijecc/2023/v13i92237

Lalander, C., Komakech, A. J., & Vinnerås, B. (2015). Vermicomposting as manure management strategy for urban small-holder animal farms—Kampala case study. *Waste Management, 39*, 96–103. https://doi.org/10.1016/j.wasman.2015.02.009

Lim, P. N., Wu, T. Y., Clarke, C., & Daud, N. N. N. (2014a). A potential bioconversion of empty fruit bunches into organic fertilizer using Eudrilus eugeniae. *International Journal of Environmental Science and Technology, 12*(8), 2533–2544. https://doi.org/10.1007/s13762-014-0648-2

Lim, S. L., Wu, T. Y., Lim, P. N., & Shak, K. P. Y. (2014b). The use of vermicompost in organic farming: Overview, effects on soil and economics. *Journal of the Science of Food and Agriculture, 95*(6), 1143–1156. https://doi.org/10.1002/jsfa.6849

Masín, C. E., Lescano, M., Rodríguez, A. R., Godoy, J. L., & Zalazar, C. S. (2018). Earthworms to assess the innocuousness of spent biomixtures employed for glyphosate degradation. *Journal of Environmental Science and Health Part B, 53*(8), 519–525. https://doi.org/10.1080/03601234.2018.1462922

Masri, I. N., Sadi, T., Karam, D. S., Abdul Sukor, A. S., Basir, M. H., Rajoo, K. S., & Abdu, A. (2022). Effects of Gliricidia sepium residue vermicompost on the yield and dry matter biomass of organic choy sum mustard. *Malaysian Applied Biology, 51*(2), 87–93. https://doi.org/10.55230/mabjournal.v51i2.2163

Maulina, L., Karnan, K., & Raksun, A. (2023). The effect of vermicompost on growth of shallots (Allium ascalonicum L.). *Jurnal Pijar Mipa, 18*(2), 265–273. https://doi.org/10.29303/jpm.v18i2.3448

Mhamdi, R. (2024). *The rise of the worm: Bibliometric insights into the growing significance of vermicomposting for sustainable agriculture and environmental management*. https://doi.org/10.21203/rs.3.rs-5521622/v1

Mishra, M., Pande, R. K., & Ray, S. (2022). A comprehensive review on earthworms' vermicompost: A strategy for sustainable waste management. *ECS Transactions, 107*(1), 20101–20108. https://doi.org/10.1149/10701.20101ecst

Mohite, D. D., Chavan, S., Jadhav, V. S., Kanase, T., Kadam, M. A., & Singh, A. S. (2024). Vermicomposting: A holistic approach for sustainable crop production, nutrient-rich bio fertilizer, and environmental restoration. *Discover Sustainability,5*(1). https://doi.org/10.1007/s43621-024-00245-y

Moledor, S., Chalak, A., Fabian, M., & Talhouk, S. N. (2016). Socioeconomic dynamics of vermicomposting systems in Lebanon. *Journal of Agriculture Food Systems and Community Development*, 1–24. https://doi.org/10.5304/jafscd.2016.064.007

Motcha, R. V. (2017). An overview: Organic waste management by earthworm. *Journal of Civil Engineering and Environmental Sciences*, 013–017. https://doi.org/10.17352/2455-488x.000015

Moustafa, Y., Mustafa, N. S., El-Dahshouri, M. F., Sameh Mohamed Mohamed El-Sawy, G., Zuhair, R., Zhang, L., & Hassoub, M. A. (2022). Influence of wastes of taro leaf, sugar beet and saw dust on physiochemical parameters of produced vermicompost. *Asian Journal of Soil Science and Plant Nutrition*, 29–40. https://doi.org/10.9734/ajsspn/2022/v8i3160

Muniya Naik, M., Chandra Mohan, K., Janardhana, K., Aruna, C., Rama Mohan, M., & Uday Kiran, V. (2024). Vermicompost (traditional methods) is an important component of agricultural productivity and an efficient tool for environmental and economic sustainability. *International Journal of Biological and Environmental Investigations, 10*(1), 55–61. https://doi.org/10.33745/ijzi.2024.v10i01.007

Murshid, N., Lamaming, J., Saalah, S., Rajin, M., & Yaser, A. Z. (2024). A review of vegetable waste bio-processing techniques in rural areas. *International Journal of Recycling of Organic Waste in Agriculture, 13*(2).

Nejatzadeh, F. (2024). Effect of vermicompost and nitrogen fertilizer on the growth and production of aloe vera. *International Journal of Plant & Soil Science, 36*(4), 337–345. https://doi.org/10.9734/ijpss/2024/v36i44486

Nunes, R. R., Pigatin, L. B. F., Oliveira, T. S., Bontempi, R. M., & Rezende, M. O. O. (2018). Vermicomposted tannery wastes in the organic cultivation of sweet pepper: Growth, nutritive value and production. *International Journal of Recycling of Organic Waste in Agriculture, 7*(4), 313–324. https://doi.org/10.1007/s40093-018-0217-7

Ose, A., Andersone-Ozola, U., & Ievinsh, G. (2021). Substrate-dependent effect of vermicompost on yield and physiological indices of container-grown Dracocephalum moldavica plants. *Agriculture, 11*(12), 1231. https://doi.org/10.3390/agriculture11121231

Oyege, I., & Sridhar, B. B. M. (2023). Effects of vermicompost on soil and plant health and promoting sustainable agriculture. *Soil Systems, 7*(4), 101. https://doi.org/10.3390/soilsystems7040101

Pathma, J., & Sakthivel, N. (2014). *Microbial diversity of vermicompost bacteria that exhibit useful agricultural traits and waste management potential* (pp. 161–208). https://doi.org/10.1201/b17368-12

Pereira, M. M. A., Moraes, L. C., Mogollón, M. C. T., Borja, C. J. F., Duarte, M., Buttrós, V. H., Luz, J. M. Q., Pasqual, M., & Soares, J. D. R. (2022). Cultivating biodiversity to harvest sustainability: Vermicomposting and inoculation of microorganisms for soil preservation and resilience. *Agronomy, 13*(1), 103. https://doi.org/10.3390/agronomy13010103

Prakash, S., Sikdar, S., & Singh, A. K. (2024). Socio-economic, socio-personal and demographic profiling of KVK-based vermicompost training participants: Evidence from Bihar, India. *Asian*

Journal of Agricultural Extension Economics & Sociology, 42(6), 293–306. https://doi.org/10.9734/ajaees/2024/v42i62492

Pramanik, P., & Chung, Y. R. (2010). Efficacy of vermicomposting for recycling organic portion of hospital wastes using Eisenia fetida: Standardization of cow manure proportion to increase enzymatic activities and fungal biomass. *The Environmentalist, 30*(3), 267–272. https://doi.org/10.1007/s10669-010-9273-x

Qasim, M., Ju, J., Zhao, H., Bhatti, S. M., Saleem, G., Memon, S. P., Ali, S., Younas, M. U., Rajput, N., & Jamali, Z. H. (2023). Morphological and physiological response of tomato to sole and combined application of vermicompost and chemical fertilizers. *Agronomy, 13*(6), 1508. https://doi.org/10.3390/agronomy13061508

Rahman, S., & Barmon, B. K. (2019). Greening modern rice farming using vermicompost and its impact on productivity and efficiency: An empirical analysis from Bangladesh. *Agriculture, 9*(11), 239. https://doi.org/10.3390/agriculture9110239

Ramnarain, Y. I., Ansari, A. A., & Ori, L. (2018). Vermicomposting of different organic materials using the epigeic earthworm Eisenia fetida. *International Journal of Recycling of Organic Waste in Agriculture, 8*(1), 23–36. https://doi.org/10.1007/s40093-018-0225-7

Ratnasari, A., Syafiuddin, A., Mehmood, M. A., & Boopathy, R. (2023). A review of the vermicomposting process of organic and inorganic waste in soils: Additives effects, bioconversion process, and recommendations. *Bioresource Technology Reports, 21*, 101332.

Raza, S. T., Wu, J., Rene, E. R., Ali, Z., & Chen, Z. (2022). Reuse of agricultural wastes, manure, and biochar as an organic amendment: A review on its implications for vermicomposting technology. *Journal of Cleaner Production, 360*, 132200.

Readyhough, T., Neher, D. A., & Andrews, T. (2021). Organic amendments alter soil hydrology and belowground microbiome of tomato (Solanum lycopersicum). *Microorganisms, 9*(8), 1561. https://doi.org/10.3390/microorganisms9081561

Ritu, & Sudesh, R. (2022). Vemicomposting of different organic matters and its effect on growth of mung bean, Vigna radiata. *International Journal of Zoological Investigations, 08*(02), 469–474. https://doi.org/10.33745/ijzi.2022.v08i02.058

Rosa, L. O. d., Souza, K. F. d., Costa, M. M., Martins, G. A., Coelho, C. W. M., Burin, P., Fick, R. B., Mendes, A. C. M., Corrêa, L. B., & Corrêa, É. K. (2023). *Vermicompost process as treatment of organic waste generated in vegetables*. https://doi.org/10.56238/devopinterscie-052

Rose, E. N., Pasion, B. J., Camille, E., Lourdes, E., & Marites, C. (2022). Production of vermicast using various organic waste materials by African Night Crawler (Eudrilus eugeniae) cultured on portable bin. *The Quest Journal of Multidisciplinary Research and Development,1*(2). https://doi.org/10.60008/thequest.v1i2.17

Rupani, P. F., Embrandiri, A., Ibrahim, M. H., Shahadat, M., Hansen, S. B., & Mansor, N. N. A. (2017). Bioremediation of palm industry wastes using vermicomposting technology: Its environmental application as green fertilizer. *3 Biotech,7*(3). https://doi.org/10.1007/s13205-017-0770-1

Saha, P., Barman, A., & Bera, A. (2022). *Vermicomposting: A step towards sustainability*. https://doi.org/10.5772/intechopen.102641

Singh, A., Tiwari, R., Chandrahas, & Dutt, T. (2020). Augmentation of farmers' income in India through sustainable waste management techniques. *Waste Management & Research the Journal for a Sustainable Circular Economy, 39*(6), 849–859. https://doi.org/10.1177/0734242x20953892

Singh, D. P., Prabha, R., Shukla, R., Sahu, P. K., & Singh, V. (2019). Agrowaste bioconversion and microbial fortification have prospects for soil health, crop productivity, and eco-enterprising. *International Journal of Recycling of Organic Waste in Agriculture, 8*(S1), 457–472. https://doi.org/10.1007/s40093-019-0243-0

Sun, B., Wang, S., Zhang, Y., Chen, B., Li, P., Zhang, X., Wang, Y., Zhao, M., Zhang, Y., & Xie, H. (2024). Enhancement of Rhizoma atractylodis quality, soil nutrients, and microbial characters of vermicompost preparations from spent mushroom and cow dung. *Agronomy, 14*(7), 1384. https://doi.org/10.3390/agronomy14071384

Suthar, S. (2008). Microbial and decomposition efficiencies of monoculture and polyculture vermireactors, based on epigeic and anecic earthworms. *World Journal of Microbiology and Biotechnology, 24*(8), 1471–1479.

Suthar, S., & Singh, S. B. (2007). Vermicomposting of domestic waste by using two epigeic earthworms (Perionyx excavatus and Perionyx sansibaricus). *International Journal of Environmental Science and Technology, 5*(1), 99–106. https://doi.org/10.1007/bf03326002

Tiwari, S. R., Stephen, A. J., & Kumar, A. (2023). A study on consumer buying behavior of vermicompost in Lucknow, Uttar Pradesh, India. *Asian Journal of Agricultural Extension Economics & Sociology, 41*(11), 112–120. https://doi.org/10.9734/ajaees/2023/v41i112267

Vuković, A., Velki, M., Ečimović, S., Vuković, R., Čamagajevac, I. Š, & Lončarić, Z. (2021). Vermicomposting—Facts, benefits and knowledge gaps. *Agronomy, 11*(10), 1952. https://doi.org/10.3390/agronomy11101952

Yatoo, A. M., Ali, M. N., Baba, Z. A., & Hassan, B. (2021). Sustainable management of diseases and pests in crops by vermicompost and vermicompost tea. A review. *Agronomy for Sustainable Development, 41*(1). https://doi.org/10.1007/s13593-020-00657-w

Zaman, A. M. K., & Yaacob, J. S. (2022). Exploring the potential of vermicompost as a sustainable strategy in circular economy: Improving plants' bioactive properties and boosting agricultural yield and quality. *Environmental Science and Pollution Research, 29*(9), 12948–12964. https://doi.org/10.1007/s11356-021-18006-z

Zhang, M., Liu, Y., Wei, Q., Liu, L., Gu, X., Gou, J., & Wang, M. (2023). Ameliorative effects of vermicompost application on yield, fertilizer utilization, and economic benefits of continuous cropping pepper in Karst areas of Southwest China. *Agronomy, 13*(6), 1591. https://doi.org/10.3390/agronomy13061591

Smart Farming: AI-Powered Mechanization for Sustainable Composting Applications

Turnad Lenggo Ginta, Tri Martini, Elisabeth Srihayu Harsanti, Muizuddin Azka, Hens Saputra, Agung Setyo Darmawan, and Nur Islahudin

Abstract This study investigates the application of artificial intelligence (AI) in optimizing composting processes, focusing on enhancing efficiency, consistency, and quality. Traditional composting methods rely on manual observations and are often prone to errors, inconsistency, and inefficiency. AI-driven systems, leveraging sensor data and machine learning algorithms, can automate real-time adjustments to key environmental parameters such as temperature, moisture, and aeration, ensuring optimal conditions for microbial activity. The research utilizes computational models to predict compost maturity time, optimize environmental factors, and reduce resource waste. Data from real-time sensors, including temperature, moisture, and nutrient levels, were used to develop predictive models that assess compost quality consistency and maturity. The methodology integrates experimental sensor data with AI-driven predictive models and optimization techniques. Performance metrics, including compost maturity time and resource efficiency, were compared between AI-based and traditional methods. Results show that AI systems significantly reduce composting cycle time, improve consistency, and enhance compost quality, demonstrating the advantages of AI in accelerating decomposition while maintaining sustainable practices. In conclusion, AI-powered systems offer a transformative approach to composting, providing precise control over environmental

T. L. Ginta (✉) · M. Azka · H. Saputra
Research Center for Process and Manufacturing Industry Technology, National Research and Innovation Agency, Tangerang Selatan, Indonesia
e-mail: turn001@brin.go.id

T. Martini
Research Center for Sustainable Production System and Life Cycle Assessment, National Research and Innovation Agency, Tangerang Selatan, Indonesia

E. S. Harsanti
Cibinong Science Center, Research Center for Horticulture, National Research and Innovation Agency, West Java, Indonesia

A. S. Darmawan
Department of Mechanical Engineering, Faculty of Engineering, Universitas Muhammadiyah Surakarta, Surakarta, Indonesia

N. Islahudin
Industrial Engineering Program at Dian Nuswantoro University, Semarang, Indonesia

A. Z. Yaser et al. (eds.), *Compost and Liquid Organic Fertilizers*,
https://doi.org/10.1007/978-981-95-8561-8_8

conditions and ensuring high-quality compost. The findings highlight AI's potential in advancing sustainable waste management practices, offering both environmental and economic benefits.

Keywords Smart farming · Composting process · AI-Power mechanization

1 Introduction

Smart farming, driven by advanced technologies like artificial intelligence (AI), is transforming agriculture by enabling precise, efficient, and sustainable practices. AI-driven mechanization plays a crucial role in optimizing resource management, crop monitoring, and sustainable compost production while minimizing environmental impact (Mana et al., 2024). Beyond automation, AI utilizes sophisticated algorithms and machine learning for real-time data analysis, predictive modeling, and intelligent decision-making, helping farmers optimize soil health and resource allocation (AlZubi & Galyna, 2023; Sharma & Shivandu, 2024). AI-powered sensors and IoT devices further enhance farming efficiency by monitoring soil conditions and environmental factors, paving the way for a new era of sustainable agricultural practices (Shaikh et al., 2022).

In addition, sustainability is central to smart farming, particularly in compost applications, where traditional production methods are often resource-intensive and environmentally taxing. AI-powered mechanization addresses these challenges by streamlining processes, reducing waste, and enhancing nutrient delivery efficiency (Balyan et al., 2024). From automated compost turners that ensure uniform decomposition to precision agriculture tools that apply fertilizers only where needed, these innovations bridge the gap between productivity and environmental stewardship (Saiz-Rubio & Rovira-Más, 2020; Zul Azlan et al., 2024). Furthermore, as the global population grows and food demand intensifies, smart farming provides a scalable solution to meet these needs while preserving ecological balance. AI-driven mechanization enables farmers to produce high-quality compost with precision, ensuring optimal crop yields and long-term soil health (Zhang et al., 2021). By promoting natural nutrient cycling, improving soil fertility, and reducing reliance on chemical inputs, sustainable compost applications enhance soil structure, water retention, and microbial diversity, positioning agriculture for a more sustainable future (Mana et al., 2024).

Moreover, the importance of sustainable practices extends beyond ecological benefits to include long-term economic viability for farmers. Traditional overuse of synthetic fertilizers often results in diminishing returns as soil health deteriorates over time, requiring higher inputs for the same yields (Sheoran et al., 2019). Sustainable compost, by contrast, create a regenerative system that maintains soil health and ensures consistent productivity. These practices reduce costs by utilizing organic waste materials, converting them into valuable compost, and minimizing input wastage through precision techniques (Tahat et al., 2020). As a result, sustainable applications not only improve crop yields but also help farmers achieve resilience against climate variability and market fluctuations, making them an essential cornerstone of modern farming practices (Sathiyapriya et al., 2024).

In recent years, the integration of Artificial Intelligence (AI) in farming has gained significant momentum, offering transformative possibilities for sustainable agricultural practices. One key area where AI can contribute to environmental sustainability is in smart composting, a critical process in organic waste management (Hoseinzadeh & Astiaso Garcia, 2024). Smart composting techniques leverage advanced sensors, machine learning algorithms, and automated systems to optimize composting processes, enhancing efficiency and reducing environmental impact (Mana et al., 2024). However, the literature surrounding smart composting is still developing, with a need for more comparative studies that evaluate the performance of various AI-driven approaches across different parameters. Previous works primarily focus on isolated elements of smart composting, such as monitoring temperature or moisture levels, but few provide comprehensive insights into the efficiency, scalability, and long-term environmental impact of these systems (Linaza et al., 2021).

To address this gap, a critical review of existing literature is necessary to identify the strengths and weaknesses of various smart composting techniques. A comparison that summarizes key studies on AI-powered composting, providing insight into the methodologies used, parameters measured, and the outcomes of these studies is presented in Table 1. This comparison not only highlights the evolution of smart composting technologies but also allows for a clear identification of areas where further research and innovation are required. The following table offers a structured overview of past research, showcasing the diversity of approaches and their implications for the future of sustainable composting.

The comparison table provides a comprehensive overview of key studies in the field of AI-powered composting, focusing on the diverse AI techniques employed and the parameters monitored across different research efforts. Chen et al. utilized machine learning to predict compost maturity by monitoring temperature, humidity, and pH, although the study's small-scale trials limit its broader applicability (Chen et al., 2024). Zhang et al. applied neural networks to predict oxygen consumption rates, demonstrating effectiveness in early composting stages, but the study lacked long-term data to evaluate its full potential (Zhang et al., 2025). Fouguira et al. integrated IoT sensors with AI to optimize composting conditions in real-time, yet did not compare their system's performance with traditional composting methods (Fouguira et al., 2025). Taheri also employed deep learning for enhanced control over temperature and microbial activity, although the system's high energy consumption remains a concern (Taheri et al., 2025). Finally, Ezeorba implemented reinforcement learning to minimize energy usage while maintaining compost quality, but the system's complexity and need for extensive training data may limit its scalability (Ezeorba et al., 2024). Collectively, the table highlights the progress in AI integration into composting, while underscoring limitations in scalability, energy consumption, and long-term applicability that future research should address.

Table 1 Comparison of AI-powered composting studies

Study	AI techniques used	Parameters monitored	Key findings	Limitations
Chen et al. (2024)	Machine Learning (ML)	Temperature, Humidity, pH	Demonstrated the role of ML in predicting compost maturity, enhancing efficiency	Limited to small-scale trials; lacks scalability analysis
Zang et al. (2025)	Neural Networks (NN)	Oxygen, Temperature, Moisture	NN models effectively predicted oxygen consumption rates during composting	Focused only on initial stages of composting, no long-term data
Fouguira et al. (2025)	Internet of Things (IoT), AI	Moisture, Temperature, pH, C:N	Used IoT sensors for real-time data collection, optimizing composting conditions	Does not compare with traditional composting methods
Taheri et al. (2025)	Deep Learning (DL), Sensor Networks	Moisture, Temperature, pH	Deep learning models improved temperature control and microbial activity	High energy consumption in the system
Ezeorba et al.(2024)	Reinforcement Learning (RL)	Temperature, Moisture, pH	RL-based system minimized energy usage while maintaining compost quality	Requires extensive training data and complex setups

In addition to machine learning (ML) and artificial neural networks (ANNs), several other AI methodologies have been applied in smart composting and broader smart agriculture contexts. For instance, deep learning (DL) techniques, particularly recurrent neural networks (RNNs), have been utilized for predicting compost maturity and optimizing composting processes by analyzing temporal data from sensors monitoring parameters like temperature and humidity (Shaikh et al., 2022). Decision trees (DTs) have been employed to model and predict gas emissions such as CO_2, H_2S, NH_3, and CO during composting, aiding in emission control and process optimization (Ali et al., 2025). Autoencoders, a type of unsupervised learning model, have been applied for anomaly detection in smart farming ecosystems, identifying outliers in sensor data indicative of system faults or environmental anomalies (Qazi et al., 2022). Additionally, reinforcement learning (RL) has been explored for optimizing decision-making in composting systems, adjusting parameters like aeration and moisture levels to enhance compost quality and efficiency (Chen et al., 2022). These AI techniques, among others, contribute to the advancement of smart composting and agriculture by enabling more precise, adaptive, and sustainable practices.

Modern agriculture faces a dual challenge of meeting the rising global demand for food production while minimizing its environmental impact. Traditional farming practices, characterized by excessive use of synthetic fertilizers and inefficient composting methods, have led to significant issues such as soil degradation, nutrient leaching, water pollution, and greenhouse gas emissions (Jha et al., 2023). These practices not only jeopardize the long-term sustainability of agricultural ecosystems but also increase production costs for farmers, making farming less economically viable over time (Nakachew et al., 2024). Furthermore, the inconsistent quality of compost and fertilizer applications exacerbates inefficiencies, resulting in suboptimal crop yields and reduced soil health. In this context, there is an urgent need for innovative and sustainable approaches to composting management. The integration of AI-powered mechanization in agriculture offers a promising solution to address these challenges by optimizing compost production, enhancing nutrient delivery precision, and reducing resource wastage (Adegbeye et al., 2020). However, despite its potential, the adoption of such technologies is limited by barriers such as cost, technological literacy, and infrastructure. Bridging these gaps is critical to creating scalable, efficient, and environmentally sustainable farming systems that can meet the needs of a growing population while preserving the planet's resources (Shah and Wu, 2019).

The purpose of this article is to explore how AI-powered mechanization is revolutionizing the production and application of compost by enhancing efficiency, productivity, and environmental sustainability. Traditional farming methods often suffer from inefficiencies, waste, and ecological damage, but the integration of artificial intelligence and automated machinery provides precise and data-driven solutions. AI-driven systems optimize processes such as compost decomposition by utilizing real-time monitoring, predictive modeling, and precision delivery, resulting in reduced waste, improved nutrient management, and lower environmental impact. This article highlights how these advancements not only enhance agricultural productivity but also contribute to long-term ecological balance, making them a cornerstone of sustainable farming practices.

2 Methodology

The research methodology employed in this study integrates computational approaches to evaluate the impact of AI-driven systems on composting processes. The primary focus was to develop and validate models that optimize key parameters such as temperature, moisture, and aeration, which are critical for efficient composting. Data was generated from previous researches, capturing variables like temperature, moisture, and oxygen concentration, along with microbial activity and nutrient levels (Law et al., 2021; Pradnya et al., 2024). These data sets were analyzed using machine learning algorithms to establish predictive models for compost maturity, quality consistency, and environmental optimization, providing a foundation for AI to make real-time adjustments (Aydın Temel et al., 2023).

The computational models were constructed based on historical data with the objective of predicting compost maturity time, optimizing environmental conditions, and minimizing resource waste. The models incorporated multiple variables, such as temperature (T), moisture level (H), oxygen concentration (O), nutrient levels (N), and decomposition rate (D). To assess the models' effectiveness, a series of simulations were run under varying environmental conditions, allowing the AI algorithms to predict necessary adjustments to maintain optimal composting conditions. These models were compared with traditional methods to highlight the advantages of AI in terms of accuracy, error reduction, and cycle acceleration.

In this study, an AI-powered mechanization system for sustainable composting, leveraging machine learning algorithms to enhance the composting process was employed. The AI system integrates sensors that track real-time data, such as temperature, humidity, and moisture levels, which are critical for optimizing composting conditions. A key algorithm used is the Support Vector Machine (SVM), which classifies different stages of the composting cycle based on environmental inputs. The SVM's ability to distinguish between different states of decomposition allows for automated adjustments in composting parameters, ensuring an efficient and consistent process. Additionally, a Neural Network (NN) is employed to predict the optimum conditions for the next composting cycle, helping to accelerate the breakdown of organic materials. The algorithm works by processing historical data to refine the composting process, adjusting factors like aeration and turning frequency to meet predefined thresholds.

The AI model focuses on four critical parameters to measure the efficiency of the composting system. Time Efficiency is improved by using real-time data to reduce the need for manual monitoring and intervention. Consistency is ensured by maintaining optimal composting conditions, minimizing fluctuations in temperature and moisture that could lead to uneven decomposition. Error Reduction is achieved through the use of automated sensors and predictive algorithms that minimize human error in monitoring and adjusting composting parameters. Finally, Cycle Acceleration is facilitated by the system's ability to adjust conditions dynamically, speeding up the composting process by optimizing conditions for microbial activity. A flowchart illustrating the process is shown in Fig. 1, where data flows from sensors to the AI system, which then adjusts the parameters in real time to enhance the composting process.

The flow chart as presented in Fig. 1 illustrates the process of an AI system designed to optimize performance using Support Vector Machine (SVM) and Neural Networks. It begins with the input of raw data, which is then processed by the AI system. The system utilizes both SVM and Neural Network models, each contributing to the analysis and learning process. SVM is a powerful supervised learning model that separates data into different classes by finding the best decision boundary, while the Neural Network, with its multiple layers, captures complex, nonlinear patterns within the data. After the models process the input data, the system adjusts parameters to enhance the model's performance. This adjustment ensures the AI system improves in key areas, such as time efficiency, consistency, error reduction, and cycle acceleration. Time efficiency refers to the system's ability to process data quickly,

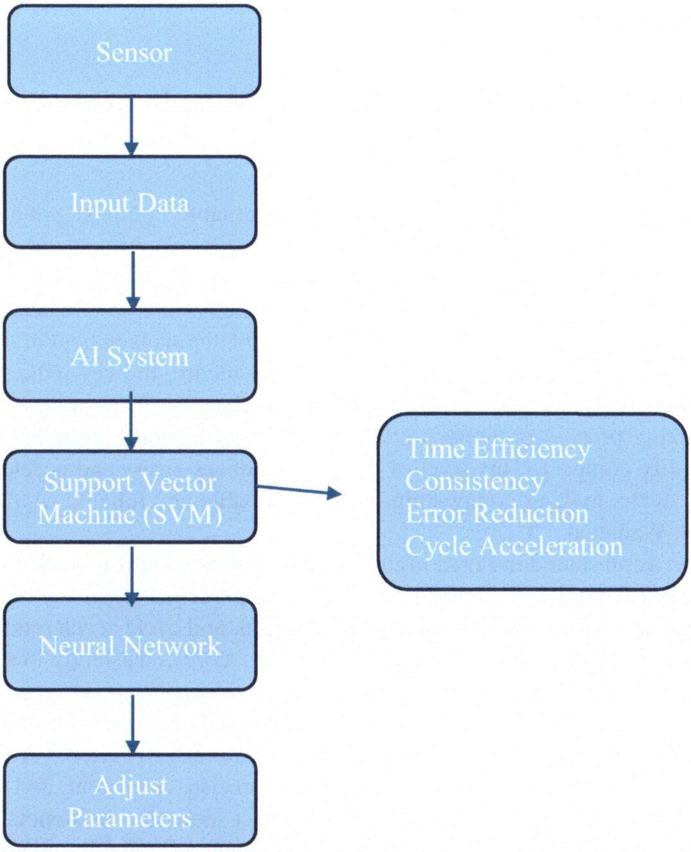

Fig. 1 AI system optimization workflow

while consistency ensures reliable, repeatable results across different datasets. Error reduction occurs through continual fine-tuning, which minimizes inaccuracies in predictions, and cycle acceleration speeds up the learning and adaptation process. Ultimately, this iterative optimization enables the AI system to become more efficient and accurate, producing faster and more reliable results.

3 Results and Discussion

3.1 The Role of AI Algorithms in Monitoring Decomposition Stages

AI algorithms play a transformative role in sustainable compost production by enabling precise monitoring of decomposition stages. Traditional methods rely heavily on manual observations, which are often inconsistent and time-consuming (Tsui et al., 2023). With AI, data collected from sensors is analyzed to determine the exact stage of the composting process, allowing for better management of the breakdown of organic materials (Martín et al., 2023). By identifying key indicators such as microbial activity, nutrient levels, and chemical changes, AI systems ensure that the composting process progresses efficiently and consistently, ultimately leading to high-quality compost. This level of precision reduces errors and accelerates the production cycle, making composting more sustainable and reliable (Aydın Temel et al., 2023; Wan et al., 2022).

The chart as presented in Fig. 2 highlights the significant advantages of AI-based monitoring over traditional methods in composting processes, focusing on four key metrics: time efficiency, consistency, error reduction, and cycle acceleration. Traditional methods, which rely on manual observations, score consistently lower across all metrics, with particularly poor performance in error reduction and cycle acceleration (Huang et al., 2024). This reflects the challenges of human error, variability, and the time-consuming nature of manual processes. In contrast, AI-based systems excel, leveraging sensor data and algorithms to provide precise, consistent, and real-time monitoring (Ayilara et al., 2020). With a near-perfect score in consistency and error reduction, AI ensures that decomposition stages are managed reliably and without the oversight common in manual methods. Additionally, AI significantly accelerates the composting cycle by optimizing critical factors like microbial activity and nutrient levels, enabling a faster, more sustainable production process (Manea et al., 2024). Overall, the chart underscores how AI transforms composting by improving precision, reliability, and efficiency, making it a critical tool for advancing sustainable practices in organic waste management.

3.2 Optimizing Temperature, Moisture, and Aeration Using AI

One of the critical factors in compost production is maintaining optimal environmental conditions, such as temperature, moisture, and aeration, to support microbial activity (López et al., 2014). Microbial activity is essential for natural processes like bioremediation, nutrient cycling, and wastewater treatment (Wang et al., 2025).

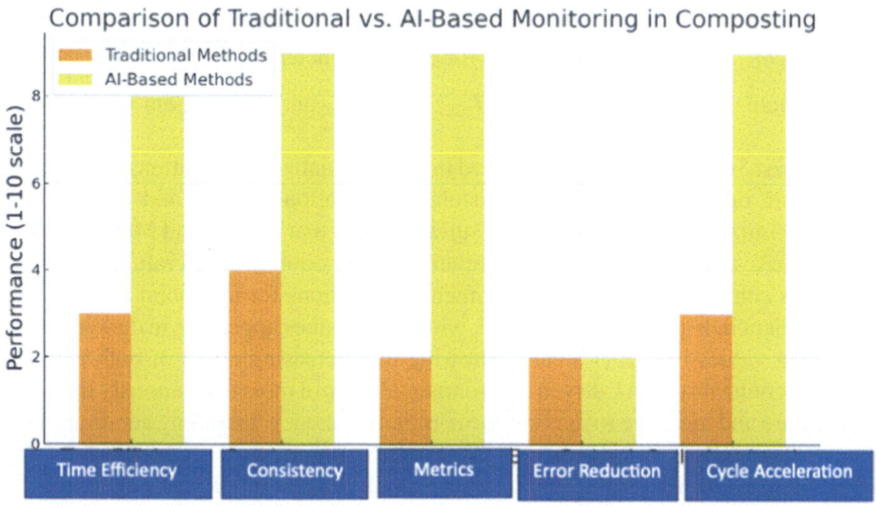

Fig. 2 Comparison of traditional methods versus AI-based monitoring in composting efficiency

Microorganisms break down pollutants, recycle nutrients, and maintain ecological balance. They are also used in industrial applications for waste treatment, biofuel production, and pharmaceutical manufacturing, contributing to sustainable, eco-friendly technologies for environmental solutions (Akinsemolu, 2023).

AI-powered systems excel in managing these parameters by continuously analyzing real-time data from integrated sensors. These systems can predict when adjustments are needed and automatically regulate conditions to ensure that microbial activity remains at its peak (Keener et al., 2005; Tomicic, 2022). For example, AI can control aerators to introduce oxygen or adjust water levels to prevent compost drying out or becoming waterlogged. By automating these processes, AI eliminates the guesswork associated with traditional composting methods, reducing resource waste and ensuring consistent compost quality. This study widely proposed models in optimizing temperature, moisture and aeration during compost production.

Environmental Optimization Model

The proposed model as presented in Eq. (1) describes how AI systems maintain optimal conditions for microbial activity by balancing temperature (T), moisture (M), and aeration (A). The model is derived as follows:

$$E_{opt} = \omega_1 + T_{opt} + \omega_2.M_{opt} + \omega_3.A_{opt} \tag{1}$$

where:

- E_{opt} : Environmental optimization score
- T_{opt}: Proximity of the temperature to the optimal range
- M_{opt}: Proximity of moisture to the optimal range

- A_{opt}: Proximity of aeration to the optimal range
- $\omega_1, \omega_2, \omega_3$: Weighting coefficients for each parameter

AI systems continuously monitor T_{opt}, M_{opt}, A_{opt} and adjust them to maximize E_{opt}.

The latest 3D surface plot as presented in Fig. 3 visualizes the relationship between temperature optimization (Topt), moisture optimization (Mopt), and the environmental optimization score (Eopt). The surface shows that as Topt and Mopt increase, Eopt also rises, indicating a direct correlation where optimal temperature and moisture levels enhance the overall environmental conditions for microbial activity. The color gradient helps illustrate this trend, with lower values appearing in darker shades and higher values in lighter shades, showing the increasing effect of both parameters. This confirms that AI-driven environmental control must continuously fine-tune temperature and moisture to maximize microbial efficiency. Including aeration (Aopt) in a further analysis would provide an even more comprehensive understanding of its impact on the system. It follows the research by Kaya that intelligent environmental control systems integrating sensors, automation, and AI are crucial for optimizing crop production in vertical farming (Kaya, 2025). Furthermore, these systems continuously monitor and adjust key parameters like temperature, humidity, and CO_2 levels to maintain optimal growing conditions. Similarly, as described by Ward et al., in anaerobic digestion, optimizing environmental conditions such as temperature and pH is essential for maximizing methane production (Ward et al., 2008). In addition, a study by Khor et al., for microalgae cultivation, controlling growth factors including pH, temperature, and light intensity is vital for enhancing biomass productivity (Khor et al., 2022).

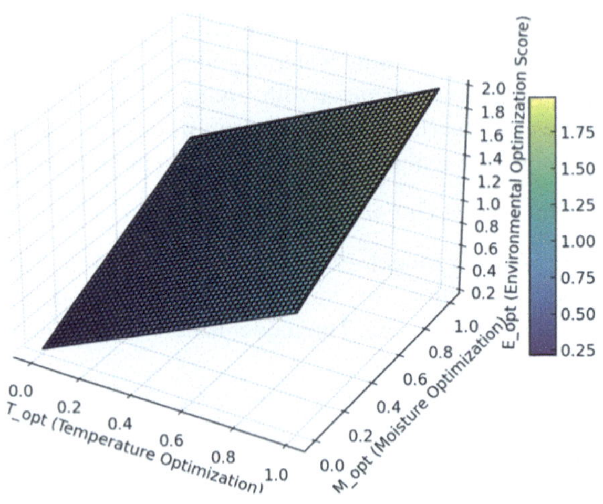

Fig. 3 Effect of temperature and moisture on environmental optimization

Predictive Adjustment Model

The proposed model as depicted in Eq. (2) predicts when adjustments are needed based on deviations from optimal environmental conditions. The model is generated as follows:

$$\Delta_{adj} = \delta_1.(T_{real} - T_{opt}) + \delta_2.(M_{real} - M_{opt} + \delta_3.(A_{real} - A_{opt}) \tag{2}$$

where:

- Δ_{adj}: Adjustment magnitude required
- $T_{real}, M_{real}, A_{real}$: Real-time measurements of temperature, moisture, and aeration
- $T_{opt}, M_{opt}, A_{opt}$: Optimal levels for each parameter
- $\delta_1, \delta_2, \delta_3$: Adjustment sensitivity coefficients

AI-powered systems use sensor data to calculate Δ_{adj} and automate adjustments. The 3D graph as presented in Fig. 4 visually represents the adjustment magnitude *(Δ_adj)* required based on deviations in temperature *(T_real)* and moisture *(M_real) from* their optimal values. As the temperature increases or decreases from its optimal level, the required adjustment magnitude changes proportionally, demonstrating a direct relationship between temperature deviations and adjustment needs. Similarly, changes in moisture levels also affect the magnitude of adjustment needed, highlighting how both factors interact in determining the adjustments required to maintain optimal conditions. The plot emphasizes that the adjustment magnitude increases as both temperature and moisture deviate from the ideal range, which is crucial for AI-powered systems in automating environmental control adjustments. This study follows the research by Alex that found that AI-driven smart irrigation systems can automatically regulate water release based on the complex relationship between soil moisture and temperature, promoting water efficiency (Alex et al., 2024). It also follows the findings by Malbog, that for indoor gardens, automatic misting systems using rule-based approaches have been developed to maintain optimal temperature ranges for plant growth (Malbog, 2020).

Resource Efficiency Model

This model as showed in Eq. (3) quantifies how AI reduces resource waste (e.g., water and energy usage) while maintaining optimal conditions. The model is constructed as follows:

$$R_{eff} = \kappa_1.W_{saved} + \kappa_2.E_{saved} - \kappa_3.A_{overhead} \tag{3}$$

where:

- R_{eff}: Resource efficiency score
- W_{saved}: Water saved through AI optimization
- E_{saved}: Energy saved through AI-controlled aeration
- $A_{overhead}$: AI system overhead (energy/resource cost)
- $\kappa_1, \kappa_2, \kappa_3$: Weighting factors for resource parameters

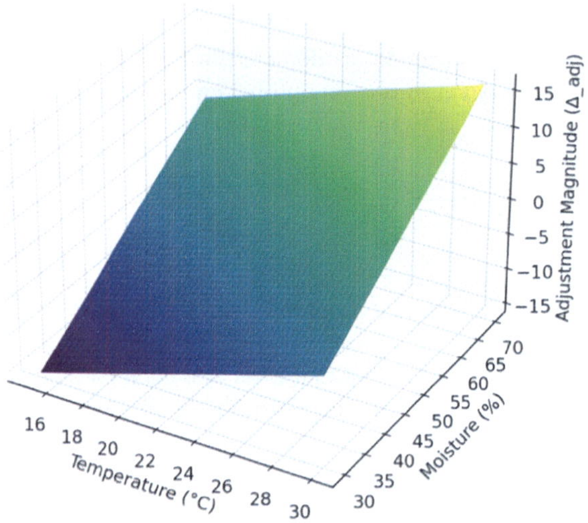

Fig. 4 Adjustment magnitude versus temperature and moisture deviations

In addition, the quadratic model can be expressed as follows:

$$R_{eff} = \kappa_1.W_{saved}^2 + \kappa_2.E_{saved}^2 - \kappa_3.A_{overhead}^2 \quad (4)$$

The 3D graph as presented in Fig. 5 represents the quadratic relationship between resource efficiency *(R_eff)*, water saved *(W_saved)*, energy saved *(E_saved)*, and AI system overhead *(A_overhead)*. As water and energy savings increase, the resource efficiency grows exponentially, as indicated by the curvature of the surface, showing a significant positive impact on efficiency with larger savings. However, the graph also demonstrates that as AI system overhead increases, it negatively affects the overall resource efficiency, with this negative impact growing quadratically. This suggests that while maximizing water and energy savings can greatly improve efficiency, managing the overhead costs associated with AI systems is crucial to maintain an overall positive outcome in resource usage. Previous studies by Lutfiani et al. and Waltermann et al. have shown that AI applications can significantly enhance energy and resource management in manufacturing, urban infrastructure, and energy companies (Lutfiani et al., 2024; Waltersmann et al., 2021). Furthermore, this model is supported by Zaharuddin et al. that implementation of AI technologies has led to increased energy efficiency, ranging from 15 to 25%, along with improvements in water management and transportation efficiency (Zaharuddin et al., 2024). Hence, in the industrial sector, AI contributes to resource efficiency through applications such as predictive maintenance, production planning, and fault detection.

Microbial Activity Optimization Model

This proposed model relates the environmental conditions to microbial activity, which is key to efficient composting. The model is proposed and constructed as follows:

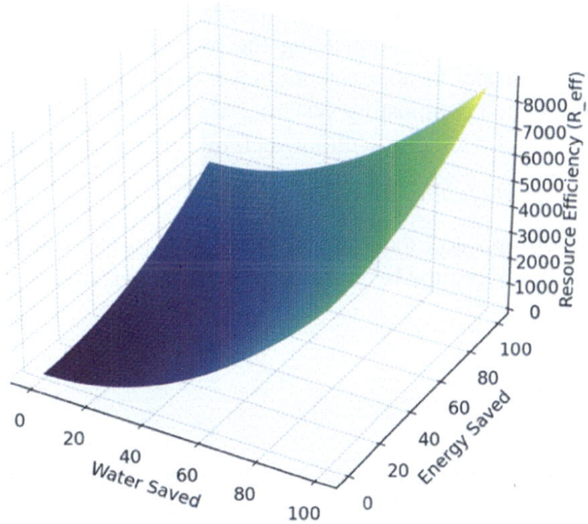

Fig. 5 Quadratic resource efficiency versus water saved, energy saved, and AI overhead

$$M_{act} = \mu.f(T, M, A) \tag{5}$$

where:

- M_{act}: Microbial activity level
- *T, M, A:* Current temperature, moisture, and aeration levels
- *f(T, M, A):* function describing the microbial response to environmental conditions (e.g., a Gaussian function with peaks at optimal conditions)
- μ; AI-driven enhancement factor accounting for real-time adjustments

The Microbial Activity Optimization Model aims to relate environmental factors, specifically temperature, moisture, and aeration, to the microbial activity during composting. In this model, M_act represents the microbial activity level, which is influenced by the current temperature *(T)*, moisture *(M)*, and aeration *(A)* levels within the composting environment. These factors are critical because they directly affect the metabolic rate and the growth of microorganisms responsible for breaking down organic matter. The function *f(T, M, A)* captures the microbial response to these conditions, which can be modeled using a Gaussian function, ensuring the model reflects the optimal environmental conditions where microbial activity is maximized. For example, each of the variables—temperature, moisture, and aeration—has an optimal range that promotes the most effective microbial activity, and the Gaussian function allows for these peaks to be captured in the model.

The parameter μ introduces an AI-driven enhancement factor to the model, allowing for real-time adjustments based on continuous monitoring of the composting environment. This AI factor adapts to fluctuations in the environment by adjusting the function to enhance microbial activity. It uses data gathered from sensors to adjust temperature, moisture, and aeration levels, ensuring that the composting process remains as efficient as possible. This model leverages machine learning algorithms to predict and adjust environmental conditions dynamically, optimizing microbial performance and improving the overall efficiency of composting. The integration of real-time AI adjustments makes this model particularly suitable for large-scale, automated composting systems, where maintaining optimal conditions continuously can be challenging without such intelligent intervention.

Compost Quality Consistency Model

The proposed model as presented in Eq. (6) evaluates the consistency of compost quality over time by minimizing fluctuations in key parameters and is constructed as follows:

$$C_{qual} = 1 - \frac{\sigma_T + \sigma_M + \sigma_A}{3} \quad (6)$$

where:

- C_{qual}: Compost quality score (closer to 1 means high consistency)
- $\sigma_T, \sigma_M, \sigma_A$: Standard deviations of temperature, moisture, and aeration over time

AI minimizes these deviations, ensuring consistent compost quality.

The 3D graph as presented in Fig. 6 illustrates the relationship between compost quality *(C_qual)* and deviations in temperature *(σ_T)* and moisture *(σ_M),* while keeping aeration *(σ_A)* fixed at a constant value. As the deviations in temperature and moisture increase, the compost quality score decreases, indicating a reduction in consistency. This demonstrates the model's focus on minimizing fluctuations in these key parameters to maintain high compost quality. The steep decline in quality with increasing deviations highlights the importance of stabilizing temperature and moisture levels to ensure consistent and optimal compost conditions. As investigated by Tilak et al. before, composting quality is heavily influenced by temperature and moisture dynamics. Maintaining stable temperature and moisture levels is crucial for optimal composting outcomes (Tilak et al., 2021). Furthermore, as studied by El Boudihi et al., temperature control is particularly important, as high temperatures during aerobic decomposition help eliminate pathogens (El Boudihi et al., 2024). It is also confirmed by Peña et al. that monitoring key parameters like pH, humidity, C/N ratio, and CO2 emissions is essential for process control and final compost quality (Peña et al., 2020).

Compost Maturity Prediction Model

This study also proposed a model in predicting the time required for compost to reach maturity based on key environmental and microbial factors, as shown in Eq. (7).

$$T_{maturity} = \alpha_1 T + \alpha_2 H + \alpha_3 O + \alpha_4 N - \alpha_5 D \quad (7)$$

where:

- $T_{maturity}$ = Predicted compost maturity time (days)
- T = Temperature (°C)

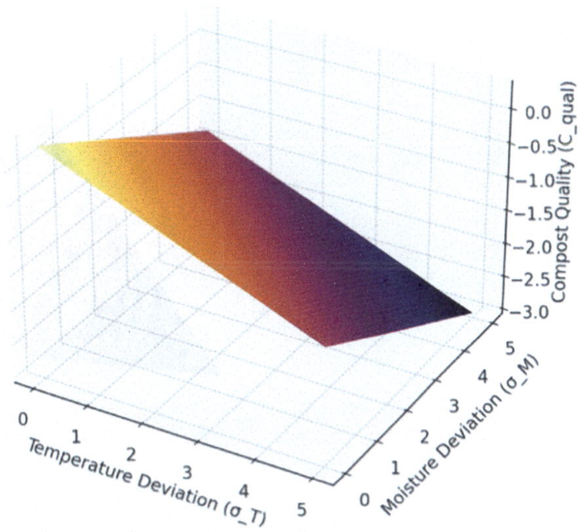

Fig. 6 Compost quality versus temperature and moisture deviations (Fixed Aeration)

- H = Moisture level (%)
- O = Oxygen concentration (%)
- N = Nutrient levels (C/N ratio, P, K)
- D = Decomposition rate (%)
- $\alpha_1, \alpha_2, \alpha_3, \alpha_4, \alpha_5$ = Weighting factors determined by AI from historical data.

This model allows AI to adjust conditions to accelerate maturity while ensuring compost quality.

The graph as presented in Fig. 7 illustrates how the predicted compost maturity time *(T_maturity)* changes with variations in temperature *(T)* and moisture *(H)*. As both temperature and moisture levels increase, the compost maturity time decreases, indicating that higher values of these factors accelerate the composting process. This model highlights the importance of optimizing temperature and moisture to reduce the time required for compost to reach maturity. It also emphasizes how AI can adjust these parameters to improve compost efficiency, ensuring quicker maturity while maintaining compost quality. This findings follow recent studies that explored the application of machine learning models to optimize composting processes and predict compost maturity. Ding et al., for instance, have employed various algorithms, including *Stacking, Extra Trees, and Gaussian Process Regression*, to analyze and predict key composting parameters such as temperature, moisture content, pH, and nutrient levels (Ding et al., 2022). Furthermore, our study could demonstrated high prediction accuracy for compost maturity indicators like seed germination index and T value.

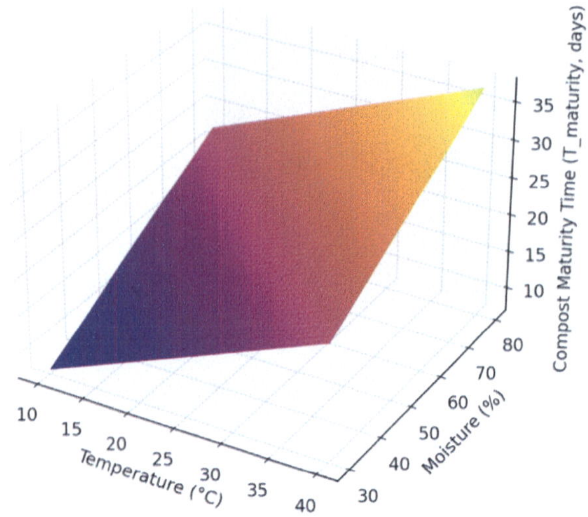

Fig. 7 Compost maturity time versus temperature and moisture

3.3 Original Contribution and Novelty of the Study

The manuscript presents an original study on the application of artificial intelligence (AI) algorithms in optimizing compost production. The methods and models discussed in the results section are directly proposed as part of our own research. Specifically, the use of AI for monitoring decomposition stages, optimizing environmental parameters such as temperature, moisture, and aeration, and improving resource efficiency during composting are all part of novel models introduced in this study. The approach is distinct from existing reviews or previous works, as it focuses on integrating AI-driven algorithms with real-time sensor data to automate and enhance various aspects of the composting process.

In contrast to traditional composting methods, which are manual and prone to errors, this study introduces a new level of precision and efficiency by leveraging AI systems. These systems are designed to predict and adjust the conditions needed for optimal microbial activity, ensuring higher-quality compost and reduced resource waste. The models discussed in the manuscript, such as the Microbial Activity Optimization Model and the Compost Maturity Prediction Model, were developed as part of this original research to optimize composting at a large scale. The discussion ties these models back to real-time data analysis and AI-powered adjustments, which are central to the study's contribution.

Furthermore, the results and discussion section emphasizes the unique contributions of this work by comparing traditional methods with AI-enhanced systems. The models proposed in this study are tested through simulations and data analysis, offering a comprehensive solution for sustainable and efficient compost production. Therefore, this manuscript is firmly rooted in the authors' own proposed work, advancing the field of AI applications in organic waste management and composting processes.

3.4 Comparison and Integration of Proposed Models for Optimizing Composting

The study introduces several AI-driven models that optimize various aspects of the composting process. The Microbial Activity Optimization Model excels at real-time adjustments of temperature, moisture, and aeration to maintain optimal microbial activity, making it ideal for large-scale composting systems. The Compost Maturity Prediction Model predicts the compost maturity time based on environmental and microbial factors, helping optimize composting duration while ensuring high-quality output. Additionally, the Resource Efficiency Model focuses on reducing resource waste, particularly water and energy, by continuously optimizing the composting conditions. The Environmental Optimization Model and Predictive Adjustment Model fine-tune system performance by maintaining ideal conditions for microbial activity and predicting when adjustments are needed.

Given their complementary nature, these models work best when used together. Integrating the Microbial Activity Optimization Model with the Compost Maturity Prediction Model accelerates the composting process while preserving quality, while combining the Environmental Optimization Model and Resource Efficiency Model enhances sustainability by minimizing resource consumption. Using all models in a unified AI-powered system allows for real-time, dynamic adjustments, offering a more efficient, sustainable, and consistent approach to compost production. This comprehensive system provides an ideal solution for optimizing organic waste management.

The distinct roles and benefits of the proposed AI-driven models in optimizing the composting process, demonstrating their utility in different stages of compost production is highlighted in Table 2. The Microbial Activity Optimization Model is crucial for maintaining optimal microbial activity by adjusting temperature, moisture, and aeration in real time, ensuring efficient decomposition. The Compost Maturity Prediction Model enhances this by predicting compost maturity time, thus optimizing the overall composting cycle and reducing unnecessary delays. The Resource Efficiency Model complements these by focusing on reducing water and energy consumption, ensuring that the composting process is sustainable. Additionally, the Environmental Optimization Model and the Predictive Adjustment Model continuously fine-tune environmental conditions to maintain consistency and optimize microbial activity. Together, these models provide a comprehensive solution that improves composting efficiency, reduces resource waste, and accelerates the production process, showcasing their effectiveness in creating a more sustainable and efficient composting system. Their integration into a unified AI-powered system

Table 2 Composting process optimization models and their application

Model	Objective	Performance & benefits	Best use case
Microbial activity optimization model	Optimizes microbial activity during composting by adjusting temperature, moisture, and aeration	– Ensures optimal microbial activity – Real-time monitoring and adjustments – Maximizes compost efficiency	Ideal for large-scale composting where continuous adjustments are needed
Compost maturity prediction model	Predicts the time required for compost to reach maturity based on environmental and microbial factors	– Reduces composting time while maintaining quality – Helps in planning the composting cycle	Best used to optimize composting duration and monitor maturity
Resource efficiency model	Quantifies how AI can reduce resource waste, focusing on water and energy usage	– Minimizes resource consumption – Ensures sustainable compost production	Effective in environments focusing on resource conservation and cost reduction
Environmental optimization model	Balances temperature, moisture, and aeration to maintain optimal environmental conditions for microbial activity	– Maintains ideal conditions for microbial activity – Reduces the guesswork in traditional composting methods	Best used in real-time environmental control systems for composting
Predictive adjustment model	Predicts when adjustments are needed for temperature, moisture, and aeration based on deviations from optimal conditions	– Automates real-time environmental control – Ensures continuous optimization for consistent compost quality	Ideal for automation in composting systems where continuous adjustments are necessary

offers a dynamic, real-time solution that addresses multiple challenges in compost production, making them highly valuable for advancing organic waste management.

4 Conclusion

The integration of AI algorithms into composting systems has revolutionized the process by offering precision and efficiency that traditional methods fail to achieve. Traditional composting relies on manual observation, which is time-consuming and prone to errors, leading to inconsistencies in compost quality. AI systems, using real-time data from sensors, continuously monitor key factors such as microbial activity, nutrient levels, and chemical changes. This automation reduces human error, accelerates the composting cycle, and improves compost quality in less time, making the process more reliable and sustainable.

AI surpasses traditional composting methods in time efficiency, consistency, error reduction, and cycle acceleration by leveraging advanced algorithms and sensor data. AI systems optimize critical environmental factors like temperature, moisture, and aeration, ensuring that microbial activity remains optimal and preventing issues such as excess dryness or wetness. The Environmental Optimization Model highlights the correlation between temperature and moisture adjustments and improved microbial conditions. Additionally, predictive modeling techniques, such as the Compost Maturity Prediction Model, use historical and real-time data to forecast when compost will mature, eliminating uncertainty and ensuring consistent quality. These advancements enhance the overall efficiency of composting systems, aligning with sustainable waste management practices.

In addition, future work could focus on further refining and integrating the proposed AI models into a fully automated composting system that combines real-time monitoring, predictive analytics, and adaptive control for even more efficient and sustainable compost production. This could involve expanding the models to include additional environmental parameters, such as pH and oxygen levels, to enhance compost quality and microbial health. Additionally, exploring the use of machine learning techniques, such as deep learning or reinforcement learning, could improve the system's ability to predict and adapt to complex, nonlinear relationships between various composting factors. The scalability of these models to different types of organic waste and composting setups, as well as their integration with broader waste management systems, would be an important area for future research. Lastly, assessing the economic and environmental impact of implementing these AI-powered models at a commercial scale could provide valuable insights into their long-term sustainability and cost-effectiveness.

References

Adegbeye, M. J., Ravi Kanth Reddy, P., Obaisi, A. I., Elghandour, M. M. M. Y., Oyebamiji, K. J., Salem, A. Z. M., Morakinyo-Fasipe, O. T., Cipriano-Salazar, M., & Camacho-Díaz, L. M. (2020). Sustainable agriculture options for production, greenhouse gasses and pollution alleviation, and nutrient recycling in emerging and transitional nations—An overview. *Journal of Clean Production, 242*, 118319.

Akinsemolu, A. A. (2023). Principles of green microbiology: The microbial blueprint for sustainable development. *Environmental Advances, 14*, Article 100440.

Alex, B., Jignasa, G., Madhubabu, K., & Gopi, A. (2024). AI-driven smart irrigation: enhancing agricultural water efficiency through intelligent valve regulation in piped and micro irrigation networks. In *2024 First International Conference on Pioneering Developments in Computer Science & Digital Technologies* (IC2SDT). IEEE, pp. 76–81.

Ali, Z., Muhammad, A., Lee, N., Waqar, M., & Lee, S. W. (2025). Artificial intelligence for sustainable agriculture: A comprehensive review of AI-driven technologies in crop production. *Sustainability, 17*, 2281.

AlZubi, A. A., & Galyna, K. (2023). Artificial intelligence and internet of things for sustainable farming and smart agriculture. *IEEE Access, 11*, 78686–78692.

Aydın Temel, F., Cagcag Yolcu, O., & Turan, N. G. (2023). Artificial intelligence and machine learning approaches in composting process: A review. *Bioresource Technology, 370*, Article 128539.

Ayilara, M., Olanrewaju, O., Babalola, O., & Odeyemi, O. (2020). Waste management through composting: Challenges and potentials. *Sustainability, 12*, 4456.

Balyan, S., Jangir, H., Tripathi, S. N., Tripathi, A., Jhang, T., & Pandey, P. (2024). Seeding a sustainable future: Navigating the digital horizon of smart agriculture. *Sustainability, 16*, 475.

Chen, L., He, P., Zhang, H., Peng, W., Qiu, J., & Lü, F. (2024). Applications of machine learning tools for biological treatment of organic wastes: Perspectives and challenges. *Circular Economy, 3*, Article 100088.

Chen, Q., Li, L., Chong, C., & Wang, X. (2022). AI-enhanced soil management and smart farming. *Soil Use Management, 38*, 7–13.

Ding, S., Huang, W., Xu, W., Wu, Y., Zhao, Y., Fang, P., Hu, B., & Lou, L. (2022). Improving kitchen waste composting maturity by optimizing the processing parameters based on machine learning model. *SSRN Electronic Journal.*

El Boudihi, H., Kifani-Sahban, F., Wahby, I., & Fihri, M. (2024). Study of temperature dynamics and influencing factors during composting process. *International Journal of Chemical Reactor Engineering, 22*, 1183–1198.

Ezeorba, T. P. C., Okeke, E. S., Mayel, M. H., Nwuche, C. O., & Ezike, T. C. (2024). Recent advances in biotechnological valorization of agro-food wastes (AFW): Optimizing integrated approaches for sustainable biorefinery and circular bioeconomy. *Bioresource Technology Report, 26*, Article 101823.

Fouguira, S., Ammar, E., Haji, M. E., & Benhra, J. (2025). Internet of things and predictive artificial intelligence for smartcomposting process in the context of circular economy. In *SMILE 2025.* MDPI, Basel Switzerland, p. 16.

Hoseinzadeh, S., & Astiaso Garcia, D. (2024). Ai-driven innovations in greenhouse agriculture: Reanalysis of sustainability and energy efficiency impacts. *Energy Conversion and Management: X, 24*, Article 100701.

Huang, L., Hou, J., & Liu, H. (2024). Machine-learning intervention progress in the field of organic waste composting: Simulation, prediction, optimization, and challenges. *Waste Management, 178*, 155–167.

Jha, A., Pathania, D., Sonu, Damathia, B., Raizada, P., Rustagi, S., Singh, P., Rani, G. M., & Chaudhary, V. (2023). Panorama of biogenic nano-fertilizers: A road to sustainable agriculture. *Environmental Research, 235*, 116456.

Kaya, C. (2025). Intelligent environmental control in plant factories: Integrating sensors, automation, and <scp>AI</scp> for optimal crop production. *Food Energy Secure*, 14.

Keener, H. M., Ekinci, K., & Michel, F. C. (2005). Composting process optimization—using on/off controls. *Compost Science Utilization, 13*, 288–299.

Khor, J. G., Lim, H. R., Chia, W. Y., & Chew, K. W. (2022). Automated cultivation system for microalgae: Growth factors and control. *Current Neutral Food Science, 18*, 776–779.

Law, L., Foscoli, B., Mastorakos, E., & Evans, S. (2021). A comparison of alternative fuels for shipping in terms of lifecycle energy and cost. *Energies (Basel), 14*, 8502.

Linaza, M. T., Posada, J., Bund, J., Eisert, P., Quartulli, M., Döllner, J., Pagani, A., G., Olaizola, I., Barriguinha, A., Moysiadis, T., & Lucat, L. (2021). Data-driven artificial intelligence applications for sustainable precision agriculture. *Agronomy, 11*, 1227.

López, M., Martinez-Farre, X., Casas, O., Quilez, M., Polo, J., Lopez, O., Hornero, G., Pinilla, M. R., Rovira, C., Ramos, P. M., Borges, B., Marques, H., & Girão, P. S. (2014). Intelligent composting assisted by a wireless sensing network. *Waste Management, 34*, 738–746.

Lutfiani, N., Puji Lestari Santoso, N., Ahsanitaqwim, R., Rahardja, U., Rahmania & Az Zahra, A. (2024). AI-based strategies to improve resource efficiency in urban infrastructure. *International Transactions on Artificial Intelligence (ITALIC), 2*, 121–127.

M. Tahat, M., M. Alananbeh, K., A. Othman, Y., & I. Leskovar, D. (2020). Soil health and sustainable agriculture. *Sustainability, 12*, 4859.

Malbog, M. A. F. (2020). MISTMATIC: Automatic misting control system for indoor garden with rule-based approach. *International Journal of Advanced Trends in Computer Science and Engineering, 9*, 6471–6476.

Mana, A. A., Allouhi, A., Hamrani, A., Rehman, S., el Jamaoui, I., & Jayachandran, K. (2024). Sustainable AI-based production agriculture: Exploring AI applications and implications in agricultural practices. *Smart Agricultural Technology, 7*, Article 100416.

Manea, E. E., Bumbac, C., Dinu, L. R., Bumbac, M., & Nicolescu, C. M. (2024). Composting as a sustainable solution for organic solid waste management: Current practices and potential improvements. *Sustainability, 16*, 6329.

Martín, A.P.-S., Marhuenda-Egea, F. C., Bustamante, M. A., & Curaqueo, G. (2023). Spectroscopy techniques for monitoring the composting process: A review. *Agronomy, 13*, 2245.

Nakachew, K., Yigermal, H., Assefa, F., Gelaye, Y., & Ali, S. (2024). Review on enhancing the efficiency of fertilizer utilization: Strategies for optimal nutrient management. *Open Agriculture*, 9.

Peña, H., Mendoza, H., Diánez, F., & Santos, M. (2020). Parameter selection for the evaluation of compost Quality. *Agronomy, 10*, 1567.

Pradnya, I. N., Hasanah, U., Asri, S., Imani, N. A. C., Auralita, K. P., & Enjelita, A. (2024). Forecasting agricultural waste composting maturity through machine learning: A mini review. *IOP Conference Series Earth Environment Science, 1381*, Article 012037.

Qazi, S., Khawaja, B. A., & Farooq, Q. U. (2022). IoT-equipped and AI-enabled next generation smart agriculture: A critical review, current challenges and future trends. *IEEE Access, 10*, 21219–21235.

Saiz-Rubio, V., & Rovira-Más, F. (2020). From smart farming towards agriculture 5.0: A review on crop data management. *Agronomy, 10*, 207.

Sathiyapriya, S., Prabhaharan, J., Sheeba, S., Anandham, R., & Ilamaran, M. (2024). Nutrient recycling through composting: Harnessing agricultural wastes for sustainable crop production. *Plant Science Today*, 11.

Shah, F., & Wu, W. (2019). Soil and crop management strategies to ensure higher crop productivity within sustainable environments. *Sustainability, 11*, 1485.

Shaikh, F. K., Memon, M. A., Mahoto, N. A., Zeadally, S., & Nebhen, J. (2022). Artificial intelligence best practices in smart agriculture. *IEEE Micro, 42*, 17–24.

Sharma, K., & Shivandu, S. K. (2024). Integrating artificial intelligence and Internet of Things (IoT) for enhanced crop monitoring and management in precision agriculture. *Sensors International, 5*, Article 100292.

Sheoran, H. S., Kakar, R., Kumar, N., & Seema. (2019). Impact of organic and conventional farming practices on soil quality: A global review. *Applied Ecology and Environmental Research, 17*, 951–968.

Taheri, M., Bigdeli, M., Imanian, H., & Mohammadian, A. (2025). An overview of machine-learning methods for soil moisture estimation. *Water (Basel), 17*, 1638.

Tilak, K. V., Buchade, P.B., & Shaligram, Dr. A.D. (2021). Building an empirical model for wet waste composting process by in situ measurement of moisture and temperature, p. 040005.

Tomicic, I. (2022). IoT-based agricultural compost monitoring system: Prototype development and sensor technology evaluation. *Compost Science Utilization, 30*, 1–14.

Tsui, T.-H., van Loosdrecht, M. C. M., Dai, Y., & Tong, Y. W. (2023). Machine learning and circular bioeconomy: Building new resource efficiency from diverse waste streams. *Bioresource Technology, 369*, Article 128445.

Waltersmann, L., Kiemel, S., Stuhlsatz, J., Sauer, A., & Miehe, R. (2021). Artificial intelligence applications for increasing resource efficiency in manufacturing companies—A comprehensive review. *Sustainability, 13*, 6689.

Wan, X., Li, J., Xie, L., Wei, Z., Wu, J., Wah Tong, Y., Wang, X., He, Y., & Zhang, J. (2022). Machine learning framework for intelligent prediction of compost maturity towards automation of food waste composting system. *Bioresour Technol, 365*, Article 128107.

Wang, Z., Li, Z., Gao, C., Jiang, Z., Huang, S., Li, X., & Yang, H. (2025). *Bacillus Subtilis* as an excellent microbial treatment agent for environmental pollution: A review. *Biotechnology Journal*, 20.

Ward, A. J., Hobbs, P. J., Holliman, P. J., & Jones, D. L. (2008). Optimisation of the anaerobic digestion of agricultural resources. *Bioresource Technology, 99*, 7928–7940.

Zaharuddin, Audiah, S., Putri Ayu Sanjaya, Y., Pertiwi Daeli, O., & Johnson, M. (2024). Transforming energy and resource management with AI: From theory to sustainable practice. *International Transactions on Artificial Intelligence (ITALIC), 2*, 158–163.

Zhang, P., Guo, Z., Ullah, S., Melagraki, G., Afantitis, A., & Lynch, I. (2021). Nanotechnology and artificial intelligence to enable sustainable and precision agriculture. *Natural Plants, 7*, 864–876.

Zhang, X., Kong, Y., Yang, Y., Liu, Y., Gao, Q., Li, J., Li, G., & Yuan, J. (2025). Using tree-based machine learning models to predict diverse compost maturity via one-hot encoding: Model deployment, experimental validation, and practical application. *Waste Management, 205*, Article 114981.

Zul Azlan, Z. H., Junaini, S. N., Bolhassan, N. A., Wahi, R., & Arip, M. A. (2024). Harvesting a sustainable future: An overview of smart agriculture's role in social, economic, and environmental sustainability. *Journal of Clean Product, 434*, Article 140338.

Sustainable Production of Compost and Liquid Fertilizers: From Waste to Resource

Intan Suhada Azmi and Mohd Jumain Jalil

Abstract The increasing generation of organic waste from agricultural and urban sources has prompted the need for sustainable waste management solutions. Composting and liquid fertilizer production offer viable methods for converting organic waste into nutrient-rich soil amendments, reducing environmental pollution while enhancing soil fertility. Unlike synthetic fertilizers, which can degrade soil health over time, compost and liquid fertilizers promote biological activity, improve nutrient retention, and support long-term soil sustainability. This chapter explores the key principles behind the transformation of organic waste into valuable agricultural inputs, focusing on microbial interactions, decomposition processes, and nutrient stabilization. By understanding the factors influencing compost and liquid fertilizer production, researchers and practitioners can optimize these processes to maximize their benefits for both agricultural productivity and environmental conservation.

Keywords Organic waste management · Composting · Liquid fertilizer production · Microbial decomposition · Sustainable agriculture

1 Introduction

The increasing accumulation of organic waste from agriculture, food processing, and households presents both an environmental challenge and an opportunity (Chew et al., 2019). When left unmanaged, organic waste contributes to greenhouse gas emissions, soil degradation, and water pollution. However, with proper treatment, these materials can be converted into nutrient-rich compost and liquid fertilizers that support soil health and plant growth. Shifting from waste disposal to resource recovery aligns with the principles of sustainability, reducing reliance on chemical fertilizers while promoting a circular approach to nutrient management (Gontard et al., 2018).

I. S. Azmi (✉) · M. J. Jalil
Faculty of Chemical Engineering, Universiti Teknologi MARA, Masai, Johor, Malaysia
e-mail: intansuhada@uitm.edu.my

A. Z. Yaser et al. (eds.), *Compost and Liquid Organic Fertilizers*,
https://doi.org/10.1007/978-981-95-8561-8_9

Composting and liquid fertilizer production offer practical and effective ways to enhance soil fertility using locally available organic materials. Composting accelerates the natural decomposition of organic matter, resulting in a stable product that improves soil structure and microbial diversity (Benyahya et al., 2022). Meanwhile, liquid fertilizers, derived from compost extracts or anaerobic fermentation, provide plants with readily available nutrients in a form that can be quickly absorbed. These organic fertilizers not only replenish essential nutrients but also contribute to long-term soil health by enhancing biological activity and moisture retention (Mengqi et al., 2023).

Despite their advantages, producing high-quality compost and liquid fertilizers requires a well-managed process. Factors such as raw material composition, microbial activity, temperature control, and moisture levels significantly affect the final product (Singh et al., 2024). Additionally, balancing nutrient content and maintaining microbial stability in liquid fertilizers can be challenging. Understanding these aspects is crucial for optimizing production efficiency and ensuring that these bio-based fertilizers meet the nutritional needs of crops (Mamatha et al., 2024). A deeper understanding of the biological and chemical processes involved in composting and liquid fertilizer production is essential for improving their efficiency and reliability. By optimizing factors such as carbon-to-nitrogen ratio, aeration, fermentation conditions, and microbial inoculation, practitioners can enhance nutrient availability and product stability (Pajura et al., 2023). Sustainable management of organic waste through these methods not only supports regenerative agricultural practices but also reduces dependence on synthetic fertilizers, making farming systems more resilient and environmentally friendly.

Recent works by Merenda et al. (2025) and Murshid et al. (2024) have explored advanced treatment methods for transforming liquid digestates and leachates into valuable fertilizers. Merenda et al. valorised liquid anaerobic digestate using a membrane bioreactor (MBR) to produce nutrient-rich liquid fertilizers, emphasizing the role of microbial communities and process control in optimizing nitrification and nutrient recovery efficiency. Meanwhile, Murshid et al. investigated the potential of converting municipal solid waste leachate into organic liquid fertilizer via acid hydrolysis and fermentation processes, highlighting its agronomic potential through physicochemical and field evaluations. In contrast, the present review offers a broader synthesis of compost and liquid fertilizer production from various organic wastes, incorporating multiple processing routes, current challenges, regulatory perspectives, and practical solutions. This chapter aims to unify the fragmented literature into a single resource and highlight current gaps to stimulate future research, positioning itself as a comprehensive and integrative reference for sustainable organic waste valorisation.

2 Organic Waste as a Resource for Compost and Liquid Fertilizer Production

The sustainable production of compost and liquid fertilizers begins with the proper selection and management of organic waste. Organic waste includes a diverse range of biodegradable materials, such as food scraps, crop residues, manure, and plant-based industrial by-products (Hiranmai et al., 2023). These materials contain essential nutrients that, when properly decomposed, can be recycled into valuable soil amendments.

2.1 Composition and Nutrient Value of Organic Waste

The quality and effectiveness of compost and liquid fertilizers are largely determined by the composition of the raw organic waste. The nutrient content, carbon-to-nitrogen (C:N) ratio, and moisture level of the feedstock influence the microbial activity and rate of decomposition (Dimkpa et al., 2023). High-nitrogen materials, such as manure and vegetable waste, decompose rapidly and provide essential macronutrients like nitrogen (N), phosphorus (P), and potassium (K), while carbon-rich materials, such as straw, wood chips, and dry leaves, contribute to the structural integrity and aeration of the composting process (Omokaro et al., 2024). The different organic materials and their nutrient contributions was shown in Table 1.

Maintaining an optimal C:N ratio (typically 25:1 to 30:1 for composting) ensures a balanced decomposition process, allowing microbes to efficiently break down

Table 1 Different organic materials and their nutrient contributions

Organic material	Carbon-to-nitrogen ratio (C:N)	Key nutrients	Additional benefits
Food waste	15:1 to 20:1	High in nitrogen, trace minerals	Requires balancing with carbon sources to avoid excessive moisture and odours
Animal manure	10:1 to 30:1 (varies by type)	Rich in nitrogen, phosphorus, and beneficial microbes	Needs proper composting to reduce pathogens and excess ammonia
Crop residues	40:1 to 80:1	High in carbon, moderate nitrogen	Enhances aeration and moisture retention in compost
Industrial by-products	10:1 to 50:1	Varies depending on source	High in organic matter, beneficial for microbial fermentation in liquid fertilizers

organic matter without excessive nitrogen loss or slow degradation (Hicks et al., 2021; Varma et al., 2024).

2.2 Pre-treatment and Waste Segregation

Efficient pre-treatment and waste segregation are essential for improving the quality of compost and liquid fertilizers (Gunina & Kuzyakov, 2022). Before organic waste can be transformed into a nutrient-rich soil amendment, it must be properly sorted to remove contaminants that can interfere with microbial activity or degrade the final product. Non-biodegradable materials such as plastics, glass, and chemically treated waste must be excluded to prevent contamination and ensure the safety of the end product (Gerke, 2022).

Once waste is sorted, size reduction plays a critical role in facilitating microbial access to organic matter. Large and fibrous materials, such as crop residues, wood chips, and fruit peels, need to be shredded or chopped into smaller particles to enhance surface area exposure. This step accelerates decomposition by improving microbial colonization and promoting uniform breakdown of organic material. For liquid fertilizer production, finely ground waste allows for better nutrient extraction during fermentation, leading to a more consistent and bioavailable final product (Allouzi et al., 2022).

Moisture regulation is another key aspect of pre-treatment. Composting requires a moisture content of approximately 50–60% to support microbial activity and prevent drying or anaerobic conditions (Waqas et al., 2023). Excessive moisture can lead to compaction and oxygen depletion, causing anaerobic fermentation that produces undesirable odors and methane emissions. In contrast, liquid fertilizer production involves controlled dilution, ensuring that organic matter remains submerged in a fermenting solution to facilitate microbial digestion and nutrient extraction. Proper moisture management optimizes microbial performance and enhances the stability of the final compost or liquid fertilizer.

The segregation and pre-treatment of waste also involve the introduction of microbial inoculants to accelerate decomposition. Beneficial microbes such as Bacillus, Lactobacillus, and nitrogen-fixing bacteria can be added during pre-treatment to jumpstart the breakdown of organic matter. In composting, these microbes enhance the degradation of complex carbohydrates, proteins, and fats, while in liquid fertilizer production, they promote fermentation and enzymatic breakdown, increasing nutrient solubility (Sánchez et al., 2017). By integrating microbial inoculation during pre-treatment, the efficiency of composting and fermentation can be significantly improved, reducing processing time and enhancing the final product's nutrient profile (Noor et al., 2024).

3 Compost and Liquid Fertilizer Production Processes

The production of compost and liquid fertilizers involves biological transformations that depend on microbial activity, nutrient composition, and environmental conditions. Composting is an aerobic process that requires oxygen to support microbial breakdown of organic matter, while liquid fertilizer production often relies on anaerobic or facultative microbial activity to extract nutrients into a soluble form (Ayilara et al., 2020). The methods used in both processes determine the nutrient composition, stability, and effectiveness of the final product.

Composting begins with the decomposition of organic material by mesophilic bacteria and fungi, which consume simple organic compounds and generate heat (Nemet et al., 2021). As the temperature increases, thermophilic microorganisms become dominant, breaking down complex molecules such as cellulose, proteins, and fats. This phase is critical as it eliminates pathogens and weed seeds, ensuring that the final compost is safe for agricultural use (Arbab et al., 2022). The process gradually transitions into a curing phase, during which microbial activity declines, and the compost stabilizes into humus-like material. Well-managed composting ensures that essential nutrients remain available while avoiding nutrient loss through leaching or volatilization.

In contrast, liquid fertilizer production relies on controlled fermentation to extract soluble nutrients from organic waste (Unnisa, 2015). Microbial activity plays a central role in this process, breaking down organic compounds into plant-available forms. The fermentation environment is carefully regulated to maintain an appropriate pH, prevent spoilage, and promote beneficial microbial growth. Different methods, such as aerated compost tea, fermented plant extracts, and fish hydrolysate production, each yield unique nutrient compositions suited for various agricultural applications (Taneja et al., 2023). Proper fermentation conditions ensure that liquid fertilizers remain stable while preserving bioactive compounds essential for plant growth. The overall transformation pathways of organic waste into compost and liquid fertilizer are summarized in Fig. 1, which illustrates the key stages involved in both aerobic and anaerobic processes leading to soil application.

The efficiency and effectiveness of compost and liquid fertilizers can be further optimized through improvements in raw material selection, microbial inoculation, and application techniques. The use of microbial inoculants, such as nitrogen-fixing bacteria and mycorrhizal fungi, can accelerate decomposition and enhance nutrient availability (Wei et al., 2024). Additionally, optimizing moisture levels and aeration during composting prevents nutrient losses, while adjusting pH and fermentation time in liquid fertilizer production ensures a more stable and effective end product.

Field applications of compost and liquid fertilizers have demonstrated significant improvements in soil structure, water retention, and crop yield. Compost serves as a long-term soil amendment that enhances soil fertility over multiple growing seasons, whereas liquid fertilizers provide immediate nutrient availability, making them ideal for fast-growing crops (Aasfar et al., 2021). By integrating both into an agricultural

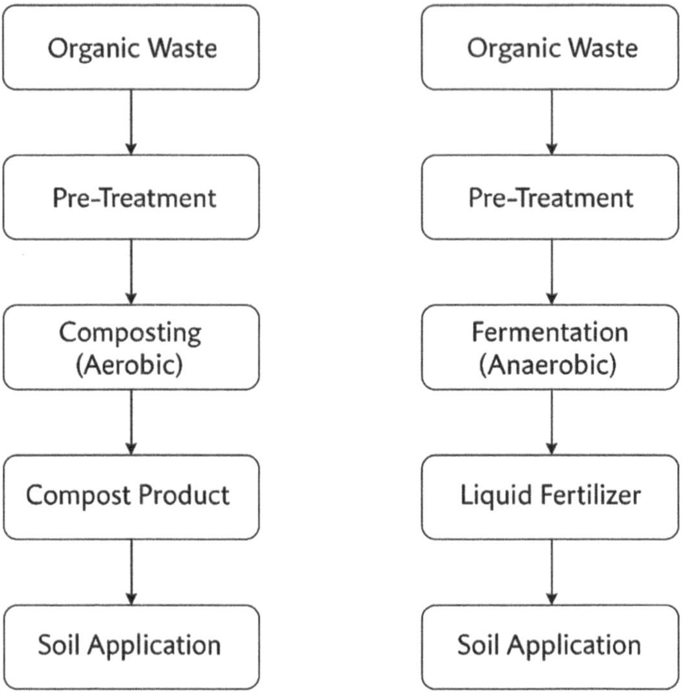

Fig. 1 Schematic flow of composting and liquid fertilizer production processes

system, farmers can achieve a balanced nutrient management approach that reduces dependency on synthetic fertilizers while promoting sustainable soil health.

4 Challenges in Compost and Liquid Fertilizer Production

Despite the numerous benefits associated with compost and liquid fertilizer production, several critical challenges continue to hinder their widespread adoption and successful implementation at larger scales. These challenges arise from both technical and non-technical factors, including variability in the quality and composition of raw organic materials, risks of contamination with non-biodegradable or hazardous substances, and inconsistent microbial activity that may affect product stability and nutrient content. In addition, issues related to waste collection, transportation, and limited processing infrastructure in certain regions can reduce overall production efficiency. Economic feasibility is also a concern, especially when organic fertilizers are compared with synthetic alternatives in terms of cost and ease of use. Furthermore, regulatory barriers such as complex certification procedures and the absence of standardized guidelines can restrict market accessibility for producers. Table 2 summarizes the major challenges along with practical strategies and approaches that can be applied to overcome them and improve the efficiency and scalability of compost and liquid fertilizer production.

One of the biggest challenges in compost and liquid fertilizer production is the inconsistency in the composition of raw materials (Sharma et al., 2024). Organic waste is highly variable in terms of nutrient content, moisture levels, and degradability, which can lead to fluctuations in product quality. Differences in feedstock types, such as food waste, crop residues, and animal manure, influence the decomposition rate and the final nutrient composition (Bremaghani, 2024). Managing these inconsistencies requires careful feedstock selection, monitoring of carbon-to-nitrogen (C:N) ratios, and proper pre-treatment methods to achieve a stable and balanced product.

Improper waste segregation poses a significant challenge, as non-biodegradable contaminants such as plastics, metals, and chemical residues can reduce the quality and safety of compost and liquid fertilizers. Contaminated feedstocks may introduce harmful substances such as heavy metals, pesticides, and pathogens, posing risks to soil health and crop safety (Bhatia & Singh, 2024). To mitigate these risks, effective waste sorting systems, public awareness campaigns, and stricter policies on organic waste collection must be implemented to ensure high-quality raw materials for composting and fermentation.

The efficiency of composting and liquid fertilizer production depends on microbial activity, which can be affected by environmental factors such as temperature, pH, and oxygen levels. Inconsistent microbial activity can lead to incomplete decomposition, nutrient imbalances, or the production of harmful by-products (Zielińska & Bułkowska, 2024). For instance, improper aeration in composting can result in anaerobic conditions, leading to methane emissions and foul odours. Similarly, liquid fertilizer fermentation can be disrupted by unwanted microbial competition, leading to spoilage. Maintaining optimal environmental conditions, such as

Table 2 Key challenges and solutions in compost and liquid fertilizer production

Challenges	Proposed solutions
Inconsistent raw material quality	Implement feedstock selection and blending strategies
Contamination with plastics/metals	Improve waste segregation and pre-treatment processes
Microbial instability	Use microbial inoculants and maintain optimal environmental conditions
Odor and methane emission	Ensure proper aeration and moisture balance
Lack of infrastructure/logistics	Develop decentralized composting units and transport systems
High production costs	Provide government incentives and promote awareness of long-term benefits
Regulatory hurdles	Standardize certification and streamline compliance procedures

controlled aeration, pH regulation, and microbial inoculation, is essential to ensure efficient decomposition and nutrient retention.

Large-scale composting and liquid fertilizer production require appropriate infrastructure for waste collection, processing, and storage. Many regions lack dedicated facilities for organic waste recycling, making it difficult to implement sustainable composting and fermentation systems. The transportation of organic waste from collection points to processing sites also poses logistical challenges, as improper handling can lead to leakage, odour issues, and nutrient losses (Fang et al., 2023). Investing in decentralized composting units, local processing centres, and improved transportation systems can help overcome these barriers and make organic waste recycling more accessible.

The cost of compost and liquid fertilizer production can be a limiting factor for widespread adoption. While organic fertilizers offer long-term benefits for soil health, their production requires labour, infrastructure, and monitoring, which can increase operational costs. Additionally, in many regions, farmers may still prefer synthetic fertilizers due to their immediate effects and ease of application. Encouraging market acceptance through subsidies, incentives for organic farming, and awareness campaigns about the long-term benefits of organic fertilizers can help address these economic challenges.

The legal framework surrounding compost and liquid fertilizer production varies across regions, with some countries imposing strict regulations on organic waste management, pathogen control, and fertilizer application (Arias et al., 2022). Obtaining certification for organic fertilizers can be a complex and time-consuming process, limiting market access for producers. Establishing clear and supportive policies, including quality control standards and incentives for organic waste recycling, is crucial for promoting compost and liquid fertilizer adoption on a larger scale.

5 Conclusion

The sustainable production of compost and liquid fertilizers represents a vital strategy for closing nutrient loops in agricultural systems. By transforming organic waste into valuable soil amendments, these practices contribute to improved soil health, reduced environmental impact, and long-term sustainability. The integration of optimized composting and fermentation processes ensures high-quality organic fertilizers that enhance nutrient efficiency and crop productivity. However, addressing challenges such as raw material variability, contamination risks, microbial inconsistencies, logistical constraints, and economic feasibility is necessary to facilitate broader adoption. With continued advancements in microbial technology, waste management strategies, and supportive policies, compost and liquid fertilizers will play a key role in promoting sustainable agriculture and environmental resilience worldwide.

References

Aasfar, A., Bargaz, A., Yaakoubi, K., Hilali, A., Bennis, I., Zeroual, Y., & Meftah Kadmiri, I. (2021). Nitrogen fixing Azotobacter species as potential soil biological enhancers for crop nutrition and yield stability. *Frontiers in Microbiology, 12*(February), 1–19. https://doi.org/10.3389/fmicb.2021.628379

Allouzi, M. M. A., Allouzi, S. M. A., Keng, Z. X., Supramaniam, C. V., Singh, A., & Chong, S. (2022). Liquid biofertilizers as a sustainable solution for agriculture. *Heliyon, 8*(12), e12609. https://doi.org/10.1016/j.heliyon.2022.e12609

Arbab, S., Ullah, H., Khan, M. I., Khattak, M. N., Zhang, J., Li, K., & Hassan, I. U. (2022). Diversity and distribution of thermophilic microorganisms and their applications in biotechnology. *Journal of Basic Microbiology, 62*(2), 95–108. https://doi.org/10.1002/jobm.202100529

Arias, O., Pulgar, J. A., & Soto, M. (2022). Application of organic wastes to soils and legislative intricacies in a circular economy context. *Clean Technologies and Environmental Policy, 24*(6), 1871–1888. https://doi.org/10.1007/s10098-022-02293-2

Ayilara, M. S., Olanrewaju, O. S., Babalola, O. O., & Odeyemi, O. (2020). Waste management through composting: Challenges and potentials. *Sustainability (Switzerland), 12*(11), 1–23. https://doi.org/10.3390/su12114456

Benyahya, Y., Fail, A., Alali, A., & Sadik, M. (2022). Recovery of household waste by generation of biogas as energy and compost as bio-fertilizer—A review. *Processes, 10*(1). https://doi.org/10.3390/pr10010081

Bhatia, T., & Singh, S. (2024). *Sustainable management of organic agricultural wastes: Contributions in nutrients availability, pollution mitigation and crop production.* Springer International Publishing. https://doi.org/10.1007/s44279-024-00147-7

Bremaghani, A. (2024). Utilization of organic waste in compost fertilizer production: Implications for sustainable agriculture and nutrient management. *Law and Economics, 18*(02), 86–98. https://journals.ristek.or.id/index.php/LE/index

Chew, K. W., Chia, S. R., Yen, H. W., Nomanbhay, S., Ho, Y. C., & Show, P. L. (2019). Transformation of biomass waste into sustainable organic fertilizers. *Sustainability (Switzerland), 11*(8). https://doi.org/10.3390/su11082266

Dimkpa, C., Adzawla, W., Pandey, R., Atakora, W. K., Kouame, A. K., Jemo, M., & Bindraban, P. S. (2023). Fertilizers for food and nutrition security in sub-Saharan Africa: An overview of soil health implications. *Frontiers in Soil Science, 3*(March), 1–18. https://doi.org/10.3389/fsoil.2023.1123931

Fang, B., Yu, J., Chen, Z., Osman, A. I., Farghali, M., Ihara, I., Hamza, E. H., Rooney, D. W., & Yap, P. S. (2023). *Artificial intelligence for waste management in smart cities: A review.* Springer International Publishing. https://doi.org/10.1007/s10311-023-01604-3

Gerke, J. (2022). The central role of soil organic matter in soil fertility and carbon storage. *Soil Systems, 6*(2). https://doi.org/10.3390/soilsystems6020033

Gontard, N., Sonesson, U., Birkved, M., Majone, M., Bolzonella, D., Celli, A., Angellier-Coussy, H., Jang, G. W., Verniquet, A., Broeze, J., & Schaer, B. (2018). A research challenge vision regarding management of agricultural waste in a circular bio-based economy. *Critical Reviews in Environmental Science and Technology, 48*(6), 614–654. https://doi.org/10.1080/10643389.2018.1471957

Gunina, A., & Kuzyakov, Y. (2022). From energy to (soil organic) matter. *Global Change Biology, 28*(7), 2169–2182. https://doi.org/10.1111/gcb.16071

Hicks, L. C., Lajtha, K., & Rousk, J. (2021). Nutrient limitation may induce microbial mining for resources from persistent soil organic matter. *Ecology, 102*(6), 1–16. https://doi.org/10.1002/ecy.3328

Hiranmai, R. Y., Neeraj, A., & Vats, P. (2023). Improvement of soil health and crop production throughutilization of organic wastes: A sustainable approach. *International Journal of Recycling Organic Waste in Agriculture, 13*(1), 132401–132402. https://doi.org/10.57647/j.ijrowa.2024.1301.01

Mamatha, B., Mudigiri, C., Ramesh, G., Saidulu, P., Meenakshi, N., & Prasanna, C. L. (2024). Enhancing soil health and fertility management for sustainable agriculture: A review. *Asian Journal of Soil Science and Plant Nutrition, 10*(3), 182–190. https://doi.org/10.9734/ajsspn/2024/v10i3330

Mengqi, Z., Shi, A., Ajmal, M., Ye, L., & Awais, M. (2023). Comprehensive review on agricultural waste utilization and high-temperature fermentation and composting. *Biomass Conversion and Biorefinery, 13*(7), 5445–5468. https://doi.org/10.1007/s13399-021-01438-5

Merenda, A., Shafaghat, A. H., Sohn, W., Seccombe, D., Phuntsho, S., & Shon, H. K. (2025). Valorisation of liquid anaerobic digestate into liquid fertilisers via membrane bioreactors: A proof-of-concept study. *Water Research*, 124026. https://doi.org/10.1016/j.watres.2025.124026

Murshid, N., Lamaming, J., Saalah, S., Rajin, M., & Yaser, A. Z. (2024). A review of vegetable waste bio-processing techniques in rural areas. *International Journal of Recycling Organic Waste in Agriculture, 13*(2), 132413–132414. https://doi.org/10.57647/j.ijrowa.2024.1302.13

Nemet, F., Perić, K., & Lončarić, Z. (2021). Microbiological activities in the composting process: A review. *Columella: Journal of Agricultural and Environmental Sciences, 8*(2), 41–53. https://doi.org/10.18380/szie.colum.2021.8.2.41

Noor, R. S., Shah, A. N., Tahir, M. B., Umair, M., Nawaz, M., Ali, A., Ercisli, S., Abdelsalam, N. R., Ali, H. M., Yang, S. H., & Ullah, S. (2024). Recent trends and advances in additive-mediated composting technology for agricultural waste resources: A comprehensive review. *ACS Omega, 9*(8), 8632–8653. https://doi.org/10.1021/acsomega.3c06516

Omokaro, G. O., Osarhiemen, I. O., Idama, V., Airueghian, E. O., West, S. T., Igbigbi, F. E., Nnake, D. C., Obolokor, E., Ahmed, A., & Omoshie, V. O. (2024). The role of organic amendments and their impact on soil restoration: A review. *Asian Journal of Environment & Ecology, 23*(11), 41–52. https://doi.org/10.9734/ajee/2024/v23i11620

Pajura, R., Masłoń, A., & Czarnota, J. (2023). The use of waste to produce liquid fertilizers in terms of sustainable development and energy consumption in the fertilizer industry—A case study from Poland. *Energies, 16*(4). https://doi.org/10.3390/en16041747

Sánchez, Ó. J., Ospina, D. A., & Montoya, S. (2017). Compost supplementation with nutrients and microorganisms in composting process. *Waste Management, 69*(26), 136–153. https://doi.org/10.1016/j.wasman.2017.08.012

Sharma, A., Soni, R., & Soni, S. K. (2024). From waste to wealth: Exploring modern composting innovations and compost valorization. *Journal of Material Cycles and Waste Management, 26*(1), 20–48. https://doi.org/10.1007/s10163-023-01839-w

Singh, N. K., Sachan, K., Bp, M., Panotra, N., & Katiyar, E. (2024). Building soil health and fertility through organic amendments and practices: A review. *Asian Journal of Soil Science and Plant Nutrition, 10*(1), 175–197. https://doi.org/10.9734/ajsspn/2024/v10i1224

Taneja, A., Sharma, R., Khetrapal, S., Sharma, A., Nagraik, R., Venkidasamy, B., Ghate, M. N., Azizov, S., Sharma, S., & Kumar, D. (2023). Value addition employing waste bio-materials in environmental remedies and food sector. *Metabolites, 13*(5). https://doi.org/10.3390/metabo13050624

Unnisa, S. A. (2015). Liquid fertilizer from food waste—A sustainable approach. *International Research Journal of Environmental Sciences, 4*(8), 22–25. www.isca.me

Varma, N., Wadatkar, H., Salve, R., & Kumar, T. V. (2024). Advancing sustainable agriculture: A comprehensive review of organic farming practices and environmental impact. *Journal of Experimental Agriculture International, 46*(7), 695–703. https://doi.org/10.9734/jeai/2024/v46i72623

Waqas, M., Hashim, S., Humphries, U. W., Ahmad, S., Noor, R., Shoaib, M., Naseem, A., Hlaing, P. T., & Lin, H. A. (2023). Composting processes for agricultural waste management: A comprehensive review. *Processes, 11*(3), 1–24. https://doi.org/10.3390/pr11030731

Wei, X., Xie, B., Wan, C., Song, R., Zhong, W., Xin, S., & Song, K. (2024). Enhancing soil health and plant growth through microbial fertilizers: Mechanisms, benefits, and sustainable agricultural practices. *Agronomy, 14*(3). https://doi.org/10.3390/agronomy14030609

Zielińska, M., & Bułkowska, K. (2024). Agricultural wastes and their by-products for the energy market. *Energies*, *17*(9). https://doi.org/10.3390/en17092099

Sustainable Liquid Fertilizers: A Review of Production from Food Waste and Fish Waste

Junidah Lamaming and Dayang Norafizan Awang Chee

Abstract The generation of food and fish waste globally poses significant environmental and economic challenges. With over 1 billion tonnes of food wasted annually and up to 50% of fish mass discarded during processing, transforming this organic waste into liquid fertilizers presents a promising solution aligned with circular economy principles. Transforming these wastes into sustainable liquid fertilizers offers a promising solution to reduce landfill burden and promote circular economy practices. This review consolidates knowledge on the nutrient composition of food and fish waste, the biological and chemical processes used to convert them into fertilizers, and the comparative advantages over synthetic fertilizers. It highlights the environmental benefits of reduced greenhouse gas emissions and improved soil health, along with economic incentives such as lower input costs and new market opportunities. The review also addresses existing barriers, such as contamination risks, regulatory hurdles, and logistical constraints and explores future directions for optimizing production technologies and standardizing quality. By integrating waste management with sustainable agriculture, food and fish waste-derived fertilizers contribute to global sustainability goals and resilient farming systems.

Keywords Food waste valorization · Fish waste liquid fertilizer · Circular economy · Sustainable agriculture · Nutrient recovery technologies

J. Lamaming (✉) · D. N. A. Chee (✉)
Faculty of Resource Science and Technology, Universiti Malaysia Sarawak, Kota Samarahan, Sarawak, Malaysia
e-mail: lajunidah@unimas.my

D. N. A. Chee
e-mail: dnorafizan@unimas.my

A. Z. Yaser et al. (eds.), *Compost and Liquid Organic Fertilizers*,
https://doi.org/10.1007/978-981-95-8561-8_10

1 Introduction

In 2022, an estimated 1.05 billion tonnes of food were wasted globally across the retail, food service, and household sectors (UNEP, 2024). About 19% of food available to consumers is wasted, with waste generated at various stages, including production, processing, retail, and consumption, as illustrated in Table 1. Food waste or loss is quantified solely for items intended for human consumption, excluding animal feed and inedible portions of products. Food losses or waste refer to the quantities of food that are lost or wasted within the segments of food chains that result in edible products intended for human consumption. Food intended for human consumption that inadvertently exits the human food chain is classified as food loss or waste, even if subsequently repurposed for non-food applications such as animal feed or bioenergy (Hoehn et al., 2023). Food waste poses distinct challenges in urban areas, where landfill pressure is often more pronounced and practices such as home composting and animal feed utilization of food scraps may be less common. In parallel, fish processing industries contribute significantly to organic waste, with nearly 30–50% of total fish mass (including heads, skins, bones, and viscera) discarded as byproducts (Bhatia et al., 2023; Kim & Mendis, 2006). These enormous waste quantities represent a critical environmental burden but also offer a valuable resource for sustainable use.

Fish waste, in particular, presents an underutilized opportunity because of its high content of essential nutrients, such as proteins, oils, and minerals (Arvanitoyannis & Kassaveti, 2008; Xia et al., 2024). Similarly, food waste is rich in organic matter suitable for conversion into fertilizers, making it a promising raw material for liquid fertilizer production. The continuous rise in global population and food demand further exacerbates waste generation, thus emphasizing the urgent need for sustainable waste management strategies (Xu et al., 2017; Yaser et al., 2022).

Improper disposal of organic waste contributes significantly to environmental degradation. Decomposing organic waste in landfills releases methane, a greenhouse gas 25 times more potent than carbon dioxide (IPCC, 2014). Moreover, leachate from improperly managed landfills contaminates groundwater, affecting ecosystems and human health. Economically, the global cost of food waste is estimated at $940 billion annually, including losses associated with resource wastage such as water, energy, and labor (EPA, 2025). Besides environmental costs, waste management imposes financial burdens on municipalities and governments. Transporting and landfilling

Table 1 Estimates of global food waste in 2022 (UNEP, 2024)

Sector	Global average (Kg/Capita/Year)	2022 Total (Million Tonnes)
Household	79	631
Food service	36	290
Retail	17	131
Total	132	1052

food and fish waste require significant resources, while landfill spaces are increasingly limited. In this context, transforming waste into valuable products like liquid fertilizers provides a dual solution—mitigating environmental impacts and recovering economic value, thus supporting the transition toward more sustainable and circular economies (European Commission, 2020).

Sustainable agricultural practices aim to maintain crop productivity without degrading the natural resource base. Integrating waste-derived liquid fertilizers into farming practices enhances soil health by increasing organic matter content and supporting beneficial microbial communities (Ahmed et al., 2023; Riddech et al., 2025). Unlike synthetic fertilizers that may lead to nutrient leaching and soil degradation, organic fertilizers provide slow-release nutrients, improving nutrient-use efficiency and soil structure. Moreover, utilizing liquid fertilizers derived from organic waste reduces the dependency on chemical fertilizers, whose production is energy-intensive and reliant on non-renewable resources such as natural gas and phosphate rock (Chojnacka et al., 2024; Tilman et al., 2002). Therefore, food and fish waste-derived fertilizers offer an environmentally friendly alternative that aligns with global goals for sustainable agriculture, such as those outlined in the United Nations' Sustainable Development Goals (SDG 12: Responsible Consumption and Production).

While previous reviews have discussed organic waste valorization or composting in general, few have specifically focused on the liquid fertilizer potential of food and fish waste, particularly in the context of their chemical composition, processing pathways, and agronomic relevance. This review stands out by providing a comparative overview of food and fish waste as dual feedstocks, an area often treated separately in existing literature. Furthermore, it uniquely combines environmental, economic, and safety considerations alongside technological insights, offering a holistic framework for evaluating the feasibility of waste-to-fertilizer systems. By identifying critical knowledge gaps and proposing future research directions, this review contributes meaningfully to the development of scalable, sustainable agricultural inputs from underutilized bioresources.

2 Composition of Food Waste and Fish Waste (Macro- and Micronutrients)

The composition of food waste varies significantly depending on the source, type of food, and region. For instance, fruit and vegetable wastes are typically rich in carbohydrates and moisture, whereas meat and dairy wastes are higher in proteins and fats (Xu et al., 2017). Such variability impacts the nutrient content and behavior of the waste during fertilizer production, necessitating customized approaches for different feedstocks. Similarly, fish waste composition is influenced by species, processing methods, and seasonal factors. Fish byproducts often contain high levels of nitrogen and phosphorus, crucial nutrients for plant growth (Arvanitoyannis &

Kassaveti, 2008; Zhang et al., 2023). However, variations in oil content and potential salt residues from marine fish can complicate the use of fish waste without proper pretreatment. Hence, comprehensive characterization of raw materials is essential to ensure consistent quality of the final fertilizer product.

The chemical composition of food waste varies significantly depending on its source (e.g., restaurant, hotel, institutional). Crude protein content ranges from 15 to 31.3%, fat content from 9.12 to 28.8%, and ash content from 3 to 14.75%. Mineral contents such as calcium, phosphorus, potassium, and magnesium also show variability as shown in Table 2. Fish waste, particularly fish waste meal, exhibits high crude protein content (49.22–57.92%), substantial fat content (7.16–19.10%), and significant ash content (21.79–30.16%). It also contains notable amounts of minerals like calcium (5.80%) and phosphorus (2.04%). These compositions highlight the potential of both food and fish waste as valuable resources for applications such as animal feed and fertilizer production.

Food and fish waste-derived fertilizers are rich in macronutrients, particularly nitrogen (N), phosphorus (P), and potassium (K), as well as micronutrients like calcium, magnesium, iron, and zinc (Jain et al., 2020; Khalil et al., 2019). Fish waste, in particular, provides high concentrations of nitrogen and phosphorus due to its protein and bone content. Similarly, vegetable and fruit waste contribute potassium and trace elements essential for balanced plant nutrition.

These organic fertilizers often have a broader spectrum of nutrients compared to synthetic fertilizers, which typically focus on NPK alone. The presence of amino acids, organic acids, and beneficial microbes enhances nutrient uptake efficiency in plants (Xu et al., 2017). Moreover, slow and steady nutrient release helps to maintain soil fertility over a longer period, reducing the risk of nutrient leaching and environmental pollution.

3 Comparative Analysis of Liquid Fertilizer from Fish and Food Waste Versus Conventional Fertilizer

Conventional fertilizers are effective at increasing crop yields, but they come with a range of environmental and economic challenges. These include high production costs, the potential for groundwater contamination, and disruption of soil pH balance. Such issues can ultimately reduce soil fertility and pose risks to human health due to the presence of toxic substances. Liquid fertilizers made from fish and food waste provide a sustainable option, following the principles of a circular economy by transforming organic byproducts into valuable agricultural inputs. The growing interest in organic agriculture for sustainability is driving efforts to develop and implement alternative methods that can replace synthetic inputs in soil and crop fertilization. The excessive use of inorganic fertilizers deteriorates soil quality, pollutes surface and groundwater, and contributes to increased greenhouse gas emissions (Chatzistathis et al., 2021). It is estimated that up to 50% of nitrogen and 90% of phosphorus

Table 2 Composition and bioactive compound of food waste and fishwaste

Component	Food waste		Fish waste	
	Sources	References	Sources	References
Crude protein (%)	15–23% (Restaurant waste)	Georganas et al., (2020), Myer et al., (2000)	56.4% (Fish waste meal)	Safitri et al., (2017)
	22.0% (Ground restaurant waste)	Georganas et al., (2020), Kwak and Kang (2006)	57.92 ± 5.26% (Fish waste)	Georganas et al., (2020), Kwak and Kang (2006)
	27.5% (Processed restaurant waste)	Garcia et al., (2005), Kandyliari et al., (2020)	49.22–57.92% (Fish by-products)	Dueñas et al., (2020), Kandyliari et al., (2020), Safitri et al., (2017)
	18–20% (Kitchen and plate waste)	Georganas et al., (2020), Myer et al., (2000)		
	31.3% (Leftover food)	Asar et al., (2018), Georganas et al., (2020)		
	25% (Dried food waste)	Chae et al., (2000), Georganas et al., (2020)		
	27.6% (Commercial and residential food waste)	Castrica et al., (2018), Dueñas et al., (2020)		
	18.9% (University dining hall food waste)	Fung et al., (2018), Georganas et al., (2020)		
	22% (Leftover food)	Cho et al., (2004), Georganas et al., (2020)		
Fat (Ether extract) (%)	17–24% (Restaurant waste)	Georganas et al., (2020), Myer et al., (2000)	13.1% (Fish waste meal)	Safitri et al., (2017)
	23.9% (Ground restaurant waste)	Georganas et al., (2020), Kwak and Kang (2006)	19.10 ± 6.06% (Fish waste)	Georganas et al., (2020), Kwak and Kang (2006)
	28.8% (Processed restaurant waste)	Garcia et al., (2005), Kandyliari et al., (2020)	7.16–19.10% (Fish by-products)	Safitri et al., (2017), Dueñas et al., (2020), Kandyliari et al., 2020)

(continued)

Table 2 (continued)

Component	Food waste		Fish waste	
	Sources	References	Sources	References
	24–26% (Kitchen and plate waste)	Garcia et al., (2005), Kandyliari et al., (2020)		
	26.08% (Leftover food)	Asar et al., (2018), Georganas et al., (2020)		
	17.3% (Dried food waste)	Chae et al., (2000), Georganas et al., (2020)		
	9.12% (Commercial and residential food waste)	Castrica et al., (2018), Dueñas et al., (2020)		
	13.58% (University dining hall food waste)	Fung et al., (2018), Georganas et al., (2020)		
	10.66% (Leftover food)	Cho et al., (2004), Georganas et al., (2020)		
Ash (%)	3–6% (Restaurant waste)	Georganas et al., (2020), Myer et al., (2000)		
	12.6% (Ground restaurant waste)	Georganas et al., (2020), Kwak and Kang (2006)		
	14.75% (Leftover food)	Asar et al., (2018), Georganas et al., (2020)		
	5–6% (Kitchen and plate waste)	Georganas et al., (2020), Myer et al., (2000)		
	14.56% (Commercial and residential food waste)	Castrica et al., (2018), Dueñas et al., (2020)		

(continued)

Table 2 (continued)

Component	Food waste		Fish waste	
	Sources	References	Sources	References
	5.01% (University dining hall food waste)	Fung et al., (2018), Georganas et al., (2020)		
Calcium (Ca, %)	0.42–1.7% (Processed restaurant waste)	Garcia et al., (2005), Kandyliari et al., (2020)	5.80 ± 1.35% (Fish waste)	Georganas et al., (2020), Kwak and Kang (2006)
	0.5–0.8% (Kitchen and plate waste)	Georganas et al., (2020), Myer et al., (2000)		
	1.09–1.25% (Commercial and residential food waste)	Castrica et al., (2018), Dueñas et al., (2020)		
	0.04–0.46% (University dining hall food waste)	Fung et al., (2018), Georganas et al., (2020)		
Phosphorus (P, %)	0.81–2.07% (Processed restaurant waste)	Garcia et al., (2005), Kandyliari et al., (2020)	2.04 ± 0.64% (Fish waste)	Georganas et al., (2020), Kwak and Kang (2006)
	0.3–0.8% (Kitchen and plate waste	Georganas et al., (2020), Myer et al., (2000)		
	0.16–0.30% (Commercial and residential food waste)	Castrica et al., (2018), Dueñas et al., (2020)		
	0.23–0.37% (University dining hall food waste)	Fung et al., (2018), Georganas et al., (2020)		
Potassium (K, %)	0.56–0.76% (Commercial and residential food waste)	Castrica et al., (2018), Dueñas et al., (2020)	0.68 ± 0.11% (Fish waste)	Georganas et al., (2020), Kwak and Kang (2006)

(continued)

Table 2 (continued)

Component	Food waste		Fish waste	
	Sources	References	Sources	References
Magnesium (Mg, %)	0.1–0.2% (Commercial and residential food waste)	Castrica et al., (2018), Dueñas et al., (2020)	0.17 ± 0.04% (Fish waste)	Georganas et al., (2020), Kwak and Kang (2006)
Crude fiber (%)	Not available		1.19 ± 1.21% (Fish waste)	Georganas et al., (2020), Kwak and Kang (2006)
Total digestible nutrients (TDN, %)	63.3–66.8% (Vegetable waste)	Castrica et al., (2018), Dueñas et al., (2020)		

from chemical fertilizers are lost from agricultural fields. This loss contributes to greenhouse gas emissions, water eutrophication, and soil salinization (Lin et al., 2020).

Liquid fertilizers made from fish waste, a byproduct of the fishing industry, offer a promising alternative, especially in organic farming. These biofertilizers are created through the microbial breakdown of organic matter, providing a dual benefit: they help reduce environmental pollution and lower the costs associated with mineral fertilizers (Andrade et al., 2023). Fish waste typically undergoes either an enzymatic or chemical hydrolysis process to release nutrients, resulting in a nutrient-rich liquid. This liquid can either be applied directly to crops or further processed into a more concentrated form. The composition of fish-based fertilizers can vary, but they generally contain essential macronutrients such as nitrogen, phosphorus, and potassium, along with micronutrients and amino acids. All of these components contribute to improved plant growth (Prayogo et al., 2021). The benefits of processing fish waste include streamlined production, reduced material costs, and lower energy consumption. Aquaponics serves as a prime example of this method by integrating aquaculture and hydroponics; it harnesses fish waste to nourish plants, creating a closed-loop system that effectively minimizes water usage and waste discharge (Nichols & Savidov, 2012). Furthermore, food waste presents a considerable environmental challenge due to its substantial volume and the potential for methane emissions in landfills. However, it can be transformed into a valuable fertilizer through composting or anaerobic digestion.

Composting food waste produces a nutrient-rich humus that enhances soil structure, improves water retention, and increases nutrient availability. On the other hand, anaerobic digestion generates biogas, a renewable energy source, and digestate, which is a nutrient-rich liquid fertilizer. To effectively implement technologies utilizing biological resources, it is necessary to establish small-scale waste solubilization or fertilizer production facilities at the site where waste is generated. This approach helps address issues related to waste transportation and sanitary hazards.

The composition of fertilizers derived from food waste can vary significantly based on the source and processing method. However, these fertilizers typically contain a balanced mix of macronutrients, micronutrients, and organic matter, all of which contribute to improved soil health. Anaerobic digestion has been used for many years to treat the odorous, pathogenic, and low-oxygen characteristics of both human and livestock wastewater (Kumar, 2013). The byproduct, known as digestate, is nutrient-rich and can be utilized as fertilizer, which helps reduce reliance on chemical fertilizers (Hegde & Trabold, 2019). Food waste is particularly suitable as a feedstock for anaerobic digestion due to its high volatile solids content and its ability to biodegrade rapidly (Kumar, 2013).

Liquid fertilizers derived from fish and food waste offer several advantages over conventional fertilizers. First, they are environmentally sustainable, as they reduce the reliance on synthetic inputs and minimize waste generation. Second, they enhance soil health by increasing organic matter content, improving water retention, and promoting beneficial microbial activity. Third, these fertilizers can be produced locally, which lowers transportation costs and supports local economies. Finally, the organic nature of these fertilizers may lead to improved crop quality and nutritional value compared to crops grown with conventional fertilizers. Anaerobic digestion is recognized as one of the most promising technologies for managing food waste. It offers a dual benefit: generating renewable energy in the form of biogas and producing digestate, a nutrient-rich fertilizer (Morales-Polo et al., 2018). This process not only contributes to energy production but also helps reduce pollution and greenhouse gas emissions. By integrating waste management practices with energy and fertilizer production, anaerobic digestion provides a sustainable pathway toward a circular economy, where resources are used efficiently and waste is minimized. It is important to consider the seasonal and geographical variations in food waste composition, which can impact the biogas potential and the nutrient content of the resulting fertilizer (Bong et al., 2017). Although anaerobic digestion of organic waste produces biogas, a renewable source of energy, the digestate, the solid residue of the process, is rich in nutrients such as nitrogen, phosphorus, and potassium, making it a valuable biofertilizer (Bong et al., 2017).

4 Overview of Production Techniques

Liquid fertilizers from food and fish waste can be produced through several biological and chemical methods. Anaerobic digestion involves the microbial breakdown of organic materials in the absence of oxygen, generating a nutrient-rich digestate that can be processed into liquid fertilizers (Appels et al., 2008). Anaerobic digestion (AD) is one of the most established technologies for converting organic waste into liquid fertilizers. It involves the microbial degradation of organic matter under anaerobic (oxygen-free) conditions, producing biogas and a nutrient-rich digestate as a byproduct. This digestate is often used as a liquid fertilizer due to its high concentrations of nitrogen (N), phosphorus (P), and potassium (K). AD is considered both scalable and cost-effective, especially in municipal and agricultural applications. However, maintaining optimal environmental conditions such as temperature, pH, and organic loading rate is critical to ensure stable microbial activity and high nutrient recovery (Khalid et al., 2011a, 2011b; Yaser et al., 2022). Poor operational control can lead to process instability and reduced fertilizer quality due to the accumulation of inhibitors like ammonia. Anaerobic digestion is generally more scalable and cost-effective for large volumes of waste, but it requires strict control of operational parameters such as temperature and pH to maintain microbial activity (Khalid et al., 2011a, 2011b; Yaser et al., 2022).

Fermentation offers another promising method for producing liquid fertilizers by utilizing specific strains of beneficial microbes to decompose organic waste. Unlike anaerobic digestion, fermentation typically focuses on enriching the fertilizer product with microbial communities that can promote plant health. For example, lactic acid fermentation of vegetable waste enhances nitrogen solubility and microbial availability, thereby improving nutrient uptake when applied to soil (Zhou et al., 2019a, 2019b). This process is especially useful for creating biofertilizers that not only supply nutrients but also improve soil microbial ecology. The efficiency of fermentation is influenced by factors such as substrate composition, moisture content, and microbial inoculum selection (Turan et al., 2023). Fermentation, especially lactic acid fermentation, offers better preservation of nutrients and the possibility to produce fertilizers with added probiotic properties, but may involve longer processing times. Microbial fermentation enhances nutrient solubilization and stabilizes organic matter, improving fertilizer quality (Murindangabo et al., 2023).

Enzymatic hydrolysis represents a more targeted technique where specific enzymes are introduced to break down complex organic molecules such as proteins, cellulose, or fats into simpler, bioavailable compounds, improving nutrient release and reducing odor. Although enzymatic hydrolysis can produce high-quality fertilizers with controlled nutrient profiles, it is often more expensive due to the cost of enzymes and processing requirements (Chen et al., 2006a, 2006b; Hopkins et al., 2025). Nevertheless, this method is advantageous in cases where precision and nutrient customization are desired. Importantly, microbial activity is central to all three methods, whether through endogenous populations in AD, selected strains in fermentation, or enzyme-producing microbes in hydrolysis, playing a vital role in converting organic waste into stable, effective, and environmentally friendly liquid fertilizers (Mao et al., 2015). Enzymatic hydrolysis allows precise nutrient targeting but remains less economical at large scales due to enzyme production costs.

Microbial activity is crucial in all processes, as microbes facilitate the breakdown of complex organic compounds into simpler, plant-available nutrients. For instance, proteolytic bacteria release amino acids, while phosphorus-solubilizing microorganisms enhance the bioavailability of phosphorus (Chen et al., 2006a, 2006b). Optimizing microbial consortia and operational conditions is vital to maximize nutrient extraction and ensure the stability and quality of the resulting liquid fertilizer. Each production method offers distinct advantages and limitations. Anaerobic digestion is widely used for its energy co-production (biogas), while fermentation can produce more specialized fertilizers rich in beneficial microbes. Enzymatic hydrolysis, though efficient in nutrient recovery, can be costly due to the need for specific enzymes. Therefore, selecting the appropriate technique depends on the nature of the waste material, intended application, cost considerations, and local infrastructure capabilities.

5 Environmental and Economic Impacts

Utilizing food and fish waste for fertilizer production significantly reduces the volume of organic waste directed to landfills, thus mitigating methane emissions from anaerobic decomposition (IPCC, 2014). Studies show that diverting organic waste streams to valorization processes like fertilizer production can cut greenhouse gas emissions by up to 50% compared to traditional landfill disposal (Möller & Müller, 2012). Moreover, recycling nutrients back into agricultural systems reduces the reliance on mineral fertilizers, whose production is energy-intensive and carbon-emitting (Tilman et al., 2002; Zhang et al., 2023). By closing nutrient loops, waste-derived fertilizers contribute to broader climate change mitigation strategies and align with sustainable waste management goals under the circular economy model (European Commission, 2020).

Liquid fertilizers from organic waste improve plant growth by providing readily available nutrients while simultaneously enhancing soil microbial activity. They contribute to better root development, increased biomass production, and higher crop yields compared to untreated controls (Abdullah et al., 2017). In addition, the organic matter content in these fertilizers promotes soil structure improvement, water retention, and erosion resistance. Compared to synthetic fertilizers, waste-derived organic fertilizers are less prone to causing nutrient imbalances or soil acidification (Diacono & Montemurro, 2010; Duan et al., 2023). Furthermore, the natural composition of these fertilizers fosters a healthy rhizosphere, enhancing symbiotic relationships such as mycorrhizal associations, which are vital for long-term soil productivity. Thus, the adoption of waste-based liquid fertilizers offers a pathway to more resilient and sustainable farming systems.

From an economic perspective, producing fertilizers from waste can lower operational costs for waste management while creating new revenue streams for waste producers and processors (Zhou et al., 2019a, 2019b). Organic fertilizers often have lower production costs than synthetic fertilizers when waste sources are locally available, reducing dependence on imported agrochemicals and stabilizing input costs for farmers. Additionally, integrating waste valorization into local economies fosters green jobs, boosts rural development, and enhances food security (FAO, 2013). Through value recovery from "waste," the agricultural sector can move towards a regenerative model where productivity, environmental stewardship, and economic viability are mutually reinforcing.

6 Challenges and Future Perspectives

Despite the rich nutrient content, using food and fish waste presents several challenges. One major concern is contamination, which may include pathogens, heavy metals, or chemical residues that can compromise the safety of the resulting fertilizers (Jain et al., 2020). Additionally, heterogeneous composition can lead to inconsistent nutrient profiles, affecting fertilizer efficacy.

Logistical challenges also arise in the collection, transportation, and storage of perishable organic waste. Without proper handling, waste can quickly decompose, leading to odor problems, pathogen proliferation, and nutrient loss. Therefore, implementing controlled collection systems and developing effective stabilization techniques are critical for the successful utilization of food and fish waste in fertilizer production (Möller & Müller, 2012).

Despite the potential benefits, regulatory challenges hinder the widespread adoption of waste-derived fertilizers. Stricter standards for pathogen and contaminant levels must be met to ensure product safety, necessitating consistent quality control measures (Chen et al., 2006a, 2006b). Regulatory uncertainty and lack of standardized protocols can discourage investment and innovation in this sector. Public perception also plays a role, as consumers may view waste-derived products with skepticism regarding safety and efficacy (Khalid et al., 2011a, 2011b). Education campaigns, transparent certification schemes, and demonstration projects are essential to build confidence among farmers and end-users about the benefits and safety of organic waste-based fertilizers.

Future developments should focus on improving production technologies, such as integrating enzymatic and microbial approaches to optimize nutrient recovery and reduce production costs. Advances in biorefinery systems could allow simultaneous production of fertilizers, biogas, and high-value biochemicals, enhancing the economic viability of waste valorization (Samoraj et al., 2024).

Further research is needed to close knowledge gaps regarding the long-term impacts of waste-derived fertilizers on different soil types and crops. Developing international standards for liquid fertilizer quality and safety will facilitate market expansion. Strategies to enhance nutrient use efficiency, such as adding controlled-release formulations and microbial inoculants, could further improve the sustainability and competitiveness of these products in the global fertilizer market.

7 Conclusion

Transforming food and fish waste into liquid fertilizers offers a viable pathway toward sustainable agriculture, environmental protection, and circular economy practices. These organic fertilizers deliver multiple advantages: they are rich in essential nutrients, improve soil health, reduce dependency on synthetic inputs, and lower greenhouse gas emissions. However, challenges such as contamination risks, variable composition, and regulatory uncertainty must be addressed to ensure safety and consistency. Advancing production technologies, establishing standardized quality frameworks, and enhancing public awareness are crucial for scaling adoption. With appropriate innovation, infrastructure, and policy support, waste-derived fertilizers can play a pivotal role in mitigating food and organic waste while supporting more sustainable and resilient agroecosystems.

References

Ahmed, T., Noman, M., Qi, Y., Shahid, M., Hussain, S., Masood, H. A., Xu, L., Ali, H. M., Negm, S., El-Kott, A. F., Yao, Y., Qi, X., & Li, B. (2023). Fertilization of microbial composts: A technology for improving stress resilience in plants. *Plants, 12*(20), 3550. https://doi.org/10.3390/plants12203550

Andrade, L. A. de, Santos, C. H. B., Frezarin, E. T., Sales, L. R., & Rigobelo, E. C. (2023). Plant growth-promoting Rhizobacteria for sustainable agricultural production [Review of plant growth-promoting Rhizobacteria for sustainable agricultural production]. *Microorganisms, 11*(4), 1088. Multidisciplinary Digital Publishing Institute. https://doi.org/10.3390/microorganisms11041088.

Appels, L., Baeyens, J., Degrève, J., & Dewil, R. (2008). Principles and potential of the anaerobic digestion of waste-activated sludge. *Progress in Energy and Combustion Science, 34*(6), 755–781. doi.org/https://doi.org/10.1016/j.pecs.2008.06.002

Arvanitoyannis, I. S., & Kassaveti, A. (2008). Fish industry waste: Treatments, environmental impacts, current and potential uses. *International Journal of Food Science & Technology, 43*(4), 726–745.

Asar, E., Raheem, H., & Daoud, J. (2018). Using dried leftover food as nontraditional feed for Muscovy duck diet. *Assiut. Vet. Med. Journal, 64*, 107–114.

Bhatia, L., Jha, H., Sarkar, T., & Sarangi, P. K. (2023). Food waste utilization for reducing carbon footprints towards sustainable and cleaner environment: A Review. *International Journal of Environmental Research and Public Health, 20*(3), 2318. https://doi.org/10.3390/ijerph20032318

Bong, C. P. C., Lim, L. Y., Lee, C. T., Klemeš, J. J., Ho, C. S., & Ho, W. S. (2017). The characterisation and treatment of food waste for improvement of biogas production during anaerobic digestion—A review [Review of The characterisation and treatment of food waste for improvement of biogas production during anaerobic digestion—A review]. *Journal of Cleaner Production, 172*, 1545. Elsevier BV. https://doi.org/10.1016/j.jclepro.2017.10.199

Castrica, M., Tedesco, E. A. D., Panseri, S., Ferrazzi, G., Ventura, V., Frisio, G. D., & Balzaretti, M. C. (2018). Pet food as the most concrete strategy for using food waste as feedstuff within the European context: A feasibility study. *Sustainability, 10*, 2035. https://doi.org/10.3390/su10062035

Chae, B. J., Choi, S. C., Kim, Y. G., Kim, C. H., & Sohn, K. S. (2000). Effects of feeding dried food waste on growth and nutrient digestibility in growing-finishing pigs. *Asian-Australasian Journal of Animal Sciences, 13*, 1304–1308. https://doi.org/10.5713/ajas.2000.1304

Chatzistathis, T., Kavvadias, V., Sotiropoulos, T., & Papadakis, I. E. (2021). Organic Fertilization and Tree Orchards. *Agriculture, 11*(8), 692. https://doi.org/10.3390/agriculture11080692

Chen, Y., Inbar, Y., & Hadar, Y. (2006a). Composted organic wastes as components of organic substrates for container-grown plants. *Horticultural Reviews, 20*, 63–112.

Chen, Y., Cheng, J. J., & Creamer, K. S. (2006). Inhibition of anaerobic digestion process: A review. *Bioresource Technology, 99*(10), 4044–4064. doi.org/https://doi.org/10.1016/j.biortech.2007.01.057

Cho, Y., Lee, G., Jang, J., Shin, I., Myung, K., Choi, K., Bae, I., & Yang, C. (2004). Effects of feeding dried leftover food on growth and body composition of broiler chicks. *Asian-Australasian Journal of Animal Sciences, 17*, 386–393. https://doi.org/10.5713/ajas.2004.386

Choe, J., Moyo, K. M., Park, K., Jeong, J., Kim, H., Ryu, Y., Kim, J., Kim, J. M., Lee, S., & Go, G. W. (2017). Meat quality traits of pigs finished on food waste. *Korean Journal of Food Science and Animals, 37*, 690–697. https://doi.org/10.5851/kosfa.2017.37.5.690

Chojnacka, K., & Moustakas, K. (2024). Anaerobic digestate management for carbon neutrality and fertilizer use: A review of current practices and future opportunities. *Biomass and Bioenergy, 180*, Article 106991. https://doi.org/10.1016/j.biombioe.2023.106991

Diacono, M., & Montemurro, F. (2010). Long-term effects of organic amendments on soil fertility: A review. *Agronomy for Sustainable Development, 30*(2), 401–422.

Duan, C. X., Li, J. B., Zhang, B. B., Wu, S. F., Fan, J. L., Feng, H., He, J. Q., & Siddique, K. H. M.(2023). Effect of bio-organic fertilizer derived from agricultural waste resources on soil properties and winter wheat (Triticum aestivum L.) yield in semi-humid drought-prone regions. *Agricultural Water Management, 289*, 108539, https://doi.org/10.1016/j.agwat.2023.108539

Dueñas, M., & García-Estévez, I. (2020). Agricultural and food waste: Analysis, characterization and extraction of bioactive compounds and their possible utilization. *Foods, 9*(6), 817. https://doi.org/10.3390/foods9060817

European Commission. (2020). Circular Economy Action Plan. https://ec.europa.eu/environment/circular-economy/

FAO. (2013). Food Wastage Footprint: Impacts on Natural Resources. Food and Agriculture Organization of the United Nations.

Fung, L., Urriola, P. E., Baker, L., & Shurson, G. C. (2018). Estimated energy and nutrient composition of different sources of food waste and their potential for use in sustainable swine feeding programs. *Translational Animal Science, 3*, 359–368. https://doi.org/10.1093/tas/txy099

Garcia, A. J., Esteban, M. B., Marquez, M. C., & Ramos, P. (2005). Biodegradable municipal solid waste: Characterization and potential use as animal feedstuffs. *Waste Management, 25*, 780–787. https://doi.org/10.1016/j.wasman.2005.01.006

Georganas, A., Giamouri, E., Pappas, A. C., Papadomichelakis, G., Galliou, F., Manios, T., Tsiplakou, E., Fegeros, K., & Zervas, G. (2020). Bioactive compounds in food waste: A review on the transformation of food waste to animal feed. *Foods (Basel, Switzerland), 9*(3), 291. https://doi.org/10.3390/foods9030291

Hegde, S., & Trabold, T. A. (2019). Anaerobic digestion of food waste with unconventional co-substrates for stable biogas production at high organic loading rates. *Sustainability, 11*(14), 3875. https://doi.org/10.3390/su11143875

Hoehn, D., Vázquez-Rowe, I., Kahhat, R., Margallo, M., Laso, J., Fernández-Ríos, A., Ruiz-Salmón, I., & Aldaco, R. (2023). A critical review on food loss and waste quantification approaches: Is there a need to develop alternatives beyond the currently widespread pathways? Resources. *Conservation and Recycling, 188*, Article 106671.

Hopkins, D. T., Berrué, F., Khiari, Z., Khiari, Z., & Hawboldt, K. A. (2025). Valorization of fisheries by-products via enzymatic protein hydrolysis: A review of operating conditions, process design, and future trends. *Process Biochemistry, 149*, 306–320. https://doi.org/10.1016/j.procbio.2024.12.024

IPCC. (2014). Climate Change 2014: Synthesis Report. Intergovernmental Panel on Climate Change.

Jain, S., Jain, S., Wolf, I. T., Lee, J., & Tong, Y. W. (2020). A comprehensive review on operating parameters and different pretreatment methodologies for anaerobic digestion of municipal solid waste. *Renewable and Sustainable Energy Reviews, 108*, Article 104429.

Kandyliari, A., Mallouchos, A., Papandroulakis, N., Golla, J. P., Lam, T. T., Sakellari, A., Karavoltsos, S., Vasiliou, V., & Kapsokefalou, M. (2020). Nutrient composition and fatty acid and protein profiles of selected fish by-products. *Foods, 9*(2), 190. https://doi.org/10.3390/foods9020190

Khalid, A., Arshad, M., Anjum, M., Mahmood, T., & Dawson, L. (2011a). The anaerobic digestion of solid organic waste. *Waste Management, 31*(8), 1737–1744.

Khalid, A., Arshad, M., Anjum, M., Mahmood, T., & Dawson, L. (2011b). The anaerobic digestion of solid organic waste. *Waste Management, 31*(8), 1737–1744. https://doi.org/10.1016/j.wasman.2011.03.021

Khalil, A., Thakur, R., & Sohail, M. (2019). Nutrient potential and microbial load of food waste compost. *Journal of Soil Science and Plant Nutrition, 19*(2), 347–357.

Kim, S. K., & Mendis, E. (2006). Bioactive compounds from marine processing byproducts—A review. *Food Research International, 39*(4), 383–393.

Kumar, K. (2013). Utilizations of food waste as an anaerobic digester feedstock. *Journal of Energy and Natural Resources, 2*(1), 1. https://doi.org/10.11648/j.jenr.20130201.11

Kwak, W. S., & Kang, J. (2006). Effect of feeding food waste-broiler litter and bakery by-product mixture to pigs. *Bioresource Technology, 97*, 243–249. https://doi.org/10.1016/j.biortech.2005.02.008

Lin, Y., Zhao, X., Bao, E., Li, J., Zou, Z., & Cao, K. (2020). Bio-organic fertilizer with reduced rates of chemical fertilization improves soil fertility and enhances tomato yield and quality. *Scientific Reports, 10*(1). https://doi.org/10.1038/s41598-019-56954-2

Mao, C., Feng, Y., Wang, X., & Ren, G. (2015). Review on research achievements of biogas from anaerobic digestion. *Renewable and Sustainable Energy Reviews, 45*, 540–555. https://doi.org/10.1016/j.rser.2015.02.032

Möller, K., & Müller, T. (2012). Effects of anaerobic digestion on digestate nutrient availability and crop growth: A review. *Engineering in Life Sciences, 12*(3), 242–257.

Morales-Polo, C., Castro, M. del M. C., & Soria, B. Y. M. (2018). reviewing the anaerobic digestion of food waste: from waste generation and anaerobic process to its perspectives. *Applied Sciences, 8*(10), 1804. https://doi.org/10.3390/app8101804

Murindangabo, Y. T., Kopecký, M., Perná, K., Nguyen, T. G., Konvalina, P., & Kavková, M. (2023). Prominent use of lactic acid bacteria in soil-plant systems. *Applied Soil Ecology, 189*, Article 104955. https://doi.org/10.1016/j.apsoil.2023.104955

Myer, R. O., Brendemuhl, J. H., & Johnson, D. D. (2000). Dehydrated Restaurant Food Waste as Swine Feed. In M. L. Westendorf (Ed.), *Food Waste to Animal Feed* (pp. 113–144). Iowa State University Press; Iowa City.

Nichols, M., & Savidov, N. (2012). Aquaponics: protein and vegetables for developing countries. *Acta Horticulturae, 958*, 189. https://doi.org/10.17660/actahortic.2012.958.22

Pandey, A., Soccol, C. R., Nigam, P., Brand, D., Mohan, R., & Roussos, S. (2000). Biotechnological potential of coffee pulp and coffee husk for bioprocesses. *Biochemical Engineering Journal, 6*(2), 153–162.

Prayogo, P., Agustono, A., Rahardja, B. S., & Amin, M. (2021). Growth performance and nutrient composition of mustard green (Brassica juncea) cultured in aquaponics systems and hydroponic system. *Journal of Aquaculture and Fish Health, 10*(3), 373. https://doi.org/10.20473/jafh.v10i3.26593

Riddech, N., Theerakulpisut, P., Ma, Y. N., & Sarin, P. (2025). Bioorganic fertilizers from agricultural waste enhance rice growth under saline soil conditions. *Scientific Reports, 15*, 8979. https://doi.org/10.1038/s41598-025-93619-9

Safitri,M., Noviandi, C.T., & Utomo, R. (2017).Chemical composition and in Sacco degradation of fish waste meal as alternative protein feed source. In *International Seminar on Tropical Animal Production (ISTAP) Proceedings*.

Samoraj, M., Çalış, D., Trzaska, K., Mironiuk, M., & Chojnacka, K. (2024). Advancements in algal biorefineries for sustainable agriculture: Biofuels, high-value products, and environmental solutions. *Biocatalysis and Agricultural Biotechnology, 58*, Article 103224. https://doi.org/10.1016/j.bcab.2024.103224

Tilman, D., Cassman, K. G., Matson, P. A., Naylor, R., & Polasky, S. (2002). Agricultural sustainability and intensive production practices. *Nature, 418*(6898), 671–677.

Turan, V., Schnug, E., & Onay, T. T. (2023). Biofertilizer production from organic waste via microbial fermentation and enzymatic hydrolysis: A review. *Current Research in Environmental Sustainability, 5*, Article 100198. https://doi.org/10.1016/j.crsust.2023.100198

USEPA, (2025). International Efforts on Wasted Food Recovery. U.S. Environmental Protection Agency. https://www.epa.gov/international-cooperation/international-efforts-wasted-food-recovery

Xia, L. W. F., Supri, S., Djamaludin, H., Nurdiani, R., Leong, S. L., Wee, Y. K., & Rovina, K. (2024). Turning waste into value: Extraction and effective valorization strategies of seafood by-products. *Waste Management Bulletin, 2*, 84–100.

Xu, F., Li, Y., Ge, X., Yang, L., & Li, Y. (2017). Anaerobic digestion of food waste—challenges and opportunities. *Bioresource Technology, 247*, 1047–1058.

Yaser, A., Lamaming, J., Suali, E., Rajin, M., Saalah, S., Kamin, Z., Safie, N., Ajie, N., & Wid, N. (2022). Composting and anaerobic digestion of food waste and sewage sludge for campus sustainability: A review. *International Journal of Chemical Engineering, 2022*(1), 1–14.

Zhang, J., Akyol, C., & Meers, E. (2023). Nutrient recovery and recycling from fishery waste and by-products. *Journal of Environmental Management, 348*, Article 119266.

Zhou, X., Zhang, L., Zhang, Z., & Liu, G. (2019a). Microbial community evolution and dynamic evaluation during food waste composting. *Bioresource Technology, 285*, Article 121347.

Zhou, X., Zeng, L., Liu, C., Tang, H., Li, Y., & Xu, F. (2019). Lactic acid fermentation of food waste for nitrogen stabilization and recovery. *Waste Management, 98*, 87–96. https://doi.org/10.1016/j.wasman.2019.08.029

Application of Liquid Organic Fertilizer in Organic Farming

Suhana Mohamed Sultan and Norahim Ibrahim

Abstract This chapter explores the development and application of liquid organic fertilizers (LOFs) derived from agricultural and food waste for use in organic and natural farming systems. Emphasis is placed on three main types of fermented organic fertilizers: Fermented Plant Juice (FPJ), Fermented Fruit Juice (FFJ), and Fermented Fish Amino Acid (FAA). These formulations are produced using simple fermentation techniques involving plant materials, fruits, and fish waste mixed with brown sugar or molasses, promoting the activity of beneficial microorganisms. Detailed procedures for preparing each type of juice are described, along with their chemical compositions, including essential macro- and micronutrients. The study highlights the role of these fertilizers in enhancing soil fertility, plant nutrition, and crop yields, while minimizing environmental impacts commonly associated with synthetic fertilizers. Field application methods and dilution recommendations are also provided. The results underscore the potential of LOFs as sustainable, eco-friendly alternatives for improving agricultural productivity in both small- and large-scale farming systems.

Keywords Liquid Organic Fertilizer (LOF) · Fermented Plant Juice (FPJ) · Fermented Fruit Juice (FFJ), · Fermented Fish Amino Acid (FAA) · Sustainable agriculture

S. M. Sultan
Department of Electronic and Computer Engineering, Faculty of Electrical Engineering, Universiti Teknologi Malaysia, Johor Bahru, Johor, Malaysia
e-mail: suhanasultan@utm.my

N. Ibrahim (✉)
Department of Biosciences, Faculty of Science, Universiti Teknologi Malaysia, Johor Bahru, Johor, Malaysia
e-mail: norahim@utm.my

A. Z. Yaser et al. (eds.), *Compost and Liquid Organic Fertilizers*,
https://doi.org/10.1007/978-981-95-8561-8_11

1 Introduction

Organic farming and natural farming are environmentally friendly farming systems. The methods used in these farming systems are both free of chemicals and toxic or harmful substances (Tahat et al., 2020). Both systems help farmers avoid the use of chemical fertilizers, pesticides or herbicides on crops or in any agricultural practices (Urra et al., 2019). Fertilizers are very important in agriculture as they promote plant growth and increase soil fertility (Alley & Vanlauwe, 2009). However, it is not the same as inorganic fertilizer. The overuse of inorganic fertilizers, such as chemical fertilizers, causes serious problems for present and future generations, such as air, water and soil pollution, degraded land, depleted soils and the greenhouse effect. These inorganic fertilizers not only have a negative impact on the environment, but also on humans, animals and other living organisms. This is because the excessive use of inorganic fertilizers can leave chemical residues in the soil that are hazardous to health (Sarkar et al., 2003; Sharma, 2017).

Farmers are encouraged to use local seed breeds and indigenous varieties of vegetables, cereals, dry seeds and other crops in both farming methods. Non-chemical and home-made methods of pest control are promoted through organic and natural farming methods. "Natural farming is a farming method that imitates the way of nature," say Masanobu Fukuoka and Mokichi Okada, who founded this method in Japan. It is also referred to as "natural farming" or "doing nothing", as the farmer is only seen as a facilitator and the actual work is done by nature itself (Fukuoka & Metreaud, 1993; Miyake & Kohsaka, 2020).

In natural farming, neither chemical nor organic fertilizers are added to the soil. In fact, no external fertilizers are introduced into the soil or given to the plants (Tahat et al., 2020). Moreover, in natural agriculture, the decomposition of organic matter by microbes and earthworms on the soil surface itself is encouraged, gradually enriching the soil with nutrients over time (Fukuoka & Metreaud, 1993). In organic farming, organic fertilizers and manures such as compost, vermicompost and cow dung are used and added to the farmland from external sources. Organic farming still requires basic agricultural practices such as plowing, tilling, mixing fertilizers and weeding.

In natural farming, there is no plowing, turning over the soil or applying fertilizers, as this is simply done as it would be in natural ecosystems. Organic farming is still expensive compared to natural farming, as it requires large amounts of fertilizers and has an ecological impact on the environment, while natural farming is a very cost-effective farming method. Organic farming generally causes less soil and water pollution and lower greenhouse gas emissions (Adnan & Zohaib Anjum, 2020).

Organic and natural farming have always been valued for their focus on sustainability and working in harmony with the environment. For example, Kumar and Kumar (2020) emphasize traditional practices such as crop rotation, green manuring, and composting as key ways to reduce chemical inputs and support biodiversity. They also point out how important farmer knowledge and cultural traditions are in keeping these methods alive. Still, their work does not provide much in terms of scientific analysis or experimental data, especially when it comes to the actual composition of organic fertilizers.

In a similar way, Enemali et al. (2023) highlight the potential of turning plant and animal waste into soil-enriching liquid organic fertilizers. Their review stresses the role of microbes, composting, and effective microorganisms (EM) in improving soil health. It even touches on some nutrient values like nitrogen, phosphorus, and potassium (NPK). However, the paper remains quite general as it doesn't compare different fermentation methods in detail, nor does it give clear practical guidance on how farmers should apply these fertilizers in real situations.

This chapter builds on those gaps by offering a much more detailed and hands-on approach. It explains how to prepare three types of fermented fertilizers—fermented plant juice (FPJ), fermented fruit juice (FFJ), and fish amino acid (FAA) with clear, step-by-step methods. Unlike earlier works, it doesn't stop at concepts. It provides detailed nutrient composition tables and practical monitoring guidelines so farmers and researchers alike can reproduce and trust the results.

Another strength of this work is the side-by-side comparison of fertilizers made from plants, fruits, and fish waste. The chapter also sets out clear instructions for how much to use and how to apply them, such as dilution ratios (for example, 1:500). By situating these findings within the Malaysian agricultural context, it speaks directly to local needs by making use of common waste materials, considering farmer practices, and focusing on crops that matter most to the region. This makes the work especially relevant to tropical and developing countries facing similar challenges.

Overall, the novelty of this chapter lies in its integrated treatment of FPJ, FFJ, and FAA, complete with illustrated preparation methods and solid chemical data to back them up. By showing how agricultural and kitchen waste can be recycled into valuable fertilizers, it not only promotes sustainable farming but also supports the broader idea of a circular economy. While earlier studies laid out the principles of organic fertilizers by Kumar and Kumar (2020) and Enemali et al. (2023), this chapter provides a practical, evidence-based, and locally grounded guide that can actually be put into practice by farmers.

2 Liquid Organic Fertilizer

2.1 Types of Fermented Juice

Organic soil improvers derived from agricultural waste have been shown to play a number of important roles in soil health and crop yield (Sayara et al., 2020; Urra et al., 2019). According to Urra et al. (2019), not all agricultural wastes have good chemical and biological properties to maintain soil health due to differences in wastes or residues. Some of them can even be harmful to plant growth and soil health. Therefore, it is important to characterize and evaluate agricultural wastes before using them as soil conditioners in order to make informed decisions about their potential in agriculture. Another important factor in convincing farmers to use these residues and residues as soil conditioners is the easy availability and abundance of agricultural wastes. This could significantly reduce their dependence on inorganic fertilizers and the associated environmental problems caused by either the overuse of chemical fertilizers or the improper disposal of agricultural waste (Sayara et al., 2020). It is important to emphasize that the use of organic soil improvers in crop production must be based on specific and precise production standards in order to achieve optimal agroecosystems that are socially, environmentally and economically sustainable (Sulok et al., 2021)

In Malaysia, agricultural wastes such as tapioca, bamboo, leaves from vegetable and banana plants, grass chaff, rotten fruits, rice straw and corn stalks can be used to produce organic soil improvers (Ibrahim et al., 2016). The conversion of agricultural wastes into fermented juices, compost and biochar is simple and acceptable techniques for local farmers to utilize their agricultural wastes as organic soil improvers in their farms. This study is conducted with the main objective of characterizing the chemical and biological properties of recycled agricultural wastes to use them as organic amendments for soil health and help local farmers to increase their yields (Zamora et al., 2016).

Liquid organic fertilizers consist of beneficial microorganisms and essential plant nutrients that recycle organic material (Phibunwatthanawong & Riddech, 2019). Liquid organic fertilizer is a good choice to improve soil fertility. It has been proven that liquid organic fertilizer is environmentally friendly and increases soil fertility. The production of liquid organic fertilizer involves effective microorganisms that play an important role in the fermentation process. There are some types of fertilizers made from fermented juices that are counted as organic liquid fertilizers, such as fermented plant juice (FPJ), fermented fruit juice (FFJ) and fermented fish amino acid (FAA).

Liquid organic fertilizer is classified as organic fertilizer because it consists of natural materials of plant or animal origin, including livestock manure, green manure, crop residues, household wastes and compost, and acts directly as a source of plant nutrients and indirectly affects the physical, chemical and biological properties of the soil (Basel & Sami, 2014). Organic fertilizers have the following benefits for improving soil fertility: (i) Increasing soil organic matter content improves soil structure, creates more air space and water retention in the soil, and increases soil nitrogen content, nutrient availability and improves nutrient mobilization (Roba et al., 2018). (ii) Organic fertilizers increase root growth by improving soil structure, promoting soil aggregates and improving cation exchange capacity (Lal, 2006) (iii) Organic fertilizers are the fastest growing area of agriculture and their main objective is to establish a balance between the interconnected systems such as soil organisms, plants, animals and humans (Berova et al., 2010).

2.1.1 Fermented Plant Juice (FPJ)

Fermented plant juice (FPJ) is used in solutions for seed and soil treatment and for plant nutrition. It consists of the young shoots of vigorously growing plants, which are usually fermented with brown sugar or molasses for 7 days. The brown sugar extracts the juices from the plant material by osmosis and also serves as a food source for the microbes that carry out the fermentation process. The weak alcohol produced during fermentation extracts chlorophyll (soluble in ethanol) and other plant components. It is non-toxic and edible.

The most important requirement when selecting plants for the production of FPJ is the use of shoot tips from fast-growing plant species. Flowers, flower buds and unripe fruits can also be used. Hard or woody parts of the plant release little or no sap. The plants should have grown vigorously at the time of collection. Plant parts should be harvested when the plants are in respiration mode (before sunrise) and not in photosynthesis mode (during daylight), as these processes affect plant chemistry. Avoid collecting plant parts during or after rainfall and do not rinse the collected plant parts to preserve the microbial populations on the surface, which are lactic acid-producing bacteria and yeasts that carry out the fermentation process. Low levels of these microbes lead to improper fermentation and low production of plant sap (Miller et al., 2013).

2.1.2 Factors that Affect the Quantity and Quality of FPJ

The most important requirement when selecting plants for the production of FPJ is the use of shoot tips from fast-growing plant species. Flowers, flower buds and unripe fruits can also be used. Hard or woody parts of the plant release little or no sap. The plants should have grown vigorously at the time of collection. Plant parts should be harvested when the plants are in respiration mode (before sunrise) and not in photosynthesis mode (during daylight), as these processes affect plant chemistry. Avoid collecting plant parts during or after rainfall (ideally wait two sunny days after the rain has stopped) and do not rinse the collected plant parts to preserve their surface microbial populations (lactic acid-producing bacteria and yeasts) that will carry out the fermentation process. Low levels of these microbes will result in improper fermentation and/or low yields of sap.

2.1.3 Fermented Fruit Juice (FFJ)

Fermented fruit juice, also known as FFJ, is a nutrition-activating enzyme and very effective in natural agriculture. It is used to revitalize crops, livestock and humans. The main ingredients for the production of fermented fruit juice are all kinds of fruits such as bananas, mangosteen, santol, papaya, mango, grapes, melons, apples and many more. Fermented fruit juice (FFJ) can be used to enhance the fruit formation process of plants and increase yield. It can be used both as a foliar spray and as an indirect application in the soil, where it nourishes the microbiome and improves soil health (Reickenberg & Pritts, 1998).

Fermented fruit juice (FFJ) is used as a foliar fertilizer that helps to increase and improve crop yield and quality. Foliar fertilization is immediately available to the plants as this form of nutrient application is better than direct fertilization (Naz et al., 2011). Young leaves take up the nutrient solution via trichomes on the leaf surface and stomata. Most of the uptake occurs by diffusion through the cuticle (Salisbury and Ross, 1992). This is in contrast to fertilizers that are applied through the soil. These are usually in solid form, e.g. powder or granules, and must be dissolved by moisture from rainfall or irrigation to be available to the plants. Fertilizers must dissolve in the soil solution to be available. If the fertilizer applied to the soil is not readily available or is insufficient, foliar fertilization is usually practiced or used as a supplement (Abbas & Ali, 2011). Foliar fertilizer application is an effective way to make up for nutrient deficiencies in the soil when plants cannot absorb them directly from the soil (Liang & Silberbush, 2002). Foliar fertilizer application results in a faster response and is more effective than soil application for some nutrients (Jamal et al., 2006).

2.1.4 Fermented Fish Amino Acid (FAA)

Fish emulsions have been shown to promote seedling growth (Murray & Anderson, 2004), fruiting (Aung & Flick, 1980) and soil microbial activity (El-Tarabily et al., 2003). One of these emulsions, fish amino acid (FAA), is produced by the fermentation of fresh fish by-products (bones, head, skin and other parts of the carcass) with brown sugar. FAA is used in conjunction with other Natural Farming products and applied either as a light spray on the leaves or as a soil fertilizer to maximize uptake and minimize run-off or leaching, providing the plant with just enough N for optimal uptake and production of chlorophyll to maintain plant health.

Organic fertilizer is a natural fertilizer produced by the decomposition of animal and plant waste. Organic fertilizer is also environmentally friendly as it is produced without the use of chemicals. In general, it proves to be a suitable means of maintaining soil fertility and plant growth (Ji et al., 2017). Fish amino acid (FAA) is one of the organic preparations used to improve plant growth development (Priyanka et al., 2019a, 2019b). FAA is produced by mixing fish waste with brown sugar and leaving it for a fermentation period of at least three months. It is of great importance for the growth of plants and microorganisms as it contains various nutrients and amino acid species (Ramesh et al., 2020). During the early or vegetative stage of development, FAA is used as a source of nitrogen to improve growth and size (Weinert et al., 2014). Application of fish amino acids through the foliage or soil can increase uptake and reduce runoff or leaching and provide the plant with sufficient N for chlorophyll production to maintain plant health (Priyanka et al., 2019a, 2019b). Commercial fertilizers are an inexpensive way to add nitrogen (N) to the soil to promote plant growth and achieve high crop yields. However, improper or excessive use of N fertilizers can lead to nitrate contamination of ground or surface water. (Foley et al., 2012; Weinert et al., 2014). Due to this issue, many researchers have been looking for an organic fertilizer that can be used by farmers without harming the environment and ecosystem. Therefore, the aim of the study is to determine the effect of fish amino acid (FAA) on plant growth and development of okra at different sampling times.

The fish amino acid was prepared from fish waste obtained from the local fish market or from kitchen waste. The collected fish waste should be mixed or blended with the activated EM, which consists of EM.1, molasses and dechlorinated water. The materials were mixed well and stored in a cool, dry place. It should be protected from direct sunlight. After 7 to 10 days, the liquid portion was filtered and used for spraying (Maghirang, 2011). The final product was a viscous liquid and had a sweet odor.

3 Production of Liquid Organic Fertilizer

3.1 Preparation of Fermented Plant Juice

Figure 1 shows the procedures in making Fermented Plant Juice (FPJ).

Step 1: Collection of plant material.

The plant was collected before sunrise, e.g. pandan leaves, mulberry leaves and any kind of plant you can select for fermentation. Avoid collecting during or after rains.

Step 2 Cut and weighing of the plant material.

To preserve the microorganisms, do not rinse the collected plant material. Record the weight of the plant material. Cut the leaves into 2–3 inch long pieces if the leaves are long, and weigh the plant parts before or after cutting, whichever is easier, as shown in Fig. 1b.

Step 3 Addition of molasses.

Weigh out an amount of Effective Microorganism.1 (EM.1) and molasses equal to the weight of the plant material and mix them in a high-powered blender (Fig. 1c). Cover as much surface area of the plant material as possible with sugar to speed up the osmotic process and draw out the plant juices.

Step 4 Pouring the mixture of plant material and brown sugar into a container.

Choose a clear or plastic container. Metal containers should not be used as they will react with the solution. Fill the container with the mixture of plant material and brown sugar or activated effective microorganisms until the container is almost full, i.e. 2/3 of the container (Fig. 1e). Close the opening of the container tightly with the lid to prevent air exchange and to keep out pests and other contaminants. Store the covered container in a well-ventilated place, away from artificial or natural light and extreme heat or cold.

Step 5 Checking the container after 24 h.

For the fermentation process to work properly, the volume of the mixture of plant material and brown sugar should settle to 2/3 of the container after 24 h. If the container is too full, the microbes will not have enough air to ferment properly. Remove some of the plant material until the container is no more than 2/3 full. If the container is less than 2/3 full, add more dechlorinated water to prevent mould

Fig. 1 a–f Steps in making Fermented Plant Juice (FPJ) for fertilizer

growth. Not all plants settle in the same way, so you will need to check and adjust the volume after the first 24 h.

Step 6: Fermentation process.

The fermentation process is dependent on the ambient temperature, but it usually only takes up to 7 days for the fermentation process to complete. The fermentation process begins when bubbles start to form in the container, which is usually on the second day. Ideally, fermentation should not take longer than 7 days, as the quality of FPJ seems to decline after that. Fermentation is complete when (1) the plant material floats and the liquid settles to the bottom (note: if too much brown sugar has been used, this separation will not be clear), (2) a slight alcohol odour can be detected due to the breakdown of chlorophyll, and (3) the liquid tastes sweet and not bitter.

Step 7: Filtration of the liquid from the solids.

After fermentation is complete, which is usually after 7 days, separate the plant material from the liquid using a sieve. The liquid is the fermented plant juice (FPJ), which can be used immediately or stored in a closed container.

Step 8: Storage of FPJ.

Pour the FPJ into a closed plastic container. Preferably store it in a dark or tinted container to avoid sunlight. The microorganisms in the solution are alive and will continue to produce gases during the fermentation process. Therefore, the lid must be tightly closed. Open the lid to remove the gas trapped in the container, otherwise the container may explode. The container should be stored in a dark place to avoid sunlight.

Preparation of Fermented Fruit Juice (FFJ).

Figure 2 shows the process for producing fermented fruit juice (FFJ). The steps for the preparation of FFJ are the same as in Sect. 3.1. In this case, the raw materials used are santol fruit, mulberries, mangosteen and bananas.

3.2 Preparation of Fermented Fish Amino Acid (FAA)

Figure 3 shows the process for producing fermented fish amino acid (FAA). The steps for the preparation of FFJ are the same as in Sect. 3.1. In this case, the raw materials used are fish wastes.

3.3 Preparation of Mixed Fermented Juices

Figure 4a–f shows the materials and steps in the production of mixed fermented juice containing plant, fruit and fish parts or fish waste. The steps for the preparation of FFJ are the same as in 3.1. In this case, the raw materials used are fish wastes, rotten bananas, and mulberry leaves.

3.4 Characterization of FPJ, FFJ and FAA

The chemical composition of the fermented fruit juices produced from banana peels is shown in Table 1. The batches of FPJ, FFJ and FAA were analysed for N, P, K, Mg and B using standard laboratory procedures to evaluate the chemical composition of the fermented juices. pH and electrical conductivity are classified as physical parameters and used as indicators of the progress of fermentation.

Fig. 2 **a–f** Steps in making Fermented Fruit Juice (FFJ) for fertilizer

Potassium (K) increases crop yield and improves quality. It is required for numerous plant growth processes (Yasav et al., 2017). Banana peels have a high K content. Table 1 shows that the potassium content in fermented banana peels is the highest compared to other chemical compositions and is 12%. Fruit contains a large amount of antioxidants, which are beneficial to our health in many ways (Faria et al., 2006). Although liquid fertilizers are not necessarily high in essential nutrients, Yadav et al. (2017) report that liquid fertilizers can be important bioresources for agriculture, as their beneficial microbes can improve plant growth and nutrient uptake through solubilization of P, K and Zn, as well as N fixation. Therefore, it is possible that the activities of these beneficial microbes can increase crop yield,

Fig. 3 **a–f** Steps in making Fermented Fish Amino Acid (FAA)

eliminate all pollutants, inhibit the activity of pathogens, and produce fixed N or other substances for plant growth and development. The effectiveness of fermented fruit juice application on plant growth was proven in a study by Ali et al. (2013), who found that foliar application of fermented fruit juice improved the growth, flowering and marketable yield of tomatoes.

Yadav et al (2017) reported that stimulation of growth by plant microbiomes may be a consequence of biological N-fixation, production of plant growth regulators such as indoleacetic acids (IAAs), gibberellic acids, cytokines and biocontrol of phytopathogens. This occurs through the production of antibiotics, antifungal or antibacterial agents, Fe chelating agents, nutrient competition, induction of acquired

Fig. 4 **a–f** Main ingredients and steps in the production of mixed fermented fish amino acid, plant and fruit juice

host resistance or improvement of mineral bioavailability. These microbes may also play a role in nutrient cycling and increasing soil biodiversity, which are important components of soil health (Urra et al., 2019). These potential positive effects on soil biological health need to be evaluated in the field through soil amendments in cropping systems.

The K requirement of fruit crops is particularly high. A high K content is required not only to achieve the highest overall fruit production, but also to produce the highest percentage of fruit suitable for marketing. Potassium is often referred to as the quality element for crop production (Usherwood, 1985) and it is generally

Table 1 Chemical composition of fermented fruit juice (banana peels)

S. no	Parameters	Unit	Result
Physical parameters			
1	pH		7 + −0.01
2	Electrical conductivity	μS/cm	6010 + −1.45
Chemical parameters			
3	Total Nitrogen (Kjedahl)	%	0.13
4	Total Phosphate as P2O5 #	%	2.75
5	Potassium as K2O #	%	12
6	Sodium as Na #	mg/kg	475
7	Calcium as Ca #	mg/kg	1170
8	Magnesium as Mg #	mg/kg	185
9	Mangenese as Mn #	mg/kg	< 0.5
10	Iron as Fe #	mg/kg	< 0.5

recognized that it plays a crucial role in many crop quality parameters. The size, appearance, color, soluble solids, acidity, vitamin content, flavor and shelf life of fruit are significantly affected by adequate potassium supply. These properties are influenced by photosynthesis, the translocation of photosynthates, the regulation of stomata, the activation of enzymes and many other processes. The role of potassium in plant water regulation and tolerance to environmental stresses such as drought, excess water, wind, and high and low temperatures is related to tree productivity and fruit quality (Ganeshamurthy et al., 2011). Siti et al. (2017) stated in their journal that fermented fruit juice, which can be made from any type of fruit, fresh or rotten fruit, is used to improve soil and increase the production of leafy plants in gardens.

Table 2 shows the chemical composition of fermented fish amino acid, which is mainly extracted from fish waste to produce this liquid organic fertilizer. The nitrogen content in fermented fish waste fertilizer is 0.7%, which is the highest compared to other macronutrients. Compared to synthetic fertilizers, organic fertilizers contain relatively low concentrations of actual nutrients, but perform important functions that synthetic fertilizers do not. As Seran et al. (2011) explained in their report, liquid organic fertilizers such as fermented fish amino acid improve the physical structure of the soil, allowing more air to reach the plant roots and increasing bacterial and fungal activity in the soil. Fermented fish waste fertilizer, like other organic fertilizers, is environmentally friendly, does not leach and remains in the soil for a long time, so it does not pollute ecosystems. Many studies have shown that fermented fish waste fertilizer increases production later in life. Tomatoes fertilized with fish waste fertilizer grew better later in life than those fertilized with chemical fertilizer. Therefore, growers should take some measures to make the nutrients in fish waste fertilizer available before the plants start to grow (Shankara et al., 2005).

Nitrogen is an essential element for plants. It consists of amino acids, which catalyze chemical reactions and transport electrons, and chlorophyll, which enables

Table 2 Chemical composition of fermented fish amino acid (fish waste)

S. no	Parameters	Unit	Result
Physical parameters			
1	pH		4.25 + −0.2
2	Electrical conductivity	μS/cm	29,800 + −1.45
Chemical parameters			
3	Total Nitrogen (Kjedahl)	%	0.7
4	Total Phosphate as P2O5 #	%	0.2
5	Potassium as K2O #	%	0.6
6	Sodium as Na #	mg/kg	4153.4
7	Calcium as Ca #	mg/kg	0.6
8	Magnesium as Mg #	mg/kg	0.5
9	Mangenese as Mn #	mg/kg	37.6
10	Iron as Fe #	mg/kg	100.9

photosynthesis in the plant. Nitrogen plays a very important role in various physiological processes. It gives plants a dark green color and promotes the growth and development of leaves, stems and other vegetative parts of the plant. It also stimulates root growth. Nitrogen causes rapid early growth, improves fruit quality, promotes the growth of leafy vegetables, increases the protein content of forage crops; it promotes the uptake and utilization of other nutrients such as potassium and phosphorus and controls the overall growth of the plant (Bloom, 2015). Nitrogen deficiency leads to reduced growth, chlorosis (change of green color to a yellow color of the leaves) and red and purple spots on the leaves and restricts the growth of lateral buds (from which leaves, stems and branches develop). Usually, the deficiency symptoms first appear on older leaves (Bianco et al., 2015), then leaf senescence begins and excessive nitrogen supply has a negative effect on plant growth by promoting an additional dark green color on the leaves, making the overall growth more succulent and favoring a lower fruit quantity with lower quality.

Table 3 shows the chemical composition of the fermented plant juice produced from water spinach. Green leafy vegetables such as water spinach are considered a rich source of micronutrients (Kala & Prakash, 2004). Table 3 above shows that the potassium content in fermented plant juice is 0.5%. The total chemical composition of sodium in fermented plant juice is 4384.3 mg/kg. The comparison of micronutrients contained in fermented plant juice has shown that water spinach has a high content of micronutrients. According to Siti et al. (2017), fruit juice from any type of fruit, whether fresh or rotten, is used to improve soil and increase the production of leafy plants in gardens. Plants require small amounts of sodium as the element serves as a micronutrient for the plant to support metabolism and the production of chlorophyll. The availability of sodium (Na^+) has been reported to stimulate the growth of many crops, especially when there is a deficiency of potassium (K^+) ions (Barker & Pilbeam, 2007; Subbarao et al., 2003).

Table 3 Chemical composition of fermented plant juice (water spinach)

S. no	Parameters	Unit	Result
Physical parameters			
1	pH		7.03
2	Electrical conductivity	μS/cm	2001
Chemical parameters			
3	Total Nitrogen (Kjedahl)	%	0.1
4	Total Phosphate as P2O5 #	%	0.01
5	Potassium as K2O #	%	0.5
6	Sodium as Na #	mg/kg	4384.3
7	Calcium as Ca #	mg/kg	0.6
8	Magnesium as Mg #	mg/kg	0.3
9	Mangenese as Mn #	mg/kg	28.6
10	Iron as Fe #	mg/kg	30.5

Fermented plant juices (FPJ) are suitable for use on vegetable crops. According to Borhan (2011), water spinach can be a good growth promoter as it grows quickly and contains many natural growth hormones such as auxin. This is supported by the result of the study conducted by Siti et al. (2017) on the application and effectiveness of fermented plant juice. The plant height, leaf area of the tomato, fresh weight and dry weight of the shoot and root were measured. The result, which shows that fermented plant juice can increase the production of tomatoes and supply the plant with nutrients, can be seen in Fig. 5.

The total nitrogen observed in mixed fermented juice is 0.2%. Mixed fermented juice contains three types of fermented juice, namely fermented plant juice, fermented fruit juice and fermented fish amino acid. The nutrient content extracted from the different types of materials used in this fermented juice is expected to be higher than the specific type of fermented juice, because in this study the nutrients were extracted from fish waste, mulberry leaves and banana peels.

Table 4 shows that the total phosphate content of the mixed fermented juice is 0.1, which is considered low nutrient content. Compared to the secondary data from a study by Ramesh et al. (2020), the phosphate content in fermented crab waste contains only 0.41%, while in a study by Sahu et al. (2014), the phosphate content in fermented fish waste is only 0.52%. The above table shows that the potassium content is also low, as the potassium content of the prepared mixed fermented juice is 0.4%. There are some factors that influenced the decrease of nutrient content in fermented plant juice. The decrease in nutrient content was reportedly caused by the high rate of nutrient uptake by microbial fermentation, which is associated with increased activity. Hoorman and Islam (2010) previously reported that the nutrient content of material undergoing decomposition decreases with the increase in microbial populations. In addition, the decrease in the amount of N contained in the fermentation also led to changes in the form elements in the gas fraction leaving the fermentation system. The biological impurities in the fermented materials were influenced by the

Table 4 Chemical composition of mixed fermented juice (FPJ, FFJ and FAA)

S. no	Parameters	Unit	Result
Physical parameters			
1	pH		8.65 + -0.01
2	Electrical conductivity	μS/cm	6300 + -1.45
Chemical parameters			
3	Total Nitrogen (Kjedahl)	%	0.2
4	Total Phosphate as P2O5 #	%	0.1
5	Potassium as K2O #	%	0.4
6	Sodium as Na #	mg/kg	3413.0
7	Calcium as Ca #	mg/kg	0.6
8	Magnesium as Mg #	mg/kg	0.3
9	Mangenese as Mn #	mg/kg	15.2
10	Iron as Fe #	mg/kg	19.0

amount of carbon contained. Therefore, the enrichment process using the fermentation of organic substances and microbial functions is proposed to improve the quality and effectiveness of the developed fertilizer.

3.5 *Application of Fermented Juice on Plants*

According to Miller et al (2013), FPJ is generally used at a concentration of 1 part per 500 parts water (1:500). In the following circumstances, a more dilute solution is required (1:800 to 1:1,000) to avoid plant damage such as leaf-burn: When more than three fermented sap ingredients are combined, the fermented plant sap, fruit juice or fish amino acid is applied in hot weather, or the fermented sap has been stored for more than one year and has become more concentrated as a result, it is used. All three types of fermented juice (FPJ, FFJ or FAA) can be applied to the plants according to the formula given in the table below.

The solution can be poured onto the plants or into the soil, or applied as a foliar spray. The nutrient solution is administered once a week and is adapted to the stages of the life cycle and the vegetative and reproductive phases of the plant.

4 Conclusion

In summary, the results of this research provide a number of important applications for future practice, especially information on the production of fermented fruit juice, unsellable waste fruit in the form of organic fertilizer, to advance in the realization of sustainable agriculture The use of liquid organic fertilizer such as FPJ, FFJ and FAA also appears to be a promising alternative to replace chemical fertilizer.

References

Abbas.MK and Ali A.S., 2011. "Effect of foliar application of NPK on some growth characters of two cultivars of rosette (Hisbiscus sabdariffa L.)." Am.J.Plant physiol.,6, 220–227.

Adnan, M., & Zohaib Anjum, M. (2020). Back to Past. *Organic Agriculture. Acta Scientific Agriculture, 5*(2), 01–02.

Alley, M., & Vanlauwe, B. (2009). *The role of fertilizers in integrated plant nutrient management.* IFA, TSBF-CIAT.

Ali, A., Ahmad, A., & Akhtar, J. (2013). Effect of organic and inorganic fertilizers on soil properties and crop yield. *Journal of Agricultural Science, 5*(5), 222–229

Aung, L. H., & Flick, G. C. (1980). The Influence of Fish Solubles on Growth and Fruiting of Tomato. *HortScience, 15*(1), 32–33.

Barker, A. V., & Pilbeam, D. J. (2007). *Handbook of plant nutrition.* CRC Press.

Basel, N., & Sami, M. (2014). Effect of Organic and Inorganic Fertilizers Application on Soil and Cucumber (Cucumis sativa L.) Plant Productivity. *International Journal of Agriculture and Forestry, 4*, 166–170.

Berova, M., Karanatsidis, G., Sapundzhieva, K., & Nikolova, V. (2010). Effect of organic fertilization on growth and yield of pepper plants (Capsicum annuum L.). *Folia Horticulturae, 22*(1), 3–7.

Bianco, M. S.,CecílioFilho A.B., and Carvalho L.B., 2015. "Nutritional status of the cauliflower cultivar Verona grown with omission of out added macronutrients." Plos One, 10(4).

Bloom, A. J. (2015). The increasing importance of distinguishing among plant nitrogen sources. *Current Opinion in Plant Biology, 25*, 10–16.

Borhan, M., 2011. Beneficial and Effective Microorganism (EM): Their Applications in Sustainable Agriculture, Syarikat Surya MajuSdnBhd, Brunei Darussalam.

El-Tarabily, K.A., A.H. Nassar, G.E.S.J. Hardy, and K. Sivasit-Hamparam. 2003. Fish emulsion as a food base for rhizo-bacteria promoting growth of radish (Raphanus sativus L. var. sativus) in a sandy soil. Plant and Soil, 252(2), 397–411.

Enemali, S. I., Alfa, J., Yusuf, O. Y., & Muazu, I. (2023). Liquid organic fertilizer availability and utilization in Nigeria: A review. *World Journal of Advanced Research and Reviews, 20*(2), 820–833.

Faria, A., Calhau, C., de Freitas, V., & Mateus, N. (2006). Procyanidins as Antioxidants and Tumor Cell Growth Modulators. *Journal of Agricultural and Food Chemistry, 54*(6), 2392–2397.

Foley, K. M., Doniger, A. R., Shock, C. C., Horneck, D. A., & Welch, T. K. (2012). *Nitrate pollution in groundwater* (p. 137). Oregon State University, Department of Crop and Soil Science Ext/CrS.

Fukuoka, M. and Metreaud, F., 1993. The natural way of farming. Madras: Bookventure.

Ibrahim, I., Suhaimee, S. and Abdul Wahab, A., 2016. Organic Agriculture in Malaysia. [online] FFTC Agricultural Policy Platform (FFTC-AP). Available at: <https://ap.fftc.org.tw/article/1010> [Accessed 29 September 2021].

Ganeshamurthy, A. N., Reddy, K. S., & Srinivasarao, C. (2011). Organic matter management for improving soil quality and crop productivity. *Indian Journal of Fertilisers, 7*(4), 56–61.

Hoorman, J. J., & Islam, R. (2010). Understanding soil microbes and nutrient recycling. Ohio State University Extension Fact Sheet (SAG-16-10).

Jamal, Z., Hamayun, M., Ahmad, N., & Fayyaz Chaudhary, M. (2006). Effects of Soil and Foliar Application of Different Concentrations of NPK and Foliar Application of $(NH_4)_2$ SO_4 on Different Yield Parameters in Wheat. *Journal of Agronomy, 5*(2), 251–256.

Ji, R., Dong, G., Shi, W., & Min, J. (2017). Effects of Liquid Organic Fertilizers on Plant Growth and Rhizosphere Soil Characteristics of Chrysanthemum. *Sustainability, 9*(5), 841.

Kala, A., & Prakash, J. (2004). Nutrient Composition and Sensory Profile of Differently Cooked Green Leafy Vegetables. *Food Properties, 7*(1), 659–669.

Kumar, K., & Kumar, M. (2020). Significance of liquid organic manures in Indian agriculture – A review. *International Journal of Current Microbiology and Applied Sciences, 9*(1), 17–24.

Lal, R. (2006). Enhancing crop yields in the developing countries through restoration of the soil organic carbon pool in agricultural lands. *Land Degradation & Development, 17*(2), 197–209.

Liang, F., & Silberbush, M. (2002). Response of maize to foliar vs. soil application of nitrogen, phosphorus potassium fertilizer on plant. *Journal of Plant Nutritional, 25*(11), 2333–2342.

Maghirang, R. G. (2011). Composting and its role in sustainable agriculture. Kansas State University Extension Publication.

Miller, S. A., Ikeda, D. M., Weinert, E., Jr., Chang, K. C., McGinn, J. M., Keliihoomalu, C., & DuPonte, M. W. (2013). Natural Farming: Fermented plant juice. *Tropical Agriculture and Human Resouces, 2*, 1–7.

Miyake, Y. and Kohsaka, R., 2020. History, ethnicity, and policy analysis of organic farming in Japan: when "nature" was detached from organic. Journal of Ethnic Foods, 7(1).

Murray, R. and R. G. Anderson. 2004. Organic fertilizers and composts for vegetable transplant production. University of Kentucky, Greenhouse use of organic fertilizers and composts – 3 Floriculture Research Report, 17–04.

Naz F., Haq JU., Asghar S., Sharih AS and Rahman A., 2011. "Studies on growth, yield and nutritional composition of different tomato cultivars in Battal Valley of district Mansehra, Khyber Pakhtunkhwa, Pakiston", 27, 569–571.

Phibunwatthanawong, T., & Riddech, N. (2019). Liquid organic fertilizer production for growing vegetables under hydroponic condition. *International Journal of Recycling of Organic Waste in Agriculture, 8*(4), 369–380.

Priyanka, B., Anoob, D., Gowsika, M., Kavin, A., Sri, S. K., Kumar, R. V. K., Gomathi, R. S., Sivamonica, B., Devi, G. V., & Theradimani, M. (2019b). Effect of Fish Amino Acid and Egg Amino Acid as Foliar Application to Increase the Growth and Yield of Green Gram. *The Pharma Innovation Journal, 8*(6), 684–686.

Priyanka, B., T. Ramesh, S. Rathika and Balasubramaniam, P. 2019. Foliar Application of Fish Amino Acid and Egg Amino Acid to Improve the Physiological Parameters of Rice. Int.J.Curr.Microbiol.App.Sci. 8(2), 3005-3009.

Ramesh, T., Amuthavalli, A., & Boopathy, R. (2020). Analysis of fermented liquid fertilizer from marine crab waste. *International Journal of Environment, Agriculture and Biotechnology, 5*(3), 636–642.

Reickenberg, R. L., & Pritts, M. P. (1996). Dynamics and nutrient uptake from foliar fertilizers in red raspberry (Rubusidaeus L.). *J. Am. Soc. Hort. Sci., 121*, 158–168.

Roba, T. (2018). Review on: The Effect of Mixing Organic and Inorganic Fertilizer on Productivity and Soil Fertility. *Oalib, 05*(6), 1–11.

Sahu, R. K., et al. (2014). Application of organic waste for soil fertility improvement. *International Journal of Recycling of Organic Waste in Agriculture, 3*(3), 1–9.

Salisbury, F. B., & Ross, C. W. (1992). Plant Physiology (4th ed.). Belmont, CA: Wadsworth Publishing Company

Sarkar, S., Singh, S., & Singh, R. (2003). The effect of organic and inorganic fertilizers on soil physical condition and the productivity of a rice–lentil cropping sequence in India. *The Journal of Agricultural Science, 140*(4), 419–425.

Sayara, T., Basheer-Salimia, R., Hawamde, F., & Sánchez, A. (2020). Recycling of Organic Wastes through Composting: Process Performance and Compost Application in Agriculture. *Agronomy, 10*(11), 1838.

Seran, T., Srikrishnah, S. and Ahamed, M., 2011. Effect of different levels of inorganic fertilizers and compost as basal application on the growth and yield of onion (Allium cepa L.). Journal of Agricultural Sciences, 5(2), 64.

Shankara N, de Jeude JL, Marja G, Martin H, Barbara D., 2005. Cultivation of tomato. production, processing and marketing. Agromisa Foundation and CTA, Wageningen..

Sharma, A., 2017. A Review on the Effect of Organic and Chemical Fertilizers on Plants. International Journal for Research in Applied Science and Engineering Technology, V(II), 677–680.

Siti Zaharah, A. R. Nur Azwani, J. Abdul Shukor, A. Md Amirul, and A. Farzad, 2017. "Effects of fermented plant juice and fruit juice on growth and yield of tomato for sustainable practices," Bangladesh J. Bot., 46 (1), 405–412.

Subbarao, G. V., Ito, O., Berry, W. L., & Wheeler, R. M. (2003). Sodium: A functional plant nutrient. *Critical Reviews in Plant Sciences, 22*(5), 391–416.

Sulok, K. M., Ahmed, O. H., Majid, N. M. A., & Kharuddin, Z. (2021). Effect of organic amendments on soil fertility and crop performance: a review. *Sustainability, 13*(4), 1–15.

Tahat, M., M. Alananbeh, K., A. Othman, Y. and I. Leskovar, D., 2020. Soil Health and Sustainable Agriculture. Sustainability, 12(12), 4859.

Urra, Alkorta and Garbisu, 2019. "Potential Benefits and Risks for Soil Health Derived from the Use of Organic Amendments in Agriculture." Agronomy, 9(9), 542.

Usherwood, N. R. (1985). The role of potassium in crop quality. *Better Crops with Plant Food, 69*(1), 6–8.

Weinert, E. J., Miller, S. A., Ikeda, D. M., Chang, K. S., McGinn, J. M., & DuPonte, M. W., 2014. Natural Farming: Fish Amino Acid. Sustainable Agriculture, SA-12. University of Hawai'i, College of Tropical Agriculture and Human Resources

Yadav, A. N., Kumar, R., Kumar, S., Kumar, V., Sugitha, T., Singh, B., Chauahan, V. S., Dhaliwal, H. S., & Saxena, A. K. (2017). Beneficial microbiomes: Biodiversity and potential biotechnological applications for sustainable agriculture and human health. *J. Appl. Biol. Biotechnol., 5*, 45–57.

Zamora, O. B., Salcedo, J. N., & Sombilla, M. A. (2016). Organic waste utilization for sustainable agriculture. *Journal of Environmental Science and Management, 19*(2), 45–56.

Effect of Preserved Compost Tea Microbial Isolates on the Growth Performance of Mustard Plants

Husnul Azan Tajarudin, Ng Chi Yan, Charles Ng Wai Chun, Yusuf Wibisono, and Rimi Akter

Abstract Compost tea contains a diverse consortium of beneficial microorganisms that enhance plant growth and soil health, yet its use is often limited by the rapid decline in microbial viability during storage. This study aimed to overcome this limitation by isolating compost tea microorganisms and evaluating two preservation strategies—freeze-drying and cold storage—to maintain long-term microbial activity. Microbial cultures harvested at the 12th hour of the exponential phase exhibited the highest post-rehydration viability after freeze-drying (1000 CFU/mL), confirming the superior stability of this method compared to cold storage at 4 °C. The preserved microorganisms were subsequently reactivated and applied to mustard (Brassica juncea) to assess their growth-promoting effects. Mustard plants treated with 30%, 50%, and 70% formulations demonstrated higher growth rates of 0.0043, 0.0037, and 0.0070 day^{-1}, respectively, compared to 0.0021 day^{-1} in the untreated control, indicating enhanced plant performance associated with microbial inoculation. Overall, the findings highlight the effectiveness of freeze-drying as a preservation technique for maintaining compost tea microbial isolates and support their potential use as a stable, sustainable bio-stimulant for crop production.

Keywords Compost tea · Separated · Preserved · Mixed culture microorganism · Freeze drying · Growth

---

H. A. Tajarudin (✉) · N. C. Yan · C. N. W. Chun · R. Akter
Division of Bioprocess Engineering Technology, School of Industrial Technology, Universiti Sains Malaysia, USM, George Town, Pulau Pinang, Malaysia
e-mail: azan@usm.my

Y. Wibisono
Program Studi Teknologi Bioproses, Fakultas Teknologi Pertanian, Universitas Brawijaya, Malang, Indonesia

A. Z. Yaser et al. (eds.), *Compost and Liquid Organic Fertilizers*,
https://doi.org/10.1007/978-981-95-8561-8_12

1 Introduction

Compost tea has emerged as a promising tool in sustainable agriculture, offering an eco-friendly solution for crop production and soil health management. It is a liquid extract derived from well-matured compost that harbors a complex community of beneficial microorganisms, enzymes, and bioactive compounds that provide nutrients and help plant growth capable of supplying nutrients, enhancing plant growth, and suppressing certain soil-borne diseases. As a natural biostimulant, compost tea contributes to the improvement of soil fertility, nutrient cycling, and overall crop performance, thereby presenting an alternative to chemical fertilizers and pesticides (Ng et al., 2020).

Despite its advantages, one of the major limitations of compost tea lies in its short shelf life. The microbial communities within compost tea are highly dynamic, and their viability decreases significantly during storage. As a result, the effectiveness of compost tea as a plant growth promoter is often inconsistent when used over extended periods. To overcome this problem Thus, developing strategies for the separation, preservation, and reactivation of compost tea microorganisms becomes essential to maintain their long-term functionality.

Preservation of microorganisms is a well-established practice in microbiology, particularly in the production of probiotics, starter cultures, and biofertilizers. Techniques such as freeze-drying (lyophilization), cryopreservation, and cold storage are widely used to extend the viability of microbial cultures (Wai Chun et al., 2021). Among these, freeze-drying is considered one of the most effective methods as it allows long-term storage with minimal loss of viability. Cold storage, on the other hand, provides a simpler option but is limited to short-term preservation due to gradual declines in microbial activity.

In this context, the present study aims to investigate the separation, preservation, and reactivation of microorganisms from compost tea, and to evaluate their subsequent effects on the growth of mustard (Brassica juncea). Mustard was chosen as a model crop due to its economic importance and rapid growth cycle, making it an ideal candidate for assessing plant growth-promoting effects. By comparing freeze-drying and cold storage as preservation methods, this work provides insights into the importance of microbial preservation for extending the shelf life and functionality of compost tea in sustainable agricultural systems.

2 Materials and Methods

2.1 Preparation, Isolation, Characterization and Preservation Microorganism in the Compost Tea

Compost tea was prepared by steeping well-decomposed compost in distilled water under aerobic conditions. The mixture was continuously aerated and agitated for **24–48 h** to enhance microbial release into the liquid extract. After brewing, the suspension was filtered to remove coarse particles, leaving a microbial-rich aqueous solution. Figure 1 shows the illustration schematic diagram to making compost tea.

(a) **Isolation and Characterization of Microorganisms**

Microorganisms from compost tea were isolated using the **serial dilution and spread plate technique** on nutrient agar and potato dextrose agar to capture both bacterial and fungal communities. Distinct colonies were selected and purified through repeated streaking. Morphological characterization (colony color, shape, margin, and elevation) and basic biochemical tests were conducted to identify the predominant microbial groups.

(b) **Preservation Techniques**

Two preservation strategies were evaluated:

Freeze-drying (lyophilization): Microbial pellets were harvested by centrifugation, resuspended in protective agents (e.g., skim milk or trehalose), frozen at –80 °C, and subsequently lyophilized. The resulting powders were sealed in sterile vials for long-term storage.

Fig. 1 Schematic diagram to prepare compost tea

Cold storage: Microbial cultures were stored at 4 °C in nutrient broth or agar slants, serving as a short-term preservation method.

(c) **Reactivation of Preserved Microorganisms**

Preserved inocula were rehydrated (in the case of freeze-dried cultures) or sub-cultured (in the case of cold-stored cultures) prior to experimental application. Viability was assessed by colony-forming unit (CFU) counts and visual growth on nutrient agar plates.

(d) **Growth Promotion Assay on Mustard (Brassica Juncea)**

Mustard seeds were surface sterilized with 1% sodium hypochlorite, rinsed thoroughly with sterile distilled water, and germinated in sterile soil under greenhouse conditions. Reactivated microbial inocula were applied as soil drench treatments. Control plants received sterile distilled water without microbial inoculation. Growth parameters measured included germination percentage, shoot and root length, number of leaves, and total biomass. Data were collected at fixed intervals and statistically analyzed to determine significant differences between treatments.

2.2 *Analysis of the Formulated Effluent of Anaerobic Digestion*

The effluent of anaerobic digestion provided by the USM@Sains at Balik Pulau was used as the medium to be formulated with the freeze dried preserved microorganism, imitating as compost tea in determining the effect of compost tea on the growth of the mustard. Therefore, analysis of BOD_5, COD, nitrogen and phosphorus concentration were performed on the filtered sterile effluent formulated with different concentrations of preserved microorganism in 30% (w/v), 50% (w/v) and 70% (w/v) as compared to the effluent before formulation.

(a) **Day Biological Oxygen Demand (BOD) Test**

The 5-Day BOD test of the formulated effluent with different concentration of preserved microorganism was carried out as specified by the 5-Day BOD protocol (APHA, 1995). Blank was prepared with 300 mL of dilution water. Initial DO of the blank and formulated effluents was measured using dissolved oxygen meter (YSI 5100, USA) and the readings were recorded. The formulated effluents and the blank were incubated at 20 °C ± 1 °C for 5 days. The final OD of the formulated effluents and the blank were measured after 5 days. The BOD_5 of the formulated effluents and blank were determined by Eq. (1):

$$\mathrm{BOD_5},\ \mathrm{mg/L} = \frac{D1 - D2}{P} \tag{1}$$

where:

D_1 = DO of diluted sample immediately after preparation, mg/L,
D_2 = DO of diluted sample after 5 d incubation at 20 °C, mg/L,

P = decimal volumetric fraction of sample used.

(b) **Chemical Oxygen Demand (COD) Test**

The COD Test was carried out as specified as in the Method 8000 in Hach (2008). Two mL of the formulated effluent with suitable dilution were pipetted into the HR vial (Hach, USA) at an angle of 45 degrees. Blank was prepared by pipetting 2 mL deionized water into the HR vial. The cap of the vials was closed tightly and the vials were inverted gently several times to mix well by holding the cap as the vials get very hot during mixing. After that, the vials were placed in the preheated DRB200 reactor (Hach, USA) for 2 h. After 2 h, the power of the reactor was switched off and the vials were left to be cooled in the reactor for at least 20 min. The COD value of the blank and the formulated effluents were obtained by measuring with DR 2800 with the program 435 COD HR. The COD value for effluents obtained in mg/L COD were recorded.

(c) **Determination of Nitrogen content using Kjeldahl method**

The determination of nitrogen content in the formulated effluent was performed using Kjeldahl method employed from the standard method in AOAC Official Method 955.04 for Nitrogen (Total) in Fertilizers. In the stage of digestion, 3 mL of the formulated effluent for each concentration, one tablet of the mercury catalyst and 20 mL H_2SO_4 were added into a micro Kjeldahl tube and blank was prepared without the addition of formulated effluent. The digestion process was carried out for 2 h and the content in the micro Kjeldahl tubes were left to cooled. Forty % NaOH was added into the micro Kjeldahl tube containing digested sample until the volume of the content in the tube reached the indicator level of 160 mL. The distillation of the solution into the boric acid in the receiving Erlenmeyer flask for 4 min. The solution in the boric acid receiving flask was titrated with standard 0.1 M HCl solution to the first trace of pink. The volume of the HCl used was recorded and the nitrogen concentration in the effluent was determined by Eq. (2)

$$\text{Nitrogen (mg/L)} = \frac{(\text{mL HCl of sample} - \text{mL HCl of blank})\text{N x 14 x 1000}}{\text{mL of sample}} \quad (2)$$

(d) **Determination of Total Phosphorus content**

The determination of the total phosphorus content of the formulated effluent was carried based on the standard protocol as in Method 365.2 (USEPA, 1971) with the modification of the usage of PhosVer 3 Phosphate Reagent Powder in the determination of orthophosphate. Preliminary persulfate digestion of the formulated effluent was performed to convert other phosphorus compound into orthophosphate. Fifty mL of the formulated effluent, one mL of 11 N H_2SO_4 and 0.4 g ammonium persulfate was transferred into a 125 mL Erlenmeyer flask. The mixture was then heated in an autoclave at 121 °C for 30 min and left to be cooled before diluted to the final volume of 40 mL with the addition of distilled water prior further analysis steps.

Ten mL of the formulated effluent obtained after persulfate digestion was pipetted into a test tube and one packet of PhosVer 3 Phosphate Reagent Powder was added into the test tube. The solution in the test tube was vortexed to mix well and left to react for 5 min. The absorbance reading of the solution was measured with spectrophotometer at 650 nm and the reading was recorded. The experiment was repeated with the formulated effluent that was diluted for 10 times dilution. The phosphorus concentration in the effluent was determined based on the phosphorus standard curve.

2.3 Evaluation of Plant Growth with Preserved Microorganism

The mustard plants used in the evaluation of the effect of freeze dried preserved microorganism formulated in effluent of anaerobic digestion were planted 2 months prior the evaluation stage. Each mustard plant was planted in the homogenized soil and compost and was given similar amount of 25 mL of water and the formulated effluent on a daily basis. Ten mL of the sterile effluent of anaerobic digestion was added with the freeze dried preserved microorganism powder in the concentration of 30% (w/v), 50% (w/v) and 70% (w/v). Each of the formulated effluent with different concentration of freeze dried mixed cultured microorganism was transferred into a spray bottle and spray on the leaves and the soil of the mustards. Control was prepared as the mustard plants were given only water. Water and the formulated effluent with different concentration of freeze dried mixed culture microorganism were given to the sample plants. The height of the plants was measured and recorded on a daily basis for 30 days. Three triplicates for each concentration of formulated effluent were performed.

2.4 Statistical Analysis

One-way analysis of variance (ANOVA) was performed using Statistical Package for Social Sciences (SPSS) to compare the means of the viable cell counting of the rehydrated freeze dried mixed culture microorganism that harvested at different time interval. Apart from that, one-way ANOVA was also performed to compare the means of different concentration of the freeze dried mixed culture microorganism on growth rate based on the height of the mustard plants after 30-days monitoring period. Statistical significance was obtained at 95% confidence level ($\alpha = 0.05$). The mean comparison was assessed by Tukey's test.

3 Results and Discussion

3.1 Separation of the Microorganism from Compost Tea

The growth profile of the mixed culture microorganism was obtained from compost tea brewed with 50 mL/min aeration in the nutrient broth containing 10 g/L glucose as shown in Fig. 2. Four distinct phases observed in a bacterial growth curve, and the four phases are distinguished by the distinct growth rate exhibited in the bacterial culture over time. In the growth profile of the mixed culture microorganism obtained shown in Fig. 3.1, the exponential phase of the mixed culture microorganism in the compost tea was determined as started from the fourth hour to the 12th hour of the compost tea brewing.

The determination of the exponential phase of the growth of mixed culture microorganism was done to deduce the harvest time of the microorganism in the compost tea. According to Morgan et al. (2006), microorganism harvested at late exponential phase a have better survival rate after the process of freeze drying as mature cells grown near to the early stationary phase are capable of protecting themselves from extreme condition by activating starvation responses. Therefore, sample was harvested at two hours interval between the fourth hour and 12th hour of the determined exponential phase of the mixed culture microorganism in the compost tea and subjected to freeze drying.

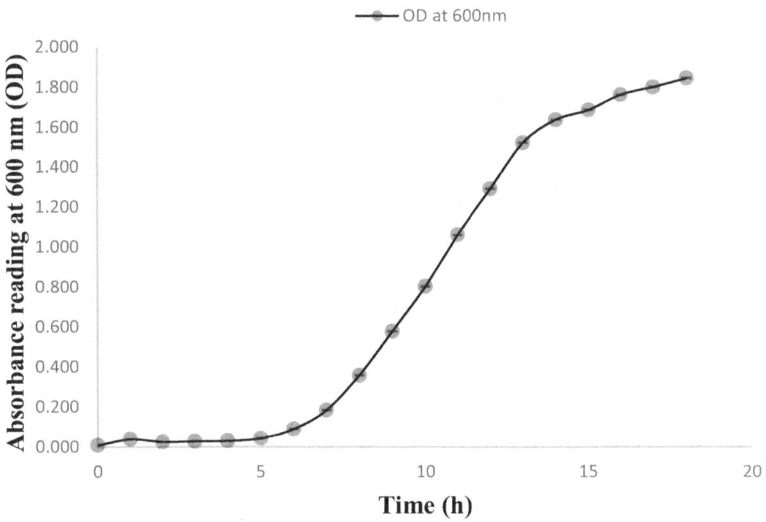

Fig. 2 Growth profile of the mixed culture microorganism in the compost tea. Error bars represent standard deviations of N = 3

3.2 Preservation of the Harvested Microorganism from Compost Tea

Viability of the preserved microorganism harvested in different time interval was analyzed to determine the optimum harvest of the microorganism in the compost tea for maximum viability of the preserved microorganism after freeze drying.

Viable cell counting was used to analyze the viability of the preserved microorganism harvested in different time interval between the fourth hour and 12th hour of the compost tea brewing using plate count method. The viable cell concentration was then expressed as colony-forming units per milliliter and the result was shown in Fig. 3. From the result obtained, there was a statistically significant difference between the viable cell count for the rehydrated freeze dried mixed culture microorganism that harvested at different time interval as determined by one-way ANOVA ($p < 0.05$).

Referring to Fig. 3, the viable cell count obtained for the rehydrated freeze dried microorganism harvested at the 12th hour of the compost tea brewing process possessed the highest viable cell concentration of 1000 CFUs/mL as compared to the viable cell concentration obtained for the rehydrated freeze dried microorganism harvested at other time interval. This outcome was supported by Morgan et al. (2006) as per mentioned that microorganism harvested at late exponential phase had a better survival rate after freeze drying process. Therefore, optimum time to harvest and separate the mixed culture microorganism from the compost tea for maximum viability

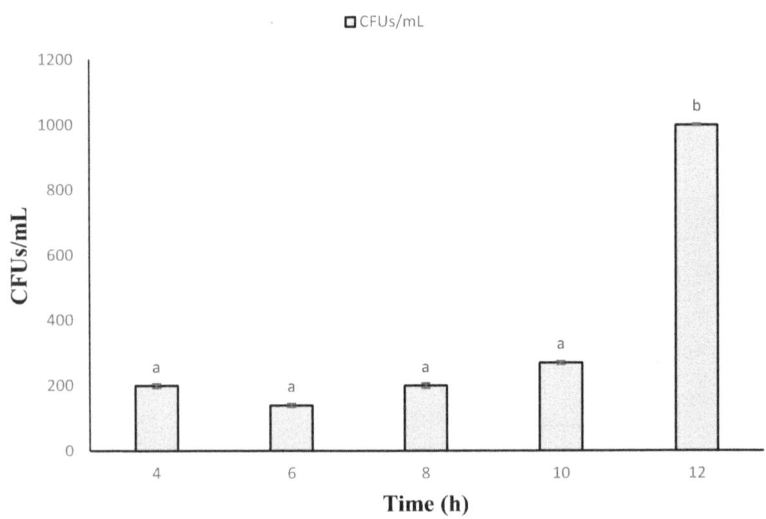

Fig. 3 Viable cell count of rehydrated freeze dried microorganism in CFUs/mL harvested at different time interval. Error bars are mean ± standard deviation of N = 3. The different variables that bear the same superscript letter were not significantly different ($p > 0.05$)

of the freeze dried presercved microorganism was determined at the 12th hour of compost tea brewing as it possessed the highest viability after rehydration.

3.3 Analysis of the Formulated Effluent of Anaerobic Digestion

The analysis of the effluent was carried out for the effluent formulated with the different concentration of freeze dried preserved microorganism in 30% (w/v), 50% (w/v) and 70% (w/v) and compared the effluent before formulation. From the results obtained and illustrated in Fig. 4, the BOD_5, COD, nitrogen and phosphorus concentration in the effluent formulated with preserved microorganism in different concentration were comparatively higher than the effluent before formulation. Referring to Fig. 4, an upward trend was observed in the BOD_5, COD, nitrogen and phosphorus concentration in the effluent formulated with different concentration of preserved microorganism. This indicated that the effluent formulated with higher concentration preserved microorganism contained higher BOD_5, COD, nitrogen and phosphorus concentration. The drastic increase in the COD, nitrogen and phosphorus concentration in the formulated effluent was explained with the presence of the carbon source from the sucrose and the nitrogen and phosphorus source from the milk protein and phosphorus that originated from the skim milk in lyoprotectant (Zayed & Roos, 2004).

3.4 Evaluation of Plant Growth with Preserved Microorganism

The effect of effluent formulated with different concentration of the preserved microorganism on the growth promotion of mustards was observed as the imitation of the effect of the compost tea on the growth of plant. Since the beneficial effects of the compost tea were reported to be contributed by the microbial community present in the compost tea, the mixed culture microorganism harvested from the compost tea played the same role as in the compost tea. The growth of the mustard plants was observed in terms of the height of the mustard plants and converted into respective growth rate. The result of the effect of different concentration of the preserved microorganism on the growth promotion of mustards was summarized and shown in Table 1. There was statistically significant difference of the effect of the freeze dried mixed culture microorganism on the growth rate of mustard plants and on the length of the root ($p < 0.05$).

The growth rate based on the height of the mustard plants did increase variedly with the increasing concentration of freeze dried mixed culture microorganism in effluent but with a low growth rate. Referring to Table 1, growth rate of the mustard

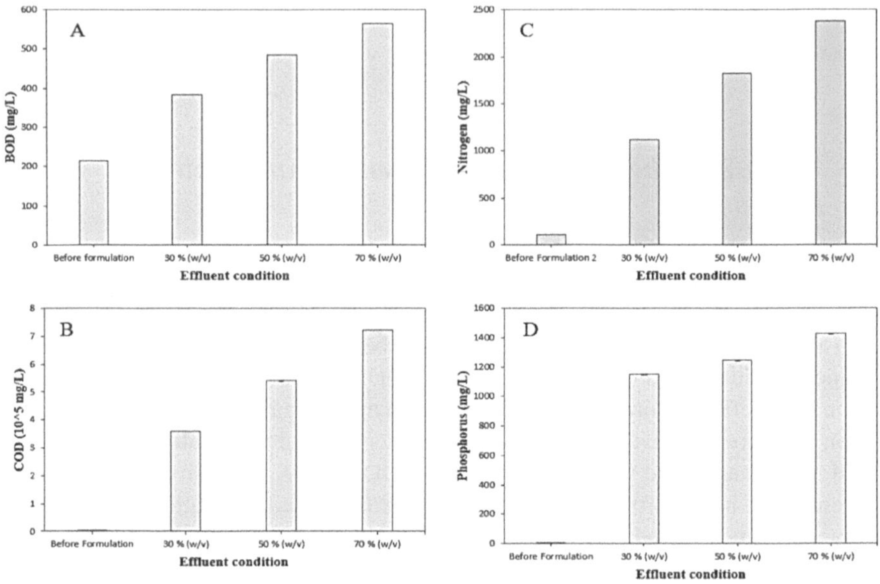

Fig. 4 **a** BOD_5, **b** COD, **c** nitrogen and **d** phosphorus concentration in the effluent before formulation and after formulated with 30% (w/v), 50% (w/v) and 70% (w/v) of preserved microorganism. Error bars represent mean ± standard deviation of N = 3. From the result obtained for COD, nitrogen and phosphorus concentration of the formulated effluent, C:N:P ratios derived for the effluent formulated was shown in Fig. 5. Referring to Fig. 5, the C:N:P ratio for effluent formulated with 30% (w/v), 50% (w/v) and 70% (w/v) of freeze dried mixed culture microorganism were 312:1:1, 434:1:1 and 507:2:1 respectively. These ratios also indicated that the formulated effluents contain excessive carbon content that contributed by the carbon source of the sucrose and skim milk in lyoprotectant used in freeze drying of the mixed culture microorganism

plants applied with effluent formulated with 30% (w/v), 50% (w/v) and 70% (w/v) were 0.0043, 0.0037 and 0.0070 day^{-1} respectively. This result did not possess a uniform upward trend where the growth rate of the mustard plants was not increased along with the increase in the concentration of the preserved microorganism in the formulated effluent. Figure 6 shows mustard plants grown under different concentrations of preserved microorganisms after the treatment period: (a) Control, (b) 30% concentration, (c) 50% concentration, and (d) 70% concentration.

As reported in several studies that the beneficial effect of the compost tea was closely related to the microbial community in the compost tea, the interaction within the microbial community in different concentration may affected the growth rate of the mustard plants differently (Koné et al., 2010; Palmer et al., 2010; Scheuerell & Mahaffee, 2006). But, the microbial interaction and the mechanisms involved in the efficacy of the compost tea application was still not fully understood and remained uncertain in some studies (Scheuerell & Mahaffee, 2006; St. Martin & Brathwaite, 2012). Mechanism and the interaction of the microbial community in the compost tea

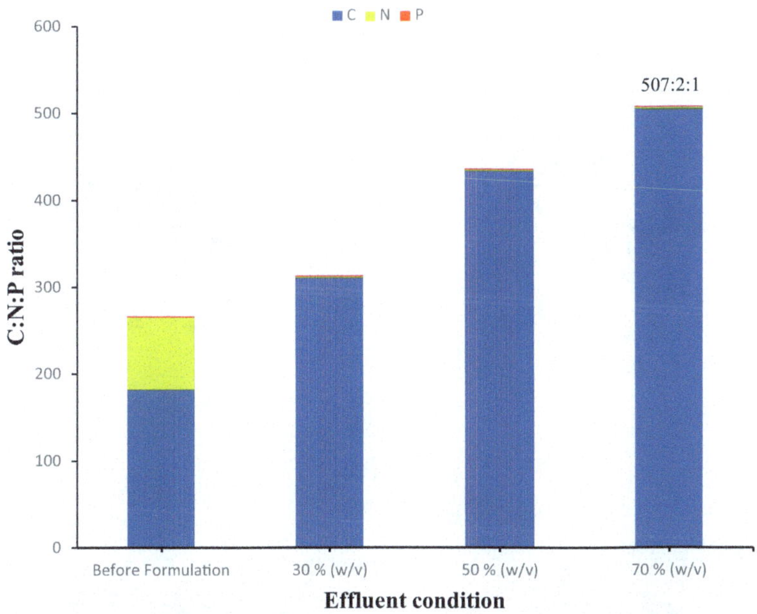

Fig. 5 C:N:P ratio for effluent before formulation and after formulated with 30% (w/v), 50% (w/v) and 70% (w/v) of freeze dried mixed culture microorganism

Table 1 Effect of different concentration of the preserved microorganism on the growth of mustard plants in terms of growth rate based on height of plants throughout the observation period. Data are mean ± standard deviation (N = 3)

Concentration of freeze dried mixed culture microorganism	Growth rate
	Height (day^{-1})
Control	0.0021 ± 0.001
30%	0.0043 ± 0.002
50%	0.0037 ± 0.002
70%	0.0070 ± 0.003

affected the effect of the formulated effluent with different concentration of freeze dried mixed culture microorganism on the growth of the plant.

Apart from that, C:N:P ratio of the effluent affected the growth of the plants in terms of affecting the soil environment and the microbial metabolism of the soil microbes as well (Li et al., 2012). Thus, the interaction of the microbial community may be affected by the C:N:P ratio of the formulate effluent and therefore created a varied efficacy effect on the growth of the mustards. According to Eschen et al. (2006), excessive carbon content in the soil affected the nitrogen absorption of the plant and therefore affect the plant growth. From the results shown in Table 1, the application of the formulated effluent did possess an observable positive effect on

Fig. 6 Mustard plants grown under different concentrations of preserved microorganisms after the treatment period: **a** Control, **b** 30% concentration, **c** 50% concentration, and **d** 70% concentration

the growth in terms of height of the mustard plants. This result may contributed by the presence of the microbial community within the compost tea, the effect of C:N:P ratio of the formulated effluent and the complication between the microbial interaction in different environment created by the effluent formulated with different concentration freeze dried mixed culture microorganism.

4 Conclusion

The present findings demonstrated that the application of the formulated effluent of anaerobic digestion with 30% (w/v), 50% (w/v) and 70% (w/v) of freeze dried mixed culture microorganism on mustard plants had yielded growth rate of 0.0043,

0.0037 and 0.0070 day^{-1} respectively as compared to 0.0021 day^{-1} in control. All in all, the result showed that the application of freeze dried mixed culture microorganism formulated in effluent of anaerobic digestion had positive effect on the growth promotion of the mustards. The result also suggested that the lyoprotectant used in the formulation for freeze drying process may also affect the growth promotion of the mustard.

References

Apha. (1995). Standard Methods for the examination of water and wastewater. In *American Public Health Association, American Water Works Association, Water Environment Federation, Washington* (19th Ed.). Washington, D.C.

Eschen, R., Müller-Schärer, H., & Schaffner, U. (2006). Soil carbon addition affects plant growth in a species-specific way. *Journal of Applied Ecology, 43*(1), 35–42.

Koné, S. B., Dionne, A., Tweddell, R. J., Antoun, H., & Avis, T. J. (2010). Suppressive effect of non-aerated compost teas on foliar fungal pathogens of tomato. *Biological Control, 52*(2), 167–173.

Li, Y., Wu, J., Liu, S., Shen, J., Huang, D., Su, Y., & Syers, J. K. (2012). Is the C: N: P stoichiometry in soil and soil microbial biomass related to the landscape and land use in southern subtropical China? *Global Biogeochemical Cycles, 26*(4).

Morgan, C. A., Herman, N., White, P., & Vesey, G. (2006). Preservation of micro-organisms by drying: A review. *Journal of Microbiological Methods, 66*(2), 183–193.

Ng, C. W., Ismail, A. F., Mohd Zaini Makhtar, M., Jamaluddin, M. N. F., & Tajarudin, H. A. (2020). Conversion of food waste via two-stage fermentation to controllable chicken feed nutrients by local isolated microorganism. *International Journal Of Recycling Of Organic Waste In Agriculture, 9*, Article 3359.

Palmer, A., Evans, K., & Metcalf, D. (2010). Characters of aerated compost tea from immature compost that limit colonization of bean leaflets by Botrytis Cinerea. *Journal of Applied Microbiology, 109*(5), 1619–1631.

Scheuerell, S. J., & Mahaffee, W. F. (2006). Variability associated with suppression of gray mold (Botrytis Cinerea) on geranium by foliar applications of nonaerated and aerated compost teas. *Plant Disease, 90*(9), 1201–1208.

St. Martin, C., & Brathwaite, R. (2012). Compost and compost tea: principles and prospects as substrates and soil-borne disease management strategies in soil-less vegetable production. *Biological Agriculture & Horticulture, 28*(1), 1–33.

Wai Chun, C. N., Tajarudin, H. A., Ismail, N., Azahari, B., & Mohd Zaini Makhtar, M. (2021). Elucidation of mechanical, physical, chemical and thermal properties of microbial composite films by integrating sodium alginate with bacillus subtilis Sp. *Polymers, 13*(13), 2103.

White, P., & Brown, P. (2010). Plant nutrition for sustainable development and global health. *Annals of Botany, 105*(7), 1073–1080.

Zayed, G., & Roos, Y. H. (2004). Influence of Trehalose and moisture content on survival of Lactobacillus Salivarius subjected to freeze-drying and storage. *Process Biochemistry, 39*(9), 1081–1086.

Compost Leachate from Food and Vegetable Wastes as Alternative Fertilizers for Brazilian Spinach (*Alternanthera sissoo*)

Abu Zahrim Yaser, Nursyafieqah Basri, Nur Amirah Sakirah Barli, Nur Fadilah Ramli, Sariah Saalah, Ahmad Safwan Ismail, Azreen Ibrahim, Nurhazwani Yusoff Azudin, and Mariani Rajin

Abstract This study evaluated the feasibility of using compost leachate as a substitute for conventional commercial fertilizers to improve the growth of Brazilian spinach (*Alternanthera sissoo*). Three treatments were prepared with replicates: food-waste compost leachate (FWCL), diluted vegetable-waste compost leachate (DVWCL), and diluted commercial fertilizer (DCF). The nitrogen level for each plant was maintained at 5 mg throughout 4 weeks cultivation period. Growth parameters, including leaf, stem, and root development were measured using ImageJ software after 4 weeks. One-way ANOVA was performed to statistically analyze growth data. The results showed that Brazilian spinach treated with FWCL and DVWCL showed similar growth responses to those treated with DCF, with no significant differences among treatments ($p > 0.05$). Dilution of VWCL and CF effectively reduced electrical conductivity (EC) and heavy metal concentrations to within permissible limits. These findings demonstrate the potential of compost leachates (FWCL and DVWCL) as alternatives to commercial fertilizers for Brazilian spinach cultivation. This study also highlights that proper dilution is essential to ensure heavy metal levels remain safe for agricultural use.

Keywords Compost leachate · Brazilian Spinach (*Alternanthera sissoo*) · Dilution · Growth performance · Commercial fertilizer

---

A. Z. Yaser (✉) · N. Basri · N. A. S. Barli · N. F. Ramli · S. Saalah · A. S. Ismail · A. Ibrahim · N. Y. Azudin · M. Rajin
Faculty of Engineering, Universiti Malaysia Sabah, Sabah, Malaysia
e-mail: zahrim@ums.edu.my

A. Z. Yaser et al. (eds.), *Compost and Liquid Organic Fertilizers*,
https://doi.org/10.1007/978-981-95-8561-8_13

1 Introduction

Compost leachate (CL) is a liquid by-product generated during microbial degradation of organic compound in biowaste, resulting in stabilized carbon compounds (Casini et al., 2021; Smith & Collins, 2006). However, CL may also contain heavy metals and phytotoxic substances, including ammonia, low-molecular weight organic compounds, and salts (Zhang et al., 2009), which can pose significant risks of water and soil contamination. Tahsini et al. (2024) reported that a significant amount (75–100 L) of CL can be produced per ton of waste, particularly when the waste has a moisture content exceeding 50%. Therefore, effective management strategies, such as dilution, biological or physicochemical treatments, are essential before recycling or releasing the CL into the environment.

Various treatment methods have been explored for CL management. Biological approaches, such as anaerobic reactors, membrane bioreactors, biofilters, and constructed wetlands, are commonly used due to their operational simplicity and cost-effectiveness (Tahsini et al., 2024). However, elevated concentrations of pollutants in CL can inhibit microbial activity, reducing treatment efficiency. Physicochemical methods, such as coagulation-flocculation, have also been studied using different coagulants and flocculants. For example, Kiani Feizabadi et al. (2015) tested ferrous sulfate, ferric chloride, poly ferric sulfate, alum, and polyaluminum chloride along with two commercial co-coagulants (K350CF and LT25, representing anionic and cationic types), in removing chemical oxygen demand (COD) and total suspended solids (TSS) from leachate generated at a composting plant. In addition, the study by Zahrim et al. (2017) demonstrated that treated water obtained after the coagulation process showed improved phytotoxicity levels. Similarly, coagulation treatment of biologically treated landfill leachate using aluminum chlorohydrate, PAC, and tannin achieved over 80% color removal and yielded germination indices above 80%, suggesting that the treated leachate was free of phytotoxins and could be safely reused for agricultural purposes (Ibrahim & Zahrim, 2025).

Recently, increasing interest has been given to the use of treated and non-treated CL as sustainable liquid fertilizer, offering a potential alternative to synthetic fertilizers that are associated with long-term environmental concerns (Ayesha, 2020). CL contains essential plant nutrients such as nitrogen, phosphorus, potassium, and trace elements that can enhance soil properties and support plant growth (Sall et al., 2019). A review by Safie et al. (2025) further highlighted that diluted vegetable compost leachate, with reduced COD, TN, and heavy metal contents, effectively improves plant growth and yield compared to raw leachate, demonstrating that a simple dilution method offers an economical and sustainable approach for utilizing vegetable waste–derived leachate as liquid fertilizer.

Brazilian spinach (*Alternanthera sissoo*) is a nutrient-dense leafy vegetable rich in vitamin C, vitamin K, folic acid, iron, calcium, and carotenoid (Alam et al., 2022). It belongs to the Amaranthaceae family (Khairul Ikram et al., 2022). Unlike most spinach varieties, *A. sissoo* produces non-viable seeds but can be effectively propagated through stem or terminal cuttings (Muda et al., 2022). The presence of leaves on cuttings enhances rooting success, sprouting, and root development (Belniaki et al., 2018). Due to its ease of cultivation and low maintenance requirements, Brazilian spinach is an ideal candidate for research on crop growth in leachate conditions. Alam et al. (2022) also reported that the species requires minimal fertilizer inputs, making it well-suited for sustainable cultivation. Furthermore, Mohd Hussain and Amir (2022) observed that Brazilian spinach is relatively less susceptible to common garden pests, futher increasing its suitability for experimental studies and home gardening.

This study aims to evaluate the potential of compost leachate as a liquid fertilizer for Brazilian Spinach by comparing its effect on plant growth with those of a commercial fertilizer under equal nitrogen supply. This study also investigates the role of dilution in optimizing nutrient composition and ensure the safe use of CL in agriculture.

2 Materials and Methods

2.1 Research Location

This study was conducted at Taman Indah Permai residential area, Kota Kinabalu Sabah, Malaysia (6.0603° N, and 116.1510° E) from February to March 2025.

2.2 Materials

Food-waste compost leachate (FWCL) was generated during the composting of a mixture of kitchen waste and water apple at the Imbak Canyon Studies Centre (ICSC) and was used without dilution. Vegetable-waste compost leachate (VWCL) was produced from a mixture of vegetable waste, chicken manure, and rice husk, sourced from Kundasang, Sabah, Malaysia (Rajin et al., 2025).The commercial liquid fertilizer (CF) used as a control treatment was Apple Green Biohybrid Organic Liquid Fertilizer (Selangor, Malaysia).

2.3 Physichochemical Analyses

Samples of VWCL, FWCL, and CF were analysed for physicochemical, including ammoniacal nitrogen, total nitrogen, Chemical Oxygen Demand (COD), orthophosphate, potassium, magnesium, calcium, copper, zinc, and iron. All analyses were carried out at the Water Research Unit, Universiti Malaysia Sabah (UMS).

3 Preparation of Treatments and Experimental Setup

Diluted VWCL (DVWCL) was prepared by mixing VWCL with water at a 1:9 ratio (v/v). A diluted commercial fertilizer (DCF) was prepared following the manufacturer's recommended dilution ratio of 1:499 (CF: water, v/v). The typical dilution ratio of compost leachate (CL) used for agricultural applications ranges from 1:10 to 1:100 (Yaser, 2022).

Although Faria et al. (2023) recommended 135 mg N per pot per week for spinach, the limited availability of FWCL restricted nitrogen application to 1.25 mg per plant per week. Each treatment (FWCL, DVWCL and DCF) consisted of four replications and was applied to the Brazilian spinach in 15 cm × 15 cm pot. The applications were performed on days 0 and 14. Over the 4-week cultivation period, each plant received a total of 5 mg of N.

4 Preparation of Cutting and Transplanting

Healthy, freshly harvested Brazilian spinach stem cuttings (7 cm in length), each bearing two leaves, were prepared. Excess foliage was removed to reduce transpiration. For each treatment, two pots (15 cm × 15 cm) were prepared, with each pot containing two cuttings, resulting in four cuttings per treatment. Each cutting was inserted to one-third of its length into the growing medium, watered immediately, and maintained under moist conditions throughout the growth period (Alam et al., 2022).

Irrigation was performed based on daily weather conditions, with approximately 200 mL of water manually applied to each pot. A homemade biopesticide was sprayed as soon as possible to prevent disease development and promote healthy growth (Alam et al., 2022). The organic pesticide was prepared from a blend of onion and peppers mixed with cooking oil and dish soap. Harvesting was carried out using garden shears by cutting the plant at root level, ideally four weeks after planting when the plants were well established.

5 Plant Growth Measurement

At week 4, all plants were harvested, and images of the entire plant including leaves, stems and roots were captured from multiple angles to ensure complete visual coverage for accurate morphological measurements. For analysis in ImageJ, the captured images were opened via the software toolbar, and a calibration scale was established using a ruler placed adjacent to a reference part of the plant during imaging. The scale was standardized to 1 cm, and the corresponding number of pixels was manually set as one unit.

For leaf area determination, the image was opened in ImageJ (File > Open) and cropped to remove extraneous elements (Image > Crop). The image was converted to grayscale (Image > Type > 8-bit), and the contrast was adjusted using the threshold function (Image > Adjust > Threshold) to isolate the leaf area. Total leaf area was calculated using Analyze > Analyze Particles., and the results were validated using a leaf area meter (LI-COR LI3100, 1996).

To measure stem length using ImageJ, the Segmented Line Tool was selected from the toolbar to trace the stem by creating a series of connected line segments along its path. By clicking at regular intervals along the stem and double-clicking at the endpoint, the tracing was completed, and the total stem length was calculated using the 'Measure' function.

Root parameters, including length, diameter, and volume, were analyzed ImageJ, in combination with SmartRoot plugin (Lobet et al., 2011). After calibration, roots were traced using Segmented Line tool, and the software automatically calculated the length of each traced line. Total root length was determined by summing the measurements of the primary, secondary, and minor roots (Albert et al., 2023).

6 Statistical Analyses

Statistical analysis and data processing, including the calculation of mean and standard deviation values, were performed using Microsoft Excel. One-way analysis of variance (ANOVA) was also conducted in Microsoft Excel to determine significant differences among treatments.

7 Results and Discussion

7.1 Physicochemical Analysis of DVWCL and DCF

Table 1 shows the physicochemical characteristics of undiluted VWCL and CF, where both exhibited electrical conductivity (EC) and heavy metals concentration exceeding the permissible limits. However, after the dilution, the characteristics of DVWCL and DCF were theoretically within acceptable ranges based on regulatory standards and previous studies (Table 2). Both DVWCL and DCF displayed neutral to alkaline pH, which is favorable for agricultural applications. In contrast, acidic fertilizers can be detrimental to soil health as they promote the proliferation of soil-borne pathogens (Zhang et al., 2022). The EC values for DVWCL and DCF were below 13,600 μS/cm, indicating their suitability for agricultural use. Romero et al. (2013) reported that EC values above this threshold can reduce crop yield. Excessive EC can disrupt physiological functions and cause toxicity due to ion accumulation (Butcher et al., 2016). Furthermore, the concentration of Cu, Zn and Fe for DVWCL

Table 1 Physicochemical properties of VWCL, FWCL and CF

Parameter	VWCL	FWCL	CF	Test method
pH	8.9	7.1	ND	pH Meter
EC (μS/cm)	24,800	1310	370,000	EC Meter
NH_4– N (mg/L)	1350	7.4	29,614	HACH8155
TN (mg/L)	2356	25.04	44,797	In-house Method
COD (mg/L)	21,050	2625	10,337	HACH 8000
P (mg/L)	99	3.2	39,934	HACH 8048
K (mg/L)	7000	8.6	54,800	APHA 3210 B
Mg (mg/L)	187	2.6	1370	APHA 3210 B
Ca (mg/L)	446	8.7	1150	APHA 3210 B
Cu (mg/L)	0.02	<0.01	7.95	APHA 3210 B
Zn (mg/L)	4.8	<0.01	18.3	APHA 3210 B
Fe (mg/L)	11.9	0.3	50.2	APHA 3210 B

ND = Not Determined

and DCF were below the recommended limits set by the FAO (Ayers & Westcot, 1985; Jeong et al., 2016), indicating no potential risk.

7.2 *Growth and Yield of Brazilian Spinach*

Plant morphology was assessed on Day 28 to monitor the growth of Brazilian Spinach, as shown in Fig. 1. The assessment focused on morphological differences in leaves, stems, and roots in response to different fertilizer treatments.

In this study, the nitrogen level was maintained at 5 mg per plant for all treatments (FWCL, DVWCL, and DCF) throughout the 4-week cultivation period. One-way ANOVA showed no significant differences ($P > 0.05$) in the growth of leaves, stems and roots across the different treatments (Tables 3, 4, 5, 6, 7 and 8). These findings are align with those of Sulistiani et al. (2023), who observed no significant difference in Brazilian Spinach growth when nitrogen levels were maintained using liquid organic fertilizer derived from pineapple peel and leftover rice waste. Similarly, Jama et al. (2023) reported no significant difference in leaf numbers when comparing green manure and urea applications at a fixed 2.3×10^7 mg N/ha. The lack of significant growth may be attributed to the limited availability of nutrients (Sulistiani et al., 2023) (Tables 4, 6 and 8).

Table 2 Physicochemical properties of DVWCL & DCF in relation with permissible limits for crop nutrient management

Parameter	FWCL	DVWCL	DCF	Permissible Limit	References
pH	7.1	8.5	6.7	>4.5	Department of Malaysia (2014)
EC (μS/cm)	1310	2480	740	<13,600	Cáceres et al. (2015), Das and Kirkland (2008), Gutiérrez-Miceli et al. (2017), Romero et al. (2013)
NH_4-N (mg/L)	7.4	135	59.23	ND-69000	Cáceres et al. (2015), Das and Kirkland (2008), Gutiérrez-Miceli et al. (2017), Romero et al. (2013)
TN (mg/L)	25.04	235.6	89.59	ND	ND
COD (mg/L)	2625	2105	20.67	ND	ND
P (mg/L)	3.2	9.9	79.87	1.6–39,000	Cáceres et al. (2015), Das and Kirkland (2008), Gutiérrez-Miceli et al. (2017), Romero et al. (2013)
K (mg/L)	8.6	700	109.6	359–9000	Cáceres et al. (2015), Das and Kirkland (2008), Gutiérrez-Miceli et al. (2017), Romero et al. (2013)
Mg (mg/L)	2.6	18.7	2.74	100	Ayers and Westcot (1985)
Ca (mg/L)	8.7	44.6	2.3	230	Ayers and Westcot (1985)
Cu (mg/L)	<0.01	0.002	0.02	0.2	Ayers and Westcot (1985)
Zn (mg/L)	<0.01	0.48	0.04	2.0	Ayers and Westcot (1985)
Fe (mg/L)	0.3	1.19	0.1	5.0	Jeong et al. (2016)

Note ND = Not Determined

Fig. 1 Brazilian Spinach plants after 28 days of growth with different treatments: (**a**) Food-waste compost leachate (FWCL), (**b**) Diluted vegetable-waste compost leachate (DVWCL), (**c**) Diluted commercial fertilizer (DCF)

Table 3 Effect of different treatments on leaf growth of Brazilian Spinach

Treatment	Parameter				
	Number of leaves	Leaf area (cm^2)	Leaf fresh weight (g)	Leaf dry weight (g)	Moisture content (%)
FWCL	$42.00 \pm 10.10_a$	$78.54 \pm 13.43_a$	$2.00 \pm 0.43_a$	$0.22 \pm 0.04_a$	$88.74 \pm 3.98_a$
DVWCL	$41.00 \pm 15.78_a$	$53.62 \pm 40.78_a$	$1.45 \pm 0.87_a$	$0.19 \pm 0.09_a$	$86.03 \pm 3.36_a$
DCF	$28.00 \pm 5.48_a$	$56.61 \pm 25.98_a$	$1.53 \pm 0.82_a$	$0.20 \pm 0.10_a$	$86.45 \pm 1.23_a$

Note The standard error of mean value (n = 4) with the same letter had no significant difference at $p < 0.05$

Table 4 Variance analysis (Leaf area)

Source of variance	SS	df	MS	F	P-value	F crit
Between group	1482.033	2	741.0163	0.882379	0.446758	4.256495
Within group	7558.139	9	839.7932			
Total	9040.17	11				

8 Conclusions

This study demonstrated that dilution of VWCL and CF effectively reduced electrical conductivity and heavy metal concentrations to levels comply with permissible agricultural limit. Over a 4-week cultivation period with a fixed nitrogen application of 5 mg per plant, FWCL and DVWCL supported Brazilian spinach growth

Table 5 Effect of different treatments on stem growth of Brazilian Spinach

Treatment	Parameter				
	stem length (cm)	Stem fresh weight (g)	Stem dry weight (g)	Moisture content (%)	Stem diameter (cm)
FWCL	$12.22 \pm 1.06_a$	$2.3 \pm 0.33_a$	$0.48 \pm 0.10_a$	$79.05 \pm 2.01_a$	$0.42 \pm 0.07_a$
DVWCL	$11.90 \pm 1.00_a$	$1.97 \pm 0.60_a$	$0.38 \pm 0.08_a$	$80.20 \pm 2.61_a$	$0.32 \pm 0.06_a$
DCF	$7.43 \pm 4.14_a$	$1.97 \pm 0.67_a$	$0.44 \pm 0.15_a$	$77.63 \pm 1.62_a$	$0.24 \pm 0.09_a$

Note The standard error of mean value (n = 4) with the same letter had no significant difference at $p < 0.05$

Table 6 Variance analysis (Stem Fresh Weight)

Source of variance	SS	df	MS	F	P-value	F crit
Between group	0.28385	2	0.141925	0.465671	0.64203	4.256495
Within group	2.742975	9	0.304775			
Total	3.026825	11				

comparable to DCF, with no significant differences in leaf, stem, or root development. These results suggest that FWCL and DVWCL can be used as alternatives to commercial fertilizers. Proper dilution is essential to ensure nutrient balance and reduce potential risks. Future studies should test higher nutrient levels, longer cultivation periods and field conditions to further confirmed their effectiveness and safety for crop production.

Table 7 Effect of different treatments on root growth of Brazilian Spinach

Treatment	Parameter					
	root length (cm)	Root fresh weight (g)	Root dry weight (g)	Moisture content (%)	Root diameter (cm)	Root volume (cm^3)
FWCL	$26.09 \pm 9.79_a$	$0.63 \pm 0.05_a$	$0.16 \pm 0.02_a$	$75.30 \pm 2.63_a$	$0.12 \pm 0.04_a$	$0.03 \pm 0.02_a$
DVWCL	$18.62 \pm 14.24_a$	$0.76 \pm 0.34_a$	$0.18 \pm 0.07_a$	$76.39 \pm 2.00_a$	$0.10 \pm 0.07_a$	$0.03 \pm 0.05_a$
DCF	$29.35 \pm 2.73_a$	$1.03 \pm 0.25_a$	$0.26 \pm 0.08_a$	$75.24 \pm 1.85_a$	$0.11 \pm 0.04_a$	$0.07 \pm 0.08_a$

Note The standard error of mean value ($n = 4$) with the same letter had no significant difference at $p < 0.05$

Table 8 Variance analysis (Root length)

Source of variance	SS	df	MS	F	P-value	F crit
Between group	242.026	2	121.013	1.18654	0.34885	4.25649
Within group	917.893	9	101.988			
Total	1159.92	11				

Acknowledgements The research was funded by Universiti Malaysia Sabah (UMS) through Dana Living Lab DLV2302 Vegetable Waste Composting: Leachate Processing.

References

Alam, M. A., Rahmat, N. A., Mijin, S., Rahman, Md. S., & Hasan, M. M. (2022). Influence of Palm Oil Mill Effluent (POME) on growth and yield performance of Brazilian spinach (Alternanthera sissoo). *Journal of Agrobiotechnology, 13*(1), 40–49. https://doi.org/10.37231/jab.2022.13.1.287

Albert, D., Nanaiah, K., & Sheshshayee, M. S. (2023). A novel automated method for screening rice genotypes for root growth rate using ImageJ software. In *The Mysore Journal of Agricultural Sciences.*

Anggraeni, L., Latifah, E., Rahmawati, D., Khamidah, A., Yustina, I., & Zubaidi, T. (2023, May). Plant architectural arrangements to improve production and quality of Brazilian Spinach. In *3rd International Conference on Smart and Innovative Agriculture (ICoSIA 2022)* (pp. 635–645). Atlantis Press.

Ayers, R. S., & Westcot, D. W. (1985). *Water quality for agriculture* (Vol. 29). Food and agriculture organization of the United Nations Rome.

Ayesha, A. (2020). Comparison of crop production efficiency of compost leachate with chemical fertilizer and evaluating its effect on germination and growth of wheat crop. *African Journal of Biotechnology, 19*(5), 282–286. https://doi.org/10.5897/ajb2020.17091

Belniaki, A. C., Rabel, L. A. D. N., Gomes, E. N., & Zuffellato-Ribas, K. C. (2018). Does the presence of leaves on coleus stem cuttings influence their rooting? *Ornamental Horticulture, 24*(3), 206–210. https://doi.org/10.14295/oh.v24i3.1204

Butcher, K., Wick, A. F., Desutter, T., Chatterjee, A., & Harmon, J. (2016). Soil salinity: A threat to global food security. In *Agronomy Journal* (Vol. 108, Issue 6, pp. 2189–2200). American Society of Agronomy. https://doi.org/10.2134/agronj2016.06.0368

Cáceres, R., Magrí, A., & Marfà, O. (2015). Nitrification of leachates from manure composting under field conditions and their use in horticulture. *Waste Management, 44*, 72–81. https://doi.org/10.1016/j.wasman.2015.07.039

Casini, D., Barsali, T., Rizzo, A. M., & Chiaramonti, D. (2021). Production and characterization of co-composted biochar and digestate from biomass anaerobic digestion. *Biomass Conversion and Biorefinery, 11*(6), 2271–2279. https://doi.org/10.1007/s13399-019-00482-6

Das, K. C., & Kirkland, J. T. (2008). Quantification of water extractable contaminants from food waste and biosolids blends at different stages of composting. *Compost Science and Utilization, 16*(3), 200–206. https://doi.org/10.1080/1065657X.2008.10702378

Department of Standard Malaysia. (2014). Malaysian Standard Liquid Fertilizers-Specification.

Faria, L. O., de Mello, M. C., & Pelá, A. (2023). Doses and timing of nitrogen application on spinach productivity. *Acta Agronomica, 72*(2), 139–145. https://doi.org/10.15446/acag.v72n2.100895

Giménez, A., Gómez, P. A., Bustamante, M. Á., Pérez-Murcia, M. D., Martínez-Sabater, E., Ros, M., Pascual, J. A., Egea-Gilabert, C., & Fernández, J. A. (2021). Effect of compost extract addition to different types of fertilizers on quality at harvest and shelf life of spinach. *Agronomy, 11*(4). https://doi.org/10.3390/agronomy11040632

Gutiérrez-Miceli, F. A., García-Gómez, R. C., Oliva-Llaven, M. A., Montes-Molina, J. A., & Dendooven, L. (2017). Vermicomposting leachate as liquid fertilizer for the cultivation of sugarcane (Saccharum sp.). *Journal of Plant Nutrition, 40*(1), 40–49. https://doi.org/10.1080/01904167.2016.1193610

Ibrahim, A. & Zahrim, A. Y. (2025) Treatment of biologically treated landfill leachate by coagulation for possible reuse in agriculture. In A. Z. Yaser, J. Lamaming, D. Johnson & N. Hilal (eds.), *Nutrients and Colored Compounds in Wastewater.* Elsevier, pp. 529–538.

Jeong, H., Kim, H., & Jang, T. (2016). Irrigation water quality standards for indirect wastewater reuse in agriculture: A contribution toward sustainablewastewater reuse in South korea. *Water (Switzerland), 8*(4). https://doi.org/10.3390/w8040169

Khairul Ikram, E. H., Wan Mohd Nasir, W. D. N., & Khairul Ikram, N. K. (2022). Antioxidant activity and total Phenolics content of Brazilian Spinach (Alternantherasissoo) and Spinach Cultivarin Malaysia. *Malaysian Journal of Medicine and Health Sciences, 18*(8), 221–229. https://doi.org/10.47836/mjmhs18.8.29

LI-COR LI 3100. (1996). *Instruction Manual LI-3100 Area Meter* ®. http://www.licor.com

Lobet, G., Pagès, L., & Draye, X. (2011). A novel image-analysis toolbox enabling quantitative analysis of root system architecture. *Plant Physiology, 157*(1), 29–39. https://doi.org/10.1104/pp.111.179895

Mahvi, A. H., Kiani Feizabadi, G., Dehghani, M. H. & Mazloomi, S. (2015). Efficiency of different coagulants in pretreatment of composting plant leachate. *Journal of Biodiversity and Environmental Sciences, 6*(6), pp. 21–28http://www.innspub.net

Mohd Hussain, N. H., & Amir, A. F. (2022). Preference of common garden pest towards edible plants. *Malaysian Journal of Sustainable Environment, 9*(1), 36. https://doi.org/10.24191/myse.v9i1.17291

Montemurro, F., Tittarelli, F., Lopedota, O., Verrastro, V., & Diacono, M. (2015). Agronomic performance of experimental fertilizers on spinach (Spinacia oleracea L.) in organic farming. *Nutrient Cycling in Agroecosystems, 102*(2), 227–241. https://doi.org/10.1007/s10705-015-9691-8

Muda, S. A., Lakitan, B., Wijaya, A., & Susilawati, S. (2022). Response of Brazilian spinach (Alternanthera sissoo) to propagation planting material and NPK fertilizer application. *Pesquisa Agropecuaria Tropical, 52*,. https://doi.org/10.1590/1983-40632022v5272730

Rajin, M., Saalah, S., Atinin, E., Shafizan, S., Azmier, M., & Yaser, A. (2025). *Mesophilic Composting of Vegetable Waste Using a Passive Aerated Static Bioreactor: Simultaneous Production of Compost and Leachate* (pp. 341–353). https://doi.org/10.1007/978-981-97-8931-3_19

Romero, C., Ramos, P., Costa, C., & Carmen Márquez, M. (2013). Raw and digested municipal waste compost leachate as potential fertilizer: Comparison with a commercial fertilizer. *Journal of Cleaner Production, 59*, 73–78. https://doi.org/10.1016/j.jclepro.2013.06.044

Safie, N. N., Lajakaya, N. A., Marnoh, N., Md. Ruzaman, N. F. & Yaser, A. Z. (2025). A review of the utilization of vegetable waste compost leachate as fertilizer by dilution method. In A. Z. Yaser, J. Lamaming, D. Johnson & N. Hilal (eds.) *Nutrients and Colored Compounds in Wastewater.* Elsevier, Chapter 8.

Sall, P. M., Antoun, H., Chalifour, F. P., & Beauchamp, C. J. (2019). Potential use of leachate from composted fruit and vegetable waste as fertilizer for corn. *Cogent Food and Agriculture, 5*(1). https://doi.org/10.1080/23311932.2019.1580180

Smith, J. L., & Collins, H. P. (2006). Management of organisms and their processes in soils. In *Soil Microbiology, Ecology and Biochemistry: Third Edition* (pp. 471–502). Elsevier. https://doi.org/10.1016/b978-0-08-047514-1.50021-4

Sulistiani, L., P. Negara, Z., Adriansyah, F., Gustiar, F., Halimi, E. S., Sodikin, E., & Ali Muda, S. (2023). The effects of shading and organic domestic waste on Brazilian Spinach growth. *Jurnal*

Lahan Suboptimal : Journal of Suboptimal Lands, 12(1), 52–61. https://doi.org/10.36706/jlso.12.1.2023.623

Tahsini, M. J., Nikaeen, M., & Nafez, A. H. (2024). Biological treatment of compost leachate: Assessing the efficacy of composting process and bioaugmentation of composting piles. *Environmental Technology and Innovation, 36*,. https://doi.org/10.1016/j.eti.2024.103859

Yaser, A. Z. (Ed.). (2022). *Pengkomposan: Prinsip, Teknik, dan Amalan.* Penerbit Universiti Malaysia Sabah.

Zahrim, A. Y., Dexter, Z. D., Joseph, C. G., & Hilal, N. (2017). Effective coagulation-flocculation treatment of highly polluted palm oil mill biogas plant wastewater using dual coagulants: Decolourisation, kinetics and phytotoxicity studies. *Journal of Water Process Engineering, 16*, 258–269.

Zhang, T., Ding, L., & Ren, H. (2009). Pretreatment of ammonium removal from landfill leachate by chemical precipitation. *Journal of Hazardous Materials, 166*(2–3), 911–915. https://doi.org/10.1016/j.jhazmat.2008.11.101

Zhang, Y., Ye, C., Su, Y., Peng, W., Lu, R., Liu, Y., Huang, H., He, X., Yang, M., & Zhu, S. (2022). Soil Acidification caused by excessive application of nitrogen fertilizer aggravates soil-borne diseases: Evidence from literature review and field trials. *Agriculture, Ecosystems & Environment, 340*, 108176. https://doi.org/10.1016/j.agee.2022.108176

Global Research Trends and Technical Insights in Biogas Production from Organic Waste

Nadia Isa and Pramila Tamunaidu

Abstract The increasing accumulation of organic waste and the environmental challenges posed by saturated landfills have intensified global interest in sustainable waste-to-energy solutions. Biogas production via anaerobic digestion offers a dual benefit of renewable energy generation and waste valorization, making it a central technology in circular bio-economy and low-carbon energy strategies. This chapter provides a comprehensive analysis of biogas research by integrating bibliometric mapping of global research trends (2000–2022) with a technical overview of biogas production processes. Using the Web of Science database, bibliometric analyses in this chapter identify publication patterns, leading countries and keyword co-occurrence, revealing hotspots of research activity and emerging thematic clusters. In parallel, the chapter examines the biological and biochemical principles of anaerobic digestion, the role of microbial communities, operational optimization strategies (temperature, retention time, feedstock composition) and pre-treatment techniques to enhance substrate accessibility and improve methane yields. The synthesis of these perspectives highlights how bibliometric insights inform technical and operational research priorities, guiding the development of efficient, sustainable and scalable biogas systems. The chapter concludes with an outlook on future research directions, including co-digestion strategies, biogas upgrading to biomethane and policy frameworks supporting energy security and climate change mitigation.

Keywords Biogas · Anaerobic digestion · Bibliometric analysis · Renewable energy · Waste-to-energy

N. Isa (✉) · P. Tamunaidu
Chemical Process Engineering Department, Malaysia-Japan International Institute of Technology (MJIIT), Universiti Teknologi Malaysia (UTM), Kuala Lumpur, Malaysia
e-mail: nadiaisa@unikl.edu.my

P. Tamunaidu
e-mail: pramila@utm.my

N. Isa
Process Engineering Technology Department, Universiti Kuala Lumpur Malaysian Institute of Chemical and Bioengineering Technology (UniKL MICET), Melaka, Malaysia

A. Z. Yaser et al. (eds.), *Compost and Liquid Organic Fertilizers*,
https://doi.org/10.1007/978-981-95-8561-8_14

1 Introduction

Nowadays, the world's economies are driven by consumer-based lifestyles, where producing waste is the most evident and unfavourable resource intensive by-products (Moya et al., 2017). With the low-income nation producing the highest organic waste, biogas production has become the preferred technology in dealing with global energy crisis. The accumulation of organic waste and the overexploitation of landfill capacities represent significant environmental and socio-economic challenges globally. Biogas is formed in the process of anaerobic fermentation, which disinfects the waste. Unlike bio-diesel and bio-ethanol, biogas is mainly produced from organic waste and does not caused problems such as food shortage or loss of biodiversity (Nevzorova & Kutcherov, 2019a, 2019b). Therefore, it is considered as a holistic and flexible energy resource that generates electricity, heat and power but also portrays environmentally friendly properties (Segun Giwa et al., 2020). The integration of biogas systems into circular economy frameworks aligns with the Sustainable Development Goals (SDGs), including SDG 7 (Affordable and Clean Energy), SDG 12 (Responsible Consumption and Production), and SDG 13 (Climate Action), underscoring its dual role in energy generation and environmental stewardship (IEA, 2023)

Understanding global trends in biogas research is critical for guiding future scientific inquiry and policy development. Bibliometric analyses offer a systematic approach to map research outputs, identify influential authors, journals, and countries and uncover thematic and collaborative patterns within a research field (Donthu et al., 2021). By synthesizing large volumes of bibliographic data, such analyses provide insights into the evolution of research priorities, highlight emerging areas of interest and reveal potential knowledge gaps that warrant further investigation. In the context of biogas production, bibliometric studies can elucidate how research has progressed across geographic regions, institutions, and topical areas, thereby informing both technical and strategic decision-making.

2 Global Research Landscape

The past two decades have witnessed a marked increase in research activity on biogas production, reflecting the growing recognition of organic waste valorization as a cornerstone of sustainable energy systems. Bibliometric analysis provides a systematic approach to quantify and visualize these developments, offering insights into publication trends, leading contributors and research hotspots geographic and institutional contexts (Donthu et al., 2021).

2.1 Publication Trends

The study is focusing on the specific time frame for the articles published between 1971 until 2022. This duration of year was selected in order to investigate the paradigm shift of the biogas production throughout the years. It was found that the earliest publication related to biogas production was in 2004 which is 34 years later with one publication. This may be due to lack of awareness and lack of interest towards the research of biogas technology as the source of energy worldwide still available and the abundance of waste still under control. Hence, the research started to grow from 2010 until today which reaches up to only 84 publications in 2021. Therefore, this factor could be one of the novelty of the research area. Figure 1 shows the annual numbers of research papers on biogas production from 2004 until 2022. The year 2021 contributing to the highest numbers of published articles followed by year 2020, 2019, 2018, 2017 and 2015. It is anticipated that the annual publication will continue to rise as up to March 2022 as 8 research papers were already been published.

Biogas technology area are extensive and many research groups worldwide are working actively in these areas. Analysis on subject area showed that energy and fuel are the major concern in biogas technology. This is proved by the total publications classified under the following research directions: Energy Fuels (161 articles), Engineering (129 articles), Environmental Sciences and Ecology (121 articles), Biotechnology Applied Microbiology (93 articles), Agriculture (73 articles), Science and Technology-Other Topics (64 articles) and Chemistry (32 articles). Biogas technology research indeed a multidisciplinary area as one of the publications was categorized under Economics research direction as shown in Fig. 2. Results also showed that the research articles in this area were published in various document types. Majority publications were from research articles with 287 papers followed by 51 review papers and 11 proceeding papers.

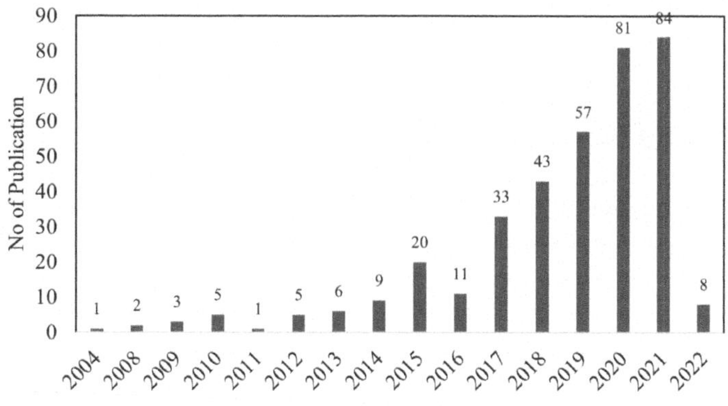

Fig. 1 Annual numbers of research papers on biogas production indexed in WOS

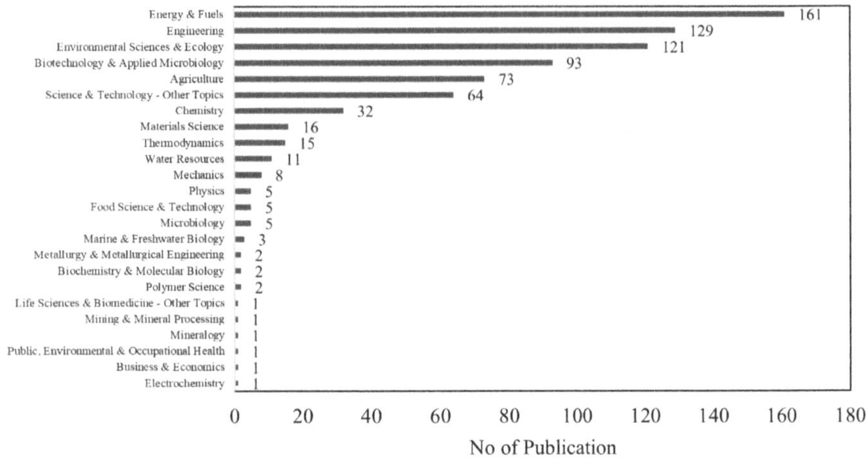

Fig. 2 Research directions in the area of biogas production

Bioresource Technology is seen to be the major source of production in this field with an h-index of 22 and total 1,794 citations for the papers within this sample. Most other top sources are having h-indices between 5 and 9. As for the cited references for the sample 369 articles, top cited articles were mostly from the Bioresource Technology supporting the importance of the source within this area of research which covers all area concerning biomass, biological waste treatment, bioenergy, bio-transformations and bio-resource system analysis and technologies associated with conversion or production. This most productive journal which is Bioresource Technology consisted of 57 research article covering 15.45% of the total publications extracted from March 2022. The article publications followed by Journal of Cleaner Production (16, 4.34%), Renewable & Sustainable Energy Reviews (14, 3.79%), Waste Management (14, 3.79%) and Science of the Total Environment (13, 3.52%) as shown in Fig. 3.

2.2 Leading Publications by Countries

From the global trend of productive countries in Fig. 4, it was found that the top two countries that contributed to global publications are China and United States of America (USA). This indicated that these two countries are the key players in the biogas research technology. China was the leading country with 125 publications followed by USA, Germany, India and United Kingdom with 50, 40, 30 and 23 respectively.

China is considered as the pioneer in the utilisation and development of biogas plant, therefore the growth of the biogas research in China is believed to occur due to

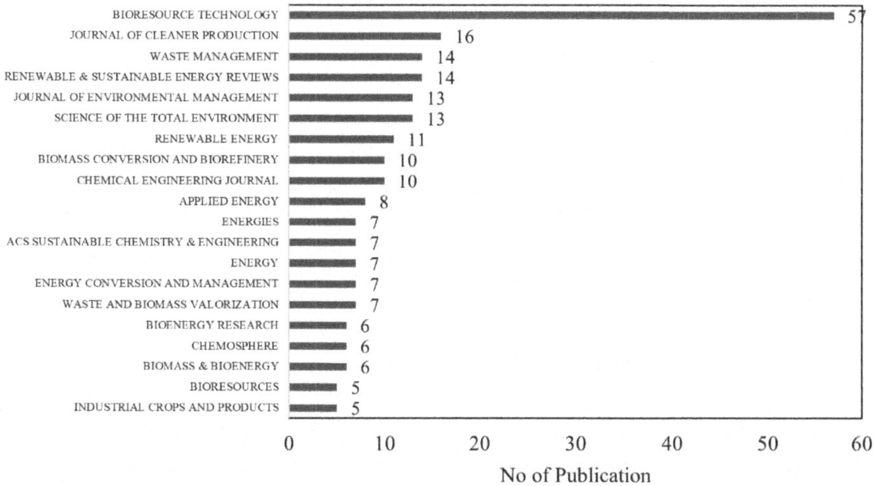

Fig. 3 List of most productive journals on research of biogas production

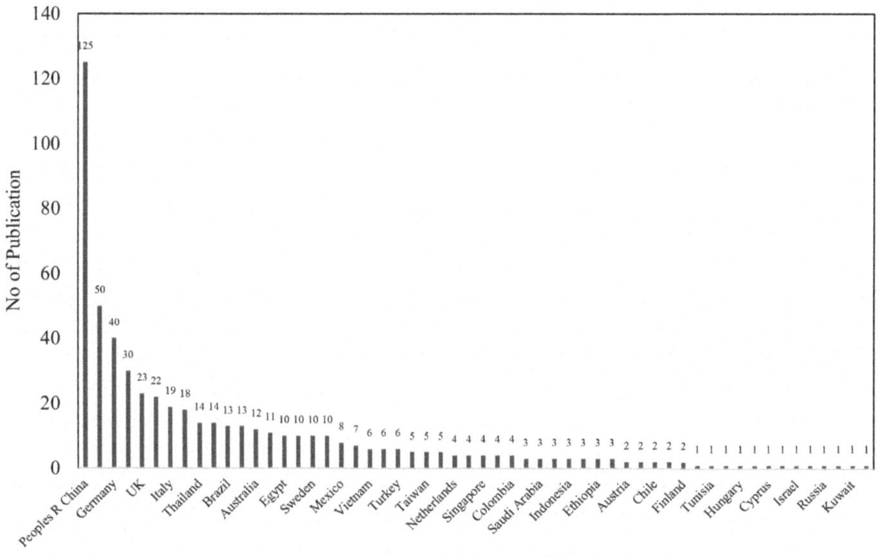

Fig. 4 Global productive countries in biogas research publications

the total energy demand in China and the country experienced accelerated industrialisation after 1990s. Moreover, the China's government has promoted the importance of research in the biogas technology by introducing programmes, provided financial grants along with biogas improvement projects which led to the biogas utilisation in the start of twenty-first century (Segun Giwa et al., 2020).

In the USA, 12% of biogas plants been built are farm based digestion plants using livestock manure. This explained the reason for the highest result of publications in America as well. Although Germany is the 3rd country with the highest article publications, the type of feedstock in the use for biogas production is an issue for the mentioned country. This is because, the use of energy crops for energy production received a bad reputation in public opinions and governments as it created a negative impact on food security, increased food prices and made inefficient usage of land for crops production (Nevzorova & Kutcherov, 2019a, 2019b).

2.3 Author Keywords and Terminology

A total of 3,258 keywords was recorded among which the keywords were used more than 20 times from year 2018 to 2021 which is ‘biogas production’. This shows that the researches related to biogas production comprises of anaerobic digestion, biomass, methane production, hydrothermal pre-treatment and lignocellulosic biomass are the most frequent keywords identified in WOS database since the research of biogas production started in 2004.

The keywords are later analysed in terms of terminology and themes being clustered in order to understand the transformation of biogas technology over the years among different countries. Most of these keywords are favourably used and studied by the China’s researchers as this country leads the world in domestic biogas technology (Zheng et al., 2020). Based on the results explanation, numerous of authors from China conducted biogas researches consisting keywords of ‘biogas production’, ‘anaerobic digestion’ and ‘hydrothermal pre-treatment’.

Our results show that three distinctive clusters were identified here (Fig. 5). The first cluster consists of 33 elements, with the most frequent terms being ‘digestion’ (246 occurrences), ‘pretreatment’ (97), ‘methane yield’ (90) and ‘hydrothermal pretreatment’ (74). The second cluster has 30 items and includes ‘biogas’ (101 occurrences), ‘energy’ (78), and ‘product’ (77). The third cluster identified 8 other words that occurred most often, with the most common being ‘digestate’ (44), followed by ‘hydrothermal carbonization’ and ‘carbon’ (with 43 and 35, respectively).

Although most frequent terms fall under Cluster 1 and Cluster 2, the terms under Cluster 3, although not as frequent as the other 2 clusters, needs to be analyzed further to see the trend of research focus in this cluster in particular. As the terms in each cluster showed trends in studies in this particular area, overlay visualization can help in identifying the recent terms being used within the clusters respectively. ‘Solubilization’, ‘methanogenesis’, ‘gompertz model’ and ‘mesophilic condition’ are among the few recently used term for Cluster 1, whereas ‘biorefinery’ looks prominent for Cluster 2. Cluster 3 is relatively new in this area and ‘sewage sludge’ and ‘co-digestion’ seem to be the recently used terms within this cluster. These could be examined in a future study as various topics can be developed based on these keywords. The terms in each cluster represents a research stream are presented in Table 1. It is interesting to see how a particular term is established.

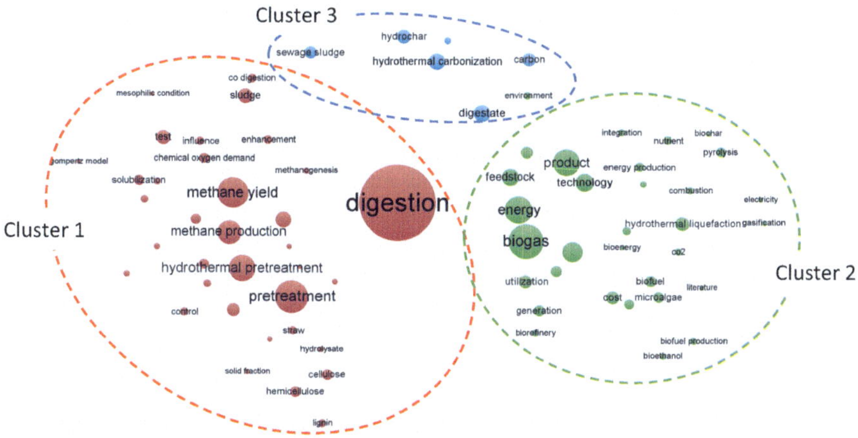

Fig. 5 Clusters based on the most frequent terms

Table 1 Most frequent terms under respective clusters

Cluster 1		Cluster 2		Cluster 3	
Term	Occurrences	Term	Occurrences	Term	Occurrences
Digestion	246	Biogas	101	Digestate	44
Pretreatment	97	Energy	78	Hydrothermal carbonization	43
Methane yield	90	Product	77	Carbon	35
Hydrothermal pretreatment	74	Conversion	54	Hydrochar	35
Methane production	66	Feedstock	46	Sewage sludge	33
Hydrolysis	41	Hydrothermal liquefaction	37	Co digestion	21
Sludge	38	Chemical	30		
Degradation	32	Property	30		
Cellulose	29	Generation	28		
Hemicellulose	29	Biofuel	27		
Solubilization	29	Microalgae	27		
Biodegradability	26				
Chemical oxygen demand	26				

3 Technical Fundamentals of Biogas Production

Biogas production is primarily achieved through anaerobic digestion, a biologically mediated process in which microorganisms decompose organic matter in the absence of oxygen, generating a methane rich biogas and nutrient rich digestate. The process is a cornerstone of sustainable organic waste management, offering both renewable energy generation and environmental benefits by mitigating greenhouse gas emissions from landfills (Li et al., 2022). Understanding the biological and biochemical principles of anaerobic digestion is critical for optimizing system performance and aligning technological innovation with emerging research priorities identified through bibliometric analysis. Anaerobic digestion proceeds through four primary stages:

1. Hydrolysis: Hydrolysis involves the enzymatic breakdown of complex organic polymers such as carbohydrates, proteins and lipids into soluble monomers like sugars, amino acids, and fatty acids. This stage frequently represents the rate-limiting step in anaerobic digestion, particularly when lignocellulosic or recalcitrant feedstocks are used (Zhang et al., 2021).
2. Acidogenesis: Hydrolyzed monomers are metabolized by acidogenic bacteria into volatile fatty acids (VFAs), alcohols, hydrogen, and carbon dioxide. This stage establishes the chemical environment that supports subsequent microbial activity.
3. Acetogenesis: Acetogenic bacteria convert VFAs and alcohols into acetate, hydrogen, and carbon dioxide, producing intermediates that are directly usable by methanogenic archaea.
4. Methanogenesis: Methanogenic archaea convert acetate, hydrogen, and carbon dioxide into methane and water. This stage determines the overall methane yield and is highly sensitive to environmental conditions, such as pH, temperature and inhibitory compounds.

Microbial communities are central to anaerobic digestion performance. The process relies on a complex consortium of bacteria and archaea that operate synergistically. Hydrolytic and acidogenic bacteria initiate substrate breakdown, while acetogens and methanogens drive methane production. The stability and composition of these communities are influenced by feedstock characteristics, operational parameters, and environmental conditions (Li et al., 2022).

The bibliometric analysis presented in Sect. 2 indicates that research interest in anaerobic digestion has grown significantly, particularly in Asia and Europe, with a strong focus on process optimization, microbial engineering, and feedstock management. These trends align with the technical challenges discussed here, such as improving hydrolysis efficiency, stabilizing microbial consortia, and enhancing methane yields through pre-treatment methods. By integrating bibliometric insights with technical understanding, researchers can identify high-impact areas for innovation and address gaps in knowledge or application.

4 Value of Microbial Responses on the Production of Biogas: Operating Strategies

The design and configuration of biogas reactors play a fundamental role in determining anaerobic digestion efficiency, process stability and methane yield. Effective reactor selection is largely governed by the system's capacity to sustain high organic loading rates (OLR), maintain short hydraulic retention times (HRT) and generate substantial volumes of high quality biogas factors that collectively reduce digester volume requirements while enhancing conversion performance (Li et al., 2022). Reactor performance is also influenced by internal engineering controls, including temperature regulation, feedstock distribution, mixing dynamics and microbial community stability (Angelidaki & Ahring, 1993). Although construction cost, fabrication materials and design complexity differ across regions, successful anaerobic digesters universally exhibit several operational pre-requisites like stable methane production, consistent treatment of high strength organic substrates and efficient hydraulic management (Khanal, 2008). In addition, bioreactors must support critical operational functions such as ensuring thorough mixing, regulating temperature, pH, pressure and nutrient availability, maintaining sterility where relevant and enabling controlled withdrawal of digestate to facilitate continuous or semi-continuous operation.

Preparation of feedstock to achieve optimum biogas digester performance requires sufficient amounts of water. The water demand depends on the moisture content of the feedstock and its organic loading rates. According to Tucho (2016), the water to fresh inoculum ration recommended is within 1:1 or 1:2 based on the moisture contents of the feedstock and optimum OLR. Several factors influence biogas production by anaerobic digestion are discussed below in detail:

(a) pH

The efficiency of biogas generation through anaerobic digestion is closely linked to the pH of the substrate, as microbial communities involved in the process are highly sensitive to hydrogen ion concentration. Excessive acidity can inhibit metabolic activity, suppress microbial growth, and impede the overall digestion process. A decline in pH often results from the accumulation of volatile fatty acids (VFAs), which disrupts methanogenesis and ultimately reduces biogas yield (Khorshidi et al., 2021). For stable digestion performance, the optimum pH typically ranges between 7.0 and 8.5, with values near 7.0 generally considered ideal for maintaining balanced microbial activity and maximizing methane production (Banerjee et al., 2022).

(b) Temperature

Anaerobic bacteria demonstrate peak metabolic activity within the mesophilic (30–40 °C) and thermophilic (50–60 °C) temperature ranges, making thermal regulation a critical factor governing hydrolysis, acidogenesis and methanogenesis. Temperature strongly influences reaction kinetics, microbial growth rates, and overall biogas productivity, with an optimal operational point for mesophilic digestion typically around 35 °C (Li et al., 2011). Thermal conditions inside the digester are modulated by several operational parameters, including the heat released from biochemical reactions, mechanical agitation through impellers, and engineered heat-exchange systems. These systems commonly employing hot water or steam enable effective heat transfer to the feedstock, ensuring uniform temperature distribution and maintaining the nutrient environment required for stable anaerobic microbial activity (Mao et al., 2015).

(c) Waste characteristics and particle size

The physicochemical characteristics of the feedstock, particularly its composition and particle size, exert a significant influence on both the quality and volume of biogas produced. Biomass provides essential nutrients and carbon sources required for sustained microbial activity, thereby supporting a long term stability of the anaerobic digestion process. Selection of suitable biomass is typically guided by operational factors, including waste characteristics and biodegradability (Banerjee et al., 2022). Pre-treatment steps such as mechanical size reduction are commonly employed to enhance substrate accessibility, after which co-digestion may be introduced to further elevate methane yield. Reducing particle size increases the substrate's surface area, improving microbial attachment and hydrolysis efficiency, which collectively enhance biodegradation rates and biogas generation. In contrast, overly large particles pose operational challenges, such as clogging of the reactor. Consequently, lignocellulosic materials with high lignin content, such as woody biomass, are generally unsuitable for biogas production due to their low biodegradability (World Bioenergy Association, 2013).

(d) Hydraulic Retention Time (HRT)

Biogas production often exhibits a time lag at the initial stages of digestion. Longer hydraulic retention times (HRT) enhance biogas yield as well as the reduction of total solids (TS) and volatile solids (VS). This improvement is attributed to prolonged exposure of the feedstock to microbial activity, which increases the efficiency of the anaerobic digestion process (Khorshidi et al., 2021).

(e) Carbon Nitrogen Ratio (C/N ratio)

The carbon-to-nitrogen (C/N) ratio is a critical parameter governing anaerobic digestion performance, with an optimal range typically around 30:1 for stable microbial activity. Carbon acts as the primary energy source for the microbial consortia, whereas nitrogen is required for protein synthesis and cellular growth among methanogenic populations. When nitrogen availability is limited (a high C/N ratio), microbial growth becomes constrained, slowing the degradation of excess carbon and subsequently reducing biogas yield. Conversely, when the C/N ratio falls below the optimal range, surplus nitrogen is converted into ammonia, elevating system pH beyond 8.5 and inhibiting methanogenic activity. This inhibition leads to process instability and diminished gas production (Banerjee et al., 2022).

(f) Inhibition factors

There are factors that contributed to the inhibition of the biological process of biogas plants and have consequences towards its production capacity and profitability. Ammonia concentration more than 3000 mg/L could cause toxic condition to the process. Therefore, in order to maintain the concentration within 50 to 200 mg/L, the organic loading rate should be reduced.

Alkalinity is the capacity of water to neutralize the acids in the biogas plant and should be maintained between 1500 to 5000 mg/L. Small pH variations can affect methanogens and influence the stability of the digestion process. Volatile fatty acids (VFA) represent the health of the digester. The production of VFA varies according to the quantity of solids supplied to the digester and should be below 2000 ppm. Concentration exceeding 300 ml/L can overloaded the digester and off balance the biology of the system and become toxic.

(g) Co-digestion

The application of co-digestion can significantly enhance biogas production by improving the nutrient balance of the feedstock. Specifically, co-digesting cow dung with crop residues provides a more favorable C/N ratio, resulting in biogas yield increases ranging from 40 to 100% compared to mono-digestion of cow dung alone (Tucho et al., 2016). Cow dung typically exhibits a low C/N ratio relative to crop residues; thus, supplementing it with plant materials rich in carbon helps mitigate ammonia accumulation and prevents excessive pH increases. This nutritional adjustment represents a key advantage of co-digestion over mono-digestion, in which nutrient balance cannot be optimized and only the organic loading rate can be managed through dilution. Consequently, optimizing both nutrient composition and operational parameters is critical for enhancing methane content and overall biogas quality. Co-digestion therefore promotes higher biogas yields with improved methane concentrations (Tucho et al., 2016).

5 Techniques to Increase Microbial Responses on Production of Biogas: Pre-Treatments

Thermal, chemical, mechanical and biological processes can be applied to speed up the degradation process for biogas production. With high energy requirement for thermal pre-treatment could promote the hydrolysis process of organic compounds compared to chemical pre-treatment whereby strong acids, alkalis, oxidizing agents and ionic liquids required. Furthermore, mechanical pre-treatment enhances the specific surface area as well as lowered the degree of polymerization and cellulose crystallinity (Zhang et al., 2019). According to Ren et al. (2018), by inoculating microorganisms and enzymes will promote the hydrolysis of substrate and increase the AD rate as well.

The effect of pre-treatment techniques on AD has only recently been investigated and it is still necessary to optimize these techniques in terms of efficiency, cost and application range (Achinas et al., 2017). Table 2 shows the comparisons of various

Table 2 Comparison of pre-treatment techniques for organic waste

Techniques	Advantages	Disadvantage	Application
Physical	– Improve the accessible surface area – Crack the lignin hemicellulosic complex structure – Increase the available cellulosic content	– High energy consumption	Biogas
Chemical	– Increase surface area and porosity – Reduce the cellulose crystallinity – Reduce fermentation time	– Expensive	Bioethanol Biogas
Biological	– Environmentally feasible solution – Microorganism tolerant to extreme conditions	– Reduce surface area for the accessibility of microbes	Bioethanol Biogas
Physicochemical	– Low energy input and negligible environmental effects – Enhance the cellulose hemicellulose solubilisation – Without chemicals – Low operating cost – Short period of time	– Generates some inhibitory compounds – Seldom break down all lignin	Bioethanol Biogas

pre-treatment techniques for organic materials whereas Table 3 shows the list of advantages and disadvantages of the physicochemical pre-treatment methods.

Bibliometric findings in Sect. 2 (Fig. 5) indicated that 'hydrothermal' emerged as one of the dominant keywords associated with biogas research between 2014 and 2022. Hydrothermal pre-treatment typically involves treating biomass in water at elevated temperatures (150–230 °C) and pressure, which facilitates the solubilization of hemicellulose and disrupts lignin cellulose linkages, thereby enhancing the accessibility of polysaccharide fractions for subsequent conversion (Martín et al., 2022). This process increases enzymatic digestibility and accelerates methane production during subsequent digestion.

6 The Value of Biogas as the Source of Energy

Biogas production is more versatile than many other biofuels, as it can be derived from carbohydrates, proteins, fats, and a variety of wet or secondary organic materials. Serving as a renewable energy source, biogas offers a sustainable solution to addressing rising energy demands (Monjurul Hasan et al., 2022). It can be directly

Table 3 Advantages and disadvantages of the physicochemical pre-treatment methods

Pre-treatment	Advantages	Disadvantages
Steam Treatment (Autohydrolysis) and Steam Explosion	– Low energy input and negligible environmental effects – Hydrolyses some of the hemicellulose and alters the lignin structure – Increase the surface area	– Produces some inhibitory compounds – Does not always break down all lignin
Ammonia Fiber Explosion (AFEX)	– Effective for pre-treatment of low lignin substrate	– Does not solubilize much of the hemicellulose – usage of toxic chemical
Hydrothermal Pre-treatment (Subcritical Condition)	– Increase the cellulose hemicellulose solubilisation – Usage of water as the reactant is safe, non-toxic, inexpensive and readily available – Lower production costs, mild operating conditions and shorter production duration	– Decrease in pH of the reaction medium

combusted to generate heat and electricity or upgraded to high-purity methane—commonly termed biomethane or renewable natural gas—through the removal of water, carbon dioxide, hydrogen sulfide, and other trace impurities. This upgraded biogas is chemically analogous to conventional natural gas and can be used as a transportation fuel in either compressed or liquefied form.

"Empowering communities by powering their homes"

Biogas plants offer a wide range of environmental benefits at the household, local, national and global levels. By diverting organic waste from landfills, they help reduce contamination of groundwater, surface water and other natural resources. Anaerobic digestion efficiently inactivates harmful pathogens, including *E. coli* and *Mycobacterium avium* subsp. *paratuberculosis*. Furthermore, the effluent from biogas digesters can be utilized as a high-quality organic fertilizer, thereby reducing the need for imported or synthetic nitrogen-based fertilizers (World Bioenergy Association, 2013). Biogas also has the potential to contribute in mitigating greenhouse gas emissions and improving the utilisation of crop nutrients (Monjurul Hasan et al., 2022). Methane combustion produces very low soot and particulate emissions because its simple molecular structure leads to more complete oxidation, resulting in cleaner flame behavior that minimizes fouling in boilers and related equipment (Bakkaloglu & Hawkes, 2024 Consequently, systems operating on methane- or biogas-based fuel often require less frequent maintenance due to lower accumulation of soot and cinders. Additionally, promoting biogas production from organic waste and sustainably produced feedstock leads to the creation of permanent jobs and regional development. The involvement of many interested parties in planning, construction and cost estimation, production, control and distribution is important to ensure the successful development of biogas technology (World Bioenergy Association, 2013).

The development of biogas constitutes a strategically significant step toward reducing reliance on fossil fuels. As a renewable energy source, it supports the establishment of a sustainable energy supply and enhances long-term energy security (World Bioenergy Association, 2013). Globally, the deployment of domestic biogas technologies has been most successful in countries where governments actively participate in providing subsidies, and in the planning, design, construction, operation, and maintenance of biogas plants (World Bioenergy Association, 2010). Although over the last two decades, the stakeholders have become motivated to produce biogas from waste and this trend increase globally, however, in comparison with other renewable energy resources such as wind and solar, the information and sharing of biogas production is still lacking in quantity and quality (Monjurul Hasan et al., 2022).

In low and middle income countries such as Bangladesh, where agricultural residues, animal manure and the organic fraction of municipal waste constitute the primary feedstocks for domestic biogas plants, there exists significant potential for both biomass gasification and biogas based power generation. However, actual achievements in these sectors have remained below their projected potential. This is due to six major barriers which covers the technical aspect (lack of waste treatment and storage facilities, lack of feedstock supply and planning and installation issues), economic (lack of investments incentives, uncertainties on return investment and competition with other renewable energy options), lack of social awareness (resistance to change), government's policies (insufficient attention from the government) and the final factor is immature biogas market in the country (Monjurul Hasan et al., 2022).

The successful deployment of biogas systems utilizing agricultural residues has been widely demonstrated in several developing countries including China, India, Nepal and Kenya despite their differing climatic conditions, technological capacities, resource availability and socio-economic contexts. In China, in particular, agricultural waste based biogas initiatives have played a significant role in accelerating national biogas adoption. The expansion of industrial scale anaerobic digesters has further driven the rapid growth of biogas production, contributing to both rural energy security and large scale waste management (Segun Giwa et al., 2020). China is considered as the pioneer in the utilisation and development of biogas plant, therefore the growth of the biogas research in China is believed to occur due to the total energy demand in China and the country experienced accelerated industrialisation after 1990s. Moreover, the China's government has promoted the importance of research in the biogas technology by introducing programmes, provided financial grants along with biogas improvement projects which led to the biogas utilisation in the start of twenty-first century (Segun Giwa et al., 2020). Despite the many successes of biogas deployment, several critical challenges remain, including weak policy frameworks, limited access to advanced pre-treatment technologies, immature industrial value chains, lack of standardized digester specifications and technical issues associated with low quality digester construction (Saidmamatov et al., 2021).

Among the African countries, since installation of biogas plants in Kenya started in 2009, it has recorded as the highest African country to have total installed biogas plants amounting 6,749 plants up to year 2012 (Patinvoh & Taherzadeh, 2019). However, it still faces some issues regarding energy, climate and food. Kenya heavily relies on fuels such as biomass, charcoal and dung to for their energy supply due to the fact that Kenya does not have crude oil and coal reserves. Unfortunately, although Kenya been introduced to biogas technology in the 1950s, the technology did not achieve its success due to high construction and maintenance costs, absence of technical knowledge in maintaining the digesters accordingly (Saidmamatov et al., 2021).

For countries in South East Asia like Vietnam, biogas technology has high potential with huge annual discharge of about 85 million tons of livestock wastes. However, the potential has not been duly met because of challenges associated with technical services, funding, inadequate operational process and lack of adequate information. Different with Indonesia and Malaysia which are the key producers of palm oil. Malaysia's renewable energy is estimated to increase from 2 to 20% by 2025. Despite the potential and opportunities, biogas implementation in Indonesia and Malaysia faces a series of challenges associated with funding, government coordination and synergy between programs and planning (Patinvoh & Taherzadeh, 2019).

7 Conclusion and Future Directions

This chapter has examined the global landscape of biogas production through an integration of bibliometric analysis and technical review. The findings reveal that biogas research has experienced substantial growth over the past decade, reflecting increasing recognition of its potential as a sustainable energy source and waste management solution. The bibliometric data demonstrated a clear upward trend in publication output since 2014, with China, the United States, Germany and India emerging as leading contributors. The predominance of journals such as *Bioresource Technology* and *Journal of Cleaner Production* highlights the interdisciplinary nature of biogas research, spanning environmental engineering, microbiology, and renewable energy systems.

From a technical perspective, the chapter has underscored the fundamental role of microbial dynamics, operational optimization and pre-treatment strategies in enhancing biogas productivity. The efficiency of anaerobic digestion depends on maintaining balanced microbial interactions and optimizing key operational parameters such as temperature, retention time, and feedstock composition. Among various pre-treatment methods, hydrothermal technique has shown the greatest potential in improving substrate solubilization and methane yield while aligning with sustainability considerations. These findings correspond with bibliometric trends that indicate a growing research focus on microbial ecology, process integration and bio-based circular systems.

In the broader context of energy transition, biogas stands as a critical component of the renewable energy mix. It contributes not only to decarbonization and energy diversification, but also to resource recovery and environmental protection. The valorization of organic waste through anaerobic digestion exemplifies circular economy principles by converting waste liabilities into energy and nutrient assets. Moreover, biogas supports decentralized energy systems, particularly in developing regions, by improving energy access and fostering local economic resilience.

References

Achinas, S., Achinas, V., & Euverink, G. J. W. (2017). A technological overview of biogas production from biowaste. *Engineering, 3*, 299–307.

Angelidaki, I., & Ahring, B. K. (1993). Thermophilic anaerobic digestion of livestock waste: The effect of ammonia. *Applied Microbiology and Biotechnology, 38*, 560–564.

Bakkaloglu, S., & Hawkes, A. (2024). A comparative study of biogas and biomethane with natural gas and hydrogen alternatives. *Energy & Environmental Science, 17*, 1482–1496.

Banerjee, S., Prasad, N., & Selvaraju, S. (2022). Reactor design for biogas production: A short review. *Journal of Energy and Power Technology, 4*, 1–14.

Donthu, N., Kumar, S., Mukherjee, D., Pandey, N., & Lim, W. M. (2021). How to conduct a bibliometric analysis: An overview and guidelines. *Journal of Business Research, 133*, 285–296.

International Energy Agency (IEA). (2023). *Net Zero Roadmap: A Global Pathway to Keep the 1.5°C Goal in Reach (2023 Update)*. IEA, Paris.

Khanal, S. K. (2008). *Anaerobic Biotechnology for Bioenergy Production: Principles and Applications*. CRC Press.

Khorshidi, M. A., Beiki, H. & Kanvisi, M. (2021). Using a novel continuous bioreactor in enhancing the biogas production. *Research Square*. [Preprint].

Li, C., He, P., Peng, W., Lu, F., Du, R., & Zhang, H. (2022). Exploring available input variables for machine learning models to predict biogas production in industrial scale biogas plants treating food waste. *Journal Cleaner Production, 380*, Article 135074.

Mao, C., Feng, Y., Wang, X., & Ren, G. (2015). Review on research achievements of biogas from anaerobic digestion. *Renewable and Sustainable Energy Reviews, 45*, 540–555.

Martín, C., Dixit, P., Momayez, F., & Jönsson, L. J. (2022). Hydrothermal Pre-treatment of Lignocellulosic Feedstocks to Facilitate Biochemical conversion. *Frontiers in Bioengineering and Biotechnology, 10*, Article 846592.

Monjurul, Hasan, A. S. M. M., Kabir, M. A., Hoq, M. T., Johansson, M. T. & Thollander, P. (2022). Drivers and barriers to the implementation of biogas technologies in Bangladesh. *Biofuels, 13*, 643–655.

Moya, D., Aldas, C., Lopez, G., & Kaparaju, P. (2017). Municipal solid waste as a valuable renewable energy resource: A worldwide opportunity of energy recovery by using waste-to-energy technologies. *Energy Procedia, 134*, 286–295.

Nevzorova, T., & Kutcherov, V. (2019a). Barriers to the wider implementation of biogas as a source of energy: A state of the art review. *Energy Strategy Reviews, 26*, Article 100414.

Nevzorova, T., & Kutcherov, V. (2019b). Barriers to the wide implementation of biogas as a source of energy: A state of the art review. *Energy Strategy Reviews, 26*, Article 100414.

Patinvoh, R. J., & Taherzadeh, M. J. (2019). Challenges of biogas implementation in developing countries. *Current Opinion in Environmental Science & Health, 12*, 30–37.
Ren, Y., Yu, M., Wu, C., Wang, Q., Gao, M., Huang, Q., & Liu, Y. (2018). A comprehensive review on food waste anaerobic digestion: Research updates and tendencies. *Bioresource Technology, 247*, 1069–1076.
Saidmamatov, O., Rudenko, I., Baier, U., & Khodjaniyazov, E. (2021). Challenges and solutions for biogas production from agricultural waste in the Aral Sea Basin. *Processes, 9*(2), 199.
Segun Giwa, A., Ali, N., Ahmad, I., Asif, M., Guo, R. B., Li, F. L., & Lu, M. (2020). Prospects of China's biogas: Fundamentals, challenges and considerations. *Energy Reports, 6*, 2973–2987.
Tucho, G. T., Moll, H. C., Uiterkamp, A. J. M. S., & Nonhebel, S. (2016). Problems with biogas implementation in developing countries from the perspective of labor requirements. *Energies, 9*(9), 750.
World Bioenergy Association. (2010). *Biogas Factsheet*. World Bioenergy Association. Accessed. https://www.worldbioenergy.org/uploads/Factsheet%20-%20Biogas.pdf
World Bioenergy Association. (2013). *Biogas–An Important Renewable Energy Source*. http://www.worldbioenergy.org. Accessed 8 Dec 2022.
Zhang, L., Loh, K. C., & Zhang, J. (2019). Enhanced biogas production from anaerobic digestion of solid organic wastes: Current status and prospects. *Bioresource Technology Reports, 5*, 280–296.
Zhang, X., Li, Q., Wang, S., & Zhu, J. (2021). Lignocellulosic biomass pretreatment for enhanced bioenergy recovery: Effect of lignocellulose recalcitrance and enhancement strategies. *Frontiers in Energy Research, 9*, Article 646057.
Zheng, X., Li, G., Li, Y., & Wang, X. (2020). Development and trends of biogas technology: A global perspective. *Renewable and Sustainable Energy Reviews, 118*, Article 109524.

Optimizing Food Waste Anaerobic Digestion: Substrate Composition, Total Solids, Bulking Agents and Digestate as a Source of Fertilizer

Boanita Susanto Putri, I. D. A. A. Warmadewanthi, and Talent Nia Pramestyawati

Abstract Food waste has rapidly emerged as a critical environmental challenge that is closely linked to both climate change and the depletion of natural resources. Anaerobic digestion (AD) offers a strategic approach to address this issue by simultaneously reducing organic waste and generating value in the form of renewable energy and nutrient-enriched digestate. This chapter examines how substrate characteristics, total solids (TS), and the addition of bulking agents (BAs) influence the performance and resilience of AD in food-waste management. The composition of carbohydrates, proteins, and lipids defines not only the potential biogas yield but also the operational limitations, while the carbon-to-nitrogen (C/N) ratio remains a crucial indicator for process balance. The TS content further determines microbial activity, methane productivity, and digestate quality, with semi-dry conditions (approximately 15–20% TS) often representing the optimal range. Moreover, bulking agents such as straw or sawdust enhance porosity, regulate moisture levels, and create favorable microbial habitats, thereby ensuring greater process stability and improved methane generation. By considering these interconnected factors, AD can be positioned not merely as a waste treatment method, but as a cornerstone of the circular economy—transforming food waste into clean energy and sustainable biofertilizer, while contributing to broader environmental and resource sustainability goals.

Keywords Anaerobic digestion · Biogas production · Bulking agents · Food waste · Total solid

---

B. S. Putri · I. D. A. A. Warmadewanthi (✉) · T. N. Pramestyawati
Department of Environmental Engineering, Institut Teknologi Sepuluh Nopember (ITS), Surabaya, Indonesia
e-mail: warma@its.ac.id

A. Z. Yaser et al. (eds.), *Compost and Liquid Organic Fertilizers*,
https://doi.org/10.1007/978-981-95-8561-8_15

1 Introduction

Food waste (FW) generation has become a critical environmental and economic concern worldwide, particularly in rapidly urbanizing regions where consumption patterns are shifting. Food waste is rich in biodegradable organic matter but poses severe environmental risks when landfilled, including methane emissions and leachate generation. To address these challenges, anaerobic digestion (AD) has gained recognition as a sustainable biotechnological approach that reduces waste volume and recovers energy and nutrients.

In Malaysia, the highest composition in solid waste is also food waste (63%). Households are the biggest source of food waste, but higher-education institutions are also among the main contributors. Food waste (FW) has a high moisture content (70–80%), C/N ratio between 14 and 27, is acidic (pH 4–7), and is rich in N, P, and K. Methane production is enhanced during the anaerobic co-digestion process, and it is recommended to apply anaerobic co-digestion while adjusting the carbon-to-nitrogen (C/N) ratio by altering the proportions of Food Waste (FW):Waste Water (WW) or Animal Manure (AM):Waste Water (WW). The co-substrates SS and FW have the highest biogas capacity according to the analysis, with ratios of 60:40, 50:50, 70:30, and 66.6:33.3 (Paranjpe et al., 2023).

Some universities have adopted AD methods for their waste. University of Florida use food waste from dining hall to produce biogas and reduce landfill dependence. AD of food waste in Asian Institute of Technology-Thailand produced 85–123 m^3 biogas/day. Although AD is promising for energy generation, its high capital costs limit its adoption on campuses (Yaser et al., 2022).

In addition to bioenergy recovery, the integration of food waste and sewage sludge through anaerobic co-digestion supports circular economy principles by transforming waste streams into renewable energy and biofertilizers. As biogas is a renewable energy source, its production has the potential to replace fossil fuels owing to its environmental and economic benefits. Recently, many studies have been conducted on the application of anaerobic co-digestion of food waste to increase biogas production (Chew et al., 2021). The resulting digestate may replace synthetic fertilizers and contribute to soil health and long-term carbon sequestration.

Although extensive research has been conducted on anaerobic digestion of FW, most studies have discussed co-digestion, microbial processes, or operational factors in isolation. Few reviews have integrated the effects of substrate composition, total solids (TS), and bulking agents (BAs) on methane yield, process stability, and digestate quality. This review addresses these gaps by synthesizing recent findings on the biochemical characteristics of FW, the operational role of TS, and the application of BAs. Together, these perspectives provide a clearer and more holistic framework for optimizing anaerobic digestion compared to previous studies.

2 Food Waste Generation

The latest research from the FAO in 2019 shows that 13.8% of food produced in 2016 was lost from farms to dining tables, excluding the retail and household stages in the global food supply chain (Despoudi et al., 2021). ASEAN countries contribute significantly to this crisis while simultaneously facing unique regional challenges in food security and waste management infrastructure development. The increasing food waste generation across some countries correlates strongly with economic development, urbanization, and changing consumption patterns. As nations experience rapid economic growth and urban migration, traditional food systems undergo significant transformation, often resulting in increased food waste at both retail and consumer levels.

Research about food waste of Indonesian took placed in Bogor Regency. Household food waste in Bogor Regency averages 77 kg per capita per year, with urban areas generating more waste (79.4 kg/cap/year) compared to rural areas (45.8 kg/cap/year). Major contributors to edible food waste include cereals, tubers, and vegetables, while fruits are the main contributors to inedible food waste. It shown by Fig. 1.

Food waste contributes significantly to greenhouse gas emissions, primarily methane, when decomposed in landfills. This exacerbates climate change and represents a loss of the energy used in food production. Waste valorisation is the process of converting waste materials into valuable products or energy, aiming to reduce waste and improve resource efficiency. This approach supports sustainable waste management and promotes a circular economy by transforming waste. Several techniques are employed in waste valorisation:

1. Recycling:Collecting and processing waste materials such as plastics, paper, and metal to produce new products. This reduces landfill waste and conserves natural resources

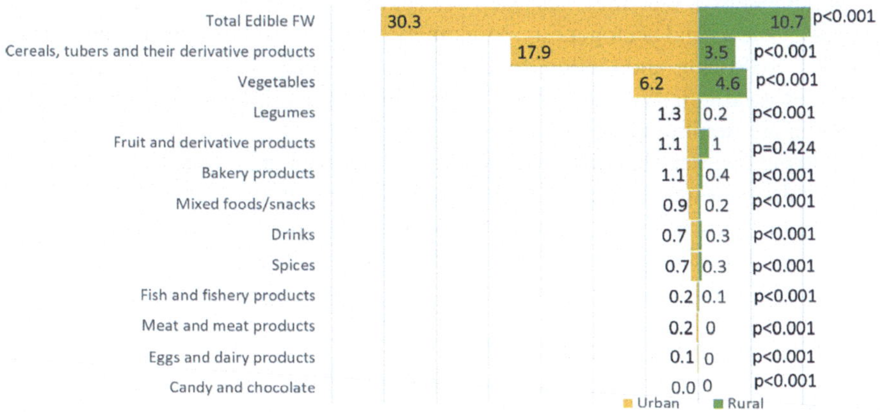

Fig. 1 Food waste composition in urban and rural areas (kg/cap/year) (Martianto et al., 2024)

2. Bioconversion: Using microorganisms to convert organic waste into biofuels, bioplastics, and other value-added products, reducing greenhouse gas emissions and dependence on fossil fuels.

3 Anaerobic Method for Food Waste Treatment

Anaerobic digestion (AD) is a biological process that decomposes organic matter in the absence of oxygen, resulting in the production of biogas and digestate. This process involves multiple stages and a diverse community of microorganisms working together to break down complex organic materials. AD is increasingly recognized as a cornerstone of sustainable waste management and the circular economy (Hussain et al., 2020). Anaerobic digestion (AD) is a widely accepted and effective technology for converting food waste into biogas, primarily methane, which can be used for heat and electricity generation. Unlike traditional landfilling, which generates uncontrolled methane emissions and toxic leachate, AD offers a cleaner, more economically viable alternative by transforming waste into valuable resources. AD produces biogas, which is rich in methane (60–70%) and can be used as a renewable energy source (Mishra et al., 2025).

3.1 The Biological Process

Compared to incineration and composting, AD has a lower environmental impact, reducing air and solid waste pollution (Chew et al., 2021):

a. Hydrolysis: This initial stage breaks down large polymers (like proteins, carbohydrates, and lipids) into smaller monomers (sugars, amino acids, and fatty acids) with the help of enzymes. This step determines the overall AD rate.
b. Acidogenesis and acetogenesis: Monomers are converted to short-chain organic acids, hydrogen, and carbon dioxide. If not properly managed, an accumulation of volatile fatty acids (VFAs) during this phase can lower the reactor's pH, potentially inhibiting the next stage.
c. Methanogenesis: Methanogenic bacteria, which are highly sensitive to environmental conditions, convert acetogenesis products into methane (CH_4) and CO_2. The end products are biogas (a mixture of methane and CO_2) and digestate (solid and liquid residue).Flow diagram AD process show by Fig. 2.

3.2 Operational Factors in AD

Operational factors can be a crucial role in determining the efficiency and stability of anaerobic digestion (AD) for food waste treatment. Maintaining optimal operational conditions is essential to ensure process stability, reduce the risk of system

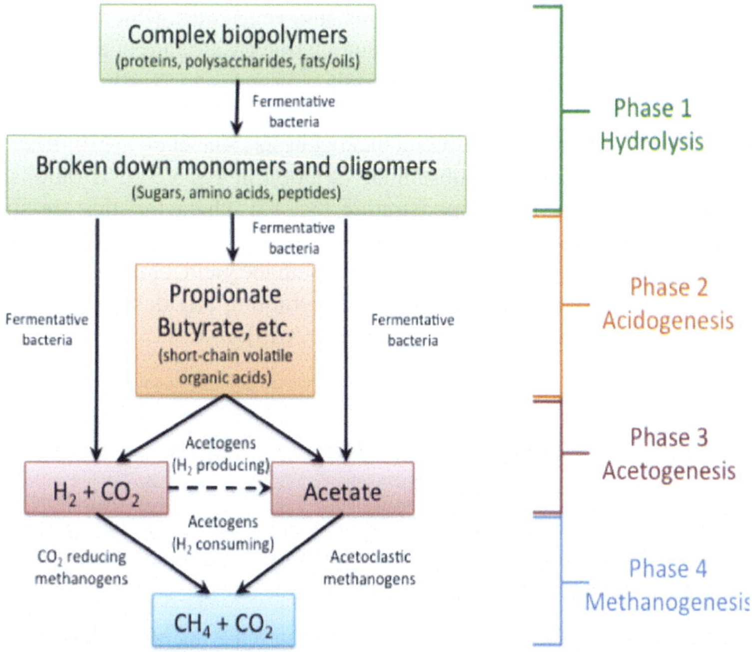

Fig. 2 Flow diagram AD process (Karuppiah & Ebenezer Azariah, 2019)

disturbances, and maximize methane production efficiency in anaerobic digestion systems. These are the operational factors in AD:

a. **Temperature**: Optimal temperatures in mesophilic condition, It is around 35–37 °C (Dinh et al., 2025). Continuous anaerobic digestion operated under thermophilic conditions has been identified as the most effective approach for achieving optimal performance in high-solids anaerobic digestion processes (Zamri et al., 2021).
b. **pH**: Maintaining a stable pH is critical for microbial activity, with an optimal range typically between 6.5 and 7.5 (Mu et al., 2023)
c. **Mixing**: Proper mixing enhances contact between microorganisms and substrates, improving digestion efficiency (Induchoodan et al., 2022).

4 Substrate Composition

The effectiveness and stability of anaerobic digestion (AD) for food waste are fundamentally determined by compositing the substrate. Food waste composition varies widely across regions, diets, and socio-economic contexts, causing significant differences in anaerobic digestion (AD) performance. FW from Indonesia and other ASEAN countries typically contains higher moisture and carbohydrate-rich fractions duet o tropical diets dominated by rice, fruits, and vegetables, which promote rapid hydrolysis but increase the risk of acid accumulation (Martianto et al., 2024; Yaser et al., 2022). In contrast, FW from East Asia, Europe, and North America often has higher protein, dat, and processed food content, leading to elevated ammonia or long-chain fatty acids that can inhibit methanogenesis (Xue et al., 2020; Rhee et al., 2023). Global analyses confirm substantial variation in TS, VS, C/N ratio, soluble COD, and ammonium levels, demonstrating that FW characteristics are strongly region-dependent (Ahmad et al., 2024; Chew et al., 2021). These differences influence microbial pathways, process stability, and methane yield, underscoring the need to characterize local FW properties before determining AD design and operating strategies (Shabbirahmed et al., 2024; Wu et al., 2025).

4.1 The Crucial Role of Organic Macronutrients

The three primary organic components in food waste—carbohydrates, proteins, and lipids—are the main energy sources for the anaerobic microbial community. Each component offers distinct advantages and challenges during AD.

1. Carbohydrates found in fruits, vegetables, and starches are highly biodegradable. Hydrolytic bacteria easily break them down into simple sugars, resulting in a rapid initial biogas production rate. However, this rapid degradation can cause a sharp drop in the reactor's pH due to the accumulation of volatile fatty acids (VFAs). This drop in pH is dangerous because it inhibits methanogenic bacteria, which are highly sensitive to acidic conditions.
2. Proteins from meat, fish, and dairy products have a high methane yield potential, making them a valuable component for maximizing energy output. Protein degradation releases ammonia nitrogen (NH_4-N). Unionized FAN at high concentrations becomes toxic to methanogenic bacteria, potentially causing process inhibition and even complete reactor failure. Their degradation releases ammonia, which can be both a nutrient and an inhibitor (Costa-pinto, 2020), managing a high protein content is one of the biggest challenges in AD of food waste.
3. Lipids (fats and oils) offer the highest CH_4 yield per unit of mass at approximately 1,014 mL CH_4/g VS. While great for energy recovery, lipid hydrolysis is a slow process. The resulting long-chain fatty acids (LCFAs) can accumulate and stick to microbial cells, thereby reducing mass transfer and inhibiting biological activity. Therefore, the lipid ratio in the substrate must be carefully controlled.

Organic macronutrients such as carbohydrates, proteins, and lipids play a crucial role in anaerobic substrates by directly influencing the biochemical degradation pathways that lead to methane production during anaerobic digestion. A study show biogas potential assessment of lipid-rich substrates were high in CH4 potential (640 mL/g VS on average) compared to the protein-rich substrates (301.1 mL/g VS on average) (Rhee et al., 2023), Their chemical characteristics, ratios, and interactions with microbial consortia determine hydrolysis efficiency and metabolic fluxes, while environmental and operational variables like pH, temperature, and organic loading rate further modulate these processes.

4.2 The Importance of C/N Ratio

The carbon to nitrogen (C/N) ratio is a critical parameter in the anaerobic digestion (AD) process, significantly influencing microbial activity, process stability, and biogas production. An optimal C/N ratio, typically ranging from 20:1 to 30:1(Panigrahi et al., 2022), ensures a balanced food source for both acid-forming and methane-producing bacteria. Ratios outside this range can lead to process inefficiencies and inhibition.

Food waste alone often has a low C/N ratio (typically 10:1 to 15:1) due to its high protein content. A low C/N ratio can lead to excessive ammonia levels, which inhibit the degradation of organic matter and hinder the methanogenesis process. This condition, known as the "inhibited steady-state," is characterized by a shift from aceticlastic to hydrogenotrophic methanogenesis pathways (Zheng et al., 2021). A high C/N ratio can lead to insufficient nitrogen availability, limiting microbial growth and reducing biogas production. High carbon content without adequate nitrogen can also result in the accumulation of volatile fatty acids (VFAs), further inhibiting the AD process (Ahmad et al., 2024). Table 1 show methane yield from various substrate.

Table 1 Comparison of methane yields from food waste from different sources

Food waste sources	Characteristic	Methane yield (mL CH_4/g Vs)	Reference
Household food waste (China)	C/N 20–25	380–520	Xue et al. (2020)
Household food waste (Thailand)	High carbohydrate and moisture content	350–480	Yaser et al. (2022)
Campus dining hall waste (USA)	High protein and lipid content	450–620	Ma et al. (2020)
Mixed food waste (Korea)	Protein- rich → higher NH_4-N accumulation risk	301.1	Rhee et al. (2023)
Fruit and vegetable waste (Global studies)	High carbohydrate, low protein	280–420	Chew et al. (2021)
Rice-based Asian food waste (ASEAN)	High starch, moderate C/N	330–470	Martianto et al. (2024)

4.3 The Role of Total Solids (TS) in Anaerobic Digestion

The total solids (TS) content is a fundamental parameter that determines the efficiency and stability of AD in transforming substrate into biogas and nutrient-rich digestate. TS represents the proportion of organic and inorganic matter remaining after water removal from the substrate. Its concentration directly influences microbial activity, reactor design, leachate management, and the final biogas and digestate characteristics. Classification and the TS operating regimes. Anaerobic digestion processes are typically categorized according to TS content as follows:

1. Wet digestion: TS < 15%
2. Semi-dry digestion: TS 15–20%
3. Dry or solid-state anaerobic digestion (SS-AD): TS > 20%

Wet AD conditions offer easier mixing and more stable microbial environments, but they require larger reactor volumes and produce large quantities of effluent, which increases digestate handling requirements. In contrast, semi-dry systems often offer a balance between manageable rheology, good mass transfer, and higher methane yield. Dry AD (>20% TS) reduces water consumption and typically increases volumetric methane productivity; however, elevated TS levels can introduce challenges such as substrate compaction, reduced free water, inhibited hydrolysis, and limited diffusion of nutrients and metabolites (Gao et al., 2023).

Recent studies confirm that TS levels between 15 and 20% tend to be optimal for food waste (FW) digestion, because FW typically contains high moisture and easily degradable carbohydrates that require sufficient water for microbial mobility and buffering capacity. Similarly, mesophilic digestion of FW between 5–20% TS showed improved volatile solids removal up to ~ 15% TS, but further increases reduced microbial diversity and slowed methanogenesis (Yi et al., 2014). Co-digestion studies also demonstrate that increasing TS toward 20% can shift the methanogenic pathway from acetoclastic to hydrogenotrophic methanogenesis, reducing specific methane yield due to mass transfer limitations (Liu et al., 2020). Table 2 summarizes representative findings on the relationship between TS content and methane yield.

The TS content in anaerobic digestion substrates plays a crucial role in determining methane yield. While moderate TS levels (10%-20%) are optimal for maximizing

Table 2 Representative studies on TS content and methane yield

Substrate	TS (%)	Methane yield	References
Kitchen waste (dry AD)	15–25%	Highest: 405.7 mL CH4/g VS at TS 20%	Liu et al. (2025)
Food waste (mesophilic AD)	5–20%	Increased performance up to 15% TS	Yi et al. (2014)
Food waste + pig manure	5–20%	Higher TS (20%) shifted pathway, reduce yield	Liu et al. (2020)

methane production, higher TS levels (30% and above) can inhibit the process due to physical and chemical limitations. TS between 15%-20% can be used as recommended value in anaerobic digestion engineering design to treat Food waste (Zhao et al., 2021).

TS also determines the physical form and nutrient profile of digestate. Higher TS digestion produces a solid or semi-solid digestate with more concentrated nutrients, making it easier to transport and apply as organic fertilizer. However, very high TS or pretreatments (e.g., thermal hydrolysis) can alter nitrogen speciation, reducing bioavailability in soils. On the other hand, moderate TS levels (15–20%) usually yield digestate with balanced moisture content, facilitating both nutrient recovery and agronomic application.

For food waste AD, maintaining semi-dry TS levels (15–20%) is widely recommended as it balances methane yield, digestate handling, microbial stability, and resource efficiency (Mirmasoumi et al., 2023). Dry AD conditions (>20% TS) can be applied in water-scarce regions, but additional measures—such as bulking agents, leachate recirculation, or intermittent mixing—are necessary to maintain porosity and avoid mass transfer limitations (Gao et al., 2023). Ultimately, optimizing TS requires consideration of substrate type, reactor design, local water availability, and the desired balance between biogas production and digestate quality.

5 Impact on Microbial Community

TS levels not only affect physical treatment but also influence the dynamics of microbial communities. High TS environments tend to reduce microbial diversity and favor hydrogenotrophic methanogens over acetoclastic pathways, influencing both methane yield and process resilience. At low TS, hydrolysis is rapid, but volatile fatty acid (VFA) accumulation can destabilize the system if methanogens cannot match the acidogenic activity.

Microbial community analysis with dairy manure and cucumber residues as substrate revealed that during the peak methane production phase, Firmicutes and *Methanoculleus* were the predominant bacterial and archaeal phyla, respectively. The microbial community structure changed with different TS levels (Wang et al., 2025). The fibrous structure of lignocellulosic BAs provides a solid surface that supports the attachment of microbial biofilms. This enhances syntrophic interactions between hydrolytic, acidogenic, and methanogenic communities, resulting in improved degradation efficiency and higher system resilience against organic loading fluctuations.

Microorganism has significant roles in AD. Anaerobic condition make microorganism degrade the substrate and produce stable sludge (digestate) and biogas. Several environmental condition such as pH, temperature can influence microorganism metabolism. A study about AD use percentage and frequency of methanogen bacteria (MCB) as variable of research. MCB is added every 10 and 15 days with

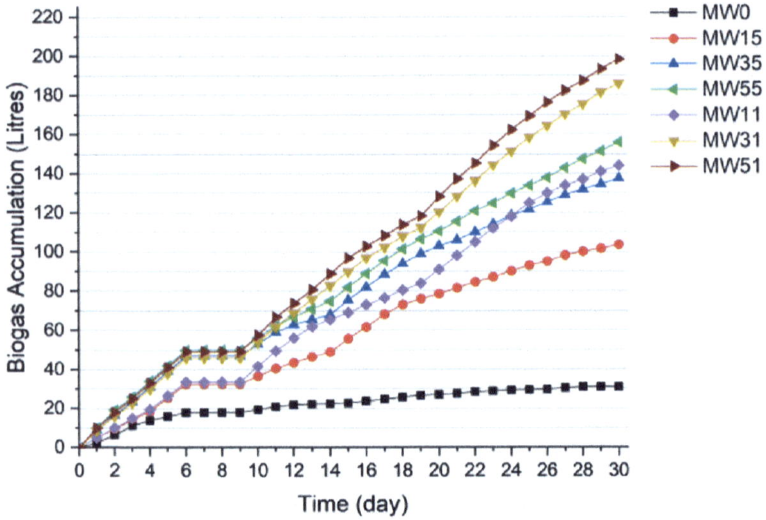

Fig. 3 Cumulative biogas production (Khairunnisa et al., 2025)

30 days required in the AD process, so there were 3 reactors that were given MCB twice in 1 batch. Figure 3 show cumulative biogas production during AD process.

The highest cumulative biogas production was achieved at MW51, where MCB was added at a rate of 50% every 10 days. The cumulative biogas produced was 198.3 L during the 30-day degradation process. At the same time, the lowest production was achieved at MW0 or the control reactor, with 30.8 L of biogas. This indicates that an addition of 50% MCB every 10 days can enhance biogas production by up to 6.43 times compared to the control.

6 Current Challenges

Despite its maturity, the AD of food waste presents unique challenges due to the high variability of the feedstock. High protein content, for instance, can lead to ammonia buildup, which is toxic to methanogenic bacteria and can inhibit the process (Wu et al., 2025). Anaerobic digestion (AD) is a well-established technology for waste management and bioenergy production, but it faces several challenges and opportunities for innovation. The following are some of the challenges:

a. Low Biogas Yield and Process Instability

Lignocellulosic biomass in agricultural waste causes challenges because it resists degradation, leading to low biogas yield and low buffering capacity (He et al., 2022). Instability issues arise from volatile fatty acid accumulation, foaming, and low buffer capacity, leading to reduced methane yields and efficiency

b. High Costs and Economic Viability

The economic feasibility of AD systems is often hindered by high capital and operating costs, as well as the costs associated with digestate disposal. Regulatory concerns and requirements also add to the financial burden (Ge et al., 2024)

c. Feedstock Variability

The variable composition of feedstocks, such as food waste, affects the stability and efficiency of the AD process. Factors like moisture content, nutritional content, and inhibitory substances (e.g. lipids, high ammonia) contribute to process instability (Shabbirahmed et al., 2024).

7 Future Innovations

Anaerobic digestion encounters considerable obstacles, continuous research and technological innovation provide promising pathways to improve its efficiency, economic feasibility, and environmental sustainability. Future developments in resolving current limitations and realizing the full potential of AD technology. The following are future innovations in AD technology:

7.1 Co-Digestion

Co-digestion by mixing food waste with other organic materials is a powerful strategy for overcoming this imbalance. The C/N ratio can be adjusted to the optimal range by combining food waste with a substrate rich in carbon and low in nitrogen (such as agricultural waste, yard trimmings, or sewage sludge). Co-digestion not only improves process stability and biogas yield but also allows the simultaneous treatment of multiple waste streams, resulting in a more integrated and efficient waste management system. Co-digestion with complementary substrates and pretreatment techniques can improve substrate degradability and process stability (Shabbirahmed et al., 2024).

7.2 Pretreatment

Pre-treatment of substrates before anaerobic digestion (AD) is crucial for enhancing biogas production and improving the efficiency of the digestion process. Various methods have been explored and can be broadly categorized into physical, chemical, biological, and combined pre-treatments. Mechanical treatment such as grinding, sieving, and using knife mills can increase the surface area of the substrate, making it more accessible to microbial action. For instance, mechanical treatments of raspberry extrudate resulted in a higher methane yield due to increased solubilization of phenolic compounds and sugars (Trujillo-Reyes et al., 2022).

7.3 Bulking Agents (BAs)

Bulking agents (BAs) such as biochar and lignocellulosic residues play a critical role in improving the performance of food waste anaerobic digestion by enhancing substrate porosity, preventing compaction, and promoting more efficient mass transfer of water, nutrients, and gases (Pei et al., 2024). Biochar, in particular, has demonstrated strong performance-enhancing effects; when added at optimal doses (e.g., 10 g/L). Moreover, co-pyrolysis biochars produced from mixtures of rice straw and digestate further enhance methane production—often exceeding 30% improvement—by supporting direct interspecies electron transfer (DIET) and stabilizing syntrophic microbial communities (Yang et al., 2025). In contrast, agricultural residues such as rice husk and wood chips primarily function by increasing structural integrity and internal aeration, which is beneficial in regions with abundant biomass, though these materials typically provide smaller gains in methane yield compared to pyrolytic biochars (Gao et al., 2023). Selecting an appropriate BA type and dosage must therefore account for local food waste characteristics—including moisture, lipid content, and particle structure—to maximize methane production, minimize inhibition, and improve overall reactor stability.

8 Anaerobic Digestate as a Source of Fertilizer

Anaerobic digestate has emerged as a valuable biofertilizer due to its rich nutrient content and contribution to sustainable agriculture. Recent research highlights its ability to enhance soil fertility, improve nutrient cycling, and reduce dependence on synthetic fertilizers. Digestate application not only supplies essential macronutrients like nitrogen, phosphorus, and potassium but also improves soil microbial activity and carbon sequestration. Digestate derived from food waste can enhance crop yields while minimizing greenhouse gas emissions associated with mineral fertilizer production.

The quality of digestate is closely linked to the stability of methanogenesis. Under optimal organic degradation, volatile fatty acids (VFAs) are efficiently consumed, preventing pH reduction and making the digestate safe for crop application (Wang et al., 2021). Conversely, VFA accumulation caused by microbial community imbalance can decrease pH and increase the risk of toxicity (Prasanna Kumar et al., 2024).

9 Conclusion

The study of food waste generation and its management through anaerobic digestion (AD) reveals a multidimensional issue that intersects environmental sustainability, energy recovery, and circular economy principles. Globally, food waste represents both a significant environmental burden and an untapped resource. Findings from Bogor Regency exemplify this duality: urbanization and changing consumption patterns accelerate waste generation, while local management systems struggle to adapt. With household food waste averaging 77 kg per capita per year, strategic valorisation becomes essential.

Anaerobic digestion emerges as a key sustainable solution, effectively converting organic waste into renewable biogas and nutrient-rich digestate. Compared to traditional disposal methods, AD minimizes greenhouse gas emissions, improves waste stabilization, and contributes to energy self-sufficiency. The biochemical stages—hydrolysis, acidogenesis, and methanogenesis—demonstrate the intricate microbial ecology required for efficient methane production. Critical operational parameters such as temperature (35–37 °C), pH (6.5–7.5), and mixing play decisive roles in maintaining process stability and optimizing yield.

The substrate composition profoundly influences AD performance. Balancing carbohydrates, proteins, and lipids ensures both stability and productivity, as imbalances lead to inhibitory conditions like ammonia toxicity or volatile fatty acid accumulation. Similarly, maintaining an optimal C/N ratio (20–30:1) is vital for microbial metabolism and process equilibrium. Semi-dry digestion systems with total solids (15–20%) offer the best compromise between biogas yield, reactor stability, and digestate manageability.

Emerging research underscores the importance of microbial community management, where methanogenic consortia (e.g., *Methanoculleus*) and bacterial phyla (e.g., *Firmicutes*) synergistically drive methane formation. Experimental results show that enhancing methanogen activity such as with 50% MCB additions—can increase cumulative biogas yields by over sixfold compared to control systems.

Nevertheless, challenges remain. Process instability, high operational costs, and feedstock variability continue to hinder AD's scalability and economic viability. Innovations such as co-digestion, pre-treatment technologies, and bulking agents represent promising solutions. Co-digestion balances C/N ratios and enhances degradability, pretreatments increase substrate accessibility, and bulking agents improve porosity and digestate quality—collectively strengthening the resilience and efficiency of AD systems.

In conclusion, tackling food waste through optimized anaerobic digestion aligns environmental stewardship with economic opportunity. By integrating technological innovation, biochemical optimization, and policy support, AD can transform food waste from an ecological liability into a cornerstone of renewable energy and sustainable resource management to supporting global transitions toward low-carbon, circular economies.

References

Ahmad, R. M. et al. (2024). Optimizing biogas production and digestive stability through waste co-digestion. *Sustainability (Switzerland), 16*(7). https://doi.org/10.3390/su16073045

Chew, K. R., et al. (2021). Effects of anaerobic digestion of food waste on biogas production and environmental impacts: A review. *Environmental Chemistry Letters, 19*(4), 2921–2939. https://doi.org/10.1007/s10311-021-01220-z

Costa-pinto, R. (2020). Macronutrients, minerals, vitamins and energy. *Anaesthesia and Intensive Care Medicine, 21*(3), 157–161. https://doi.org/10.1016/j.mpaic.2019.12.006

Despoudi, S. et al. (2021). Chapter 1—Food waste management, valorization, and sustainability in the food industry. In C. M. Galanakis (ed.). San Diego: Academic Press, pp. 3–19. https://doi.org/10.1016/B978-0-12-820563-1.00008-1

Dinh, P. Van et al. (2025). A critical review on factors affecting the two-stage anaerobic digestion of biodegradable solid waste. *Vietnam Journal of Science and Technology, 63*(2), 226–248. https://doi.org/10.15625/2525-2518/19962

Gao, Q., Li, L., Wang, K., & Zhao, Q. (2023). Mass transfer enhancement in high-solids anaerobic digestion of organic fraction of municipal solid wastes: A review. *Bioengineering, 10*(9), 1084. https://doi.org/10.3390/bioengineering10091084

Ge, X. et al. (2024). Chapter One—Rethinking anaerobic digestion for bioenergy and biopolymers production: Challenges and opportunities. In B. Li (ed.), *Conversion of waste and biomass to fuels and polymers*. Elsevier, pp. 1–109. https://doi.org/10.1016/bs.aibe.2024.01.001.

He, C. et al. (2022). Chapter 3—Conversion of agricultural wastes to bioenergy and biochemicals via anaerobic digestion. In A. Pandey et al. (Eds.). Elsevier, pp. 45–68. https://doi.org/10.1016/B978-0-323-90633-3.00007-9

Hussain, Z., Mishra, J., & Vanacore, E. (2020). Waste to energy and circular economy: The case of anaerobic digestion. *Journal of Enterprise Information Management, 33*(4), 817–838. https://doi.org/10.1108/JEIM-02-2019-0049

Induchoodan, T. G., Haq, I., & Kalamdhad, A. S. (2022). Factors affecting anaerobic digestion for biogas production: A review. In C. Hussain & S. B. T.- A. O. W. M. Hait (Eds.). Elsevier, pp. 223–233. https://doi.org/10.1016/B978-0-323-85792-5.00020-4.

Karuppiah, T., & Ebenezer Azariah, V. (2019). Biomass pretreatment for enhancement of biogas production. *Anaerobic Digestion*. https://doi.org/10.5772/intechopen.82088

Khairunnisa, C. F., Warmadewanthi, I. D. A. A., & Hanh, D. V. B. (2025). Biogas production from modification of food waste anaerobic digestion process. *BIO Web of Conferences, 157*. https://doi.org/10.1051/bioconf/202515705009

Liu, Y., Dong, H., & Zhu, H. (2020). Impact of total solids content on anaerobic co-digestion of pig manure and food waste: Insights into shifting of the methanogenic pathway. *Waste Management, 114*, 96–106. https://doi.org/10.1016/j.wasman.2020.06.048

Liu, L., Zheng, J., & Liu, J. (2025). Enhancing methanogenic efficiency in dry anaerobic digestion of kitchen waste through optimization of total solids content. *Environmental and Green Processess, 13*(8), 2332. https://doi.org/10.3390/pr13082332

Ma, G., et al. (2020). Methane yields during anaerobic co-digestion of animal manure with other feedstocks : A meta-analysis. *Science of the Total Environment, 728*, Article 138224. https://doi.org/10.1016/j.scitotenv.2020.138224

Martianto, D. et al. (2024). The quantity and composition of household food waste: Implications for policy. *PLoS ONE, 19*(6 June), 1–11. https://doi.org/10.1371/journal.pone.0305087

Mirmasoumi, S., Taherzadeh, M. J., & Horváth, I. S. (2023). Role of moisture and solids content in food waste anaerobic digestion: A critical review. *Waste Management, 160*, 121–134. https://doi.org/10.1016/j.wasman.2023.04.028

Mishra, S., et al. (2025). Microbial process in anaerobic digestion of food wastes for biogas production: A review. *Systems Microbiology and Biomanufacturing, 5*(2), 531–549. https://doi.org/10.1007/s43393-024-00303-6

Mu, L., et al. (2023). Emerging strategies for enhancing propionate conversion in anaerobic digestion: A review. *Molecules, 28*(9), 1–24. https://doi.org/10.3390/molecules28093883

Panigrahi, S., Sharma, H. B., & Dubey, B. K. (2022). Optimization of F/M ratio during anaerobic codigestion of yard waste with food waste: Biogas production and system stability BT—treatment and disposal of solid and hazardous wastes. In D. Sengupta, B. K. Dubey, & S. Goel (Eds.). Cham: Springer International Publishing, pp. 185–192. https://doi.org/10.1007/978-3-030-29643-8_9

Paranjpe, A., Saxena, S., & Jain, P. (2023). A review on performance improvement of anaerobic digestion using co-digestion of food waste and sewage sludge. *Journal of Environmental Management, 338*(December 2022), 117733. https://doi.org/10.1016/j.jenvman.2023.117733

Pei, S., Fan, X., Qiu, C., Liu, N., Li, J., Qi, L., & Wang, S. (2024). Effect of biochar addition on the anaerobic digestion of food waste : Microbial community structure and methanogenic pathways. *Water Science and Technology, 90*(3), 894–907. https://doi.org/10.2166/wst.2024.199

Prasanna Kumar, D. J., et al. (2024). A comprehensive study on anaerobic digestion of organic solid waste: A review on configurations, operating parameters, techno-economic analysis and current trends. *Biotechnology Notes, 5*(February), 33–49. https://doi.org/10.1016/j.biotno.2024.02.001

Rhee, C. et al. (2023). Mapping microbial dynamics in anaerobic digestion system linked with organic composition of substrates : Protein and lipid TS. *Energy, 275*(December 2022), 127411. https://doi.org/10.1016/j.energy.2023.127411

Shabbirahmed, A. M. et al. (2024). *Challenges and strategies for waste food anaerobic digestion: insights and future directions, Environment, Development and Sustainability*. Springer Netherlands. https://doi.org/10.1007/s10668-024-04820-1.

Trujillo-Reyes, Á., et al. (2022). Assessment of different mechanical treatments for improving the anaerobic biodegradability of residual raspberry extrudate. *Waste Management, 139*, 190–198. https://doi.org/10.1016/j.wasman.2021.12.034

Wang, Z. et al. (2021). Effect of biochar addition on the microbial community and methane production in the rapid degradation process of corn straw. *Energies, 14*(8). https://doi.org/10.3390/en14082223

Wang, Y. et al. (2025). *Anaerobic co-digestion of dairy manure and cucumber residues : Methane production efficiency and microbial community characteristics*, pp. 1–15.

Wu, H. M., et al. (2025). Food waste anaerobic digestion under high organic loading rate: Inhibiting factors, mechanisms, and mitigation strategies. *Processes, 13*(7), 1–26. https://doi.org/10.3390/pr13072090

Xue, S., et al. (2020). Interactive e ff ects of carbohydrate, lipid, protein composition and carbon/nitrogen ratio on biogas production of di ff erent food wastes. *Bioresource Technology, 312*(April), Article 123566. https://doi.org/10.1016/j.biortech.2020.123566

Yang, Q., Liu, H., Liu, L., Yan, Z., Chui, C., Yang, N., Wang, C., Shen, G., & Chen, Q. (2025). Enhanching methane production in anaerobic digestion of food waste using co-pyrolysis biochar derived from digestate and rice straw. *Molecules, 30*(8), 1766. https://doi.org/10.3390/molecules30081766

Yaser, A. Z. et al. (2022). Composting and anaerobic digestion of food waste and sewage sludge for campus sustainability: A review. *International Journal of Chemical Engineering, 2022*. https://doi.org/10.1155/2022/6455889

Yi, J., Dong, B., & Dai, X. (2014). Effect of increasing total solids contents on anaerobic digestion of food waste under mesophilic conditions: Performance and microbial characteristics analysis. *PLoS ONE, 9*(7), Article e102548. https://doi.org/10.1371/journal.pone.0102548

Zamri, M. F. M. A. et al. (2021). A comprehensive review on anaerobic digestion of organic fraction of municipal solid waste. *Renewable and Sustainable Energy Reviews, 137*(November 2020), 110637. https://doi.org/10.1016/j.rser.2020.110637

Zhao, X. et al. (2021). *Effect of total solids contents on the performance of anaerobic digester treating food waste and kinetics evaluation*, Vol. 01026, pp. 1–8.

Zheng, Z., et al. (2021). The effects of C/N (10–25) on the relationship of substrates, metabolites, and microorganisms in "inhibited steady-state" of anaerobic digestion. *Water Research, 188*, Article 116466. https://doi.org/10.1016/j.watres.2020.116466

Enhancing Phosphorus Recovery from Septic Tank Sludge Through Anaerobic Digestion as Biofertilizer-Effect of pH and Temperature

Rafidah Selaman, Ajimi Jawan, and Dewi Farhana Nazra

Abstract Phosphorus (P) is considered a limited and non-renewable mineral source that is essential in our daily lives. It is essential nutrient required for plant growth and development. Septic tank sludge (STS) is a residual waste generated from wastewater treatment which rich in organic matter and essential nutrient such as P. Instead of being disposed of as waste, it can serve as a resource for nutrient recovery. One effective method for this is anaerobic digestion (AD). Therefore, this study carried out to investigate the highest of reactive P recovery and time digestion, with a focus on optimizing pH and temperature conditions. The physicochemical and biological of raw STS was determined, and it was reported that waste was suitable to undergo the AD process. Optimization of pH 4.0 show the highest of reactive P recovery with value 22.99 mg/L ± (0.21), followed by pH 5.0 with 20.58 mg/L ± (0.15) and the lowest recovery was pH 6.0 with 19.65 mg/L ± (0.12). All digestion day was at day 9. While in temperature optimization, resulted 55.0 °C ± (1.0) recovered the highest of reactive P with 69.95 mg/L ± (0.12). Followed by 35.0 °C ± 1.0 (25.78 mg/L ± 0.05), and 45.0 °C ± 1.0 (17.56 mg/L ± 0.13), respectively. These findings indicate optimizing of pH and temperature significantly can enhance the P recovery efficiently. Moreover, this study also provides information on new strategies used in recovering P.

Keywords Anaerobic digestion · Phosphorus recovery · Septic tank sludge · pH · Temperature

R. Selaman (✉) · A. Jawan · D. F. Nazra
Faculty of Applied Sciences, Universiti Teknologi MARA Sabah Branch, Kota Kinabalu, Sabah, Malaysia
e-mail: rafidah5045@uitm.edu.my

A. Z. Yaser et al. (eds.), *Compost and Liquid Organic Fertilizers*,
https://doi.org/10.1007/978-981-95-8561-8_16

1 Introduction

Phosphorus (P) is a critical macronutrient essential for plant growth, energy transfer and cellular function is both plants and animals. It play a key role in food production through its use in fertilizers and is therefore directly linked to global food security (Witek-Krowiak et al., 2022). However, P is a finite and non-renewable resource primarily mined from phosphate rock., with global reserves concentrated in a few countries (Selaman & Wid, 2023). At the same time, excessive P discharge into water bodies contributes to eutrophication, leading to algal blooms, hypoxia and ecosystem damage (Okunola et al., 2018). This simultaneous challenge of P scarcity and environmental pollution underscores the need for strategies that enable P recovery and reuse. Septic tank sludge (STS) is a byproduct of domestic wastewater treatment (DWT) systems, is an underutilized but promising resource for nutrients recovery. It is rich in organic matter and essential nutrients such as P (Nakarmi et al., 2023). STS is frequently considered a waste material and disposed of in ways that can cause environmental contamination. P recovery from STS would not reduce waste and mitigate pollution but also support to a circular economy by generating a renewable source of fertilizer (Rose et al., 2015; Uddin & Wright, 2023). Anaerobic digestion (AD) is a widely applied biological treatment process that breaks down organic matter with absence of oxygen (Osei-Owusu et al., 2023). This technology offers a favourable opportunity for the generation of biogas and nutrient-rich digestate. Besides that, AD can also facilitate the release of P from sludge into recoverable form (Matjuda et al., 2022; Jankauskienė et al., 2024). However, the efficiency of P release and recovery is strongly influence by operational parameters such as pH and temperature. Optimizing of these parameters is crucial for maximizing the nutrient recovery. Due of this, this study focuses on investigate the effect of pH and temperature on P recovery from STS during AD process. The physicochemical and biological properties of raw STS was examined to confirm the suitability of AD. Then, experiments were performed to identify the pH and temperature conditions that resulted in maximised the P recovery. The findings from this research offer valuable insights into effective strategies for nutrient recycling, thereby supporting sustainable waste management and contributing and environmental security.

2 Methodology

2.1 Sample Collection and Preparation

STS was collected from a domestic tank located at Taman Indah Permai, Sepanggar. The collected raw STS was immediately transfer to the clean sealed container to prevent contamination and minimized nutrient loss.

2.2 Analytical Method

The physicochemical and biological characteristics of the STS were analysed following the standard procedure by Wid et al. (2017) and Aris et al. (2023). These included the determination of total solid (TS), volatile solid (VS), pH, total volatile fatty acids (TVFAs) and nutrient concentrations, particularly reactive P and nitrogen (N). In addition, determination of facultative anaerobic bacteria (FAB) was performed to evaluate the microbial activity and organic matter degradation potential. All analysed were conducted in triplicates to ensure accuracy and reliability, and the results are expressed as mean ± standard error of the mean (SEM).

2.3 Operation Star-Up for AD

The AD experimental for reactive P recovery was carried out in a reactor which represented by Duran bottle with working volume of 400 mL. The STS sample was put in Duran bottle according to the value of TS and VS. The optimization of pH and temperature was done according to Table 1. The digestion was done for 15 days. Initial pH was controlled with 1.0 M HCl and 1.0 M NaOH (Selaman & Wid, 2023). The gas generated during this period was discharged using the water displacement method once in three days of digestion to avoid any explosion. About 10 mL of AD liquor was taken out for reactive P determination once in three days (Table 2).

3 Results and Discussion

Figure 1 shows the graph of the concentration of reactive P recovery obtained from the optimization of three different pH value which were 4.0, 5.0 and 6.0. The results show the concentration of reactive P recovery for all pH recovery increased starting from day 1 of AD until day 9 of the AD process, then gradually decrease until day 15. Among all treatment, pH 4.0 resulted the highest percentage of reactive P recovery, with 22.99 mg/L ± (0.21), representing an increase of approximately 76.90% from the initial value of raw reactive P reported in Table 1. Followed by pH 5.0 with 20.58 mg/L ± (0.15) and the lowest recovery was pH 6.0 with 19.65 mg/L ± (0.12). The slow rate of reactive P recovery during day 3 to day 6 indicates the hydrolysis stage, where complex organic matter is broken down by hydrolytic bacteria into simpler and soluble molecules like amino acids, sugars, and fatty acids (Wid & Selaman, 2025). The sudden increase in reactive P recovery is due to the high availability of soluble organic molecules, especially TVFAs, which serve as carbon sources for polyphosphate accumulating organisms (PAOs) in anaerobic conditions (Mohammad et al., 2021). According to Li et al. (2021), pH 4.0 enhanced reactive P recovery higher that other pH level could be due to the pH promotes the dissolution

Table 1 The physicochemical and biological properties of raw STS

Parameters	TS (%)	VS (%)	pH	TVFAs (mg CH_3COOH/L)	N (mg/L)	Reactive P (mg/L)	FAB (CFU/mL)
Raw STS	30.12 ± (0.06)	85.32 ± (0.13)	7.70 ± (0.03)	4.20×10^5 ± (0.24)	13.07 ± (0.18)	5.31 ± (0.21)	4.20×10^4 ± (0.14)

Table 2 The optimization of pH and temperature

Parameters	pH	Temperature
pH	4.0 5.0 6.0	Fixed
Temperature	Fixed	35 °C 45 °C 55 °C

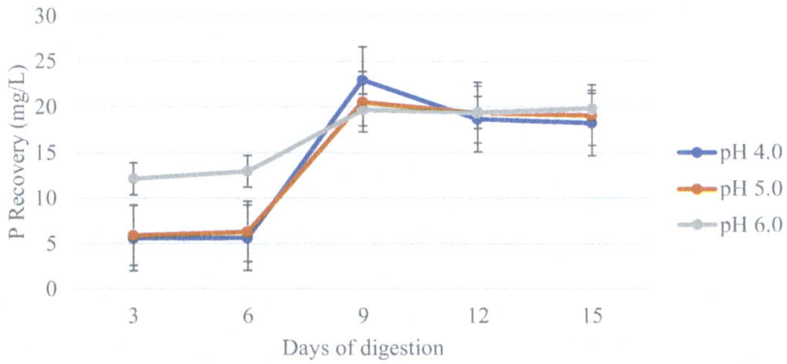

Fig. 1 The optimization of pH in 15 days of AD

of metal phosphate complexes. The increasing concentration of H^+ ions could break down the insoluble metal such as Fe, Al and Ca bound forms, releasing soluble P into the digestate liquor.

Table 3 compares the optimum pH values obtained in this study with those reported from previous studies by using different types of substrates. According to Kanellos et al. (2024), the optimum pH of FW co-digested with sewage sludge was 6.0. While study by Li et al. (2021) and Selaman and Wid (2023), observed optimum values 5.0 for WAS and POMES, respectively. In contract present study found that STS achieved its highest performance at lower pH of 4.0. This suggests that the optimum pH highly dependent on the type and composition of the substrate used. The lower pH requirement for STS may be attributed to its increase the formation of TVFAs formation which could cause the increasing of reactive P removal from STS and cause the FAB synergetic effects increasing, this increasing the reactive P recovery in anaerobic liquor (Selaman & Wid, 2023; Uddin et al., 2023). These differences highlight the importance of substrate characteristics in determining the most suitable operating condition of AD.

Temperature is a fundamental factor in AD, as it influences both bacterial growth and metabolic activity. In this study, temperature ranges corresponding to mesophilic (35.0°C ± 1.0), intermediate mesophilic (45.0°C ± 1.0), and thermophilic (55.0°C ± 1.0), conditions was used. Referring to Fig. 2, the highest concentration of reactive P recovery was achieved at 55.0°C (69.95 mg/L ± 0.12), followed by 35.0 °C (25.78 mg/L ± 0.05), and 45.0 °C (17.56 mg/L ± 0.13). Temperature of 55 °C

Table 3 Comparison the results obtained from the other studies

References	Types of substrates	Optimum pH
Li et al. (2021)	Waste activated sludge (WAS)	5.0
Selaman and Wid (2023)	Palm oil mill effluent sludge (POMES)	5.0
Kanellos et al. (2024)	FW and sewage sludge	5.0
This study	Septic tank Sludge (STS)	4.0

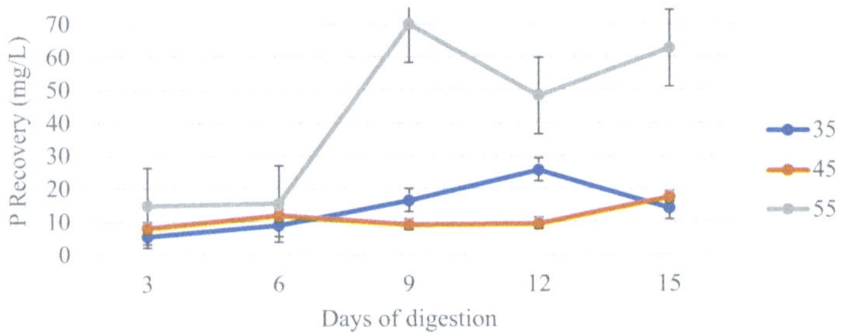

Fig. 2 The optimization of temperature in 15 days of AD

recover the highest reactive P with 92.41% increase. This could be because elevated temperatures accelerate the biochemical reactions that influence the rate of degradation, resulting in an enhanced hydrolysis rate (Singh et al., 2023). Besides that, study by Shin et al. (2022) mentioned that thermal hydrolysis occurs when optimal heat is applied, increasing the reactivity of water and allowing it to break down chemical bonds in the sludge. The rapid increase in reactive P concentration after day 6 also reflects the thermal breakdown of phosphorus-rich organic matter and microbial lysis, leading to high reactive P release into the liquid phase.

Table 4 show the optimum temperature for P recovery from STS was in this study (55.0 °C ± 1.0) is higher than those reported for other substate, which generally fall within the mesophilic range (30–38 °C). This suggested that thermophilic conditions are more favourable for enhancing reactive P solubilization and release from STS from other substrates. The higher of temperature could accelerated microbial activity and organic matter degradation then resulting a higher nutrient release (Osei-Owusu et al., 2023; Shin et al., 2022). These finding indicate that, thermophilic AD could be an effective strategy for maximizing reactive P recovery from STS although it could require additional energy input compared to mesophilic process.

Table 4 Comparison the results obtained from the other studies

References	Types of substrates	Optimum temperature (°C)
Okunola et al. (2018)	Pre-treated agro-food waste	30.0–38.0 (±0.0)
Jingjing et al. (2019)	Waste activated sludge (WAS)	35.0 ± 0.0
Wid and Selaman (2025)	Palm oil mill effluent sludge (POMES)	35.0 ± 0.0
This study	Septic tank sludge (STS)	55.0 ± 1.0

4 Conclusions

As the conclusion, this research highlights the potential of STS as a valuable source of reactive P recovery through AD. Experimental results demonstrated that both pH and temperature play a key role in enhancing P recovery efficiency. The highest reactive P recovery was achieved at pH 4.0 (22.99 mg/L ± 0.21) on day 9 of AD and temperature thermophilic condition (55.0°C ± 1.0) produced the greatest reactive P recovery (69.95 mg/L ± 0.12). These findings confirm that optimizing pH and temperatures can significantly improve reactive P recovery from STS. Hence, it can be concluded that AD is an effective method in promotes sustainable waste utilization while supporting a circular economy by transforming waste into a valuable nutrient for agricultural use.

Acknowledgements Authors would like to express sincere gratitude to the Universiti Teknologi Mara Sabah Branch for the financial support and lab assistants at the Science and Agrotechnology Complex, Universiti Teknologi Mara Sabah Branch.

Conflict of Interest The authors declare no conflict of interest.

References

Aris, A. M., Yuzir, A. M., Anuar, A. N., & Hara, H. (2023). Identification of facultative anaerobic bacteria for Low-Cost nutrient removal in treating domestic wastewater. *Chemical Engineering Transactions, 106*, 1201–1206.

Jankauskienė, J., Laužikė, K., & Kaupaitė, S. (2024). The use of anaerobic digestate for greenhouse horticulture. *Agronomy, 14*(10), 2437.

Jingjing, C., Tang, S., Yan, F., & Zhang, Z. (2019). Efficient recovery of phosphorus in sewage sludge through hydroxylapatite enhancement formation aided by calcium-based additives. *Water Research, 171*, Article 115450.

Kanellos, G., Tremouli, A., & Kondylis, A. (2024). Anaerobic co-digestion of the liquid fraction of food waste with waste activated sludge. *Waste Biomass Volar, 15*, 3339–3350.

Li, C., Sheng, Y., & Xu, H. (2021). Phosphorus recovery from sludge by pH enhanced anaerobic fermentation and vivianite crystallization. *Journal of Environmental Chemical Engineering, 9*(1), Article 104663.

Matjuda, D., Tekere, M., & Thaela-Chimuka, M. J. (2022). Characterization of the physicochemical composition of anaerobically digested (digestate) high throughput red meat abattoir waste in

South Africa and the determination of its quality as a potential biofertilizer. *Heliyon, 9*(11), Article e21647.

Mohammad, S., Baidurah, S., Kobayashi, T., Ismail, N., & Leh, C. P. (2021). A review- anaarobic digestion. *Processes, 9*(5), 739.

Nakarmi, K., Daneshvar, E., Manttari, M., & Bhatnagar, A. (2023). Removal and recovery of nutrients from septic tank wastewater using microalgae: Key factors and practical implications. *Journal of Environmental Management, 345*, Article 118922.

Okunola, A., Adewumi, A., Okonkwo, C., & Kim, D. (2018). Physico-chemical properties of various biomass used in anaerobic digestion. *International Journal of Civil Engineering and Technology, 9*(13), 1403–1410.

Osei-Owusu, B. A., Baidoo, M. F., Arthur, R., & Oduro-Kwarteng, S. (2023). Physico-chemical characterization of selected feedstocks as co-substrates for household biogas generation in Ghana. *International Journal of Sustainable Engineering, 16*(1), 117–128.

Rose, C., Parker, A., Jefferson, B., & Cartmell, E. (2015). The Characterization of feces and urine: A review of the literature to inform advanced treatment technology. *Critical Reviews in Environmental Science and Technology, 45*(17), 1827–1879.

Selaman, R., & Wid, N. (2023). Comparing phosphorus recovery and optimum time digestion from single and mixing substrates by using anaerobic mesophilic digestion technique. *Borneo Akademika, 7*(2), 70–76.

Shin, J., Lee, S., Park, H., Son, H., Raza, S., Wang, J., & Kim, Y. (2022). Effects of thermal hydrolysis on anaerobic digestion and abundance of antibiotic resistance genes during recuperative thickening digestate treatment of sewage sludge. *Chemical Engineering Journal, 450*(3), Article 138128.

Singh, R., Hans, M., Kumar, S., & Yadav, Y. K. (2023). Thermophilic anaerobic digestion: An advancement towards enhanced biogas production from lignocellulosic biomass. *Sustainability, 15*(3), 1859.

Uddin, M., & Wright, M. (2023). Anaerobic digestion fundamentals, challenges, and technological advances. *Physical Sciences Reviews, 8*(9).

Wid, N., Selaman, R., & Marcus, J. (2017). Enhancing phosphorus recovery from different wastes by using anaerobic digestion technique. *Advanced Science Letters, 23*(1), 437–1439. https://doi.org/10.1166/asl.2017.8381

Wid, N., & Selaman, R. (2025) Phosphorus recovery from different organic substrates using single and mixed anaerobic digestion. In *Nutrients and Colored Compounds in Wastewater*, pp. 103–10. Elsevier.

Witek-Krowiak, A., Gorazda, K., Szopa, D., Trzaska, K., Moustakas, K., & Chojnacka, K. (2022). Phosphorus recovery from wastewater and biobased waste: An overview. *Bioengineered, 13*(5), 13474–13506.

Decoding Veterinary Antibiotic Behavior in Aerobic Composting and Anaerobic Digestion of Manure

Kiki Gustinasari, Ahamd Rizqi Alima Fabri, Rizkiy Amaliyah Barakwan, Inggit Kresna Maharsih, Nimas Mayang Sabrina Sunyoto, and Jaisy Aghniarahim Putritamara

Abstract The common use of veterinary antibiotics on livestock means that they end up in manure, which is a big source of antimicrobial resistance (AMR) in the environment. Aerobic composting and anaerobic digestion (AD) are common ways to treat manure, but how well they get rid of antibiotics and resistance genes (ARGs) varies a lot. This review looks at how the main classes of antibiotics break down during these processes. Antibiotic breakdown is greatly affected by operational parameters and microbial communities. Aerobic composting frequently attains enhanced antibiotic elimination via thermophilic and oxidative mechanisms. In contrast, while AD is useful for energy recovery, it often leads to incomplete degradation and the persistence of transformation products (TPs). A significant discovery is the contradictory

K. Gustinasari (✉)
Study Program of Environmental Engineering, Department of Biosystem Engineering, Faculty of Agricultural Engineering, Universitas Brawijaya, Malang, Indonesia
e-mail: kikigustinasari@ub.ac.id

A. R. A. Fabri (✉)
Faculty of Mathematics and Natural Sciences, Universitas Brawijaya, Malang, Indonesia
e-mail: arizqialima@gmail.com

Universitas Islam Malang, Malang, Indonesia

R. A. Barakwan
Study Program of Environmental Engineering, Department of Biology, Faculty of Science and Technology, Universitas Airlangga, Surabaya, Indonesia

I. K. Maharsih
Department of Biosystems Engineering, Faculty of Agricultural Technology, Study Program of Bioprocess Engineering, Universitas Brawijaya, Malang, Indonesia

N. M. S. Sunyoto
Department of Agro-Industrial Technology, Faculty of Agricultural Technology, Universitas Brawijaya, Malang, Indonesia

J. A. Putritamara
Department of Livestock Socio-Economics, Faculty of Animal Science, Universitas Brawijaya, Malang, Indonesia

A. Z. Yaser et al. (eds.), *Compost and Liquid Organic Fertilizers*,
https://doi.org/10.1007/978-981-95-8561-8_17

function of these systems: they have the capacity to diminish ARGs while potentially augmenting them in suboptimal circumstances. Important knowledge gaps are the absence of field-scale studies, data from tropical areas, and risk evaluations of TPs. Future research should emphasize field validation, employ advanced analytical techniques, and incorporate multi-omics approaches to enhance the robustness and comprehensiveness of scientific findings. Ultimately, putting manure management at the top of the list of One Health strategies in a circular economy is important for stopping the spread of AMR.

Keywords Veterinary antibiotics · Manure treatment · Aerobic composting · Anaerobic digestion · ARGs · One health · Transformation products

1 Introduction

Modern food animal production increasingly relies on antimicrobial agents to maintain animal health and productivity under intensive production systems. In veterinary medicine, antibiotics are typically used for treating diagnosed infections, metaphylaxis (treating groups when some are infected), prophylaxis (preventing anticipated outbreaks), and historically in many jurisdictions growth promotion through subtherapeutic dosing. Although growth promotion has now been banned in the EU since 2006 and restricted in the U.S. and several Asian countries, it remains prevalent in regions where livestock intensification has outpaced regulatory oversight (Aidara-Kane et al., 2018; Page & Gautier, 2012).

Recent quantitative analyses highlight the scale and trajectory of antibiotic use in livestock. Van Boeckel et al. (2015) estimated global food-animal antimicrobial consumption at 63,151 tons in 2010, projecting a 67% increase by 2030, mainly driven by rapidly intensifying production in middle-income economies. The adoption of biomass corrected metrics, such as mg/PCU (milligrams per population-correction unit), has further improved cross-country comparability, revealing striking disparities in antibiotic intensity across production systems (Sabah et al., 2025).

An often-overlooked pharmacokinetic aspect is that 30–90% of administered antibiotic doses are excreted as active compounds or metabolites, making livestock manure a primary reservoir for antibiotic residues and ARGs (Zhang & He, ; He et al., 2020). When untreated manure is applied to agricultural land, these compounds persist in the soil, strongly adsorb organic fractions, leach into surface and groundwater, and affect the structure of microbial communities (Hu et al., 2017; Li et al., 2017a, 2017b; Xie et al., 2018).

Recent review papers have made significant contributions to the understanding of antibiotic degradation and resistance gene dynamics in aerobic composting. Zhao et al. (2024) describes optimizing aerobic composting for the degradation of antibiotic resistance genes by emphasizing factors such as aeration, bulking agents, and temperature-controlled fans to enhance ARG reduction. Another review by Feng et al. (2023) analyzes the behavior of antibiotics and ARGs in antibiotic fermentation waste composting, focusing primarily on microbial and resistome dynamics under aerobic conditions. While both reviews provide insights into compost-based mitigation, their coverage is limited to aerobic processes and does not include AD. This review explains the behavior, degradation mechanisms, and transformation pathways of veterinary antibiotics in both aerobic composting and AD systems. It further integrates these findings with microbial resistance dynamics and environmental considerations, framed within the One Health and circular economy concepts.

2 Veterinary Antibiotics in Livestock Manure

Against this background, we next characterise antibiotic classes, usage patterns, and their occurrence in manure.

2.1 Classes and Pattern of Use

Table 1 demonstrates how different antibiotic classes react to different treatment approaches and behave differently in manure. Tetracyclines and macrolides primarily segregate into solid fractions, but sulfonamides have greater mobility and a higher propensity to leach. Even while β-lactams break down quickly, they can have a temporary effect on the levels of ARGs. Thermophilic aerobic composting and biochar addition accelerate the breakdown of tetracyclines and macrolides. Anaerobic digestion, on the other hand, works somewhat well and only on certain compounds (Gaballah et al., 2024; Wallace et al., 2018).

Public policy influences usage patterns. The European Union banned antibiotic growth promoters on January 1, 2006; the new framework, Regulation (EU) 2019/6, which came into effect on January 28, 2022, further tightens prophylaxis/metaphylaxis and establishes a list of antibiotics reserved for human use only. Similarly, the US implemented a veterinary feed directive (VFD), ending the use of "medically important" growth promoters and requiring veterinary supervision for feed and water (since 2017). In another case, China banned colistin as a feed additive in 2017, followed by a marked decline in the prevalence of the mcr-1 gene in animals, the environment, and humans (European Commission, 2006).

2.2 Occurrence and Pathways Dissemination

Animal antibiotic residues are commonly found in raw manure from swine, poultry, and cattle. Recent field monitoring studies and meta-analyses indicate that typical concentrations range from tens to thousands of μg/kg (≈ 0.01–1 mg/kg), with some

Table 1 Veterinary antibiotic classes, environmental occurrence traits, and treatment relevance in manure processing systems

Class	Common livestock indications	Typical manure occurrence and traits (Qualitative)	Notes for treatment relevance
Tetracyclines (oxytetracycline, chlortetracycline)	Broad-spectrum therapy and metaphylaxis in swine, cattle, poultry	High excretion (>70%); strong sorption to solids; persistent in lagoons/soils	Thermophilic composting improves degradation; biochar co-composting enhances antibiotic and ARG attenuation (Tian et al., 2019; Zhang and He, 2019a, 2019b)
Sulfonamides (sulfamethazine, sulfadiazine, SMX)	Often combined with trimethoprim for enteric/respiratory infections	Moderate sorption; mobile in porewater; variable removal efficiencies in compost and AD	Suboptimal composting can temporarily enrich antibiotic-resistant bacteria; careful control of temperature and moisture is required (Wallace et al., 2018)
Macrolides (tylosin, erythromycin)	Respiratory disease therapy; historical growth promoters	High sorption to organic matter; detectable in lagoons/soils	Removal efficiency depends on thermophilic exposure, oxygen availability, and retention time (Katada et al., 2021)
Fluoroquinolones (enrofloxacin, ciprofloxacin)	Targeted therapy in swine and poultry	Strong sorption; persistent in solids; low mobility in aqueous phases	Compounds are highly recalcitrant; integrated biochar-amended AD improves partial removal (Zhao et al., 2025a, 2025b, 2025c)
β-lactams (ampicillin, cephalosporins)	Mastitis treatment, systemic infections	Rapid hydrolysis under neutral–alkaline conditions; low persistence	Although residues degrade quickly, β-lactam exposure exerts strong selective pressure on clinically critical resistance genes (Dolliver et al., 2008)

reaching tens to hundreds of mg/kg for widely used compounds/classes (e.g., tetracyclines, macrolides, ionophores in poultry). Reported values are often expressed per dry weight (dw), although some studies use wet weight (ww); this difference in basis is important when comparing values (Qian et al., 2022). β-lactams (e.g., cephalosporins) tend to be more labile in moist raw manure, so their concentrations are often low/quickly declining after administration; for example, ceftiofur degrades very rapidly during composting ($t1/2 \sim 1.1$ days), although data immediately after excretion in raw manure do indicate the presence of metabolites (Ma et al., 2024).

Persistence is determined by (i) sorption on organic/mineral materials, (ii) transformation (biodegradation, photolysis, hydrolysis) and the TPs, and (iii) manure biophysics (pH, solids, metals, dissolved organic carbon (DOC)). Tetracyclines and fluoroquinolones are highly "sticky" (strong sorption) and persist in the topsoil for weeks to months. Sulfonamides have weaker sorption properties and are therefore relatively easy to dissolve and migrate. Microlides are generally strongly bound, but their degradation varies depending on conditions (Nkoh et al., 2024).

2.3 Ecological Risks

Bacteria and cyanobacteria are the most sensitive groups to many antibiotics; exposure to low concentrations can inhibit photosynthesis/microalgal growth, as well as alter aquatic community structure. Antibiotics in soil, particularly tetracyclines and sulfonamides, can disrupt the nitrogen (N) and carbon (C) cycles, suppress denitrification/nitrification, and alter microbial diversity and ecosystem function (Guo et al., 2025). The direct and indirect toxicological impacts of human exposure through water/food production are generally low; the primary risk is promoting AMR in the environment, which then impacts human/animal health (infections that are difficult to treat).

3 Manure Treatment Pathways

With these occurrence patterns in mind, we now outline the two dominant treatment pathways.

3.1 Principles of Aerobic Composting

Composting is familiar method for processing solid manure. Meso- and thermophilic microorganisms decompose biodegradable matter, such as simple carbohydrates and amino acids, in the presence of oxygen. This process converts complex organics into simpler compounds, heat, CO_2, and water. There are two major steps in composting, namely active and maturation steps. The first step occurs when microorganisms consume oxygen and simple organic compounds simultaneously. This stage produces heat by increasing the temperature up to nearly 70 °C due to microbes metabolism and the presence of thermophilic bacteria. Afterward, the maturation step starts when the activity of microorganisms decreases indicated by decreasing in temperature and new compounds are formed, such as humification of lignocellulosic compounds.

There are three composting configurations that commonly operated, including pit process or static pile, stacking or windrow, and in-vessel systems. The differences of these three lies on the type of composting unit and aeration methods. Pit process or static pile uses a pit to pile manure statically. Ventilation is provided as a channel for airflow and drainage of leachates. Insulation is needed in this configuration to reduce excessive heat loss. In contrast, the stacking/windrow system utilizes mechanically aerated system with a blower to supply air to the system. Lastly, in-vessel composting is carried out in a rotating drum that ensures continuous forced aeration and mixing. This system allows precise control of temperature, oxygen, and moisture content to optimized the decomposition (Liu and Wang, 2019; Amuah et al., 2022).

3.2 Principles of Anaerobic Digestion

Anaerobic digestion is a conversion process of manure to biogas with high methane using anaerobic microorganisms in enclosed and oxygen-free operation, called digester. The decomposition process usually produces several intermediate chemicals, including hydrogen sulphides, methane, and organic acids. The overall reaction network including hydrolysis, acidogenesis, acetogenesis, and methanogenesis processes, as shown in Eqs. (1)–(4b) (Braeutigam et al., 2014). The hydrolysis process breaks down the complex organic matters like carbohydrates, proteins, and fats to monomers or oligomers of sugars, amino acids or fatty acids. Subsequently, in ideal condition of acidogenesis stage, fermentative bacteria degrade simpler organic compounds into organic acids and alcohols. However, usually simpler organic compounds can be directly converted into volatile fatty acids, hydrogen, and carbon dioxide. Then, the remaining organic acids and alcohols are altered to produce CO_2, H_2, HCOOH, and acetate through acetogenesis phase. Finally, during methanogenesis stage, CH_4 and CO_2 are formed and released.

$$C_xH_yO_zN\ (complex\ organic\ compounds) \xrightarrow{hydrolysis} simpler\ organic\ compounds \tag{1}$$

$$simple\ organic\ compounds \xrightarrow{acidogenesis} organic\ acids + alcohols + CO_2 + H_2 + HCOOH + Acetate \tag{2}$$

$$organic\ acids\ and\ alcohols \xrightarrow{acetogenesis} CO_2 + H_2 + HCOOH + Acetate \tag{3}$$

$$CO_2 + H_2 + HCOOH \xrightarrow{methanogenesis} CH_4 + CO_2 \tag{4a}$$

$$Acetate \xrightarrow{methanogenesis} CH_4 + CO_2 \tag{4b}$$

In downstream process, biogas can be purified to enhance the methane composition. This compound can be used to generate heat and electricity. Slurry and liquid can be used as raw materials in AD. In comparison with aerobic composting, AD process results in lower nutrient loss but requires longer time to obtain final product (Liu and Wang, 2019; Amuah et al., 2022).

3.3 Operational Factors Influencing Treatment Performance

Aerobic composting depends on temperature, aeration, moisture condition, and the carbon-to-nitrogen ratio to obtain optimum product. The optimal operating conditions are temperature of 40–65 °C, with frequent aeration to supply oxygen, removes excessive heat and CO_2, and reduce moisture to 40–60%. These conditions are associated with the most favourable environment for microbial growth and activity. Additionally, the optimal carbon-to-nitrogen ratio for aerobic composting ranges between 25:1 and 30:1. Thus, raw manure needs to be pre-treated before composting because it consists high nitrogen and moisture content.

Similar to aerobic composting, AD is strongly influenced by temperature, pH, and the carbon-to-nitrogen ratio of manure as the feedstock. The optimal operating conditions are a temperature around 35 °C, a neutral pH, and a carbon-to-nitrogen ratio of 20:1 to 30:1. A continuous digester configuration is recommended to ensure a constant supply of manure. Sudden changes in feedstock composition can alter operating conditions, which negatively affects microbial growth and activity (Liu and Wang, 2019).

4 Antibiotic Behavior in Aerobic Composting

Having set out the treatment pathways, we first examine antibiotic fate under aerobic composting.

4.1 Mechanisms of Degradation and Transformation

During aerobic composting, the concentration of antibiotics decreases due to a mix of biotic (biodegradation by microbes and their enzymes) and abiotic (hydrolysis, oxidation) processes. This is also due to the antibiotics being absorbed or sequestered into the organic matter and compost minerals, which makes them less available to living things. In the thermophilic phase, high temperatures (around 55–70 °C) speed up the metabolism of microbes (like *Bacillus* and *Thermus*) and the breakdown of several abiotic classes. Important enzymes: There are many microbial enzymes that attack antibiotic structures. These include β-lactamases (which open β-lactam rings), esterases, mono- and di-oxygenases, nitroreductases, and ligninolytic enzymes (laccases and peroxidases) that cause the nonspecific oxidation of macrolides, tetracyclines, and other antibiotics. Figure 1 shows the mechanism of antibiotic biodegradation by bacteria. Recent studies have confirmed the ability of laccases/peroxidases to decompose a wide range of antibiotics, making them a promising bioremediation strategy (Dutta et al., 2025). Thermophilic compost rapidly degrades various antibiotics. In broiler manure compost (40 days), > 99% of

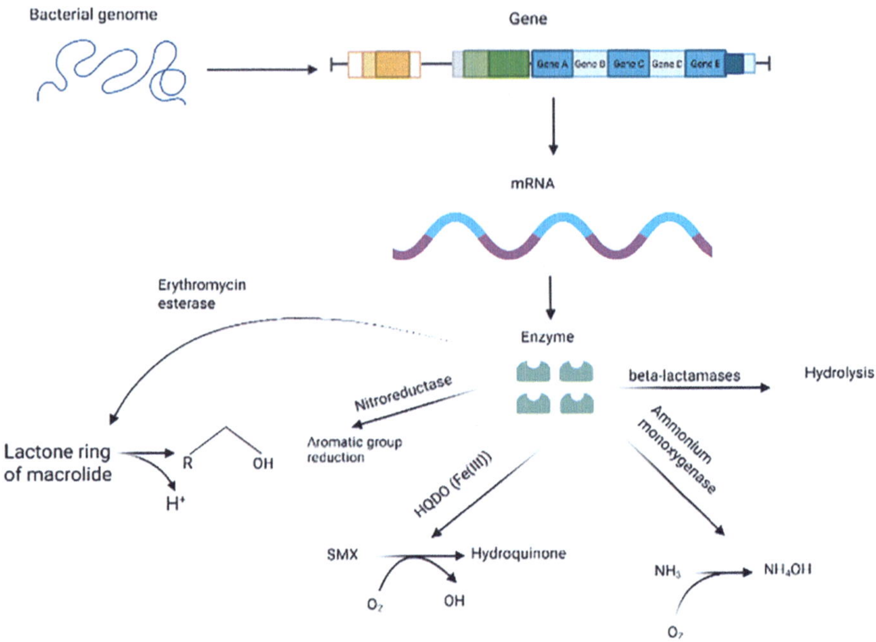

Fig. 1 The mechanism of antibiotic biodegradation by bacteria. Adopted from Dutta et al. (2025)

9 antibiotics and 1 hormone were lost with a t1/2 of 1.3–3.8 days, indicating very rapid degradation/transformation under adequate processing conditions (Ho et al., 2013).

5 Antibiotic Behavior in Anaerobic Digestion

By comparison, the following section considers fate and persistence in AD.

5.1 Transformation Mechanisms Under Anaerobic Conditions

Antibiotics in A are subject to sorption and microbial biotransformation. Unlike aerobic systems, AD achieves only partial mineralisation, with transformation pathways altering antimicrobial activity and persistence in digestate (Wallace et al., 2018). Antibiotic removal in AD is mediated by cooperative activity of fermenters, methanogens, and syntrophs, each with distinct contributions to degradation and redox balance. Their main roles are summarised in Table 2.

Table 2 Role of microbial consortia in antibiotic removal during AD (Turker et al., 2018; Ma et al., 2021; Chu et al., 2022; G. Wang et al., 2022a, 2022b; Wu et al., 2022; Trueba-santiso et al., 2025)

No.	Microbial group	Example microorganisms	Antibiotic	Mechanism of removal/impact
1	Fermenters	*Desulfovibrio vulgaris, Clostridium spp., Firmicutes*	Sulfamethoxazole, Ciprofloxacin	Supply H_2, formate, and VFAs; reductive cleavage (N–O bond reduction); ciprofloxacin alters acidogenesis and methane yield
2	Methanogens	*Methanothrix, Methanosaeta, Methanosarcina*	Ciprofloxacin, Sulfamethazine, Florfenicol	Maintain low redox; hydroxylation/side-chain modifications; acetoclastic methanogens more sensitive; > 88% ciprofloxacin removal
3	Syntrophs	*Syntrophomonas spp., Anaerolineaceae, SAO consortia*	Tetracycline, Oxytetracycline	Maintain ultra-low H_2; biochar enhances DIET, boosting tetracycline removal (24–158%); reduce ARGs

5.2 Persistence and Transformation Products in Digestate

Digestate from AD often retains residual antibiotics and TPs, reflecting incomplete biodegradation and sorption, with ecological risks upon land application. Table 3 presents a synthesis of persistence and TPs of antibiotics during AD.

Table 3 Transformation products of antibiotics in digestate and the removal efficiency

No.	Antibiotic	Metabolites in digestate	Removal efficiency (%)	References
1	Sulfamethoxazole	Aminobenzoate-type fragments	70–90	Chu et al. (2022)
2	Tetracycline	4-epi-tetracycline, iso-tetracycline	50–60	Cheng et al. (2016)
3	Ciprofloxacin	Hydroxylated CIP, de-cyclopropyl CIP	40–70	Trueba-santiso et al. (2025)
4	Chlortetracycline & Roxithromycin	Hydroxylated CTC, ROX residues	CTC 59; ROX 50	Ezzariai et al. (2018)
5	Oxytetracycline	Epimers and demethylated OTC	≈100	Lu et al. (2023)
6	Sulfonamides & macrolides	Hydroxylated and N-dealkylated metabolites	SMX/TMP 90; macrolides 50	Mejías et al. (2025a)
7	Multiple (45 PhACs incl. antibiotics)	Hydroxylated, conjugated TPs	Up to 83	Dolu and Nas (2023)

6 Comparation of Aerobic Versus Anaerobic Digestion in Remove Antibiotic

With these mechanisms established, we compare performance across antibiotic classes between composting and AD. Table 4 presents comparative performance of composting and AD of major veterinary antibiotic. Aerobic composting achieves higher removal of tetracyclines/sulfonamides and rapid β-lactam hydrolysis; macrolides/fluoroquinolones remain partially persistent. AD can accumulate TPs; ARG responses are inconsistent. Integrated AD with composting balances energy recovery with contaminant polishing.

Reported removals in composting and AD differ mainly due to:

(i) **Operating Conditions**

Sustained Thermophilic Composting:

- **Higher Attenuation**: Continuous thermophilic composting (CTC) significantly enhances the degradation of antibiotics due to the high temperatures maintained throughout the process. This method accelerates organic matter degradation and shortens the composting cycle, primarily driven by the synergism of bacterial genera (Wang et al., 2024). The thermophilic phase is crucial for the high degradation of antibiotics, as temperature is a major factor influencing the process (Selvam et al., 2013).

Mesophilic/Unstable Runs and **Short Hydraulic Retention Time (HRT)/ Solid Retention Time (SRT) in AD**:

- **Moderate Removal**: AD with HRT and SRT yields moderate removal of antibiotics. The persistence of antibiotics like fluoroquinolones in AD biosolids indicates that longer retention times or additional treatment stages may be necessary for effective removal (Liu et al., 2019).

(ii) **Matrix/Speciation (Sorption and Persistence)**:

- **Strong Sorption (TCs, FQs**: Tetracyclines (TCs) and fluoroquinolones (FQs) exhibit strong sorption to organic matter, which can mask their persistence in the environment. This sorption is influenced by factors such as pH and the presence of organic matter (Chen et al., 2023; Lou et al., 2017).
- **N-acetyl/De-acetyl (SAs**: Sulfonamides (SAs) undergo N-acetylation and de-acetylation, affecting their stability and persistence in the environment (Han et al., 2021).
- **Hydrolysis of β-lactams**: β-lactams hydrolyze readily but can leave behind bound or transient residues, complicating their complete removal (Han et al., 2021).

(iii) **Analytics of Liquid chromatography-tandem mass spectrometry (LC–MS/MS)**:

Table 4 Comparative performance of composting versus anaerobic digestion

No.	Antibiotic (class)	Key metabolites/ behaviour in Aerobic Composting	Removal (%) Composting	Key metabolites/ behaviour in Anaerobic Digestion	Removal (%) AD	Sources
1	Tetracyclines (TC, CTC, OTC)	Epimerisation (4-epi forms), iso-TC; oxidation; strong sorption to humic matter	70–95	Epimers (4-epi/iso), demethylated OTC; solids binding	40–60	Wu et al. (2011), Tong et al. (2012), Kitazono et al. (2015), Coban et al. (2016), Liu et al. (2018) and Zhang et al., (2019a, 2019b)
2	Sulfonamides (e.g., SMX, SDZ)	N-acetyl ↔ de-acetyl cycling; partial oxidation; reduced bioavailability with humification	60–85	Reductive de-amination to aminobenzoate-type products; acetyl/ de-acetyl interconversion	30–70	Shao et al. (2013), Liu et al., (2015, 2018), Jin et al. (2017), Chen et al. (2022), Province and Energy (2023) and Li et al. (2024)
3	Macrolides (e.g., ROX, ERY)	Hydroxylated/ demethylated TPs; persistence unless long high-T exposure	< 50–60	Partial transformation; residues persist in digestate	< 40–50	Zhang and Li (2018), Yan et al. (2023), Fanfoni et al. (2025) and Mejías et al. (2025b)
4	Fluoroquinolones (e.g., CIP, NOR)	Strong sorption to organic/mineral phases; limited oxidative TPs	40–70	Hydroxylation/ defluorination; piperazine cleavage; strong sorption to solids	30–50	Li et al. (2017a, 2017b), Yang, Meng and Xue (2017), Lacroix et al. (2018), Liao et al. (2019), Gaballah et al. (2023) and Sun et al. (2023)

(continued)

Table 4 (continued)

No.	Antibiotic (class)	Key metabolites/behaviour in Aerobic Composting	Removal (%) Composting	Key metabolites/behaviour in Anaerobic Digestion	Removal (%) AD	Sources
5	β-lactams (e.g., ceftiofur, penicillins)	Rapid hydrolysis; "bound residues" decline during thermophilic phase	> 90 (rapid)	Hydrolysis; protein-bound residues; short-lived activity	Rapid–partial	Khadra et al. (2019), Gurmessa et al. (2020) and Gaballah et al. (2021)
6	Trimethoprim–Sulfamethoxazole pair	Oxidative TP formation; improved removal with sustained thermophilic regime	70–90	Hydroxylated/N-dealkylated TMP; aminobenzoate-type SMX TPs	50–80 (variable)	Taylor et al. (2012), Chen et al. (2022) and Ma et al. (2025)
7	Multi-class (mixed manure)	Oxidation + humification + thermal inactivation; community succession drives attenuation	Up to > 90 (class-dependent)	Reductive biotransformation + sorption; metabolite accumulation possible	Up to ~ 70 (class-/HRT-dependent)	Liu et al. (2018), Pei et al. (2021), Wang et al., (2022a, 2022b) and Fanfoni et al. (2025)

- **Parent-only Analysis**: LC–MS/MS is the gold standard for antibiotic quantification, focusing on parent compounds. However, matrix effects and the lack of stable isotopically labeled standards can affect accuracy (Zhao et al., 2025a, 2025b, 2025 c).
- **Matrix Effects**: Matrix effects are a significant challenge in LC–MS/MS analysis, requiring robust sample preparation and the use of internal standards to mitigate their impact (Zhao et al., 2025a, 2025b, 2025c).
- **ARG Normalisation**: Normalizing ARGs in environmental samples is crucial for accurate assessment, as their abundance can vary significantly during composting and other treatment processes (Fahad et al., 2021).

A conceptual model illustrating the divergent fate of antibiotics and ARGs under these two redox conditions is presented in Fig. 2. The diagram highlights the oxidative degradation pathway in composting (left/top) versus the reductive transformation and sorption mechanisms in anaerobic digestion (right/bottom), along with their respective implications for One Health and circular economy targets.

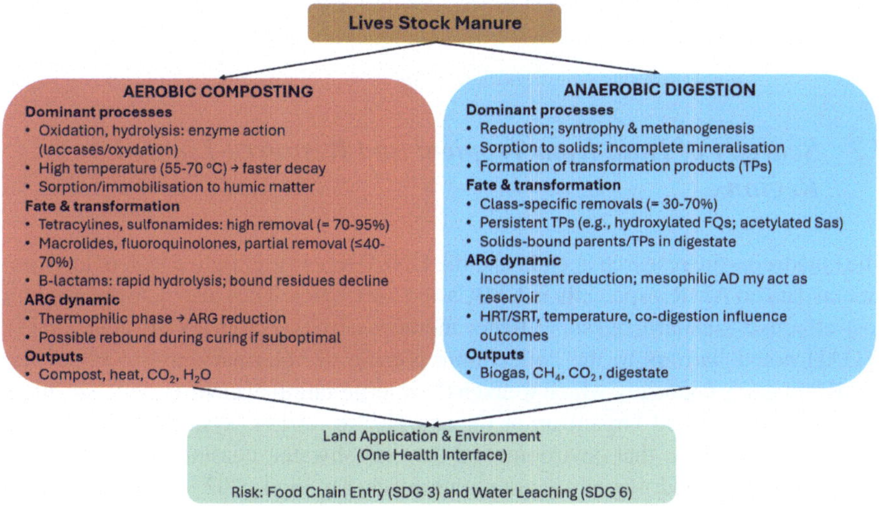

Fig. 2 Conceptual framework comparing the fate of veterinary antibiotics and ARGs in aerobic composting versus anaerobic digestion

7 Research Gaps and Future Directions

7.1 Limited Comparative and Long-Term Studies

Substantial empirical evidence demonstrates the presence of antibiotic residues and resistance genes in both manure and environmental samples. However, only a few studies compare different products, climates, or processing methods over several years. Most research has focused on laboratory or short-term studies, making it hard to understand long-term effects or create regional risk models. Recent reviews stress the need for research that covers multiple sites and years (Marutescu et al., 2022). Environmental monitoring is still limited and incomplete. The long-term effects of manure applications on antibiotic and antibiotic resistance gene (ARG) dynamics are not yet fully characterized. Numerous reviews have identified inconsistent outcomes regarding antibiotic and ARG reduction during storage, aerobic composting, and anaerobic digestion. Furthermore, there is a recognized lack of systematic comparative studies evaluating different AD configurations, including upflow anaerobic sludge blanket (UASB), continuously stirred tank reactor (CSTR), embedded digesters, and septic systems. (Zhang et al., 2021).

7.2 Scarcity of Data from Tropical and Resource-Limited Regions

The Quadrupartite research agenda and the UNEP report highlight a lack of environmental data on AMR, especially in LMICs and the tropics. They also note the need for cost-effective, standardized surveillance methods across sectors (WHO et al., 2023). WOAH noted barriers to the collection/reporting of veterinary antimicrobial use (AMU) in many countries, which weaken the interpretation of environmental residue/ARG findings (World Organization for and Health, 2019). Systematic reviews of Southeast Asia show that environmental data (freshwater, coastal, wastewater) are sporadic, and methodologies are not yet uniform (Siri et al., 2023). Typical tropical production systems employ integrated practices, such as livestock–aquaculture. The monsoon season increases the risk of runoff and percolation. These risks are rarely monitored in real-time. Therefore, practical recommendations are needed. One is cost-efficient tiered monitoring. Use Polar Organic Chemical Integrative Sampler (POCIS) for time-averaged screening in rivers and irrigation systems. Forward priority samples to LC–MS/MS at regional hub laboratories (Domínguez-García et al., 2025).

7.3 Insufficient Monitoring of Transformation Products

Most monitoring programs still focus solely on parent compounds. However, during manure processing and field application, antibiotics degrade into TPs that may be more persistent or toxic. Since these TPs often retain antimicrobial activity, current ecological risk assessments likely underestimate the true selective pressure for resistance (Zhang et al., 2025). Examples of important but often overlooked TPs. Tetracycline ⇄ 4-epimers (e.g., TC/OTC ⇄ 4-epiTC/4-epiOTC) interconvert in soil/manure and can coexist; if epimers are not monitored, estimates of "total TC" levels and impacts on the community/ARGs are biased (Lu et al., 2024). Acetylated sulfonamides (N-acetyl-SAs) are commonly excreted; they can be deacetylated back to the parent compound and exhibit different environmental behavior, but are rarely included in routine panels (Puhlmann et al., 2022). β-lactam (ceftiofur) is rapidly degraded and binds to proteins to form "bound residues"; official analysis breaks down all associated residues to desfuroyl-ceftiofur (DFC)—if only the parent is targeted, exposure is underestimated (Fsis, 2012). Fluoroquinolones undergo photolysis and biotransformation, producing various TPs through defluorination, hydroxylation, decarboxylation, and piperazine ring cleavage. These products differ in biological activity and environmental persistence (Klementová et al., 2022).

7.4 Future Direction

Future research should prioritize long-term comparative studies across diverse climates and treatment systems. There is a need to collect additional data from tropical regions and low-income countries. Monitoring antibiotic TPs is essential, as these compounds may persist or retain biological activity. Advanced analytical and molecular methods, including high-resolution mass spectrometry, non-target screening, and multi-omics integration, are necessary to link residue concentrations with ARG dynamics and environmental risks. Integrating these findings within a One Health and circular economy framework can position manure treatment strategies as both agricultural solutions and global health interventions. Figure 3 summarizes research gaps, available tools and strategies, and key outcomes to inform evidence-based policy and mitigate AMR risks.

8 One Health and Circular Economy Implications

Translating these technical findings to systems level, we position manure treatment within One Health and the circular economy. Positioning manure treatment within One Health means embedding it in national AMR action plans as a downstream control that complements antimicrobial stewardship upstream. Practically,

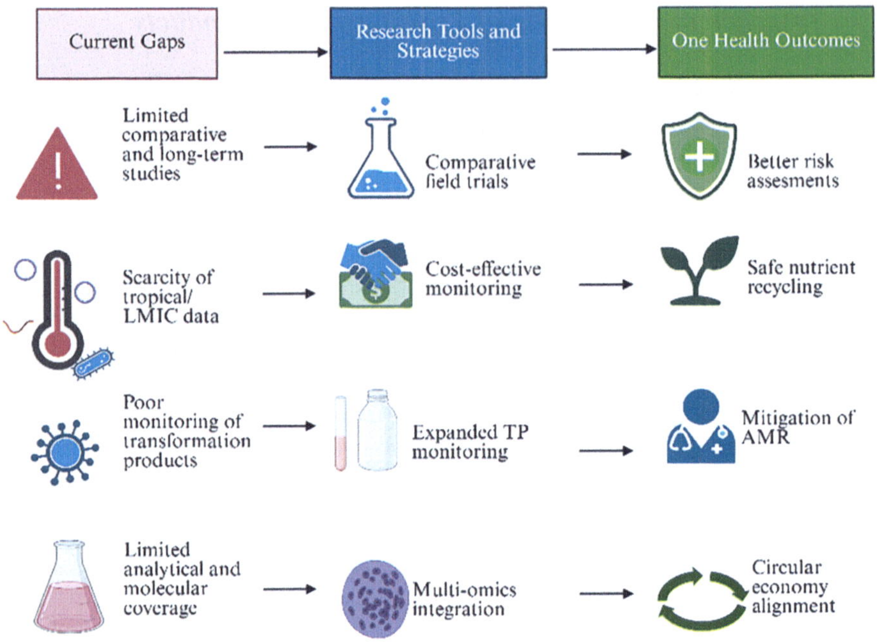

Fig. 3 Future research roadmap for the fate of animal antibiotics in livestock manure treatment systems

this requires integrated surveillance that couples chemical analytics (for parents and TPs) with genetic profiling of ARGs and mobile genetic elements. Monitoring should span the entire manure–soil–water continuum using consistent reporting metrics (e.g., dry weight). Furthermore, these data must be linked to veterinary antimicrobial-use reporting (e.g., WOAH). Regulators should establish benchmarks for compost and digestate reuse (residue and ARG thresholds) and, in high-risk contexts, require polishing steps for example, AD → thermophilic composting before land application.

In a circular-economy frame, AD delivers biogas that displaces fossil energy (SDG 7) and contributes to climate mitigation (SDG 13), while composting restores soil organic matter and nutrient cycling (SDG 2, SDG 15) and, when applied safely, protects water quality (SDG 6) under responsible production and consumption (SDG 12). By lowering environmental antibiotic loads and AMR selection pressure, these treatments also support SDG 3 (Good Health and Well-Being). Policy takeaway: use AD for energy recovery and composting for contaminant polishing—together, under a One Health governance lens, to achieve safe nutrient recycling and durable AMR mitigation.

9 Practical and Policy Implications

For rapid degradation of labile drugs while stabilizing organic matter, thermophilic aerobic composting has proven effective for example, in the case of ceftiofur in pig manure compost, which impacted the microbial community and reduced ARGs (Ma et al., 2024). On the other hand, AD provides energy recovery (biogas) while retaining N/P nutrients; antibiotic/ARG reduction performance is improved by pretreatment, co-digestion, conductive materials, or electro-assisted AD (Meng et al., 2024). To minimize release to water bodies, injection/sub-surface practices and application scheduling that avoids early rainfall have been shown to reduce antibiotic/ARG losses compared to surface broadcasts (Barrios et al., 2020; Le et al., 2018).

Policy, risk thresholds, and the One Health approach all play important roles in antimicrobial stewardship. For example, prescription rules and drug limits can include environmental monitoring that uses Predicted No Effect Concentration (PNED) thresholds to help prevent antimicrobial resistance. Regulations often rely on inter-class reference values from minimum inhibitory concentration (MIC) data (Bengtsson-Palme et al., 2023). Evidence shows that targeted interventions can reduce antimicrobial resistance markers. For example, after China implemented a colistin ban, mcr-1 prevalence declined in both animals and humans (Wang et al., 2020). Technological assessments demonstrate that the effectiveness of AD in reducing ARGs depends on system configuration and operational conditions. A recent review highlights unresolved questions and underscores the need for more rigorous monitoring (Wang et al., 2025). Furthermore, integrated omics studies of AD in combination with composting systems have identified specific antibiotic and resistome dynamics. These findings should inform the design of future monitoring protocols (Fanfoni et al., 2025).

10 Conclusion

Veterinary antibiotics present in livestock manure pose significant environmental and public health risks. Aerobic composting typically achieves greater removal of antibiotics, particularly tetracyclines and sulfonamides, compared to AD. In contrast, AD often results in lower and more variable reductions, and it can generate more persistent TPs. However, neither composting nor AD fully eliminates antibiotic resistance genes. While composting can reduce these genes during the thermophilic phase, their levels may subsequently increase. Anaerobic digestion, especially under mesophilic conditions, can serve as a reservoir for resistance genes. From a sustainability standpoint, composting offers more effective contaminant mitigation but is associated with increased greenhouse gas emissions. Anaerobic digestion facilitates renewable energy recovery, yet it introduces greater uncertainty regarding the fate of antibiotic residues and the persistence of resistance genes. Therefore, integrating both treatment systems, supported by long-term, region-specific research and advanced analytical monitoring, is essential for aligning livestock manure management with One Health and circular economy objectives. This integrated strategy promotes safe nutrient recycling and reduces the risk of antimicrobial resistance.

References

Aidara-Kane, A. et al. (2018). World Health Organization (WHO) guidelines on use of medically important antimicrobials in food-producing animals. *Antimicrobial Resistance and Infection Control*, *7*(1), 1–8. https://doi.org/10.1186/s13756-017-0294-9

Barrios, R. E. et al. (2020). Fate and transport of antibiotics and antibiotic resistance genes in runoff and soil as affected by the timing of swine manure slurry application. *Science of The Total Environment*, *712*, 136505. https://doi.org/10.1016/J.SCITOTENV.2020.136505

Bengtsson-Palme, J. et al. (2023). Towards monitoring of antimicrobial resistance in the environment: For what reasons, how to implement it, and what are the data needs? *Environment International*, *178*(july). https://doi.org/10.1016/j.envint.2023.108089

Chen, J. et al. (2022). Anaerobic sulfamethoxazole-degrading bacterial consortia in antibiotic-contaminated wetland sediments identi fi ed by DNA-stable isotope probing and metagenomics analysis *24*, 3751–3763. Accessed. https://doi.org/10.1111/1462-2920.16091

Chen, C. et al. (2023). Chemosphere biosorption and biotransformation behaviours of veterinary antibiotics under aerobic livestock wastewater treatment processes. *Chemosphere*, *335*(1), 139034. https://doi.org/10.1016/j.chemosphere.2023.139034.

Cheng, W. et al. (2016). Behavior of antibiotics and antibiotic resistance genes in eco-agricultural system : A case study *304*, 18–25.

Chu, Y. et al. (2022). Biotransformation of sulfamethoxazole by microalgae : Removal efficiency, pathways, and mechanisms. *Water Research*, *221*(February), 118834. https://doi.org/10.1016/j.watres.2022.118834

Coban, H. et al. (2016). Degradation of oxytetracycline and its impacts on biogas- producing microbial community structure, pp. 1051–1060. https://doi.org/10.1007/s00449-016-1583-z

European Commission. (2006). Ban on antibiotics as growth promoters in animal feed enters into effect. *Regulation* (December 2005), p. 1.

Dolliver, H., Gupta, S., & Noll, S. (2008). Antibiotic degradation during manure composting. *Journal of Environmental Quality*, *37*(3), 1245–1253. https://doi.org/10.2134/jeq2007.0399.

Dolu, T. & Nas, B. (2023). Journal of water process engineering full-scale anaerobic digestion of sewage sludges : Fate evaluation of pharmaceuticals and main metabolites. *Journal of Water Process Engineering*, *51*(November 2022), 103366. https://doi.org/10.1016/j.jwpe.2022.103366.

Domínguez-García, P., Almirall, X. O., & Gómez-Canela, C. (2025). A POCIS-based approach for the monitoring of pharmaceuticals in wastewater treatment plants: Calibration and deployment challenges. *Environmental Pollution*, *367*(January). https://doi.org/10.1016/j.envpol.2025.125641

Dutta, P. et al. (2025). Biotic remedies for Antibiotic pollution: A Review on Bioremediation Strategies. *Water, Air, and Soil Pollution*, *236*(7), 1–26. https://doi.org/10.1007/s11270-025-08078-9

Ezzariai, A. et al. (2018). Human and veterinary antibiotics during composting of sludge or manure: Global perspectives on persistence, degradation, and resistance genes. *Journal of Hazardous Materials*, 359, 465–481. https://doi.org/10.1016/j.jhazmat.2018.07.092.

Fahad, M. et al. (2021). Bioresource technology the fate of antibiotic resistance genes in cow manure composting : Shaped by temperature-controlled composting stages. *Bioresource Technology*, *320*(PB), 124403. https://doi.org/10.1016/j.biortech.2020.124403.

Fanfoni, E. et al. (2025). Metagenomics and untargeted metabolomics reveal antibiotic resistance dynamics in an anaerobic digestion–composting system treating organic fraction of municipal solid waste. *Environmental Microbiome*, *20*(1). https://doi.org/10.1186/s40793-025-00769-4

Feng, M. et al. (2023). Antibiotics and antibiotic resistance gene dynamics in the composting of antibiotic fermentation waste–A review. *Bioresource Technology*, *390*, 129861. https://doi.org/10.1016/J.BIORTECH.2023.129861

Fsis, U. (2012). CLG-CEF1.00 Determination for ceftiofur-related (DCA) Residues by HPLC, pp. 1–15.

Gaballah, M. S. et al. (2021). Bioresource technology a review targeting veterinary antibiotics removal from livestock manure management systems and future outlook. *Bioresource Technology*, *333*(April), 125069. https://doi.org/10.1016/j.biortech.2021.125069.

Gaballah, M. S. et al. (2023). Science of the total environment removal performance and inhibitory effects of combined tetracycline, oxytetracycline, sulfadiazine, and nor floxacin on anaerobic digestion process treating swine manure. *Science of the Total Environment*, *857*(August 2022), p. 159536. https://doi.org/10.1016/j.scitotenv.2022.159536.

Guo, K. et al. (2025). Occurrence and accumulation characteristics of antibiotics in soil and effects of carbon and nitrogen cycle. *Current Opinion in Environmental Science and Health*, *45*, 100619. https://doi.org/10.1016/j.coesh.2025.100619.

Gurmessa, B. et al. (2020). Science of the total environment manure anaerobic digestion effects and the role of pre- and post-treatments on veterinary antibiotics and antibiotic resistance genes removal efficiency. *Science of the Total Environment*, *721*, 137532. https://doi.org/10.1016/j.scitotenv.2020.137532.

Han, S. et al. (2021). Simultaneous determination of seven antibiotics and five of their metabolites in municipal wastewater and evaluation of their stability under laboratory conditions. *International Journal of Environmental Research and Public Health*, *18*(10640).

He, Y. et al. (2020). Antibiotic resistance genes from livestock waste: occurrence, dissemination, and treatment. *npj Clean Water*, *3*(1), 1–11. https://doi.org/10.1038/s41545-020-0051-0.

Ho, Y., & Bin et al. (2013). Degradation of veterinary antibiotics and hormone during broiler manure composting. *Bioresource Technology*, *131*, 476–484. https://doi.org/10.1016/j.biortech.2012.12.194.

Hu, Y., Cheng, H., & Tao, S. (2017). Environmental and human health challenges of industrial livestock and poultry farming in China and their mitigation. *Environment International*, *107*(March), 111–130. https://doi.org/10.1016/j.envint.2017.07.003.

Jin, H. et al. (2017). Distribution of sulfonamides in liquid and solid anaerobic digestates : Effects of hydraulic retention time and swine manure to rice straw ratio. *Bioprocess and Biosystems Engineering*, *40*(2), 319–330. https://doi.org/10.1007/s00449-016-1699-1

Katada, S. et al. (2021). Aerobic composting and anaerobic digestion decrease the copy numbers of antibiotic-resistant genes and the levels of lactose-degrading enterobacteriaceae in dairy farms in Hokkaido, Japan. *Frontiers in Microbiology*, *12*(September), 1–11. https://doi.org/10.3389/fmicb.2021.737420

Khadra, A. et al. (2019). Fate of antibiotics present in a primary sludge of WWTP during their co-composting with palm wastes. *Waste Management*, *84*, 13–19. https://doi.org/10.1016/j.wasman.2018.11.009.

Kitazono, Y. et al. (2015). Degradation of veterinary antibiotics during anaerobic digestion of dairy manure. *10*(3), 532–537. https://doi.org/10.2166/wpt.2015.061.

Klementová, Š. et al. (2022). Photodegradation of fluoroquinolones in aqueous solution under light conditions relevant to surface waters, toxicity assessment of photoproduct mixtures. *Environmental Science and Pollution Research*, *29*(10), 13941–13962. https://doi.org/10.1007/s11356-021-16182-6

Lacroix, Z. et al. (2018). Chemosphere a pressurized liquid extraction approach followed by standard addition method and UPLC-MS / MS for a fast multiclass determination of antibiotics in a complex matrix *211*. https://doi.org/10.1016/j.chemosphere.2018.08.021.

Le, H. T. V., Maguire, R. O. & Xia, K. (2018). Method of dairy manure application and time before rainfall affect antibiotics in surface runoff. *Journal of Environmental Quality, 47*(6), 1310–1317. https://doi.org/10.2134/JEQ2018.02.0086.

Li, N. et al. (2017). Science of the total environment partition and fate analysis of fl uoroquinolones in sewage sludge during anaerobic digestion with thermal hydrolysis pretreatment. *Science of the Total Environment, 581–582*, 715–721. https://doi.org/10.1016/j.scitotenv.2016.12.188.

Li, J. et al. (2017). Long-term manure application increased the levels of antibiotics and antibiotic resistance genes in a greenhouse soil. *Applied Soil Ecology, 121*(August), 193–200. https://doi.org/10.1016/j.apsoil.2017.10.007

Li, Y. et al. (2024). Heliyon Isolation, identification, and optimization of conditions for the degradation of four sulfonamide antibiotics and their metabolic pathways in Pseudomonas stutzeri strain DLY-21 *Heliyon, 10*(7), e29123. https://doi.org/10.1016/j.heliyon.2024.e29123.

Liao, J., et al. (2019). In fl uence of hydraulic retention time on behavior of antibiotics and antibiotic resistance genes in aerobic granular reactor treating biogas slurry. *Frontiers of Environmental Science & Engineering, 13*(3), 1–9.

Liu, B. et al. (2015). Bioresource technology effects of composting process on the dissipation of extractable sulfonamides in swine manure. *Bioresource Technology, 175*, 284–290. https://doi.org/10.1016/j.biortech.2014.10.098.

Liu, H. et al. (2018). Removal of tetracyclines, sulfonamides, and quinolones by industrial-scale composting and anaerobic digestion processes. *Environmental Science and Pollution Research, 25*(36), 35835–35844. https://doi.org/10.1007/s11356-018-1487-3

Liu, H. et al. (2019). Science of the total environment dissipation and persistence of sulfonamides, quinolones and tetracyclines in anaerobically digested biosolids and compost during short-term storage under natural conditions. *Science of the Total Environment, 684*, 58–66. https://doi.org/10.1016/j.scitotenv.2019.05.341.

Lou, Y. et al. (2017). ScienceDirect sorption behavior of tetracyclines on suspended organic matters originating from swine wastewater. *Journal of Environmental Sciences, 65*, 144–152. https://doi.org/10.1016/j.jes.2017.03.019

Lu, Y. et al. (2023). Copper in different forms and tetracycline affect behavior and risk of antibiotic resistome in thermophilic anaerobic digestion of cattle manure. *Environmental Science and Pollution Research, 30*(49), 108162–108175. https://doi.org/10.1007/s11356-023-29923-6.

Lu, C. et al. (2024). Overlooked interconversion between tetracyclines and their 4-epimers in soil and effects on soil resistome and bacterial community. *Environment International, 190*(April), 108941. https://doi.org/10.1016/j.envint.2024.108941

Ma, J. et al. (2021). Journal of environmental chemical engineering effects of different antibiotic operation modes on anaerobic digestion of dairy manure : Focus on microbial population dynamics. *Journal of Environmental Chemical Engineering, 9*(4), 105521. https://doi.org/10.1016/j.jece.2021.105521

Ma, W. et al. (2024). Ceftiofur in swine manure contributes to reducing pathogens and antibiotic resistance genes during composting. *Environmental Research, 252*(P3), 119033. https://doi.org/10.1016/j.envres.2024.119033

Ma, S. et al. (2025). Journal of environmental chemical engineering effects of high-concentration sulfamethoxazole on antibiotic resistance genes during swine manure aerobic composting. *Journal of Environmental Chemical Engineering, 13*(4), 117216. https://doi.org/10.1016/j.jece.2025.117216

Marutescu, L. G. et al. (2022). Insights into the impact of manure on the environmental antibiotic residues and resistance pool. *Frontiers in Microbiology, 13*(September), 1–18. https://doi.org/10.3389/fmicb.2022.965132

Mejías, C. et al. (2025b). Chemosphere distribution and fate of antibiotics and their main metabolites in sludge stabilization technologies. *Chemosphere, 385*(July), 144573. https://doi.org/10.1016/j.chemosphere.2025.144573

Mejías, C. et al. (2025a). Chemosphere distribution and fate of antibiotics and their main metabolites in sludge stabilization technologies. *Chemosphere, 385*(June), 144573. https://doi.org/10.1016/j.chemosphere.2025.144573

Meng, Q., & Bin et al. (2024). Roles and fates of antibiotics in anaerobic digestion of waste activated sludge: Insights to pro- and reduction of antibiotic resistance genes. *Chemical Engineering Journal, 500*, 156633. https://doi.org/10.1016/J.CEJ.2024.156633

Nkoh, J. N. et al. (2024). Antibiotics soil-solution chemistry: A review of environmental behavior and uptake and transformation by plants. *Journal of Environmental Management, 354*(October 2023), 120312. https://doi.org/10.1016/j.jenvman.2024.120312

Page, S. W. & Gautier, P. (2012). Use of antimicrobial agents in livestock. *OIE Revue Scientifique et Technique, 31*(1), 145–188. https://doi.org/10.20506/rst.31.1.2106

Pei, F. et al. (2021). Bioresource technology links between microbial compositions and metabolites during aerobic composting under amoxicillin stress was evaluated by 16S rRNA sequencing and gas chromatography-mass spectrometry : Benefit for the plant growth. *Bioresource Technology, 340*(July), 125687. https://doi.org/10.1016/j.biortech.2021.125687

Province, L. & Energy, C. (2023). Effects of residual sulfamethoxazole on biogas production characteristics and hydrolytic acidification bacteria in anaerobic digestion of chicken manure *21*(1), 753–769.

Puhlmann, N., Olsson, O., & Kümmerer, K. (2022). Transformation products of sulfonamides in aquatic systems: Lessons learned from available environmental fate and behaviour data. *Science of the Total Environment, 830.* https://doi.org/10.1016/j.scitotenv.2022.154744

Qian, X. et al. (2022). Occurrence of veterinary antibiotics in animal manure, compost, and agricultural soil, originating from different feedlots in suburbs of Shanghai, East China. *Environmental Monitoring and Assessment, 194*(5). https://doi.org/10.1007/s10661-022-10010-1

Sabah, E., Bilgin, N., & Bulut, E. (2025). Removing an antibiotic from aqueous solution using sepiolite-rich dolomite–kinetic, isotherm, and thermodynamic studies. *Water SA, 51*(3), 288–301. https://doi.org/10.17159/wsa/2025.v51.i3.4135

Selvam, A. et al. (2013). Degradation of tetracycline and sulfadiazine during continuous thermophilic composting of pig manure and sawdust. *Environmental Technology*, 37–41. https://doi.org/10.1080/09593330.2013.772644

Shao, L. et al. (2013). Bioresource technology comparison of sludge digestion under aerobic and anaerobic conditions with a focus on the degradation of proteins at mesophilic temperature. *Bioresource Technology, 140*, 131–137. https://doi.org/10.1016/j.biortech.2013.04.081

Siri, Y. et al. (2023). Antimicrobial resistance in southeast Asian water environments: A systematic review of current evidence and future research directions. *Science of the Total Environment, 896*(March), 165229. https://doi.org/10.1016/j.scitotenv.2023.165229

Sun, W. et al. (2023). Residual enrofloxacin in cattle manure increased persistence and dissemination risk of antibiotic resistance genes during anaerobic digestion. *Journal of Environmental Management, 326*(PB), 116864. https://doi.org/10.1016/j.jenvman.2022.116864

Taylor, P. et al. (2012). Journal of environmental science and health, Part B : Dissipation of sulfamethoxazole and trimethoprim antibiotics from manure-amended soils Dissipation of sulfamethoxazole and trimethoprim antibiotics from manure-amended soils. *Journal of Environmental Science and Health, 47*(4), 240–249. https://doi.org/10.1080/03601234.2012.636580.

Tian, R. et al. (2019). Preparation of biochar via pyrolysis at laboratory and pilot scales to remove antibiotics and immobilize heavy metals in livestock feces. *Journal of Soils and Sediments, 19*(7), 2891–2902. https://doi.org/10.1007/s11368-019-02350-2

Tong, Z.-L. et al. (2012). Anaerobic digestion of animal manure contaminated by tetracyclines. *Huan jing ke xue= Huanjing kexue, 33*(3), 1028–1032

Trueba-santiso, A. et al. (2025). Ciprofloxacin biotransformation during anaerobic digestion : Major transformation products elucidation and environmental proteomics *495*(April). https://doi.org/10.1016/j.jhazmat.2025.139159

Turker, G. et al. (2018). Ecotoxicology and environmental safety operating conditions in fl uence microbial community structures, elimination of the antibiotic resistance genes and metabolites during anaerobic digestion of cow manure in the presence of oxytetracycline *147*(September 2017), 349–356. https://doi.org/10.1016/j.ecoenv.2017.08.044

Van Boeckel, T. P. et al. (2015). Global trends in antimicrobial use in food animals. *Proceedings of the National Academy of Sciences of the United States of America, 112*(18), 5649–5654. https://doi.org/10.1073/pnas.1503141112.

Wallace, J. S. et al. (2018). Occurrence and transformation of veterinary antibiotics and antibiotic resistance genes in dairy manure treated by advanced anaerobic digestion and conventional treatment methods. *Environmental Pollution, 236*, 764–772. https://doi.org/10.1016/j.envpol.2018.02.024

Wang, Yang et al. (2020). Changes in colistin resistance and mcr-1 abundance in Escherichia coli of animal and human origins following the ban of colistin-positive additives in China: an epidemiological comparative study. *The Lancet Infectious Diseases*, *20*(10), 1161–1171. https://doi.org/10.1016/S1473-3099(20)30149-3

Wang, Y. et al. (2022). Science of the total environment effects of multiple antibiotics residues in broiler manure on composting process. *Science of the Total Environment, 817*, 152808. https://doi.org/10.1016/j.scitotenv.2021.152808

Wang, G. et al. (2022). Composting temperature directly affects the removal of antibiotic resistance genes and mobile genetic elements in livestock manure. *Environmental Pollution, 303*(March), 119174. https://doi.org/10.1016/j.envpol.2022.119174

Wang, S. et al. (2024). Bioresource technology continuous thermophilic composting of distilled grain waste improved organic matter stability and succession of bacterial community. *Bioresource Technology, 394*(October 2023), 130307. https://doi.org/10.1016/j.biortech.2024.130307

Wang, C. et al. (2025). Antibiotic resistance genes in anaerobic digestion: Unresolved challenges and potential solutions. *Bioresource Technology*, *419*, 132075. https://doi.org/10.1016/J.BIORTECH.2025.132075

WHO et al. (2023). *A one health priority research agenda for antimicrobial resistance, FAO publications catalogue 2022.*

World Organization for and Health, A. (2019). Annual report on antimicrobial agents intended for use in animals. *Journal of Language Relationship*, 7–8.

Wu, X. et al. (2011). Bioresource technology the behavior of tetracyclines and their degradation products during swine manure composting. *Bioresource Technology, 102*(10), 5924–5931. https://doi.org/10.1016/j.biortech.2011.03.007

Wu, Q. et al. (2022). Science of the total environment effects of antibiotics on anaerobic digestion of sewage sludge : Performance of anaerobic digestion and structure of the microbial community. *Science of the Total Environment*, *845*(May), 157384. https://doi.org/10.1016/j.scitotenv.2022.157384

Xie, W. Y., Shen, Q. & Zhao, F. J. (2018). Antibiotics and antibiotic resistance from animal manures to soil: a review. *European Journal of Soil Science, 69*(1), 181–195. https://doi.org/10.1111/ejss.12494

Yan, S. et al. (2023). Chemosphere effects of erythromycin and roxithromycin on river periphyton : Structure, functions and metabolic pathways. *Chemosphere, 316*(November 2022), 137793. https://doi.org/10.1016/j.chemosphere.2023.137793

Yang, B., Meng, L. & Xue, N. (2017). Removal of five fluoroquinolone antibiotics during broiler manure composting. *Environmental Technology*, 0(0), 1–9. https://doi.org/10.1080/09593330.2017.1301568

Zhang, X. & Li, R. (2018). Bioresource technology variation of antibiotics in sludge pretreatment and anaerobic digestion processes : Degradation and solid-liquid distribution. *Bioresource Technology*, *255*(January), 266–272. https://doi.org/10.1016/j.biortech.2018.01.100

Zhang, M., & He, L. Y., et al. (2019). Fate of veterinary antibiotics during animal manure composting. *Science of the Total Environment, 650*, 1363–1370. https://doi.org/10.1016/j.scitotenv.2018.09.147

Zhang, M., & He, L., et al. (2019). Science of the total environment Fate of veterinary antibiotics during animal manure composting. *Science of the Total Environment, 650*, 1363–1370. https://doi.org/10.1016/j.scitotenv.2018.09.147

Zhang, Q. et al. (2021). Performance of full-scale aerobic composting and anaerobic digestion on the changes of antibiotic resistance genes in dairy manure. *Bioresource Technology, 342*(September), 125898. https://doi.org/10.1016/j.biortech.2021.125898.

Zhang, Y. et al. (2025). Transformation products of antibiotics: Overlooked drivers for enhancing the environmental spread of antibiotic resistance. *Journal of Environmental Exposure Assessment, 4*(2). https://doi.org/10.20517/jeea.2024.59

Zhao, K. et al. (2024). Optimizing the management of aerobic composting for antibiotic resistance genes elimination: A review of future strategy for livestock manure resource utilization. *Journal of Environmental Management, 370*, 122766. https://doi.org/10.1016/j.jenvman.2024.122766

Zhao, Y. et al. (2025). Advancing antibiotic detection in environmental waters : Standardization of solid phase extraction procedures and development of certified reference materials. *Sustainable Horizons*, 15, 100153. https://doi.org/10.1016/j.horiz.2025.100153

Zhao, E. et al. (2025). A review on the degradation of antibiotic resistance genes during composting of livestock manure. *Toxics, 13*(8), 1–26. https://doi.org/10.3390/toxics13080667

Zhao, F. et al. (2025). Metagenomic analysis reveals the diversity, transmission and potential ecological risks of yak nasal bacteria-carried antibiotic resistance genes in the Sichuan region of Qinghai-Tibet plateau. *Journal of Hazardous Materials, 494*, 138476. https://doi.org/10.1016/j.jhazmat.2025.138476

A Review on the Isolation of Facultative Anaerobes from Gut Bacteria with Bioplastic Degradation Potential

Lilian Jose, Ikki Matsuda, Cahyo Budiman, and Vijay Kumar

Abstract The increasing accumulation of bioplastics has become a growing environmental concern, as their degradation under natural of industrial conditions remains inefficient. Gut bacteria, adapted to low-oxygen and enzyme-rich gastrointestinal environments, represent a promising yet underexplored microbial resource for enhancing bioplastic biodegradation. This review synthesizes current knowledge on gut-derived facultative anaerobes and their potential roles in enhancing anaerobic digestion processes. Representative taxa such as *Citrobacter koseri, Klebsiella oxytoca, K. pneumoniae, Enterobacter ludwigii, E. sichuanensis* and *Serratia* sp. have been associated with the degradation of various synthetic and bio-based plastics. These findings emphasize the metabolic versatility of gut microorganisms and their potential integration into sustainable waste management systems. Nevertheless, the opportunistic pathogenicity of several taxa necessitates careful biosafety evaluation before practical implementation in biodegradation and bioconversion systems.

Keywords Gut bacteria · Bioplastic biodegradation · Anaerobic digestion. Facultative anaerobes

L. Jose · C. Budiman · V. Kumar (✉)
Biotechnology Research Institute, Universiti Malaysia Sabah, Kota Kinabalu, Sabah, Malaysia
e-mail: vijay@ums.edu.my

I. Matsuda (✉)
Wildlife Research Center, Kyoto University, Kyoto, Japan
e-mail: matsuda.ikki.5a@kyoto-u.ac.jp

Chubu Institute for Advanced Studies, Chubu University, Kasugai, Aichi, Japan

Institute for Tropical Biology and Conservation, Universiti Malaysia Sabah, Kota Kinabalu, Sabah, Malaysia

A. Z. Yaser et al. (eds.), *Compost and Liquid Organic Fertilizers*,
https://doi.org/10.1007/978-981-95-8561-8_18

1 Introduction

Plastic materials, composed of synthetic or semi-synthetic polymers, are widely utilized for their durability and resistance to environmental stressors. However, their persistence in terrestrial and aquatic ecosystems presents a significant challenge for waste management. Biodegradable plastics offer a partial solution, yet their degradation particularly under anaerobic conditions remains inefficient. Anaerobic digestion (AD) is a promising strategy for plastic waste treatment, as it facilitates energy recovery while reducing environmental burden. Nevertheless, optimizing AD for bioplastic degradation is complex, with microbial composition playing a critical role in determining its efficacy. Most bacteria found to degrade bioplastics are aerobic strains isolated from environments such as soil and sewage (Kamarudin et al., 2021; Sasanami et al., 2022).

Recent studies have begun to explore facultative anaerobes, particularly those isolated from gut environments, as potential agents for enhancing AD processes (CF et al., 2021; Nyamjav et al., 2023; Soleman et al., 2022; Tan et al., 2021; Wang et al., 2024). Gut bacteria, shaped by dietary inputs and adapted to extreme gastrointestinal conditions such as low pH and limited oxygen, exhibits metabolic flexibility that may be advantageous for bioplastic degradation. Research conducted on insect gut microbiota has revealed the presence of facultative anaerobes capable of degrading contaminants, including plastics. Notably, *Citrobacter koseri* has been reported to degrade polyvinyl chloride, while a *Serratia* sp. isolated from the gut of the wood-feeding beetle *Plesiophthalmus davidis* oxidized polystyrene through C–O and C = O bond formation (Woo et al., 2020). Plastic degradation under anaerobic conditions has been demonstrated (Quecholac-Piña et al., 2020), and microbial composition has been shown to influence AD performance (Bátori et al., 2018). However, different perspectives on anaerobic digestion are needed to advance current technologies (Uddin & Wright, 2022).

To contextualize these recent developments, earlier microbiological studies established the foundational understanding of plastic degradation, predominantly under aerobic conditions, by isolating versatile taxa such as *Pseudomonas*, *Bacillus*, and *Rhodococcus* from natural environments including soil, compost, and marine systems (Shah et al., 2008). Although these pioneering works did not focus on gut-derived facultative anaerobes, they provided the methodological and enzymatic framework upon which subsequent anaerobic and gut-based research has expanded. Building upon this foundation, the discovery of PETase and MHETase in *Ideonella sakaiensis* further advanced the mechanistic understanding of polyester degradation (Yoshida et al., 2016).

This review was performed through a systematic selection of peer-reviewed publications with the focus on the isolation of facultative anaerobes from the gut bacteria. It synthesizes current literature, highlights microbial mechanisms of action, and discusses the implications for sustainable waste management. The ubiquity of microplastics, even in remote regions such as highlands (Rüthi et al., 2023; Thamsenanupap et al., 2024), has intensified interest in gut bacteria as a reservoir for plastic-degrading strains (Nyamjav et al., 2023; Tan et al., 2021; Zhao et al., 2024). Similar phenomena have been observed in marine environments, where the emergence of the plastisphere reflects microbial colonization of plastic debris and its ecological consequences (Du et al., 2022).

2 Materials and Methods

2.1 Literature Search

The literature search was conducted using Google Scholar and Scopus databases. Keywords used for the search included the combinations of "*biodegradation*", "*bioplastic degradation*", "*plastic degrading*", "*facultative anaerobes*", "*anaerobic*" and "*gut*" connected with the Boolean operators. The search was limited to peer-reviewed articles published in English between 2015 and 2025. This time frame was selected to capture the most recent decade of research, which marks the emergence of gut microbiota–based biodegradation studies and the integration of anaerobic digestion approaches into bioplastic degradation research.

2.2 Inclusion Criteria

The inclusion criteria were defined to ensure that the selected studies directly addressed the isolation or identification of facultative anaerobic strains from animal gut environments. Eligible publications (1) reported evidence of plastic or bioplastic degradation through culture-based assays, enzymatic or omics analyses, or genomic/metabolic inference, and (2) characterized bacterial taxa with facultative anaerobic potential. Studies that primarily investigated aerobic degradation were also considered when they included gut-derived facultative anaerobes. Studies from non-gut environments, reports limited to strict obligate aerobes, those focusing solely on non-polymeric additives, or those lacking organism-level attribution were excluded from this review. In this review, aerobic studies are cited as background evidence when gut-derived facultative anaerobes were characterized, whereas only anaerobic activity was considered demonstrative for the AD context.

3 Results and Discussion

In the subsequent sections, several foundational aspects on the isolation of facultative anaerobes from the gut are discussed. The section begins with an overview of the diversity of gut bacteria and their relevance to bioplastic degradation (Sect. 3.1).

It then examines the specific hosts from which bioplastic-degrading facultative anaerobes are isolated, including insects, non-insect invertebrates and vertebrates (Sect. 3.2). Detailed justifications on the contributions of these isolated facultative anaerobes to the anaerobic digestion system are provided in Sect. 3.3.

The search returned with 5,930 hits. After excluding hits related to sewage, aerobics, plants, plastic additives, non-related terms (e.g. amine, cancer) and obligate/strict anaerobes, the hits were left with 341. Studies that used bacteria not isolated from the gut were also excluded. The resulting publications were then manually screened from the title and abstract before they were selected to be included in this review. A summary of the screening outcomes and the distribution of relevant studies is provided in Table 1.

3.1 Overview of Diversity of Gut Bacteria and the Relevance to Bioplastic Degradation

Gut bacteria refer to the different species of bacteria residing in the stomach of a host and it is responsible for many aspects of the host physiology, including energy harvest, nutrition, behaviour, and immune system response (Lee et al., 2019). It is very flexible, and it can adapt to the diet of the host. Gut environment is a harsh environment, where food ingested will be digested in an environment with extreme pH and low oxygen.

In a study on Japanese macaque, it was reported that the variability of the bacterial species or also termed as 'gut bacterial composition' was higher for the individuals living in the wild, as compared to the individuals kept in captivity (i.e. zoo) and provisioned (i.e. living near humans) (Lee et al., 2019). This was likely due to the more variability of food in the wild as compared to the food given in captivity or found near human settlement area. A report on colobines, which is a group of monkeys with multichambered stomachs, found that they were able to consume bitter plant tissues such as mature leaves, that are typically toxic to other mammals. This unique dietary ability is attributed to their gut bacteria, which can detoxify harmful secondary metabolites present in these plants (Matsuda et al., 2019). Given this adaptability, the presence of bioplastic in the gut which is ingested unintentionally from the environment can act as a substrate to trigger the action of responsible bacteria to produce the necessary enzyme to degrade the bioplastic.

3.2 Isolation of Facultative Anaerobes from the Gut of Different Host

Different classes of organisms exhibit diverse dietary preferences (Lu et al., 2019; Qiu et al., 2022). However, the widespread presence of bioplastics in various ecosystems

Table 1 Summary of isolated facultative anaerobes from different organisms and the type of plastics degraded

Category of Host	Specific Organism	Type of Plastics Degraded	Facultative Anaerobes Isolated	References
Insects	Wood termites	Polyethylene (PE)	*Lysinibacillus* and *Bacillus* sp. T2	Thamil (2016)
	Black soldier fly larvae (*Hermetia illucens*)	Mix of microplastic - polyethylene terephthalate, polylactic acid (PLA), and polyhydroxybutyrate	*Methylobacterium* sp., *Lysinibacillus* sp., and *Brevundimonas* sp.	Dragone et al. (2025)
	Galleria mellonella larvae	PE	*Enterococcus innesii sp. nov*	Gooch et al. (2021)
	Galleria mellonella larvae	PE and LDPE	*Acinetobacter* sp.	Cassone et al. (2020)
	Zophobas atratus larvae	PE and PS	*Pseudomonas aeruginosa*	Zaman et al. (2024)
	Tribolium castaneum larvae	PS	*Acinetobacter* sp.	Wang et al. (2020)
	Achroia grisella larvae	LDPE	*Citrobacter freundii* and *Bacillus* sp.,	Ali et al. (2023)
	Plesiophthalmus davidis larvae	PS	*Serratia* sp.	Woo et al. (2020)
	Spodoptera frugiperda larvae	Polyvinyl chloride (PVC)	*Klebsiella* sp.	Zhang et al. (2022)
Non-insects invertebrate	*Tenebrio molitor* larvae (mealworms)	PS, PE	*Enterobacter hormaechei*	Kang et al. (2023)
	Tenebrio molitor larva	PS, PE, and PU	*Enterococcus sp., Pseudomonas sp., Chryseobacterium sp.* and *Spiroplasma*	Guo et al. (2025)
	Bombyx mori (silkworm)	PLA	*Bifidobacterium catenulatum* and *Schaalia odontolytica*	Wu et al. (2023)
Vertebrate	Human	LDPE and polypropylene (PP)	*Klebsiella oxytoca, Klebsiella pneumoniae, Enterobacter ludwigii, Enterobacter sichuanensis*	Jang et al. (2024)
	Yak	Poly(butylene adipate-co-terephthalate)	*Lactobacillus paracasei*	Zhou et al. (2024)

have introduced a new ecological niche that may influence microbial colonization. Organisms exposed to bioplastics can harbour bacteria, particularly facultative anaerobes with the potential to degrade these polymers. Such bacteria are adapted to low oxygen environments and often possess specialized enzymatic pathways that enable the metabolism of complex polymeric compounds.

3.2.1 Insects

Insect guts are increasingly recognized as reservoirs of plastic-degrading microbes, particularly facultative anaerobes (Chen et al., 2025; Yang et al., 2023). The larvae of *Galleria mellonella* (wax moth) and *Tenebrio molitor* (mealworm) have been shown to degrade polyethylene and polystyrene, respectively, with microbial activity playing a central role in the process. A study done on *Galleria mellonella* has demonstrated that the larvae could oxidize polyethylene films, and gut bacteria were implicated in the degradation pathway (Barrionuevo et al., 2022; Bombelli et al., 2017). Similarly, it has been reported that *Tenebrio molitor* larvae could mineralize polystyrene, with gut microbes contributing to the breakdown under anaerobic conditions (Tsochatzis et al., 2021). More recently, research conducted on *Zophobas atratus* revealed that the larvae could completely degrade and mineralize polystyrene, reinforcing the potential of insect gut microbiota as sources of facultative anaerobes for plastic degradation (Tan et al., 2021).

3.2.2 Non-Insect Invertebrate

Beyond insects, non-insect invertebrates such as earthworms also host gut microbes capable of degrading plastics. Earthworms play a vital role in soil ecosystems and their digestive tracts contain diverse microbial communities adapted to decomposing organic matter. Furthermore, bacteria from earthworm guts that could degrade low-density polyethylene has been isolated suggesting that soil-dwelling invertebrates may contribute to plastic breakdown in terrestrial environments (Lwanga et al., 2018). Although this study was conducted under predominantly aerobic soil conditions, it provides useful background evidence of gut-associated bacteria capable of plastic degradation.

3.2.3 Vertebrates

Vertebrate gut microbiota, particularly those of mammals, are dominated by facultative anaerobes such as *Citrobacter*, *Enterococcus*, and *Escherichia* (Jang et al., 2024; Nugrahapraja et al., 2022). While several vertebrate-associated bacteria, including strains of *Klebsiella*, *Enterobacter*, and *Lactobacillus*, have been reported to degrade synthetic or biodegradable plastics, most of these studies were conducted under aerobic conditions (Jang et al., 2024; Zhou et al., 2024). Therefore, direct evidence of anaerobic degradation by vertebrate gut isolates remains limited, and such findings should be interpreted as aerobic reference cases rather than demonstrations of anaerobic activity. Although direct evidence of plastic degradation in vertebrate gut microbes is limited, previous studies on colobine monkeys have shown microbial detoxification of plant secondary metabolites such as phenolics and tannins, indicating a broad enzymatic potential that could theoretically be harnessed for biopolymer degradation (Matsuda et al., 2019). Such metabolic versatility supports the relevance of vertebrate gut microbes in anaerobic digestion systems targeting bio-based polymers.

In a study on human faecal sample, plastic-degrading bacteria specifically for low-density polyethylene (LDPE) and polypropylene (PP) were isolated after an enrichment period of 100 days (Jang et al., 2024). The study has also captured facultative anaerobes in their experiments such as *Klebsiella oxytoca*, *Klebsiella pneumoniae*, *Enterobacter ludwigii*, and *Enterobacter sichuanensis*. However, it is important to note that several vertebrate gut isolates, including *Klebsiella* and *Enterobacter* spp., are opportunistic pathogens (Jang et al., 2024). Their use in bioplastic degradation or anaerobic digestion should therefore be approached with caution and appropriate biosafety considerations.

Through a molecular approach to detect plastic degrading genes in human gut, several genes including *cutA*, *alkA*, and *feaB* which are responsible for hydrolysing specific bonds in the plastics were identified (Nugrahapraja et al., 2022) and an expression study has been performed in gut models to examine the evidence of human adaptation to plastics (Fournier et al., 2023). This approach offers a rapid and efficient way to detect the adaptation of gut bacteria to plastic exposure without isolating every single bacterial species. A list of plastic-degrading genes is provided in Table 2.

3.3 Role of Facultative Anaerobes in Anaerobic Digestion Systems

AD is a widely adopted biological process for treating organic waste, producing biogas, and reducing environmental pollution. However, its application to plastic waste, especially bioplastics remains limited due to the recalcitrant nature of polymeric materials and the enzymatic constraints of strict anaerobes. Facultative anaerobes offer a strategic advantage in this context, as they can initiate polymer breakdown under both aerobic and anaerobic conditions, bridging the metabolic gap between oxygen-dependent degradation and methane-producing anaerobic stages (Fournier et al., 2023).

Facultative anaerobes possess versatile metabolic pathways that allow them to survive and function in fluctuating oxygen environments, making them ideal for integration into multi-phase anaerobic digestion systems. Their ability to secrete extracellular enzymes such as oxidases, hydrolases, and depolymerases enables the

Table 2 List of plastic-degrading genes from different hosts with the name of the enzymes produced

Gene	Enzyme Produced	Host	References
HGMP01	PET hydrolase	Human	Zhang et al. (2024)
PDB1_01395	Short-chain dehydrogenase	*Zophobas atratus* larvae	Zaman et al. (2024)
PDB1_01396	Esterase	*Zophobas atratus* larvae	Zaman et al. (2024)
Not provided	Laccase, alkane monooxygenase and peroxidases	*Spodoptera frugiperda*	Zhang et al. (2022)
tpx	Thiol peroxidase	*Tenebrio molitor* larvae (mealworms)	Kang et al. (2023)
ahpC	Alkyl hydroperoxide reductase C		Kang et al. (2023)
bcp	Bacterioferritin comigratory protein		Kang et al. (2023)
feaB	Phenylacetaldehyde dehydrogenase	Human	Nugrahapraja et al. (2022)
pbsA	Polyester-degrading enzyme	Human	Nugrahapraja et al. (2022)

initial fragmentation of complex polymers into smaller, more bioavailable molecules. These intermediates can then be further metabolized by strict anaerobes, enhancing overall system efficiency. Illustration of this process is shown in Fig. 1.

Studies on insect gut microbiota, such as those involving *Tenebrio molitor* and *Zophobas atratus*, have demonstrated that facultative anaerobes can degrade polystyrene and polyethylene under low-oxygen conditions (Bombelli et al., 2017; Tsochatzis et al., 2021). Similarly, earthworm-associated bacteria and human gut strains have shown potential for initiating plastic degradation in anaerobic environments (Lwanga et al., 2018). These findings suggest that facultative anaerobes could be harnessed to improve the biodegradation of bioplastics in engineered AD systems.

Moreover, the inclusion of facultative anaerobes may enhance microbial synergy within digesters, promoting stable operation and reducing lag phases during substrate adaptation. Their resilience to environmental stressors and ability to metabolize xenobiotics further supports their relevance in treating complex waste streams that include bioplastics and mixed organic materials.

The findings summarized here should be viewed as part of a continuum that began with early efforts to identify aerobic plastic-degrading bacteria in natural and engineered environments during the early 2000s (e.g., Shah et al., 2008). The transition toward gut-derived facultative anaerobes represents a natural progression of this field, extending biodegradation research from external ecosystems into internal microbial environments.

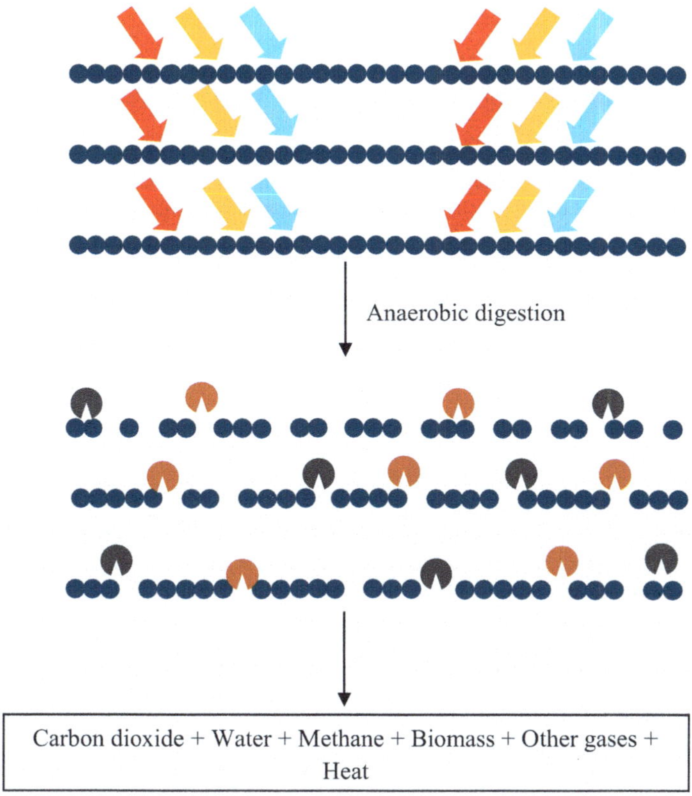

Fig. 1 Illustration of facultative anaerobes (grey circular sector) addition with strict anaerobes (orange circular sector) in anaerobic digestion of bioplastics. Light (yellow arrow), temperature (red arrow), and water (blue arrow) are required to provide an anaerobic condition suitable for the digestion of bioplastic polymer (dark blue circles)

4 Conclusions

In summary, this review underscores the potential of facultative anaerobes isolated from the gut of various organisms, including insects, non-insect invertebrates, and vertebrates for plastic degradation under anaerobic conditions. Most existing studies are focusing on conventional plastic degradation, while its potential in bioplastic breakdown remains underexplored. Facultative anaerobes could be promising biological agents in enhancing anaerobic digestion systems given their metabolic flexibility, resilience to low-oxygen environments, and enzymatic capabilities.

Isolation of the gut-derived facultative anaerobes and their integration into engineered bioreactors could contribute to the acceleration of bioplastic degradation and improving the stability of the system. Future research should explore the isolation, characterization, and optimization of the targeted bacterial strains to advance sustainable waste management technologies, particularly plastics and bioplastics.

5 Conflicts of Interest

The authors declare no conflict of interest.

Acknowledgements The authors would like to express their gratitude to the Universiti Malaysia Sabah and Kyoto University, Japan for their unwavering support and continuous encouragement in fostering the progress of our research towards contributing to microplastic research.

Funding This research is partially funded by UMSGreat Grant Code No: GUG0704-2/2024/2025, Nagao Environment Foundation Grant, UMS Grant Code LPA2502, the Japan Society for the Promotion of Science (JSPS) KAKENHI (no. 24H00774) and the JSPS Core-to-Core Programme, Asia–Africa Science Platforms (JPJSCCB20250006).

References

Ali, S. S., Elsamahy, T., Zhu, D., & Sun, J. (2023). Biodegradability of polyethylene by efficient bacteria from the guts of plastic-eating waxworms and investigation of its degradation mechanism. *Journal of Hazardous Materials, 443*(PB), 130287. https://doi.org/10.1016/j.jhazmat.2022.130287.

Barrionuevo, J. M. R., Vilanova-Cuevas, B., Alvarez, A., Martín, E., Malizia, A., Galindo-Cardona, A., de Cristóbal, R. E., Occhionero, M. A., Chalup, A., Monmany Garzia, A. C., *et al.* (2022). The Bacterial and Fungal Gut Microbiota of the Greater Wax Moth, *Galleria mellonella* L. Consuming Polyethylene and Polystyrene. *Frontiers in Microbiology, 13*(July). https://doi.org/10.3389/fmicb.2022.918861

Bátori, V., Åkesson, D., Zamani, A., Taherzadeh, M. J., & Horváth I. S. (2018). Anaerobic degradation of bioplastics: A review. *Waste Management, 80*, 406–413. https://doi.org/10.1016/j.wasman.2018.09.040

Bombelli, P., Howe, C. J., & Bertocchini, F. (2017). Polyethylene bio-degradation by caterpillars of the wax moth *Galleria mellonella. Current Biology, 27*(8), R292–R293. https://doi.org/10.1016/j.cub.2017.02.060

Cassone, B. J., Grove, H. C., Elebute, O., Villanueva, S. M. P., & LeMoine, C. M. R. (2020). Role of the intestinal microbiome in low-density polyethylene degradation by caterpillar larvae of the greater wax moth, *Galleria mellonella. Proceedings of the Royal Society b: Biological Sciences, 287*(1922), 9–11. https://doi.org/10.1098/rspb.2020.0112

CF, S. F., Rebello, S., Aneesh, E. M., Sindhu, R., Binod, P., Singh, S., & Pandey, A. (2021). Bioprospecting of gut microflora for plastic biodegradation. *Bioengineered, 12*(1), 1040–1053. https://doi.org/10.1080/21655979.2021.1902173

Chen, K., Wang, L., Liu, J., Zheng, H., Wu., X., & Liao, X. (2025). The ant that may well destroy a whole dam: a systematic review of the health implication of nanoplastics/microplastics through gut microbiota. *Critical Reviews in Food Science and Nutrition, 65*(30), 6941–6962. https://doi.org/10.1080/10408398.2025.2453632

Dragone, N. B., van Hamelsveld, S., Nazmi, A. R., Stott, M., Hatley, G. A., Moloney, K., Bohm, K., Gutierrez-Gines, M. J., & Weaver, L. (2025). Examining the potential of plastic-fed black soldier fly larvae (*Hermetia illucens*) as "bioincubators" of plastic-degrading bacteria. *Journal of Applied Microbiology, 136*(4). https://doi.org/10.1093/jambio/lxaf085

Du, Y., Liu, X., Dong, X., & Yin, Z. (2022). A review on marine plastisphere: Biodiversity, formation, and role in degradation. *Computational and Structural Biotechnology Journal, 20*, 975–988. https://doi.org/10.1016/j.csbj.2022.02.008

Fournier, E., Leveque, M., Ruiz, P., Ratel., J., Durif, C., Chalancon, S., Amiard, F., Edely, M., Bezirard, V., & Gaultier, E., et al. (2023). Microplastics: What happens in the human digestive tract? First evidences in adults using in vitro gut models. *Journal of Hazardous Materials, 442*(September 2022). https://doi.org/10.1016/j.jhazmat.2022.130010

Gooch, H. C. C., Kiu, R., Rudder, S., Baker, D. J., Hall, L. J., & Maxwell, A. (2021). *Enterococcus innesii* sp. nov., isolated from the wax moth *Galleria mellonella, 71*(12), 1–8. https://doi.org/10.1099/ijsem.0.005168

Guo, H., Zhao, X., Yang, K., & Cui, L. (2025). Optimizing mealworm rearing conditions and gut microbiome function for enhanced plastics biodegradation. *Journal of Environmental Sciences, 157*(November 2025), 417–429. https://doi.org/10.1016/j.jes.2024.12.004

Jang, Y., Nyamjav, I., Kim, H. R., Suh, D., Park, N., Lee, Y. E. & Lee, S. (2024). Identification of plastic-degrading bacteria in the human gut. *Science of the Total Environment, 929*(15 June 2024), p. 172775. https://doi.org/10.1016/j.scitotenv.2024.172775

Kamarudin, N. H. A., Noor, N. D. M., & Abd. Rahman, R. N. Z. R. (2021). Microbial degradation of polylactic acid bioplastic. *Journal of Sustainability Science and Management, 16*(7), 299–317. http://doi.org/10.46754/jssm.2021.10.021

Kang, M., Kwak, M., Kim, Y. (2023). Polystyrene microplastics biodegradation by gut bacterial *Enterobacter hormaechei* from mealworms under anaerobic conditions : Anaerobic oxidation and depolymerization. *Journal of Hazardous Materials, 459*(5 October 2023), 132045. https://doi.org/10.1016/j.jhazmat.2023.132045

Lee, W., Hayakawa, T., Kiyono, M., Yamabata, N., & Hanya, G. (2019). Gut microbiota composition of Japanese macaques associates with extent of human encroachment. *American Journal of Primatology, 81*(12), Article e23072. https://doi.org/10.1002/ajp.23072

Lu, L., Luo, T., Zhao, Y., Cai, C., Fu, Z., & Jin, Y. (2019). Interaction between microplastics and microorganism as well as gut microbiota: A consideration on environmental animal and human health. *Science of the Total Environment, 667*, 94–100. https://doi.org/10.1016/j.scitotenv.2019.02.380

Lwanga, E. H., Thapa, B., Yang, X., Gertsen, H., Salánki, T., Geissen, V., & Garbeva, P. (2018). Decay of low-density polyethylene by bacteria extracted from earthworm's guts: A potential for soil restoration. *Science of the Total Environment, 624*(15 May 2018), 753–757. https://doi.org/10.1016/j.scitotenv.2017.12.144

Matsuda, I., Grueter, C. C. & Teichroeb, J. A. (2019). *The Colobines*, In I. Matsuda, C. C. Grueter, & J.A. Teichroeb (Eds.) 1st edn, Cambridge, United Kingdom; New York, NY: Cambridge University Press 2022. https://doi.org/10.1017/9781108347150

Nugrahapraja, H., Sugiyo, P. W. W., Putri, B. Q., Ni'matuzahroh, F., Hafza, N., Götz, F., Santoso, H., Wibowo, A. T., et al. (2022). Effects of microplastic on human gut microbiome: Detection of plastic-degrading genes in human gut exposed to microplastics—preliminary study. *Environments, 9*(11), 140. https://doi.org/10.3390/environments9110140

Nyamjav, I., Jang, Y., Lee, Y. E., & Lee, S. (2023). Biodegradation of polyvinyl chloride by *Citrobacter koseri* isolated from superworms (*Zophobas atratus* larvae). *Frontiers in Microbiology, 14*(May), 1–12. https://doi.org/10.3389/fmicb.2023.1175249

Qiu, X., Qi, Z., Ouyang, Z., Liu, P., & Guo, X. (2022). Interactions between microplastics and microorganisms in the environment: Modes of action and influencing factors. *Gondwana Research, 108*(August), 102–119. https://doi.org/10.1016/j.gr.2021.07.029

Quecholac-Piña, X., del Consuelo Hernández-Berriel, M., del Consuelo Mañón-Salas, M., Espinosa-Valdemar, R. M., & Vázquez-Morillas, A. (2020). Degradation of plastics under anaerobic conditions: A short review. *Polymers, 12*(1), 1–18. https://doi.org/10.3390/polym12010109

Rüthi, J., Cerri, M., Brunner, I., Stierli, B., Sander, M., & Frey, B. (2023). Discovery of plastic-degrading microbial strains isolated from the alpine and Arctic terrestrial plastisphere. *Frontiers in Microbiology, 14*(May), 1–16. https://doi.org/10.3389/fmicb.2023.1178474

Sasanami, Y., Honda, M., Nishiki, H., Tachibana, K., Abe, H., Hokamura, A., & Yamada, M. (2022). Purification and characterization of an enzyme that degrades polyamide 4 into gamma-aminobutyric acid oligomers from. *Polymer Degradation and Stability, 197*, Article 109868. https://doi.org/10.1016/j.polymdegradstab.2022.109868

Shah, A. A., Hasan, F., Hameed, A., & Ahmed, S. (2008). Biological degradation of plastics: A comprehensive review. *Biotechnology Advances, 26*, 246–265. https://doi.org/10.1016/j.biotechadv.2007.12.005

Soleman, Z. T. M., Mantiri, F., & Rumondor, M. (2022). Isolation and characterization of potential lignocellulosic degrading bacteria from chicken manure compost. *Jurnal Bios Logos, 13*(2), 1–9. https://doi.org/10.35799/jbl.v13i2.42791

Tan, K. M., Mohd Fauzi, N. A., Mohd Kassim, A. S., A Razak, A. H., & Kamarudin, K. R. (2021). Isolation and identification of polystyrene degrading bacteria from *Zophobas morio*'s gut. *Walailak Journal of Science and Technology, 18*(8), 9118. https://doi.org/10.48048/wjst.2021.9118

Thamil, C. T. (2016). Isolation and identification of polyethylene biodegradation bacterial from the guts of plastic bags—eating damp wood termites. *Life Science Archives (LSA), 2*(2), 490–499.

Thamsenanupap, P., Pumipuntu, N., Tanee, T., Kyes, P., Karaket, A., & Kyes, R. C. (2024). First report of microplastics in wild long-tailed Macaque (*Macaca fascicularis*) Feces at Kosumpee Forest Park, Maha Sarakham, Thailand. *Veterinary Sciences, 11*(642). https://doi.org/10.3390/vetsci11120642.

Tsochatzis, E. D., Berggreen, I. E., Nørgaard, J. V., Theodoridis, G., & Dalsgaard, T. K. (2021). Biodegradation of expanded polystyrene by mealworm larvae under different feeding strategies evaluated by metabolic profiling using GC-TOF-MS. *Chemosphere, 281*. https://doi.org/10.1016/j.chemosphere.2021.130840

Uddin, M. M., & Wright, M. M. (2022). Anaerobic digestion fundamentals, challenges, and technological advances. *Physical Sciences Reviews, 8*(9), 2819–2837. https://doi.org/10.1515/psr-2021-0068

Wang, Z., Xin, X., Shi, X., & Zhang, Y. (2020). A polystyrene-degrading *Acinetobacter* bacterium isolated from the larvae of *Tribolium castaneum. Science of the Total Environment, 726*, Article 138564. https://doi.org/10.1016/j.scitotenv.2020.138564

Wang, J., Jia, M., Zhang, L., Li, X., Zhang, X., & Wang, Z. (2024) 'Biodegradable microplastics pose greater risks than conventional microplastics to soil properties, microbial community and plant growth, especially under flooded conditions. *Science of the Total Environment, 931*(2024), 172949. https://doi.org/10.1016/j.scitotenv.2024.172949

Woo, S., Song, I., & Cha, H. J. (2020). Fast and Facile Biodegradation of Polystyrene by the Gut Microbial Flora of *Plesiophthalmus davidis* Larvae. *Applied and Environmental Microbiology, 86*(18), e01361-e1420. https://doi.org/10.1128/AEM.01361-20

Wu, X., Zhang, X., Chen,., Ye, A., Cao, J., Hu, X., & Zhou, W. (2023). The effects of polylactic acid bioplastic exposure on midgut microbiota and metabolite profiles in silkworm (*Bombyx mori*): An integrated multi-omics analysis. *Environmental Pollution, 334*(2023), 122210. https://doi.org/10.1016/j.envpol.2023.122210

Yang, X.-G., Wen, P.-P., Yang, Y.-F., Jia, P.-P., Li, W.-G., & Pei, D.-S. (2023). Plastic biodegradation by in vitro environmental microorganisms and in vivo gut microorganisms of insects. *Frontiers in Microbiology, 13*, 1001750. https://doi.org/10.3389/fmicb.2022.1001750

Yoshida, S., Hiraga, K., Takehana, T., Taniguchi, I., Yamaji, H., Maeda, Y., Toyohara, K., Miyamoto, K., Kimura, Y., & Oda, K. (2016). A bacterium that degrades and assimilates poly(ethylene terephthalate). *Science, 351*, 1196–1199. https://doi.org/10.1126/science.aad6359

Zaman, I., Turjya, R. R., Shakil, M. S., Al Shahariar, M., Howlader Emu, M. R. R., Ahmed, A., & Hossain, M. M. (2024). Biodegradation of polyethylene and polystyrene by *Zophobas atratus* larvae from Bangladeshi source and isolation of two plastic-degrading gut bacteria. *Environmental Pollution, 345*(October 2023), 123446. https://doi.org/10.1016/j.envpol.2024.123446

Zhang, Z., Peng, H., Yang, D., Zhang, G., Zhang, J., & Ju, F. (2022). Polyvinyl chloride degradation by a bacterium isolated from the gut of insect larvae. *Nature Communications, 13*(1), 5360. https://doi.org/10.1038/s41467-022-32903-y

Zhang, G., Du, J., Zhang, C., Zhao, Z., Chen, Y., Liu, M., Chen, J., Fan, G., Ma, L., Li, S., et al. (2024). Identification of a PET hydrolytic enzyme from the human gut microbiome unveils potential plastic biodegradation in human digestive tract. *International Journal of Biological Macromolecules, 283*(November), Article 137732. https://doi.org/10.1016/j.ijbiomac.2024.137732

Zhao, S., Liu, R., Lv, S., Zhang, B., Wang, J., & Shao, Z. (2024). Polystyrene-degrading bacteria in the gut microbiome of marine benthic polychaetes support enhanced digestion of plastic fragments. *Communications Earth and Environment, 5*(1), 1–12. https://doi.org/10.1038/s43247-024-01318-6

Zhou, L., Li, A., Zuo, Y. F., Liu, S. P., Tan, Y. L., Li, X., Huang, Y. W., & Zou, K. (2024). A yak gut probiotic *Lactobacillus paracasei* T1–9 displays superior degradation of poly (butylene adipate-co-terephthalate) bioplastic. *Environmental Technology & Innovation, 36*(May), 1–13. https://doi.org/10.1016/j.eti.2024.103805

Controlled-Release Nanofertilizers: A Paradigm Shift in Agriculture

Nur Syafiqah Farhanah Dzulkharnien, Rosiah Rohani, and Rasul Sauket Ali

Abstract Agriculture serves as a fundamental pillar in sustaining the rapidly growing global population and driving economic prosperity. However, the excessive application of chemical fertilizers in modern farming has severely degraded soil quality by depleting essential nutrients and disrupting the delicate ecological balance. This overuse of fertilizers has not only left harmful residues on agricultural products but has also contributed to environmental pollution, disrupted agroecosystems and reduced soil fertility. Moreover, conventional fertilizers exhibit poor nutrient utilization efficiencies due to leaching. This leads to significant economic losses and long-term soil degradation. Therefore controlled-release nanofertilizers (CRNFs) are foreseen to become an alternative to replace the conventional fertilizers due to their excellent characteristic on releasing macronutrients or micronutrients in a gradual manner. Nanoparticles (NPs) of fine size, usually below 100 nm, are introduced as micronutrients or nanocarriers that are encapsulated in liquid fertilizers such as urea and ammonium nitrate mixture. Meanwhile the controlled-release behaviour of composite polymer hydrogels in the designated CRNFs assists in promoting the plant growth efficiently. These distinct characteristics are essential to facilitate and sustain the release of the essential nutrients slowly into the soil to ensure optimum absorption by the plant's uptake. The incorporation of polymer-derived nanomaterials is a recent and promising approach to enhancing fertilizer efficiency in agriculture. This sustainable and environmentally friendly strategy optimizes nutrient management, minimizes ecological impact, and promotes higher crop yields, thereby contributing to the advancement of sustainable agricultural practices.

Keywords Nanofertilizer · Controlled-release · Hydrogel · Plant growth · Agriculture

---

N. S. F. Dzulkharnien · R. Rohani (✉) · R. S. Ali
Department of Chemical and Process Engineering, Faculty of Engineering and Built Environment, Universiti Kebangsaan Malaysia, 43600 UKM Bangi, Selangor, Malaysia
e-mail: rosiah@ukm.edu.my

R. Rohani
Research Centre for Sustainable Process Technology, Faculty of Engineering and Built Environment, Universiti Kebangsaan Malaysia, 43600 UKM Bangi, Selangor, Malaysia

A. Z. Yaser et al. (eds.), *Compost and Liquid Organic Fertilizers*,
https://doi.org/10.1007/978-981-95-8561-8_19

1 Introduction

Food security has become one of the most concern topics in recent years due to the increase in food consumption but decrease in the food source. It has been reported that about 850 million of the world's populations are malnourished and most of them come from rural areas of developing countries. Several factors have influenced the production of food from the agriculture sector such as high population, increase in demand, climate change, import–export activities and so on, had a profound consequence in the agricultural economy (Oladosu et al., 2022; Syed, 2015). Hence, research on novel materials that able to improve crop production is necessary to support the demand and sustain the food supply for the consumers.

Agricultural soil encompasses various types of microbes including bacteria, fungi, protozoa, actinomyces and others. Bacteria such as *Rhizobium* sp., *Pseudomonas* sp., *Escherichia coli (E. coli), Staphylococcus aureus (S. aureus)* are common microorganisms abundantly found in the soil that contribute to the major mineral recycling such as nitrogen (N), phosphorus (P) and carbon (C), that greatly impact the soil fertility (Javed et al., 2022). For instance, *Rhizobium* sp. and *Pseudomonas* sp. play an important role in nitrogen fixation and decomposition of organic matter, respectively. Although these bacterial strains are responsible in the production of minerals, some of them are plant pathogens that could cause plant diseases (Steffan et al., 2020). Additionally, some of these pathogens also could lead to the degradation of fertilizers by changing the chemical structure of nutrients or absorbing nutrients in the soil (Wu et al., 2021). Thus, the development of an antibacterial fertilizer is foreseen to be a promising innovation in combating diseases in plantation while maintaining soil microbial activity.

Traditional fertilizers such as NPK fertilizer containing nitrogen, phosphorus and potassium have been utilized in decades to boost crop production and meet the food supply demand. The consumption of fertilizers used in agriculture has been reported to increase yearly from 40 to 130 million for the past thirty years and it is projected to increase twice by the year 2030 (Wang et al., 2021). These traditional fertilizers made from chemical materials, enriched with easily soluble nutrients, are responsible for the soil fertility that affects the crop yield and quality. Nevertheless, the traditional fertilizers are quickly leach from the soil due to their good mobility, particularly during the rainy season, can lead to environmental impact such as eutrophication, air pollution and irremediable in ecological system (Skrzypczak et al., 2021).

Liu et al. (2024) have conducted a comprehensive review of past literature on the effects of nanofertilizers on plant photosynthesis. In the review, the authors highlighted the multifaceted benefits of nanofertilizers, along with enhanced photosynthetic efficiency, biomass production in plants, modulation of pigmentation, regulation of Reactive Oxygen Species (ROS), increase in gas exchange and plant stress resistance. Nanofertilizers responsible in increasing the crop abiotic stress tolerance, promote photosynthesis, induce hormonal cascades, regulate enzymatic activity, and eventually enhance crop yield over conventional fertilizers. The amount of utilized fertilizer can be reduced as the nano-size of fertilizers are able to penetrate plant cell walls effectively contributing to maximum nutrient absorption efficiency (Sun et al., 2022). Pan et al., (2023) also recorded that the plant absorbs light effectively using low concentration of nanofertilizers, thus enhancing the electron transfer rate in photoreaction and subsequently promote the photolysis of water and reduction of $NADP^+$ to NADPH. Collectively, these actions had results to enhancement in the photosynthetic efficiency, therefore stimulate the plant growth.

Meanwhile, Haydar et al. (2024) elaborated an overview on the preparation methods of slow- (SRNFs) and controlled-release nanofertilizers (CRNFs), and discussed nutrient-release mechanisms, detection methods, development and commercialization prospects in agronomic management as well as future research prospects. The terms CRNFs and SRNFs are often interchangeable, despite having different structural and functional properties. They described that SRNFs are typically nanocomposite-based systems in which the nutrients are embedded in a polymer matrix that regulates the nutrient release at slow rate depending on soil-physic-chemical properties (pH, moisture percentage) and environmental condition temperature, rainfall, etc.). Conversely, CRNFs are generally comprising polymer coated fertilizers formulated to release nutrients in precisely control manner under the influence of external stimuli from environment (Elsayed et al., 2022).

The release mechanisms of CRNFs and SRNFs are completely distinct in nature. For CRNFs, the release profile involved three successive phases which are: lag, steady release, and decay (Shabiya et al., 2025). The lag phase is initiated by absorbing moisture from soil water into the cracked microscopic pores of the outer layer coating that eventually penetrated slowly in the CRNFs' core through water vapor pressure gradient force without releasing the nutrients yet. Next, in the steady phase, ongoing water absorption induces solid fertilizer dissolution resulting in the increment of osmotic potential in the interior and accumulation of a sufficient level of saturated solution, which contributes to continuous diffusion of nutrients out from the CNRFs through polymer membrane before undergoing decaying process (Nawadkar et al., 2024). In contrast, as SRNFs were exposed to soil moisture, the matrix began to swell and loose, consequently discharged the stored nutrients slowly from the polymer network (Lakshani et al., 2023).

The aforementioned studies primarily focused on the plant physiology response with the use of nanofertilizers and discussion on the structural and mechanistic comparison between CRNFs and SRNFs with focus on the synthesis approaches (i.e., emulsification, coacervation, nanoprecipitation, inclusion complex and coating process) and physiochemical analyses (i.e., swelling ratio, water absorption capacity, loading efficiency, entrapment efficiency, etc.). However, their discussion was still lacking with little emphasis on novel material developments or formulation strategies. Compared to the previous study mentioned before, this chapter offers a more specific examination of CRNFs in the context of polymer coating recent advances (in the past five years) and their implications for fertilizer efficiency in the supply of nutrients in agriculture.

CRNFs can be defined as nanomaterial encapsulated or coated with nutrients or liquid organic fertilizer, have been introduced as a future alternative that is expected to replace the traditional fertilizer. Researchers had tailored a variety of nano-fertilizers including nano emulsion, nanogels, solid lipid nanoparticles (NPs), dendrimers, liposomes, polymeric micelles, and nano-sized metal oxide NPs (Thangavelu et al., 2024; Verma et al., 2022). The fine size of NFs has benefits them in many ways in plantation such as effective penetration into the soil, efficient mineral transportation to the plant root system and being able to supply typical bioavailability minerals like phosphorus and zinc directly to the plant cells, thus enhancing the efficiency of nutrient absorption (Kumar et al., 2021). Concurrently, this action will reduce the possibility of excess discharged nutrients such as nitrate into the soil and minimize the amount of fertilizer used during cultivation that may cause soil pollution (Avila-Quezada et al., 2022).

Utilization of CRNFs in plantation has been seen to pave a new pathway in the agriculture sector as CRNFs has the potential to improve the quality and quantity of the crops production effectively, thus lead to sufficient food supply and demand globally (Kekeli et al., 2025). Consequently, the economy within the country will rise due to the increase in the food exports abroad and eventually attract foreign investors (Salama et al., 2022). However, CRNFs have also been seen to pose drawbacks with its usage, particularly in the ecological system. Some studies reported that the release of CRNFs into the soil affected the growth of some microorganisms that play an important role in the ecosystem (Zulfiqar et al., 2019). Wei et al. (2021) reported that the copper oxide nanoparticles (CuO NPs) have encouraged the growth of the rhizosphere microorganisms in *Salvia miltiorrhiza* after 165 days of exposure with the NPs.

In this chapter, the main aim of this review is to comprehensively elaborate a few common CRNFs available and their potential in current agricultural industries. Discussions on the types of NPs, synthesis methods the NPs, along with the CRNFs originating from composite polymer hydrogel and their mechanism for controlled-release the micronutrients for plant's uptake are elaborated further in this chapter.

2 Metal Oxide Nanoparticles as Nanofertilizers in Agriculture

Nanomaterials or NPs that function as fertilizers or nanocarriers of liquid organic fertilizers in the plantation are considered as nanofertilizers (NFs) because these nanomaterials can assist the nitrogen cycle, induce the enzymatic processes and generate plant growth-promoting microorganisms (Kalwani et al., 2022). Diverse types of NFs have been designed and developed to address different needs, including metal and metal oxide NPs such as iron oxide NPs, zinc oxide NPs, magnesium oxide NPs, copper oxide NPs etc. The following discussion in the next sub-topic will elaborate the different types and functionality of each metal oxide in the crop production.

2.1 *Iron Oxide Nanoparticles*

Iron plays an important role in many physiological processes of plants including chlorophyll biosynthesis, respiration and redox reactions. Although iron has been known as the fourth dominant element found in the earth's crust, nonetheless, most of it is fixed in the form of poorly soluble Fe^{3+} ions from iron oxides/hydroxides which cannot be absorbed by plants. Naturally, plants are only able to absorb iron nutrients in the form of Fe^{2+} ions from the soil. However, lack of Fe^{2+} ions in the soil could cause deficiency of these nutrients on the plants (Haydar et al., 2022). Meanwhile, excess of iron transport in plants will limit the plant growth and reduce the crop production due to the aerobic condition of iron that interacts with hydrogen peroxide (H_2O_2) that forms ROS through Fenton reaction (Li et al., 2023a, 2023b). Hence, the iron uptake by the plant as well as the homeostasis in the cells need to be balanced to prevent excess or deficit of iron that causes drawbacks on the yield and quality of the crops. Research on the use of iron oxide NPs, particularly maghemite (Fe_2O_3), have received great attention in agriculture as fertilizer for the plants due to their remarkable quantum confinement, high surface energy and surface area-to-volume ratio with excellent catalytic potential. Raiesi-Ardali et al. (2022) have found that the organically coated iron oxide NPs, which was humic acid coated iron oxide (Fe_3O_4@HA) NPs, have improved the growth and yield of tomato plant due to a significant increase of the iron concentration in the tomato plant shoot. Meanwhile, Manzoor et al. (2023) reported that the efficacy of iron oxide NPs in reducing the toxicity of arsenic in the soil at various concentrations (25, 50, and 100 mg kg^{-1}) have eventually enhanced the wheat (*Triticum aestivum L.*) growth performance.

2.2 *Copper Oxide*

Copper is a redox-active transition metal, which is considered as one of the essential micronutrients for plant growth during cultivation. It presents various physiological and biochemical processes in plant at optimum level such as mitochondrial respiration, in the electron transport chain, photosynthesis, cell wall metabolism, lignin synthesis, oxidative stress response and hormone signaling (Mir et al., 2021). However, excess accumulation of copper uptake by plants may bring harm and inhibit the growth of the plants due to the weak membrane integrity, decline in photosynthesis process and change the enzyme activity (Shabbir et al., 2020). Several studies have reported on the toxicity of excess copper, which have caused negative effects on seed germination and plant growth on various crops such as on pea seeds, wheat, barley and others (Kadri et al., 2022; Nagdalian et al., 2024; Thakur et al., 2021). In contrast, the low concentration of copper oxide NPs supplied has proven to promote crop growth efficiently. For instance, Essa et al. (2021) discovered that the wheat shoots and root grew about 172.78 ± 23.11 and $215.94 \pm 37.76\%$, respectively, after being treated with Cu NPs. These results conclude that the optimal concentration of copper release into the soil needs to be controlled to avoid inhibition on plant growth.

2.3 Magnesium Oxide

Major challenges faced by the farmers are to sustain sufficient and good quality crop production. These are critical requirements to meet food demand under abiotic stresses (such as climate change, heavy metals, salinity) and biotic microorganism (such as bacteria, fungi) (Ali et al., 2024). Magnesium is one of the micronutrients needed for plant growth and it poses an important role in several plant processes including photosynthesis and chlorophyll synthesis (Salman et al., 2023). Owusu Adjei et al. (2021) revealed that the chlorophyll content of *Ananas comosus* var. *bracteatus* leaves rise upon treated with a low concentration of magnesium oxide NPs (1 g/mL) and vice versa. Although magnesium oxide NPs provided sources of magnesium micronutrients for plants to increase the growth and yield of the crops, an optimum concentration of NPs used during cultivation must be regulated properly to ensure chlorophyl supply sufficient for the plant growth. Concurrently, according to Abdul-Karim (2021), magnesium oxide NPs have effectively treated tomato seedlings from the studied pathogen infection, known as *fusarium oxysporum* f.sp. *lycopersici*, which have proved the antifungal ability of the NPs. Overall, due to the fine size and antifungal property of magnesium oxide NPs, it could be used as nano fertilizers that have potential in improving the plant growth and yield by providing magnesium into the soil for plant's uptake.

2.4 Zinc oxide

Zinc (Zn) is one of the important key components for plant growth since this micronutrient is responsible for various cellular functions. For instance, it is required to enhance biophysiochemical and metabolism activities, enzyme stimulation, and ion homeostasis (Hamzah Saleem et al., 2022). Thus, in an effort to supply zinc nutrients to the plant system, zinc oxide NPs is potentially be used as a soil fertilizer, which could replace the conventional fertilizer. Like other micronutrients, excessive amounts of zinc released from the NPs will lead to chlorosis and eventually cause a serious inhibition on plant growth. Several studies have been reported that the presence of these NPs may influence soil toxicity and plant growth at certain dosages. Singh et al. (2018) have reported that high concentrations of zinc oxide NPs, at 500 and 1000 ppm, have stunted rice growth due to the toxicity effects of excessive present of zinc oxide NPs. This study proved that the toxic effects of ZnO NPs were more pronounced at high concentrations of NPs. Meanwhile, a study conducted by Abdelaziz et al. (2022) described that an optimal concentration of mycosynthesized ZnO NPs using *Penicillium expansum* (500 μg/mL) may promote the growth and metabolic activities of the infected cultivated eggplant (*Solanum melongena*) by suppressing the *Fusarium oxysporum* wilt disease on the studied plants. In general, ZnO NPs could be an alternative zinc nutrient source for the plants' uptake, to improve the plant development and inhibit the growth of the unenviable soil microorganisms at suitable concentration of NPs.

Next, several well-known methods to synthesize NPs and their advantages as well as disadvantages will be discussed in the next topic in this chapter.

3 Methods of Synthesizing Metal Oxide Nanoparticles

NPs synthesis can be achieved through two (2) primary strategies: the bottom-up and top-down approaches. A top-down approach involves breaking down large-sized materials into small nano-sized structures or particles. The top-down synthesis technique is an extension of the technique that has been used to produce nano-sized particles. The top-down approach is inherently simpler and relies on the removal or division of a large portion of the material to produce the desired structure with the appropriate properties. Examples of top-down approach techniques are high-energy wet ball milling, electron beam lithography, atomic force manipulation, gas-phase condensation, and aerosols (Balasooriya et al., 2017). In contrast, the bottom-up approach refers to the accumulation of material, which is the formation of nanostructures by combining or arranging small particles into a larger system. Among the examples of chemical methods used in the bottom-up approach are the sol–gel process, molecular reactions in thermos, microemulsion and hydrothermal and co-precipitation, in which these methods are widely used to produce NPs, nano fertilizers, nano catalysts and others (Jamkhande et al., 2019). Table 1 provides an overview of various synthesis techniques categorized under these two methodologies with their advantages and disadvantages.

The techniques listed in Table 1 can be classified into three (3) categories which are physical, chemical and biological methods. Physical methods such as hydrothermal are less suitable to use for synthesizing NPs because they require large amounts of energy, time and space (Sharma et al., 2020). Meanwhile, conventional chemical methods such as sol gel processes require the use of organic solvents that are harmful to the environment and physiological systems, formation of toxic by-products, high costs, and complicated laboratory processes (Ismail et al., 2019). Therefore, an environmentally friendly biological method such as using green synthesis technique, has developed rapidly in the field of nanotechnology due to their sustainable preparation process, less use of chemicals and production of waste, easy to prepare and cost saving (Sheldon, 2018).

Although NPs have lots of potential in diverse fields, including NFs in the agricultural sector, however, excessive release of NPs into the soil that lead to soil toxicity has become a serious concern recently. Poor control of these NPs may cause downside effects on plant growth and eventually influence crop production. In conjunction to these drawbacks, inventions with slow-release capability are needed to control the

Table 1 Comparison between the techniques commonly used to synthesize metal oxide NPs

Techniques	Advantages	Disadvantages	References
Green synthesis using plant extracts	• Can be carried out at room temperature • Does not involve toxic solvents • Low cost of raw materials • Environmentally friendly technology • Does not involve high equipment costs • Good repeatability	• Nanoparticle synthesis mechanism that is specific according to the type of plant • Different plants have different uptake rates • Difficult to develop the results of the synthesis process	(Manikandan & Lee, 2022)
Ultrasonication	• Safe and easy operation • Short synthesis time if a high ultrasound frequency is used • Low operating conditions • Cost effective process	• Materials that are sensitive to heat will not withstand acoustic cavitation • Method requires a high energy supply • Product yield per energy supply is low	(Yu et al., 2021)
Sol–gel	• Produces a high purity and homogeneous product • Economically-friendly • Easy to synthesis a uniform size nanoparticle • Facile method • Low processing temperature	• Response time is longer • Solvents used are dangerous • High permeability • Weak bonds between molecular particles • Low wear resistance	(Baig et al., 2021)
Precipitation	• Easy preparation technique and produce high yield • Short reaction time • Low processing temperature • Used environmental-friendly solvent such as water	• Difficult to control shape and size distribution of nanoparticles	(Salehipour et al., 2021)

(continued)

Table 1 (continued)

Techniques	Advantages	Disadvantages	References
Microemulsion	• Minimal agglomeration • Can easily control particle size and composition • Easy handling, inexpensive and convenience	• Requires a large amount of concentrated surfactant • External factors such as temperature and pH affect microemulsion stability • Limited solvent capacity for materials with high melting points • Involves the use of expensive autoclaves • Requires high temperature and pressure	(Salabat & Mirhoseini, 2022)
Hydrothermal	• Able to synthesize unstable NPs near to their melting point • Able to synthesize high quality nano crystal	• Expensive equipment	(Hachem et al., 2022)

release of NPs efficiently into the soil for plants' uptake. Thus, a discussion on the CRNFs made up of composite polymer hydrogels and its mechanism will be elaborated in the next topic.

4 Composite Polymer Hydrogels as Controlled-Release Nanofertilizers

One of the most challenging aspects to manage in the cultivation of plants is the nutrients deficit that affects plant growth and produces fewer nutritional crops. Fertilizers originated from NPs, have become a promising invention in the fertilizer industry due to their significant potential to promote nutrient into the soil for optimal plant growth (Cyriac et al., 2023). Nonetheless, uncontrol release of these fertilizers will lead to soil toxicity and subsequently will cause negative impact to the crop production. Recently, related research on the use of CRNFs based composite polymer hydrogels incorporating with NFs has been widely investigated to solve the nutrients deficiency issue.

Composite polymer hydrogels can be defined as a combination of two (2) different types of materials that consists of a network of hydrophilic polymer chains. These porous hydrogels are commonly posed as superabsorbent ability with high-water content (>90%) and able to hold an abundance of water in their three-dimensional (3D) structure. Due to this intrinsic property, this type of hydrogel can retain water in the soil, moisten the soil sufficiently and eventually enhance plant growth. This helps to reduce the amount of irrigation and increase the efficiency of water used in agriculture (Maksimova et al., 2023).

In addition, composite polymer hydrogels are also categorized as safe materials to control the release of nutrients over time, which are beneficial in the agricultural sector by reducing the environmental impact and increasing the crop yields and quality simultaneously. In agriculture, CRNFs made of composite polymer hydrogels are foreseen to be potential materials to substitute the traditional synthetic fertilizers. These types of CRNFs are more environmentally friendly compared to the traditional ones, as the materials used can be modified using natural based materials. Alteration on the formulation of the CRNFs are believed can improve the physical properties of the synthesized CRNFs such as water and air retention, mechanical strength, swelling properties, nutrient release ability, etc. (Tariq et al., 2023). This alteration allows CRNFs to provide a continuous source of nutrients to the plants, rather than a one-time application of synthetic fertilizer. Few examples of composite polymer hydrogels that are commonly used in the research to form CRNFs are namely: polylactic acid (PLA), polyvinyl alcohol (PVA), alginate (Alg), and others. Further elaboration on the CRNFs made up from each of these polymers will be described below.

4.1 Polylactic Acid (PLA) Nanofertilizers

Polylactic acid (PLA) originates from an aliphatic polyester of lactic acid that is derived from biodegradable materials such as maize, cassava, sugar cane and sugar beet. PLA is a well-known biocompatible and environmental-friendly polymer with hydrophobic behaviour that plays a vital role as a physical barrier against water diffusion that is essential on providing a long-term release of nutrients. However, PLA has certain limitations that hinder its usage in a large-scale application such as extreme brittleness and poor ductility (Li et al., 2023a, 2023b; Nooeaid et al., 2021). Numerous studies have reported on the potential of PLA as a CRNFs. Rubel and Wei (2022) developed a novel nitrogen biochar-based CRNFs using PLA as the coating polymer. They have found that almost 70% of nitrogen fertilizer released from 21.15% of PLA from day-2 to day-27, indicating the potential of PLA on controlling the release of fertilizer into soil. Recent reports show that this polymer has been utilized to release the NPs slowly into the soil to ensure the plant received sufficient micronutrients over time. For instance, Zilabpour et al. (2024) reported that a CRNFs composed of iron (Fe) nano carbon dot coated with PLA polymer at a concentration of 10 mg Fe kg^{-1} soil. 3 parts per thousand (ppt) of carbon dot foliar spraying has shown the best performance on sorghum cultivation with 211 mg kg^{-1} and 186 mg

kg^{-1} of iron in the steam and leaf, respectively. It is expected that the research on CRNFs combined with PLA will be significantly increased in number soon. This is because a comprehensive understanding of the controlled-release interaction of this type of CRNFs on the studied plant is acquired to be explored further.

4.2 Polyvinyl Alcohol Nanofertilizers

Polyvinyl alcohol (PVA) is a water-soluble synthetic polymer made up from vinyl acetate that acts as a protective agent for crops in agriculture. It could retain moisture in the plant by forming a thin protective layer on the surface of the leaves and it also can be used as a seed coat to protect seeds from drying out (Farias et al., 2019). Besides, few literatures also reported that PVA is commonly used as CRNFs to increase nutrient uptake by plants due to its hydrophilic nature towards water and reduced leaching. Kumar et al. (2018) reported the efficiency of Cu–Zn micronutrients carrying carbon nanofibers (CNFs) incorporated PVA-starch (PBMC) through in-situ polymerization on the chickpea plant. PBMC with 1–1 of Cu–Zn:CNF ratio exhibited the tallest grown plants (33 cm) compared to the other tested Cu–Zn:CNF ratios. The result has signified the potential of the synthesized CRNFs for the study. Meanwhile, Marzouk et al. (2022) have found that the ZnO NPs coated PVA with particle size of 33 nm exhibited the best performance in terms of height, plant haulm and number of stems on the potato plant. The obtained results remarked the important function of this CRNFs on enhancing plant vegetative growth by fully utilizing solar radiation and essential minerals through photosynthesis. In another study conducted by Channab et al. (2024), a CRNFs of triple superphosphate (TSP) fertilizers coated with a mixture of carboxymethyl cellulose (CMC) and PVA incorporated with zeolitic imidazolate type 8 (ZIF-8) NPs resulted in a 13.79% increase in leaf numbers and a 15.83% increase in plant heights of tomato plant due to the slow release of phosphate ($12.29 \pm 2.75\%$) into the soil that led to low soil toxicity. The uncontrolled release of nutrients can lead to soil toxicity, while an insufficient nutrient supply may hinder plant growth. Therefore, precise regulation of nutrient release is essential to achieving an optimal dosage that supports healthy plant development and maximizes agricultural productivity.

4.3 Alginate Nanofertilizers

Alginate (Alg) or known as alginic acid salt, is a natural anionic polysaccharide extracted from brown seaweeds that are derived β-d-mannuronic acid(M-block) bonded by C1–C4 with the blocks of α-l-guluronic acid (G-block). Excellent features of alginate such as high stability, biocompatible, biodegradable, and non-toxic have made alginate possible to be used in various industries, including agriculture

(Sahoo & Biswal, 2021). Moreover, alginate also exhibits adequate swelling properties that are important for loading and releasing activities of nutrients. Ekanayake and Godakumbura (2021) have reported on the synthesized CRNFs with nanonutrients of zinc and copper ions, which crosslinked with sodium alginate to form the alginate hydrogel (Zn-Alg and Cu-Alg, respectively). It was found that the Cu-Alg CRNFs have reduced uptake of copper ions, while the amounts of zinc ions in the tomato leaves were undetected by Flame Atomic Absorption Spectroscopy (FAAS). Next, Leonardi et al. (2021) designated a novel polyelectrolyte complex (PEC) comprising chitosan-alginate and encapsulated a CuO NPs. These 300 nm sized CRNFs exhibited an excellent efficacy on *Fortunella margarita* Swingle seeds with an evident that showed CuO/PEC_10 and CuO/PEC_50 revealed the best root and shoot growth compared to CuO/PEC_100. This result is attributed to the toxicity of high concentration CuO micronutrients in CuO/PEC_100 that reduces the root development of the plant which consequently disturbs the water and nutrient uptake that led to growth reduction. Figure 1 shows the root development of *Fortunella margarita* Swingle seeds after 30 days treatment with CuO NPs and CuO/PEC at different concentrations. Recently, Fathy et al. (2024) reported the efficiency of single-coated and double-coated CRNFs with treated nano-apatite using urea as fertilizer and Alginate/lignin as biodegradable polymer on the yield of green bean plants with a result 98% and 88% of total yield at 25% dose, respectively. This study showed that the NPs could also act as a nanocarrier containing macronutrient fertilizer, instead of being used as a fertilizer during cultivation.

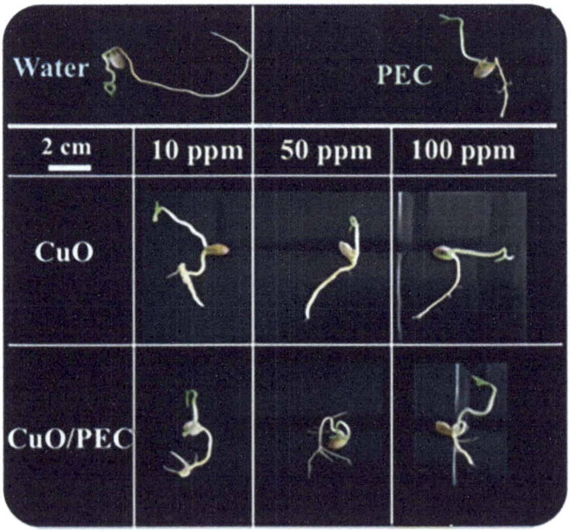

Fig. 1 Morphological characteristics of *Fortunella margarita* Swingle seeds after treated with CuO NPs and CuO/PEC at 10 ppm, 50 ppm and 100 ppm. Used with permission of Right Society of Chemistry, from Smart nanocomposites of chitosan/alginate nanoparticles loaded with copper oxide as alternative nanofertilizers, Leonardi et al., (8(1), 2021; permission conveyed through Copyright Clearance Center, Inc

5 Nutrient Transportation Mechanism from Controlled-Release Nanofertilizers

CRNFs are responsible for controlling the release of nutrients or NPs from the polymer matrix into the soil gradually over time and eventually reach to the root for maximum nutrients uptake. This action could help in the root elongation which led to optimal nutrient utilization by plant and consequently minimize nutrient loss through leaching. Figure 2 illustrates the mechanism of the CRNFs through the synergistic interaction between NPs and composite polymer hydrogel. The high surface-to-volume ratio NPs with antibacterial property serves as a crucial micronutrient for enhancing the plant growth, while the composite polymer hydrogel that encapsulates the NPs is responsible for controlling and sustaining the release of nutrients. The mechanism initiates when the composite polymer hydrogel matrix encapsulating the nanoparticles undergoes surface cracking due to exposure to sunlight and mechanical stress within the soil. According to Fig. 2, the soil water that enters the matrix through the pores and presence of soil microorganism had causes shrinkage on the hydrogel matrix. Simultaneously, the encapsulated micronutrients are dissolved into a liquid state and ready to be diffused into the soil continuously in a gradual manner to be taken by the plant root (Li & Chen, 2020; Yadav et al., 2023). The controlled and sustained release of nanoparticles (NPs) from the synthesized nano-fertilizers (NFs) presents multiple benefits, including improved nutrient uptake efficiency, minimized nutrient loss due to leaching, and prolonged nutrient availability, ultimately enhancing crop productivity while mitigating environmental impact (Kusumastuti et al., 2019). The findings of this study further emphasize that the gradual release of nutrients not only optimizes plant growth but also preserves soil health and reduces environmental degradation by minimizing nutrient runoff.

6 Conclusion

In conclusion, CRNFs hold great promises for transforming agriculture by enhancing nutrient efficiency, minimizing environmental impact, and supporting global food security. Their controlled-release mechanisms and targeted nutrient delivery offer significant advantages over conventional fertilizers. However, the challenges related to cost-effective production, optimizing nutrient release rates, and ensuring safety through well-defined regulations need to be properly addressed to realize their full potential. Collaboration between researchers, policymakers, and agricultural stakeholders is essential for integrating CRNFs into mainstream farming practices. Additionally, educating farmers on their benefits and potential risks will facilitate their adoption. Continuous research and monitoring are crucial to maximizing the benefits while mitigating risks, ultimately ensuring a more sustainable and resilient agricultural system.

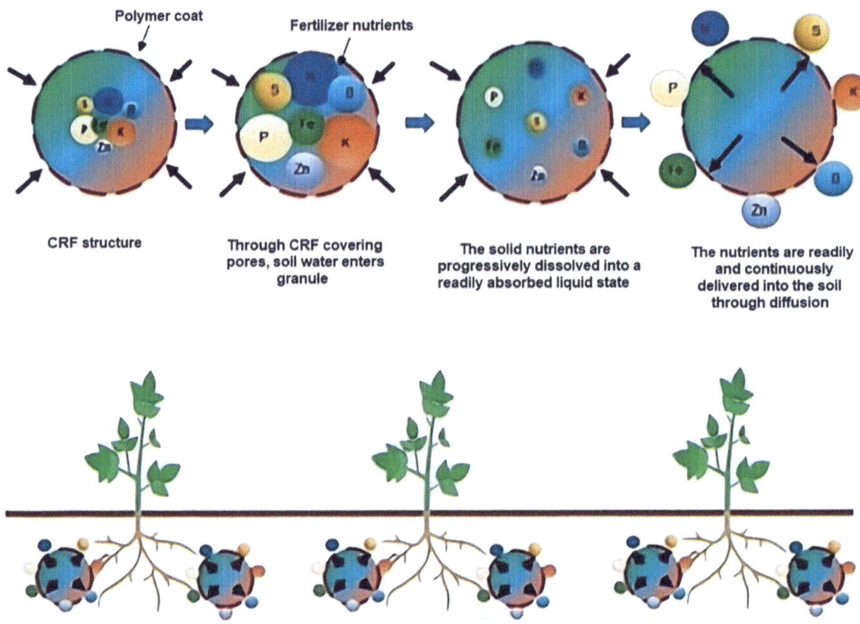

Fig. 2 Mechanism of slow release of nutrients from fertilizer. (*Source* Farias et al., 2019). CC BY 4.0

References

Abdelaziz, A. M., Salem, S. S., Khalil, A. M. A., El-Wakil, D. A., Fouda, H. M., & Hashem, A. H. (2022). Potential of biosynthesized zinc oxide nanoparticles to control Fusarium wilt disease in eggplant (Solanum melongena) and promote plant growth. *BioMetals, 35*(3), 601–616.

Abdul-Karim, E. K. (2021). The efficiency of magnesium oxide, nano magnesium oxide and cinnamon alcoholic extract in controlling Fusarium oxysporum f. sp. lycopersici which causes Fusarium wilt on tomato. *International Journal of Agricultural and Statistical Sciences, 17*(1), 1611–1618.

Ali, S., Ulhassan, Z., Shahbaz, H., Kaleem, Z., Yousaf, M. A., Ali, S., Sheteiwy, M. S., Waseem, M., Ali, S., & Zhou, W. (2024). Application of magnesium oxide nanoparticles as a novel sustainable approach to enhance crop tolerance to abiotic and biotic stresses. *Environmental Science: Nano, 11*(8), 3250–3267.

Avila-Quezada, G. D., Ingle, A. P., Golińska, P., & Rai, M. (2022). Strategic applications of nano-fertilizers for sustainable agriculture: Benefits and bottlenecks. *Nanotechnology Reviews, 11*(1), 2123–2140.

Baig, N., Kammakakam, I., & Falath, W. (2021). Nanomaterials: A review of synthesis methods, properties, recent progress, and challenges. *Materials Advances, 2*(6), 1821–1871.

Balasooriya, E. R., Jayasinghe, C. D., Jayawardena, U. A., Ruwanthika, R. W. D., Mendis de Silva, R. & Udagama, P. V. (2017) 'Honey mediated green synthesis of nanoparticles: new era of safe nanotechnology', *Journal of Nanomaterials*

Channab, B.-E., El Idrissi, A., Ammar, A., Akil, A., White, J. C., & Zahouily, M. (2024). ZIF-8 metal organic framework, carboxymethylcellulose and polyvinyl alcohol bio-nanocomposite controlled-release phosphorus fertilizer: Improved P management and tomato growth. *Chemical Engineering Journal, 495*, Article 153610.

Cyriac, J., Thomas, B., Sreejit, C.M., Yuvaraj, M., & Joseph, S. (2023) 'Chitosan zinc nanocomposite: A promising slow releasing zinc nano fertilizer', *Materials Today: Proceedings* [Preprint].

Ekanayake, S. A., & Godakumbura, P. I. (2021). Synthesis of a dual-functional nanofertilizer by embedding ZnO and CuO nanoparticles on an alginate-based hydrogel. *ACS Omega, 6*(40), 26262–26272.

Elsayed, A. A. A., Ahmed, E.-G., Taha, Z. K., Farag, H. M., Hussein, M. S., & AbouAitah, K. (2022). Hydroxyapatite nanoparticles as novel nano-fertilizer for production of rosemary plants. *Scientia Horticulturae, 295*, Article 110851.

Essa, H. L., Abdelfattah, M. S., Marzouk, A. S., Shedeed, Z., Guirguis, H. A., & El-Sayed, M. M. H. (2021). Biogenic copper nanoparticles from Avicennia marina leaves: Impact on seed germination, detoxification enzymes, chlorophyll content and uptake by wheat seedlings. *PLoS ONE, 16*(4), Article e0249764.

Farias, B. V., Pirzada, T., Mathew, R., Sit, T. L., Opperman, C., & Khan, S. A. (2019). Electrospun polymer nanofibers as seed coatings for crop protection. *ACS Sustainable Chemistry & Engineering, 7*(24), 19848–19856.

Fathy, M. A., Helal, M. I. D., Khater, H. A., Abdelkader, N. H., & Abdelgawad, K. F. (2024). Optimizing green bean yield: Controlled nitrogen release with nano-urea-modified apatite. *Scientific Journal of King Faisal University: Basic and Applied Sciences, 25*(1), 13–20.

Hachem, K., Ansari, M. J., Saleh, R. O., Kzar, H. H., Al-Gazally, M. E., Altimari, U. S., Hussein, S. A., Mohammed, H. T., Hammid, A. T., & Kianfar, E. (2022). Methods of chemical synthesis in the synthesis of nanomaterial and nanoparticles by the chemical deposition method: A review. *BioNanoScience, 12*(3), 1032–1057.

Hamzah Saleem, M., Usman, K., Rizwan, M., Al Jabri, H., & Alsafran, M. (2022) Functions and strategies for enhancing zinc availability in plants for sustainable agriculture. *Frontiers in Plant Science, 13*, p. 1033092.

Haydar, M. S., Ghosh, D., & Roy, S. (2024). Slow and controlled release nanofertilizers as an efficient tool for sustainable agriculture: Recent understanding and concerns. *Plant Nano Biology, 7*, Article 100058.

Haydar, M. S., Ghosh, S., & Mandal, P. (2022). Application of iron oxide nanoparticles as micronutrient fertilizer in mulberry propagation. *Journal of Plant Growth Regulation, 41*(4), 1726–1746.

Ismail, N. A., Shameli, K., Wong, M.M.-T., Teow, S.-Y., Chew, J., & Sukri, S. N. A. M. (2019). Antibacterial and cytotoxic effect of honey mediated copper nanoparticles synthesized using ultrasonic assistance. *Materials Science and Engineering: C, 104*, Article 109899.

Jamkhande, P. G., Ghule, N. W., Bamer, A. H., & Kalaskar, M. G. (2019). Metal nanoparticles synthesis: An overview on methods of preparation, advantages and disadvantages, and applications. *Journal of Drug Delivery Science and Technology, 53*, Article 101174.

Javed, A., Ali, E., Afzal, K. B., Osman, A., & Riaz, S. (2022). Soil fertility: factors affecting soil fertility, and biodiversity responsible for soil fertility. *International Journal of Plant, Animal and Environmental Sciences, 12*(1), 21–33.

Kadri, O., Karmous, I., Kharbech, O., Arfaoui, H., & Chaoui, A. (2022). Cu and CuO nanoparticles affected the germination and the growth of barley (Hordeum vulgare L.) seedling. *Bulletin of Environmental Contamination and Toxicology, 108*(3), 585–593.

Kalwani, M., Chakdar, H., Srivastava, A., Pabbi, S., & Shukla, P. (2022). Effects of nanofertilizers on soil and plant-associated microbial communities: Emerging trends and perspectives. *Chemosphere, 287*, Article 132107.

Kekeli, M. A., Wang, Q., & Rui, Y. (2025). The role of nano-fertilizers in sustainable agriculture: Boosting crop yields and enhancing quality. *Plants, 14*(4), 554.

Kumar, R., Ashfaq, M., & Verma, N. (2018). Synthesis of novel PVA–starch formulation-supported Cu–Zn nanoparticle carrying carbon nanofibers as a nanofertilizer: Controlled release of micronutrients. *Journal of Materials Science, 53*(10), 7150–7164.

Kumar, Y., Tiwari, K. N., Singh, T., & Raliya, R. (2021). Nanofertilizers and their role in sustainable agriculture. *Annals of Plant and Soil Research, 23*(3), 238–255.

Kusumastuti, Y., Istiani, A., Rochmadi and Purnomo, C. W. (2019). 'Chitosan-based Polyion multilayer coating on NPK fertilizer as controlled released fertilizer', *Advances in Materials Science and Engineering*, 2019(1), p. 2958021.

Lakshani, N., Wijerathne, H. S., Sandaruwan, C., Kottegoda, N., & Karunarathne, V. (2023). Release kinetic models and release mechanisms of controlled-release and slow-release fertilizers. *ACS Agricultural Science & Technology, 3*(11), 939–956.

Leonardi, M., Caruso, G. M., Carroccio, S. C., Boninelli, S., Curcuruto, G., Zimbone, M., Allegra, M., Torrisi, B., Ferlito, F., & Miritello, M. (2021). Smart nanocomposites of chitosan/alginate nanoparticles loaded with copper oxide as alternative nanofertilizers. *Environmental Science: Nano, 8*(1), 174–187.

Li, M., Watanabe, S., Gao, F., & Dubos, C. (2023a). Iron nutrition in plants: Towards a new paradigm? *Plants, 12*(2), 384.

Li, S., & Chen, G. (2020). Contemporary strategies for enhancing nitrogen retention and mitigating nitrous oxide emission in agricultural soils: Present and future. *Environment, Development and Sustainability, 4*, 2703–2741.

Li, X., Lin, Y., Liu, M., Meng, L., & Li, C. (2023b). A review of research and application of polylactic acid composites. *Journal of Applied Polymer Science, 140*(7), Article e53477.

Liu, X., Nadeem, M., & Rui, Y. (2024). Effects of Nanofertilizers on the Mechanism of Photosynthetic Efficiency in Plants: A Review. *Phyton, 93*(12), 3197.

Maksimova, Y. G., Shchetko, V. A., & Maksimov, A. Y. (2023). Polymer hydrogels in agriculture. *Sel'skokhozyaistvennaya Biol, 58*, 23–42.

Manikandan, V., & Lee, N. Y. (2022). Green synthesis of carbon quantum dots and their environmental applications. *Environmental Research, 212*, Article 113283.

Manzoor, N., Ali, L., Al-Huqail, A.A., Alghanem, S.M.S., Al-Haithloul, H.A.S., Abbas, T., Chen, G., Huan, L., Liu, Y. and Wang, G. (2023) 'Comparative efficacy of silicon and iron oxide nanoparticles towards improving the plant growth and mitigating arsenic toxicity in wheat (Triticum aestivum L.)', *Ecotoxicology and environmental safety*, 264, p. 115382.

Marzouk, N. M., Soliman, M. S., El-Tanahy, A. M. M., & Mounir, A. M. (2022). Effect of both potassium phosphate and zinc nanocomposites prepared via gamma radiation on growth and productivity of potato under new reclaimed soils. *Egyptian Journal of Chemistry, 65*(10), 285–301.

Mir, A. R., Pichtel, J., & Hayat, S. (2021). Copper: Uptake, toxicity and tolerance in plants and management of Cu-contaminated soil. *BioMetals, 34*(4), 737–759.

Nagdalian, A., Askerova, A., Blinov, A., & Shariati, M. A. (2024). Evaluation of the toxicity of copper oxide nanoparticles toward pea seeds. *World Journal of Environmental Biosciences, 13*(3–2024), 23–30.

Nawadkar, A., Chaskar, J., Tiwari, A., & Chaskar, A. (2024). Controlled release of nanofertilizers: A promising approach for efficient delivery of nutrients in plants. *Journal of Crop Improvement, 38*(3), 189–218.

Nooeaid, P., Chuysinuan, P., Pitakdantham, W., Aryuwananon, D., Techasakul, S., & Dechtrirat, D. (2021). Eco-friendly polyvinyl alcohol/polylactic acid core/shell structured fibers as controlled-release fertilizers for sustainable agriculture. *Journal of Polymers and the Environment, 29*(2), 552–564.

Oladosu, Y., Rafii, M. Y., Arolu, F., Chukwu, S. C., Salisu, M. A., Fagbohun, I. K., Muftaudeen, T. K., Swaray, S., & Haliru, B. S. (2022). Superabsorbent polymer hydrogels for sustainable agriculture: A review. *Horticulturae, 8*(7), 605.

Owusu Adjei, M., Zhou, X., Mao, M., Xue, Y., Liu, J., Hu, H., Luo, J., Zhang, H., Yang, W., & Feng, L. (2021). Magnesium Oxide nanoparticle effect on the growth, development, and microRNAs expression of Ananas comosus var. bracteatus. *Journal of Plant Interactions, 16*(1), 247–257.

Pan, X., Nie, D., Guo, X., Xu, S., Zhang, D., Cao, F., & Guan, X. (2023). Effective control of the tomato wilt pathogen using TiO 2 nanoparticles as a green nanopesticide. *Environmental Science: Nano, 10*(5), 1441–1452.

Raiesi-Ardali, T., Ma'mani, L., Chorom, M., & Moezzi, A. (2022). Improved iron use efficiency in tomato using organically coated iron oxide nanoparticles as efficient bioavailable Fe sources. *Chemical and Biological Technologies in Agriculture, 9*(1), 59.

Rubel, R. I., & Wei, L. (2022). Biochar-based controlled release nitrogen fertilizer coated with polylactic acid. *Journal of Polymers and the Environment, 30*(10), 4406–4417.

Sahoo, D. R., & Biswal, T. (2021). Alginate and its application to tissue engineering. *SN Applied Sciences, 3*(1), 30.

Salabat, A., & Mirhoseini, F. (2022). Polymer-based nanocomposites fabricated by microemulsion method. *Polymer Composites, 43*(3), 1282–1294.

Salama, D. M., Abd El-Aziz, M. E., Shaaban, E. A., Osman, S. A., & Abd El-Wahed, M. S. (2022). 'The impact of nanofertilizer on agro-morphological criteria, yield, and genomic stability of common bean (Phaseolus vulgaris L.)', *Scientific Reports, 12*(1), p. 18552.

Salehipour, M., Rezaei, S., Mosafer, J., Pakdin-Parizi, Z., Motaharian, A., & Mogharabi-Manzari, M. (2021). Recent advances in polymer-coated iron oxide nanoparticles as magnetic resonance imaging contrast agents. *Journal of Nanoparticle Research, 23*(2), 48.

Salman, J. M., Grmasha, R. A., Stenger-Kovács, C., Lengyel, E., Al-Sareji, O. J., Al-Cheban, A. M. A. A., & Meiczinger, M. (2023). 'Influence of magnesium concentrations on the biomass and biochemical variations in the freshwater algae, Chlorella vulgaris', *Heliyon*, 9(1).

Shabbir, Z., Sardar, A., Shabbir, A., Abbas, G., Shamshad, S., Khalid, S., Murtaza, G., Dumat, C., & Shahid, M. (2020). Copper uptake, essentiality, toxicity, detoxification and risk assessment in soil-plant environment. *Chemosphere, 259*, Article 127436.

Shabiya, A., Bhat, M. A., Singh, P., Hassan, S., & Hussain, N. (2025). 'Nanofertilizers: Mechanisms, efficacy, and implications for sustainable agriculture', in *Agricultural Nutrient Pollution and Climate Change: Challenges and Opportunities*. Springer, pp. 311–353.

Sharma, S., Kumar, K., Thakur, N., Chauhan, S., & Chauhan, M. S. (2020). The effect of shape and size of ZnO nanoparticles on their antimicrobial and photocatalytic activities: A green approach. *Bulletin of Materials Science, 43*(1), 1–10.

Sheldon, R. A. (2018). Metrics of green chemistry and sustainability: Past, present, and future. *ACS Sustainable Chemistry & Engineering, 6*(1), 32–48.

Singh, A., Singh, N. áB, Afzal, S., Singh, T., & Hussain, I. (2018). 'Zinc oxide nanoparticles: a review of their biological synthesis, antimicrobial activity, uptake, translocation and biotransformation in plants', *Journal of materials science, 53*(1), pp. 185–201.

Skrzypczak, D., Mikula, K., Izydorczyk, G., Dawiec-Liśniewska, A., Moustakas, K., Chojnacka, K., & Witek-Krowiak, A. (2021). New directions for agricultural wastes valorization as hydrogel biocomposite fertilizers. *Journal of Environmental Management, 299*, Article 113480.

Steffan, J. J., Derby, J. A., & Brevik, E. C. (2020). Soil pathogens that may potentially cause pandemics, including severe acute respiratory syndrome (SARS) coronaviruses. *Current Opinion in Environmental Science & Health, 17*, 35–40.

Sun, Y., Zhu, G., Zhao, W., Jiang, Y., Wang, Q., Wang, Q., Rui, Y., Zhang, P., & Gao, L. (2022). Engineered nanomaterials for improving the nutritional quality of agricultural products: A review. *Nanomaterials, 12*(23), 4219.

Syed, M. A. (2015). 'Regional analysis, import-export and related issues on food security', in *Food Security and Risk Reduction in Bangladesh*. Springer, pp. 111–134.

Tariq, Z., Iqbal, D. N., Rizwan, M., Ahmad, M., Faheem, M., & Ahmed, M. (2023). Significance of biopolymer-based hydrogels and their applications in agriculture: A review in perspective of synthesis and their degree of swelling for water holding. *RSC Advances, 13*(35), 24731–24754.

Thakur, N., Chrungoo, S., Rana, S., Kaur, S., Kaur, S., & Pathak, A. (2021). Effect of copper nanoparticles on in-vitro seed germination of wheat (Triticum Aestivum L.) varieties. *International Journal of Pharmaceutical Sciences and Research, 12*(8), 4307–4313.

Thangavelu, R. M., da Silva, W. L., Zuverza-Mena, N., Dimkpa, C. O., & White, J. C. (2024). Nano-sized metal oxide fertilizers for sustainable agriculture: Balancing benefits, risks, and risk management strategies. *Nanoscale, 16*(43), 19998–20026.
Verma, K. K., Song, X.-P., Joshi, A., Tian, D.-D., Rajput, V. D., Singh, M., Arora, J., Minkina, T., & Li, Y.-R. (2022). Recent trends in nano-fertilizers for sustainable agriculture under climate change for global food security. *Nanomaterials, 12*(1), 173.
Wang, Y., Shaghaleh, H., Hamoud, Y. A., Zhang, S., Li, P., Xu, X., & Liu, H. (2021). Synthesis of a pH-responsive nano-cellulose/sodium alginate/MOFs hydrogel and its application in the regulation of water and N-fertilizer. *International Journal of Biological Macromolecules, 187*, 262–271.
Wei, X., Cao, P., Wang, G., Liu, Y., Song, J., & Han, J. (2021) 'CuO, ZnO, and γ-Fe2O3 nanoparticles modified the underground biomass and rhizosphere microbial community of Salvia miltiorrhiza (Bge.) after 165-day exposure', *Ecotoxicology and Environmental Safety, 217*, p. 112232.
Wu, J., Sha, C., Wang, M., Ye, C., Li, P., & Huang, S. (2021). Effect of organic fertilizer on soil bacteria in maize fields. *Land, 10*(3), 328.
Yadav, A., Yadav, K., & Abd-Elsalam, K. A. (2023). Nanofertilizers: Types, delivery and advantages in agricultural sustainability. *Agrochemicals, 2*(2), 296–336.
Yu, Y., Naik, S. S., Oh, Y., Theerthagiri, J., Lee, S. J., & Choi, M. Y. (2021). Lignin-mediated green synthesis of functionalized gold nanoparticles via pulsed laser technique for selective colorimetric detection of lead ions in aqueous media. *Journal of Hazardous Materials, 420*, Article 126585.
Zilabpour, Y., Barani, M., Rafieinia, M., Bidram, E., & Yahyaabadi, M. (2024). Investigating the effect of iron nano carbon dot fertilizer application on vegetative growth parameters of sorghum in a calcareous soil. *Agricultural Engineering, 46*(4), 343–370.
Zulfiqar, F., Navarro, M., Ashraf, M., Akram, N. A., & Munné-Bosch, S. (2019). Nanofertilizer use for sustainable agriculture: Advantages and limitations. *Plant Science, 289*, Article 110270.

Bioproduct-Derived Nanofertilizers for Soil Health and Sustainable Agriculture

Effa Aida Nordin, M. N. A. Uda, Z. A. Arsat, M. K. R. Hashim, Uda Hashim, M. N. Afnan Uda, Maisara A. M. Akhir, Yuri Pamungkas, Widyastuti, Subash C. B. Gopinath, M. Isa, and N. H. Ibrahim

Abstract Nanofertilizers have arisen as new tools in agriculture, which provide efficient nutrient delivery systems for improving plant growth and yield. Nevertheless, emerging safety concerns to plant physiology and environment demand systematic investigation. This review critically discusses the uptake, translocation and accumulation mechanisms of NP in plants highlighting their agronomic implications like nitrogen use efficiency (NUE) enhancement, stress amelioration and nutrient absorption pathways optimization. The review also describes possible negative impacts such as phytotoxicity, oxidative stress and photosynthetic disorder. There is an emphasis on the effect of soil microbiota and fertility, not only in respect to their interactions with NANOPTERNs but also nanofertilizer performance considerations including particle composition, size, concentration, and plant species sensitivity. Recent developments in risk assessment models and regulatory approaches have been examined to pinpoint current deficiencies and difficulties. This review ultimately

E. A. Nordin · M. N. A. Uda (✉) · Z. A. Arsat · M. K. R. Hashim
Faculty of Mechanical Engineering & Technology, Universiti Malaysia Perlis, Kampus Tetap Pauh Putra, Arau, Perlis, Malaysia
e-mail: nuraiman@unimap.edu.my

M. N. A. Uda · S. C. B. Gopinath · M. Isa
Institute of Nano Electronic Engineering, Universiti Malaysia Perlis, Kangar, Perlis, Malaysia

M. N. A. Uda
Centre of Excellence for Biomass Utilization, Universiti Malaysia Perlis, Arau, Perlis, Malaysia

U. Hashim · M. N. Afnan Uda
Faculty of Engineering, Universiti Malaysia Sabah, Kota Kinabalu, Sabah, Malaysia

M. A. M. Akhir · N. H. Ibrahim
Faculty of Chemical Engineering & Technology, Universiti Malaysia Perlis, Arau, Perlis, Malaysia

Y. Pamungkas
Faculty of Medicine & Health, Institut Teknologi Sepuluh Nopember, Surabaya, Indonesia

Widyastuti
Department of Material and Metallurgical Engineering, Institut Teknologi Sepuluh Nopember, Surabaya, Indonesia

A. Z. Yaser et al. (eds.), *Compost and Liquid Organic Fertilizers*,
https://doi.org/10.1007/978-981-95-8561-8_20

emphasizes the need for standardized safety protocols, sustainable formulation and application strategies, and continued multidisciplinary research to guarantee the safe and responsible mixing of nanofertilizers into sustainable agricultural systems.

1 Introduction

Growing interest has been shown in the use of nano-fertilizers as an innovative method to enhance nutrient efficiency and increase agricultural yields in recent years. These tiny, nano-scale fertilizer guides are designed to absorb nutrients like a sponge placed in a glass of water or to act like a corral to keep resources in one place, so the resource is not contaminating the outside environment and is not wasted by not sticking to the target. The widespread use of chemical fertilizers, an integral component of modern farming since the Green Revolution, has presented substantial ecological and environmental challenges. The substantial removal of nutrients from arable fields through leaching and gaseous emissions is one of the most challenging issues driving environmental destruction and climate change (Kumar & Yogendra, 2021). Nanoparticles offer exciting possibilities across various fields, including agriculture, horticulture, and even pharmaceuticals, but at the same time, they raise concerns. These substances provide unknown risks to individuals and the environment, requiring caution while using them (Devi et al., 2023).

The rising cost of chemical fertilizers, along with their negative environmental effects, like chemical runoff and pollution, has pushed farmers to look for safer options. Biofertilizers are now seen as a better alternative, as they're made from specific microbes that support plant growth without harming the environment (Fatima et al., 2021). However, agrochemicals have potential health and environmental impacts (like contamination of water resources and food residue) as well (Kah, 2015).

Excessive usage of nano-fertilizers can lead to the formation of nanoparticles in soil and water, thereby harming the environment and human health. One reason is that they are SO small and have such a huge surface area, making them super reactive and they can oxidize (stress) the cells. Moreover, the chronic effects of such particles on soil microbial community dynamics and ecosystem processes, such as nutrient mineralization and decomposition, remain unexplored (Borah et al., 2025).

2 Uptake and Accumulation of Nano-Fertilizers in Plants

2.1 Mechanisms of Absorption

Plants require several essential nutrients to survive and develop properly. These nutrients are needed in sufficient quantities and are available for plant uptake. The more minerals present in the soil profile, the more minerals will be transported to the absorption system. This type of root interception technique places a great stress on the growth of roots. However, some plant physiologists consider smoked gas root interception, for soil–plant exchange, a somewhat less envisioned flow process of the soil–plant exchange (or, the slow gas flow would have otherwise not been observed) (Alaoui et al., 2022). It has been proven that nutrient uptake and transport inside the root can occur through three distinct pathways. First, concentration gradients encourage nutrient movement through diffusion, while electrochemical gradients drive transport in general.

Second, nutrients that are dissolved in water flow into and through the root via mass transport (Hopmans & Bristow, 2002). Finally, root interception is defined as the uptake of minerals from the root surface by contact with soil colloids (Alaoui et al., 2022). Water and other nutrients flow rather freely through the plant's root apoplast, while water translocation in the symplast is a structural differentiation, with separate transport channels and mechanisms (Fig. 1).

The foliar applied plant nutrients are absorbed by the leaf through the cuticle and via the stomata, and the relative importance of the two mechanisms under such conditions is still debated. Solute penetration of leaf surfaces is a passive process caused by the concentration differential between the surface and the leaf core (Wang & Liu, 2007). Water-soluble nutrients may be taken up directly from the plant cuticle, particularly if the cuticle is thin, damaged to some extent, or inherently porous. It is also frequently observed on young leaves where the cuticle has not reached full maturity. Solubilized nutrients carried by dew, rain, or irrigation water can then be taken up here into the leaf cells (Hadrami, 2024).

The major purpose of xylem is to transfer water and minerals from the roots throughout the plant. It also serves as the principal structural component of wood, providing strength to plants and trees. As a channel comprised of all the sugars and

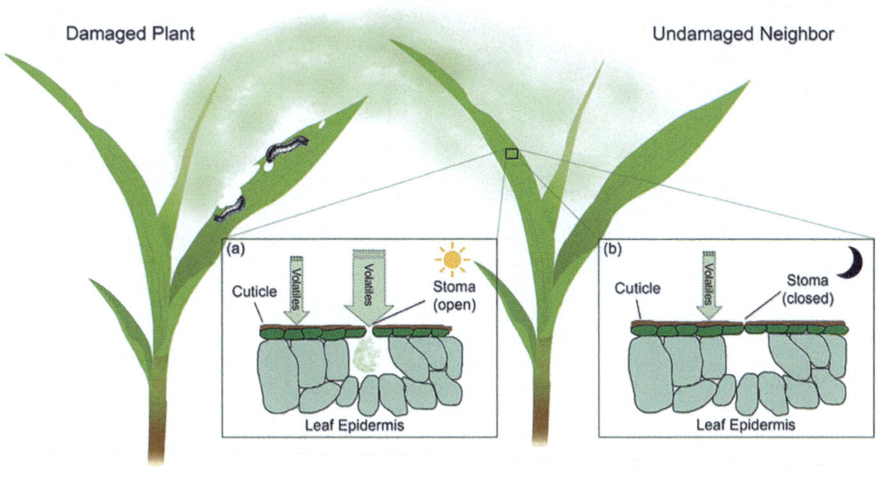

Fig. 1 Pathways of volatile entry into leaf tissues (Wang & Erb, 2022)

nutrients generated by the shoots from photosynthesis (which is transported from the shoots to other areas of the plant), phloem translocates all organs with the metabolism and energy essential to growth and function (Bieleski, 1966).

2.2 *Accumulation and Metabolism*

The role of nanoparticles in plant contamination and accumulation, particularly in seeds, has significant implications for food safety and human health. Soil NPs enter seeds based on size, type, surface charge, and root distribution. They are then distributed from the seed coat to the embryo and other sections of the plant (Malik et al., 2023). Plants can absorb NPs through root soaking, but the most common technique is foliar application. Plants absorb nanoparticles mostly through their leaves and roots (Tirkey et al., 2025). The distribution of Nanoparticles in plants varies according to their chemical composition and transport method, but can occur at the entry site or within the plant itself (Fig. 2).

To effectively use nanotechnology in agriculture, precise real-time monitoring of NP-plant interactions is crucial. This includes uptake, translocation, and accumulation, as well as potential consequences on plant growth and development (Tirkey et al., 2025). Such nano-fertilizers can also have an interference effect on a plant's life at the sub-cellular and cellular scale, when they get to interact with the metabolism process. Within the plant, after arrival, the effect of NPs at the cellular level depends

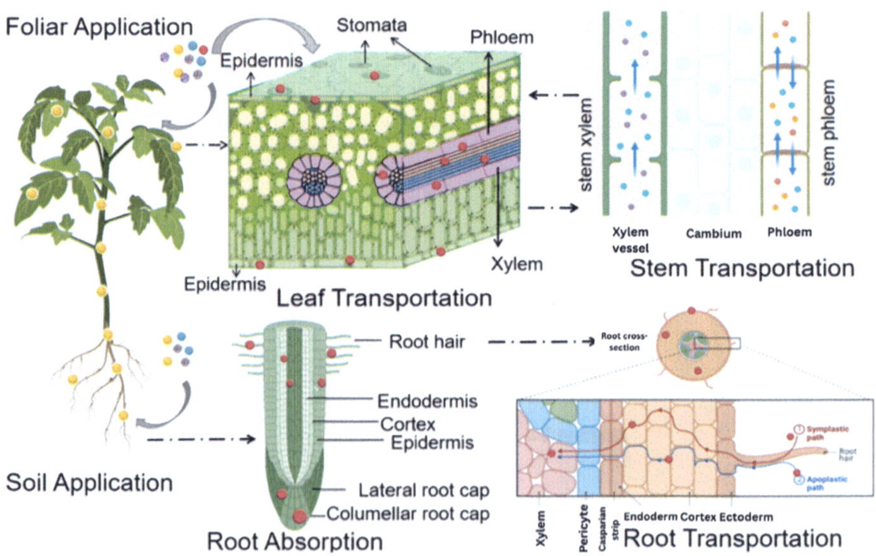

Fig. 2 A schematic diagram of the uptake and translocation of NPs in plants through foliar application or root exposure treatment (Wang et al., 2023)

on interactions with cell organelles, including mitochondria, chloroplasts, and the vacuole. Nanoparticles may also cause oxidative stress, which might impact cellular integrity and function, either positively or negatively, depending on the number and type of nanoparticles (Thavaseelan & Priyadarshana, 2021).

3 Positive Effects on Plant Health and Growth

3.1 Improved Nutrient Use Efficiency

Nanotechnology is the latest element in ensuring quantity, quality, environmental benignity, biological innocuity, and economic soundness in food production. Nano fertilizers can be classified into three types based on plant nutrient requirements: macro, micro, and nano-particulate. The nano-fertilizer might be a powder or liquid with particles smaller than 100 nm. They supply plants with plant-available nutrients, which can increase plant nutrient uptake and plant production. Previously, nanomembranes have been successfully applied for coating fertilizer particles to realize slow release for supplements. Slow-releasing nano-fertilizers to mitigate overuse of inorganic fertilizers. Due to its slow profile, it is referred to as a supplement in plant production and slow-release nano-fertilizers. As an alternative and these smart 3 ratio nano-fertilizers can be recommended, because they may be an effective source of water-soluble inorganic fertilizers. Fertilizer Particles have greater surface tension than conventional fertilizer because of the nano surface material and can control the release of nutrients. The fertilizer intensity is improved while the fertilizer amount is increased. Mannose and amino acids are frequently recognized for their role in nutritional absorption (El-Saadony et al., 2021).

3.2 Enhanced Growth and Stress Tolerance

Nanoparticles (NPs) showed substantial potential for boosting plant development and tolerance to a variety of abiotic stressors, including drought, salinity, and heat. These improvements are achieved through a variety of methods, including the management of reactive oxygen species (ROS), activation of antioxidant enzymes such as superoxide dismutase (SOD), catalase (CAT), and peroxidase (POD), and modification of stress-responsive genes. Zinc oxide (ZnO) and titanium dioxide (TiO_2) nanoparticles have been shown to reduce oxidative stress in plants by improving antioxidant defence systems. This protects plant cells from damage during stress conditions (Zaman et al., 2025). Nanoparticles also promote more nutrient intake, root development, and chlorophyll production, which combine to enhance photosynthesis and plant vigour (El-Saadony et al., 2022).

Drought stress severely limits crop growth and yield, which seriously affects food security worldwide. PGPR improves crop stress tolerance through many physiological and molecular mechanisms, including nitrogen fixation, phosphate solubilization, hormone and antioxidant synthesis, nutrient uptake, and transcript expression (Fig. 3).

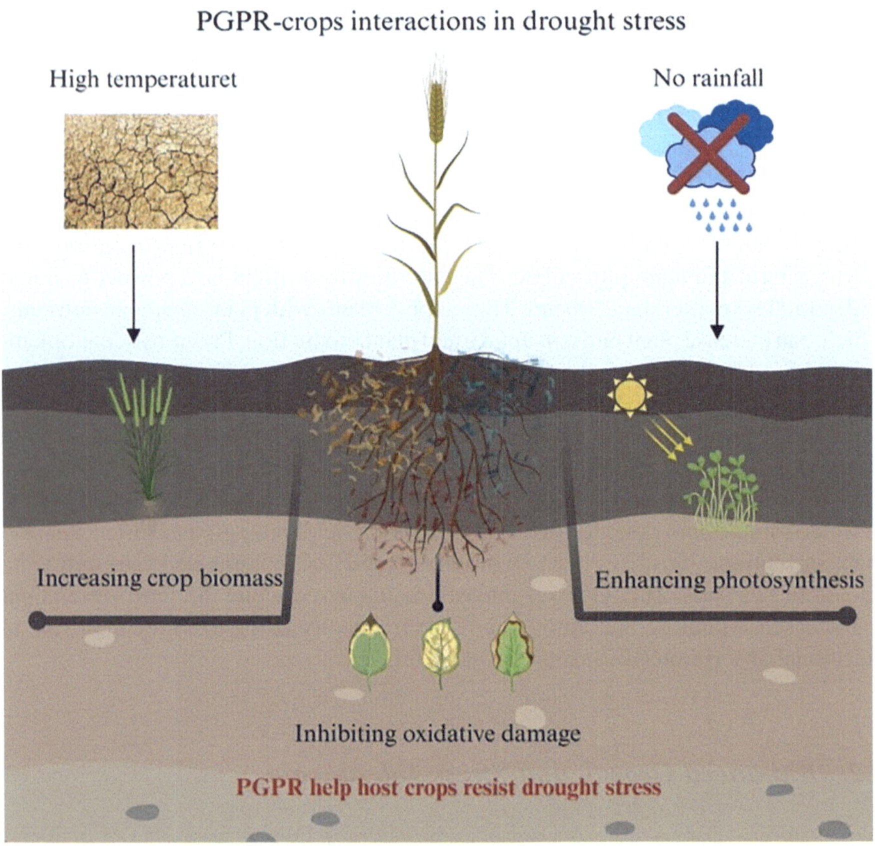

Fig. 3 Schematic presentation of PGPR-crops interaction impact on drought stress. PGPR might also be involved in increasing the drought response of plants because, under drought conditions with high temperature and depletion of rainfall, PGPR might have an apparent role in improving drought tolerance through promoting crop biomass, photosynthetic efficiency, and repressing oxidative damage (Liu et al., 2025)

4 Positive Effects on Plant Health and Growth

4.1 Case Studies and Examples

Nano-fertilizers, such as nano-Zn, nano-Fe, and nano-NPK, have been shown to improve crop health and growth, including wheat, maize, and rice. It has been demonstrated that these top-notch additions can improve yield, mass yield, as well as biomass accumulation of plants, implying that recent agriculture practices have great potential for these high-tech inputs (Table 1).

5 Potential Negative Impacts on Plants

5.1 Phytotoxicity Risks

Nanoparticles (NPs) can be beneficial for agriculture, yet also bring negative impacts on plant health at higher concentrations and certain physico-chemical properties. It is hazardous to continuously release zinc oxide nanoparticles (ZnO NPs) into the environment since this might cause a buildup of zinc in the soil, which can have a phytotoxic effect. Higher dosages of most NPs cause phytotoxicity and decreased agricultural productivity by altering the cellular structure, physiology, biochemistry, and molecular abnormalities of plants (Murali et al., 2022). The physical–chemical state of plants changed by a series of ultra-structural, morpho-physiological, and biochemical changes brought on by nanoparticle-bioreceptive bioaccumulation, including sedimentation, aggregation, and dissolution. Furthermore, it is clear from the literature that nanoparticles have two functions. For example, foliar treatment of modest concentrations of nanoparticles was found to improve general plant development characteristics. However, larger nanoparticle concentrations were found to cause phytotoxicity in many plant varieties (Gowtham et al., 2024).

5.2 Generation of Reactive Oxygen Species (ROS) and Oxidative Stress

Nanofertilizer exposure disrupts the redox equilibrium in plant cells, leading to reactive oxygen species (ROS) production, including hydrogen peroxide (H_2O_2), superoxide (O_2^-) and hydroxyl radicals (•OH). These molecules usually generate during photosynthesis and respiration however elevate significantly under nanoparticle stress leading to oxidative damage of cellular structures (Dai et al., 2024; Ghorbani et al., 2024). When accumulation of ROS exceeds the cell's antioxidant capacity, lipids, proteins and DNA become attacked by it which result in membrane damage,

Table 1 Recent case studies on nanofertilizer use in various cropping systems

Types of crop	Type of nanofertilizer & dosage concentration	Key findings	Highlights	Reference
Rice (Oryza sativa L.)	Calcium oxide nanoparticles (CaO NPs)—10 and 50 mg L^{-1} under 10 mg L^{-1} fluoride stress	Green-synthesized CaO NPs (20–25 nm, crystalline 46.7%) reduced oxidative stress (↓ MDA ≈ 42%), lowered fluoride uptake in roots (–65%) and shoots (–76%), and improved chlorophyll, sugar, and protein content	Calcium oxide nanofertilizer reduced fluoride toxicity in rice by decreasing oxidative stress, improving ion balance, and preventing fluoride transfer from roots to shoots	(Koley & Mondal, 2024)
Wheat (Triticum aestivum L.)	Nano-urea foliar spray (5%)	Elevated grain yield and nitrogen-use efficiency with reduced N losses and improved NUE	Reinforcement of nano-urea as an alternative to traditional urea for sustainable nitrogen management	(Vuswas; et al., 2024)
Savoy cabbage (Brassica oleracea L.)	Urea–hydroxyapatite (180:100:250 kg $N:P_2O_5:K_2O$ ha^{-1})	Half-rate nanofertilizer matched full NPK yield; improved P availability and soil fertility	Highlights hybrid nano-formulations that reduce nutrient inputs without yield loss	(González-Feijoo et al., 2024)
Tomato *(Solanum lycopersicum)*	ZnO NPs (100 ppm foliar)	Enhanced growth, yield (+200%), nutrient uptake, and fruit quality	Exhibits superior physiological performance and nutrient assimilation at 100 ppm ZnO-NP foliar treatment	(Ahmed et al., 2023a, 2023b)
Lettuce (Lactuca sativa L.)	Ca-based nanofertilizer (100 ppm foliar)	Synergistic improvement in nutrient uptake, photosynthetic efficiency, and tomato yield (+50%) compared to control	Demonstrates synergistic enhancement of nutrient uptake, photosynthetic efficiency, and yield using Ca + ZnO + TiO_2 nanoparticles (100 ppm each)	(Salama et al., 2024)

(continued)

Table 1 (continued)

Types of crop	Type of nanofertilizer & dosage concentration	Key findings	Highlights	Reference
Pak choi (Brassica rapa Subsp. Chinensis)	Zinc oxide nanoparticles (ZnO NPs)—100 mg/L under chromium stress	Improved spinach growth, chlorophyll, and protein levels; reduced oxidative stress and chromium accumulation in leaves	ZnO nanofertilizer helped spinach tolerate heavy metal stress and improved yield and plant health	(Guardiola-Márquez et al., 2024)
Rosemary (Rosmarinus officinalis L.)	Silver nanoparticles (200 mM L^{-1}) made from ginger extrac	Improved rosemary growth in polluted soil; increased phenolics like gallic acid, vanillic acid, and resveratrol; enhanced gene expression for secondary metabolites	AgNPs reduced wastewater stress and boosted antioxidants and plant quality	(Al Kashgry et al., 2024)
Lettuce (L. sativa L.)	Nano-zeolite ZSM-5–15 (25 mg/kg) + 50% NPK	Increased lettuce growth by about 13% but caused higher nitrous oxide (N_2O) emissions	Helped plant growth but not environmentally safe due to gas loss	(Chadwick et al., 2025)
Maize (Zea mays L.)	Nano-urea fertilizer (0.2–0.4%) with 75% NPK fertilizer	Improved maize growth and nutrient uptake compared to using only NPK. Increased yield and plant health	Nano-urea enhanced growth, saved fertilizer use, and supported sustainable farming	(Vasanth et al., 2025)
Tomato (Solanum lycopersicum L.)	Iron nanocomplex (Fe NC)—0%, 0.2%, 0.4% (w/v) with Atonik (0, 125, 250 ppm)	The result was obtained with 0.4% Fe NC and 250 ppm Atonik. This treatment improved plant growth, chlorophyll, protein, sugar, and nutrient uptake	The combination of Fe NC and Atonik enhanced tomato growth, photosynthesis, and nutrient efficiency. It acts as a sustainable nano-biofertilizer	(Tavallali & Darvishzadeh, 2025)

enzyme inactivation and photosynthetic reduction. (Chen et al., 2025a, 2025b). Heavy metal nanoparticles like ZnO, CuO, and Ag raise MDA content, lower chlorophyll concentration and inhibit carbon assimilation by disturbing electron transport chains in chloroplasts and mitochondria (Abdullah et al., 2025; Movahedi et al., 2024).

Plants prevent oxidative stress-damage by induction of antioxidant enzymes such as superoxide dismutase (SOD), catalase (CAT), peroxidase (POD) and ascorbate peroxidase (APX) that scavenge the surplus ROS (An et al., 2025; Sami et al., 2021). Non-enzymatic antioxidants including ascorbate, glutathione, carotenoids and flavonoids also add to redox balance and membrane stability. Application of nano-nutrients at low concentration being entrapped, has been proved to stimulate these defense mechanisms and developed resistance against drought, salt, temperature stress (Yadav, 2025). Leaf spray application of nano-silica and nano-zinc increases SOD and CAT activities in wheat and lettuce, stabilizing chlorophyll and photosynthesis under stress (Yadav, 2025).

Nanofertilizer seems to have a biphasic physiological role as inducers of stress and stimulator of defense response. If they are applied too much, ROS is produced and metabolic balance is disturbed, whereas an appropriate concentration can act as an antioxidant and increase the ability of plants to tolerate stress environments (Singh et al., 2024). Thus, elucidation of how NPs-mediated redox signaling occurs and development of accurate dosage regimen is crucial for the safe, efficient and sustainable application of nanofertilizers in the modern agriculture community.

5.3 *Disturbance in Cellular Processes*

Cellular processes are essential for maintaining life, and disruptions can lead to severe consequences for organisms. Two major disturbances include:

5.3.1 Disruption in Photosynthesis

Environmental stresses including high temperature, drought and salinity represent key abiotic stressors that exert additional pressure on the photosynthetic efficiency of plants. These stress generates accumulation of toxic free radicals and ROS which are involved in the degradation of thylakoid membranes (Panda et al., 2024), decreased efficiency and activity of PSII (Chen et al., 2025a, 2025b) or reduction in chlorophyll production (Nabi et al., 2025). High temperature enhances photoinhibition and enzyme denaturation (while drought decreases CO_2 assimilation by decreasing stomatal conductance and water-use efficiency (Lee et al., 2024). Similarly, saline stress causes ionic imbalance and osmotic stress resulting in carbon fixation capacity and photosynthetic pigment concentration reduction (Pacheco-Sangerman et al., 2024). Recent studies have shown that the detrimental effects could be lowered due to improved antioxidant enzyme activities (ROS scavenging), chloroplast ultrastructure stability, nutrient and water uptake, efficiency in light harvesting by the plants particularly in nano–ZnO–, nano–TiO_2– and nano-SiO 2 treated plant systems (Khan et al., 2025). These nanocatalysts act as catalytic micronutrients additives which bring redox homeostasis for membrane stability to retain photosynthetic machinery function under harsh ambient environment. Their potential to reduce oxidative stress and maintain chloroplast function provides an exciting future for the engineering of stress-tolerant crops (Semenova et al., 2024).

5.3.2 Effects on Cell Division and Growth

Cell growth and division have a variety of effects, such as changing the size and form of individual cells and the overall development of tissue and organisms. A cell's mechanical characteristics and spatial pattern depend on cell division, although its growth may be randomly regulated by its environment and/or genetic predispositions. Cell division in plants is necessary to maintain tissue organization, as it is not capable of reorganizing cells. This mechanism determines cell size and shape, and consequently, the morphography of the entire tissue (Robinson, 2021). Cell division modes significantly impact tissue growth dynamics. Random divisions increase growth heterogeneity, while divisions aligned with mechanical stress reduce variability, enhancing growth regulation (Alim et al., 2012) (Table 2).

6 Factors Affecting Nanofertilizer Safety

The safety of nano-fertilizers in agriculture is influenced by various factors, including the type of nanomaterials used, application dosage and frequency, and the specific plant species and their growth stages.

Table 2 Findings of the two major disturbances

Major disturbance	Findings	Reference
Disruption in photosynthesis	ER Stress: Induction of ER stress in Chlamydomonas reinhardtii results in reduced photosynthetic efficiency due to downregulation of photosynthesis-related proteins	(Chen et al., 2024)
	Nutrient Limitation: Nucleotide limitation impairs photosynthesis and alters gene expression, reducing energy states and increasing reactive oxygen species	(Bellin et al., 2023)
Effects on cell division and growth	Growth Impairment: Nucleotide limitation results in altered leaf morphology and reduced growth, indicating a direct link between photosynthetic capacity and growth	(Bellin et al., 2023)

6.1 Type of Nanomaterials Used

6.1.1 Metal-Based Nanoparticles

Metal-based nanoparticles (NPs) such as zinc oxide (ZnO), copper oxide (CuO), and iron oxide (Fe_3O_4) are commonly used in nanofertilizers due to their essential roles in plant nutrition. However, their safety profiles vary (Sohail et al., 2019).

1. **Zinc Oxide (ZnO) NPs**: The physical and chemical properties of zinc oxide nanoparticles (ZnO NPs) can be easily adjusted by modifying their shape or structure, which depends on the synthesis method, type of precursor, or raw materials used. One of the key advantages of ZnO nanoparticles in biomedical applications is their excellent surface characteristics, making them suitable for various treatments. Additionally, zinc oxide is well-known for its strong antimicrobial properties, making it an effective agent against a wide range of harmful microbes (Lakshmipriya & Gopinath, 2020).
2. **Copper Oxide (CuO) NPs**: CuO NPs have also been applied in agriculture as pesticides, herbicides, fertilisers, soil amendments, and growth regulators [6]. Positive as well as negative effects of CuO NPs are found on the plant. In contrast, cytotoxicity of CuO NPs at higher dosages has been observed. 1000 mg/L of CuO NPs decreased the photosynthesis rate (Chw), transpiration rate (T), and photosynthetic pigment in rice (Faraz et al., 2022).
3. **Iron Oxide (Fe_3O_4) NPs**: Iron-based nanoparticles (FeNPs) offer some unique benefits for plant development. They can influence key physiological processes like stomatal opening, seed germination, and early seedling growth through a range of complex mechanisms. These nanoparticles also appear to help plants cope with drought stress, likely by reducing the oxidative damage caused by limited water availability. In addition, plants treated with FeNPs often show increased chlorophyll production, which supports better photosynthesis. Beyond plant growth, FeNPs have also shown promise in cleaning up heavy metal contamination in both soil and water (Satya et al., 2024)

6.1.2 Carbon-Based Nanomaterials

Carbon-based nanomaterials such as CNTs have been shown to be promising for the electroanalysis of agrochemicals because of their high surface areas. Nevertheless, the direct utilization of bare carbon-based materials for agrochemicals transportation presents limitations, including low loading efficiency and failure to deliver agrochemicals to the site where they are needed promptly and smoothly. To overcome such obstacles, modification of the nanocarriers' surface has been recently introduced as a smart strategy to improve the properties of nanocarriers, which in turn convert them into intelligent cargo carriers (Hegde et al., 2024).

Table 3 Effects of low and high nanoparticle concentrations on plant growth and physiology

Low concentrations	High concentrations
Appropriate dosages can enhance nutrient availability, improve stress tolerance, and promote overall plant health. For instance, foliar application of TiO_2 NPs at 50 mg/L improved the nutritional quality of coriander without toxicity (Mandal4, 2023). Low concentrations can also improve root development and encourage microbial activity in the rhizosphere, further enhancing nutrient cycling	Excessive application can lead to nanoparticle aggregation on leaf surfaces, stomatal blockage, and oxidative stress, resulting in reduced photosynthesis and growth inhibition. For example, high concentrations of ZnO NPs have been linked to decreased pollen germination and tube elongation in lilies (Yoshihara et al., 2021). Such exposure can also disrupt hormone signaling pathways, leading to delayed flowering and abnormal plant development

6.2 Dosage and Frequency of Application

6.2.1 Low Versus High Concentration Effects

For a better understanding of the dual impact of nanoparticles on plants, it is necessary to examine their effects at low and high concentrations. Low doses can stimulate growth and improve stress tolerance, but high doses can be toxic and disrupt physiological processes. Table 3 summarises the varied results seen in plants subjected to various nanoparticle concentrations.

6.3 Plant Species and Growth Stage Sensitivity

6.3.1 Differential Uptake in Monocots versus Dicots

The translocation and distribution of nanofertilizers are quite different between monocots and dicots due to their contrasting root anatomical features as well as leaf morphology. Traits such as cuticle thickness, root type and vascular patterns determine the entry, transit and accumulation of nanoparticles in the plant tissues (Patyka et al., 2025).

Monocot plant materials, such as Oryza sativa (rice), exhibit thicker cuticles and fibrous roots with implications for NP uptake and translocation. These structural characteristics might explain why monocots are more suitable for some type of nanofertilizers (Wang et al., 2022). For instance, rice applied with nanoscale sulfur (NS) were observed to have better growth in the presence of arsenic stress and exhibited higher biomass production and yield than unapplied plants (Khepar et al., 2023). NS application also decreased As uptake and translocation, suggesting its use for enhancing production and food security in contaminated environment.

In comparison, dicot plants like lettuce (Lactuca sativa) and tomato (Solanum lycopersicum) are relatively more sensitive to nanoparticles due to their leaf-like

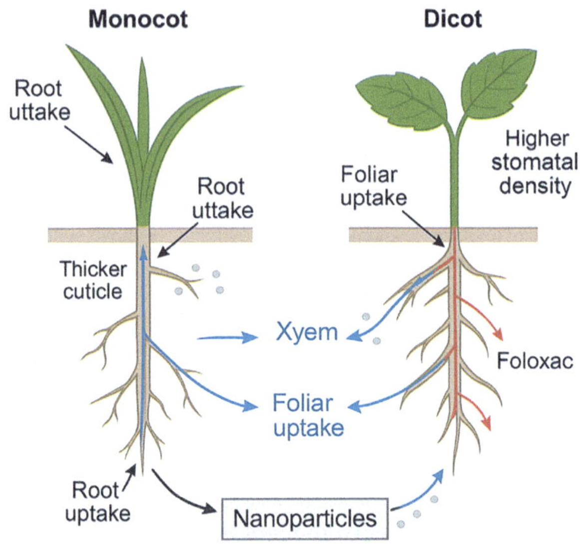

Fig. 4 Monocot and dicot absorbent is penetrated differently by nanoparticles. Blue arrows indicate nanoparticle uptake from roots and release into xylem in monocots (left) while red arrows are for entry through stomata and root followed by distribution to phloem in dicotyledons (right). The image indicates structural differences of fibrous root system and parallel venation in monocots and the taproot systems as well as reticulate venation in dicots which might be responsible for distribution and tolerance of NPs

structure harboring more stomata. Some nanomaterials, such as graphene oxide (GO), have also been shown to induce phytotoxicity (Chen et al., 2022). A study with tomato seedlings, cabbage plants and red spinach showed that, as graphene concentration increased, growth was reduced and reactive oxygen species (ROS) levels in cell were higher resulting in induced damage. It is interesting to note that lettuce was more tolerant under the same conditions, and thus not all dicots have the same response. From these findings, nanomaterials interaction with plants is intricate and species specific (Echeverría-Pérez et al., 2025; Karacan et al., 2025) (Fig. 4).

6.3.2 Sensitivity During Germination and Maturity

The developmental stage of plants influences their response to nanofertilizers. Several studies have demonstrated the beneficial effects of nanofertilizers on plant development, particularly during germination and maturity stages. During germination, nanozeolite and chitosan-based nanofertilizers have been shown to enhance nutrient absorption by increasing surface area and improving bioavailability. This leads to better seed vigor, root elongation, and early growth. A study published in International Journal of Molecular Sciences highlighted the potential of nanochitosan in sustainable agriculture, showing its role in promoting seedling development (Wang et al., 2024).Additionally, foliar applications of nanofertilizers have been found to improve reproductive success by enhancing pollen germination and tube elongation, as noted in a review on bionanofertilizers and crop development (Arora et al., 2024).

At the maturity stage, plants benefit from nanofertilizers through more efficient nutrient regulation and uptake. This is particularly evident in fruit-bearing crops like date palms, where foliar nano-fertilization has been shown to accelerate fruit ripening and improve biochemical responses (Shareef et al., 2021). Beyond plant growth, nanofertilizers also contribute to environmental sustainability by reducing nutrient leaching and enhancing microbial activity in the soil. These findings underscore the role of nano-based formulations in improving both crop productivity and soil health (Kekeli et al., 2025b).

7 Environmental and Soil Interactions

7.1 Soil Microbiota Impact

The presence of nanoparticles in agricultural soils affect not just the microbial quantity, but also the functional and biochemical integrity of the entire soil ecosystem. Metallic-NP such as Ag, CuO, ZnO can alter the diversity and activity of several important bacteria and fungi groups favoring some community shifts towards environmentally-stress-tolerant taxa like Actinobacteria or Bacillales (Afzal & Singh, 2022; Shahzadi et al., 2025). These changes are frequently associated with the decline of enzyme activities such as dehydrogenase, urease, phosphatase and related to C and N cycling (Pulikova et al., 2024). In addition, the exudation pattern of roots-induced responses can be changed due to nanoparticle exposure indirectly affecting beneficial symbionts including Rhizobium, Azospirillum and arbuscular mycorrhizal fungi which are known for mediating nutrient acquisition (Khudhur, 2025). At higher doses, certain nanoparticles inhibit organic-matter decomposition and interfere with nitrogen fixation, reducing soil fertility. However, some bio-and silica-based nanofertilizers have been shown to increase microbial respiration and the availability of nutrients when applied prudently at low dose (Nawaz et al., 2024). The results of this study together indicate that the ecological effects of nanoparticles, particularly at extremely low doses, reach far beyond microbial populations and further extend to enzyme regulation, nutrient transformation and soil–plant health in general.

7.2 Soil Health and Fertility

7.2.1 Effects on Soil Structure and Chemistry

Nanomaterials such as nano-clays and nano-silica have been shown to improve soil structure by promoting the formation of stable soil aggregates. This aggregation increases soil porosity and water-holding capacity, providing a conducive environment for root growth and microbial activity. Improved soil structure also enhances aeration, which is vital for root respiration and microbial processes (Kumar et al., 2025a, 2025b).

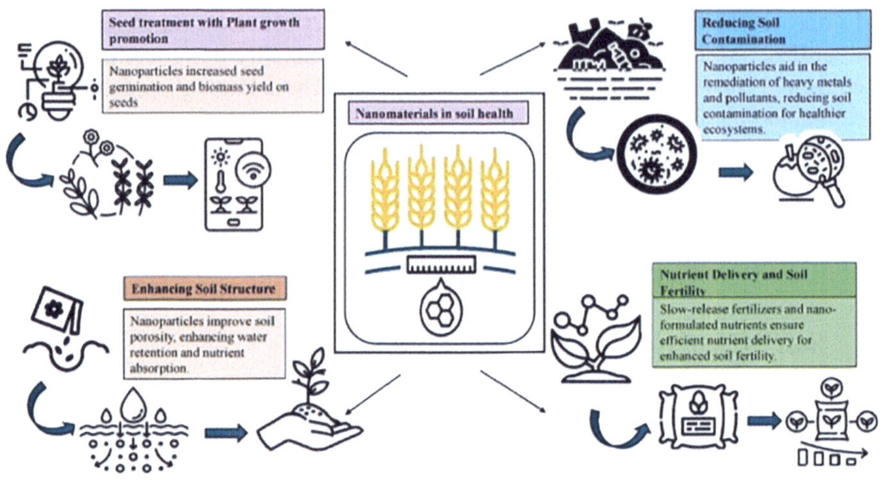

Fig. 5 Nanomaterial Interventions for Enhanced Soil Structure, Fertility, and Pollution Control (Kumar et al., 2025a, 2025b)

The application of nanofertilizers can alter soil pH and nutrient availability. For instance, nano-sulfur has been reported to lower soil pH more effectively than conventional sulfur due to its higher surface area and reactivity, leading to increased microbial oxidation to sulfuric acid. This acidification can enhance the availability of certain nutrients like phosphorus and micronutrients (Kekeli et al., 2025a).

Nanofertilizers can influence the dynamics of nutrient cycling in the soil. For example, the use of nano-phosphates has been found to enhance the bioavailability of phosphorus by increasing its solubility, which is otherwise often fixed in the soil and unavailable to plants. Similarly, nano-nitrogen fertilizers encapsulate nitrogen in nanoscale matrices, facilitating a slow release that matches plant uptake, thereby reducing losses due to volatilization and leaching (Kumar et al., 2025a, 2025b).

Nanofertilizers can influence the dynamics of nutrient cycling in the soil. For example, the use of nano-phosphates has been found to enhance the bioavailability of phosphorus by increasing its solubility, which is otherwise often fixed in the soil and unavailable to plants. Similarly, nano-nitrogen fertilizers encapsulate nitrogen in nanoscale matrices, facilitating a slow release that matches plant uptake, thereby reducing losses due to volatilization and leaching (Kumar et al., 2025a, 2025b) (Fig. 5).

7.2.2 Residual Nanoparticles and Soil Accumulation

As nanofertilizers continue to gain popularity in modern agriculture, it's important to consider the long-term impact of their residues on the soil environment. After application, nanoparticles don't just vanish—they can accumulate in the soil over time. These residual particles may eventually be absorbed by plant roots and transported to leaves, stems, and even fruits, raising concerns about their potential entry into the food chain. For example, a study by Wang et al. (2023) highlighted that metal nanoparticles like silver and gold could be absorbed and translocated within various plant tissues, which could pose health risks to consumers (Gao et al., 2023).

Beyond plants, residual nanoparticles can also affect the soil's living organisms. Beneficial microbes that support nutrient cycling and organic matter breakdown are particularly vulnerable. Some nanoparticles—especially silver nanoparticles known for their antimicrobial properties—can reduce microbial diversity and alter the balance of soil microbial communities. This was evident in a study by Siddique et al. (2023), which reported microbial disruption in soils treated with high concentrations of engineered nanomaterials (A. Ahmed et al., 2023a, 2023b).

In the long run, the accumulation of nanoparticles could change soil chemistry and reduce fertility. Persistent particles may interfere with nutrient availability, impact soil structure, or even contribute to toxic build-up, making soils less productive. As shown in another study on soil exposure to nanoparticles, continued application without careful management could lead to ecological imbalances and reduced agricultural sustainability.

8 Research Findings and Risk Assessments

The current part provides a comprehensive review on the recent scientific findings of nanofertilizer in terms of its fate, toxicity and environmental consequences. It has been restructured into two major parts to improve the clarity and critical depth: (i) Summary of Scientific Studies, which provides a metaanalysis of the current literature, and (ii) Risk Assessment Models.

8.1 Summary of Scientific Studies

Nanofertilizers have received extensive scientific investigation the last decade for their ability to enhance nutrient-use efficiency (NUE), plant growth and stress tolerance. Many experimental studies have reported that the nano–ZnO, nano–Fe_2O_3, and nano–SiO_2 based formulations boost crop productivity via controlled nutrient release, root uptake, and photosynthetic efficiency (Hudzovskyi et al., 2024; Moura et al., 2024; Wei et al., 2025). Nevertheless, the extent of these advantages depends strongly on particle size, surface functionalization, concentration and route of administration. Nanofertilizer application enhances antioxidant defense system and suppresses oxidative stress on plants at their suitable doses. In contrast, overexposed to the NFs could cause ROS accumulation, lipid peroxidation and chlorophyll degradation might be inferred as potential phytotoxicity (Pudhuvai et al., 2025).

Recent studies also show that soil microbes and nutrient cycling may be influenced by exposures to metal- and silica-based nanoparticles. Moderate bio-based nanomaterials improve soil health by promoting beneficial microorganisms and their associated enzymes in organic fertilization processes. In contrast, presence of a higher level of silver (AgNPs) or copper oxide (CuO NPs) nanoparticles restrict the nitrogen fixing species as well as hinder uniform enzymes such as urease and dehydrogenase activities (Pulikova et al., 2024). Such disturbances can decrease nutrient levels and degrade soil productivity over time, suggesting the importance of balancing agronomic incentives against ecological safety through common frameworks of risk assessment and control by government regulation.

8.1.1 Meta-Analysis of Current Research

Multiple meta-analyses agree that enhanced-efficiency fertilizers (EEFs) deliver both agronomic and environmental advantages. On average, physically coated formulations or products containing urease and/or nitrification inhibitors provide modest but significant yield gains, with the size of the response influenced by site factors such as soil pH. Second-order syntheses report that EEFs lower nitrate leaching by 17–58%, cut ammonia volatilization by 50–74%, and reduce nitrous oxide emissions by 28–49%. Pollution swapping, where nitrification inhibitors alone suppress N_2O emissions and nitrate losses but raise ammonia emissions, can be avoided by combining urease and nitrification inhibitors in the same product. These mitigation effects appear consistent across inhibitor chemistries and fertilizer sources.

Despite these documented benefits, controlled-release and stabilized formulations still account for less than 10% of the global fertilizer market, although demand for specialty products grew by about 5% in 2023. Wider adoption will therefore require continued agronomic validation, clear economic incentives for growers, and supportive policy frameworks (Maaz et al., 2025).Cost remains one of the primary barriers to the widespread adoption of enhanced-efficiency fertilizers (EEFs). Prices for EEFs can range from 30% higher to up to ten times the cost of conventional NPK fertilizers, making them less accessible to many farmers. Another concern is the environmental impact of certain formulations particularly controlled-release fertilizers that use non-biodegradable polymer coatings, which may contribute to long-term plastic pollution in agricultural soils.

8.2 Risk Assessment Models

Evaluation of the risk associated with nanomaterials must encompass traditional toxicological concepts, but also include particle size, surface reactivity and environmental fate issues specific to the nanoscale (More et al., 2021). The traditional Hazard Identification Dose Response Exposure Risk Characterization (HIDRRC) scheme still stands, with each level of the paradigm to be tailored according to well-established nano-property (Pathak et al., 2025).

Adverse effects such as oxidative stress, genotoxicity and ecotoxicity are included in the assessment of hazard identification involving in vitro and in vivo bioassays. The dose response analysis should be based not only on mass but also particle number, surface area and dissolution kinetics, because they strongly affect nanoparticle–cell interactions (Chen et al., 2023). Exposure characterization goes beyond occupational exposure and also considers environmental and dietary pathways recognising that nanoparticles retention, transformation, aggregation or disaggregation can occur in other matrices. Risk characterisation then combines hazard and exposure information by use of quantitative or probabilistic methods, such as nanospecific risk quotient (NRQ) and bayesian network models to estimate the likelihood and magnitude of adverse effects (Abidin et al., 2020; Dhall et al., 2024).

Other approaches, such as those employed in the Nano Risk Assessment (NanoRA) tool and the GUIDEnano decision-support system were recently supported by European projects under nanosafety programs to enable predictive computational modeling coupled with experimental testing for predicting nanoparticles exposure and environmental fate (Cazzagon et al., 2025). The NanoRA framework is based on a semi-quantitative scoring system that enables ranking of manufactured nanomaterial risks based on given values, while GUIDEnano uses the lifecycle-based approach to suggest prioritized risk managements during production and use and disposal (Isigonis et al., 2025). These include life-cycle assessment (LCA), connecting nanoparticle manufacture, use and end-of-life stages to identify possible intoxication hot spots. Taken together, these nano-modified risk-assessment models offer a scientifically sound basis for the safe and sustainable use of nanofertilizers in present-day farming (Moura et al., 2024).

9 Conclusion

This study has highlighted the promising role of bioproduct-derived nano fertilizers in supporting more sustainable and efficient agricultural practices. The review of scientific studies shows that nano fertilizers, when applied at appropriate concentrations, can improve nutrient use efficiency, enhance plant growth, and contribute positively to soil health. Specific findings demonstrate improved chlorophyll content, root development, and stress tolerance across various plant species. However, it was also observed that excessive or prolonged exposure to nanoparticles may lead to phytotoxic effects and negatively impact beneficial soil microorganisms. These outcomes underscore the importance of applying nano fertilizers carefully, considering factors such as particle type, dosage, plant species, and soil characteristics. Furthermore, while some countries like India have established regulatory frameworks, the global standardization of safety testing and environmental risk assessments is still limited. Ethical concerns, including public awareness and transparent labeling, must also be addressed to support broader acceptance. In conclusion, nano fertilizers derived from bioproducts hold strong potential for advancing sustainable agriculture. Still, their success depends on responsible use, continued research, and the development of clear guidelines that prioritize both productivity and environmental safety.

References

Abdullah, M., Sana, A., Bilal, Z., Amir, A., Muhammad, N., Hameed, A., & Ali, I. (2025). Effect of different levels of copper oxide (CuO) nanoparticles and biochar on soil ' s microbial activities and maize (Zea Mays L.) Growth. *Journal of Soil Science and Plant Nutrition*. https://doi.org/10.1007/s42729-025-02618-2

Abidin, M. Z., Rusli, R., & Halil, N. (2020). Conceptual framework—hazard assessment of nanomaterials using bayesian network. *Pertanika Journal of Science & Technology. 28*((S1)), 1–13.

Afzal, S., & Singh, N. K. (2022). Effect of zinc and iron oxide nanoparticles on plant physiology, seed quality and microbial community structure in a rice-soil-microbial ecosystem ☆. *Environmental Pollution, 314*(July), Article 120224. https://doi.org/10.1016/j.envpol.2022.120224

Ahmed, A., He, P., He, P., Wu, Y., He, Y., & Munir, S. (2023). Environmental effect of agriculture-related manufactured nano-objects on soil microbial communities. In *Environment International* (Vol. 173). Elsevier Ltd. https://doi.org/10.1016/j.envint.2023.107819.

Ahmed, R., Uddin, M. K., Quddus, M. A., Samad, M. Y. A., Hossain, M. A. M., & Haque, A. N. A. (2023). Impact of foliar application of zinc and zinc oxide nanoparticles on growth, yield, nutrient uptake and quality of tomato. *Horticulturae, 9*(2). https://doi.org/10.3390/horticulturae9020162.

Al Kashgry, N. A. T., Darwish, H., Aljomiha, N. A., Alharthi, S., Alayafi, A. A. M., Fallatah, A. M., El-Deeb, B. A., Abd El-Gawad, H. G., Hewidy, M., Al-Harbi, N. A., Al-Qahtani, S. M., Alotaibi, M. O., Algotiml, R. A., & Noureldeen, A. (2024). Silver nanoparticles alleviate the impact of soil contamination and wastewater irrigation on rosemary plants: modulating of gene expression and secondary metabolites. *Materials Research Express, 11*(6). https://doi.org/10.1088/2053-1591/ad5788.

Alaoui, I., Ghadraoui, O. El, Serbouti, S., Ahmed, H., Mansouri, I., Kamari, F. El, Taroq, A., Ousaaid, D., Squalli, W., & Farah, A. (2022). The mechanisms of absorption and nutrients transport in plants: a review. In *Tropical Journal of Natural Product Research* (Vol. 6, pp. 8–14). Faculty of Pharmacy, University of Benin. https://doi.org/10.26538/tjnpr/v6i1.2.

Alim, K., Hamant, O., & Boudaoud, A. (2012). Regulatory role of cell division rules on tissue growth heterogeneity. *Frontiers in Plant Science, 3*. https://doi.org/10.3389/fpls.2012.00174.

An, W., Wang, G., & Dou, J. (2025). Protective mechanisms of exogenous melatonin on chlorophyll metabolism and photosynthesis in tomato seedlings under heat stress. *Frontiers in Plant Science, February*, 1–19. https://doi.org/10.3389/fpls.2025.1519950.

Arora, P. K., Tripathi, S., Omar, R. A., Chauhan, P., Sinhal, V. K., Singh, A., Srivastava, A., Garg, S. K., & Singh, V. P. (2024). Next-generation fertilizers: the impact of bionanofertilizers on sustainable agriculture. In *Microbial Cell Factories* (Vol. 23). BioMed Central Ltd. https://doi.org/10.1186/s12934-024-02528-5.

Bellin, L., Melzer, M., Hilo, A., Amaya, D. L. G., Keller, I., Meurer, J., & Möhlmann, T. (2023). Nucleotide limitation results in impaired photosynthesis, reduced growth and seed yield together with massively altered gene expression. *Plant and Cell Physiology, 64*, 1466–1481. https://doi.org/10.1093/pcp/pcad063

Bieleski, R. L. (1966). Sites of accumulation in excised phloem and vascular tissues. *Plant Physiology, 41*, 455–466. https://doi.org/10.1104/pp.41.3.455

Borah, K. K., Goswami, Y., & Chakrabartty, I. (2025). Toxicity of nanofertilizers due to overuse. In *Nanofertilizers for Sustainable Agriculture* (pp. 31–56). Springer Nature Switzerland. https://doi.org/10.1007/978-3-031-78649-5_2.

Cazzagon, V., Vanhauten, R., Hanlon, J., Jiménez, A. S., Harrison, S., Auffan, M., Braakhuis, H., Boyles, M., Candalija, A., Katsumiti, A., Rodriguez-llopis, I., Catalan, J., Cross, R. K., Lahive, E., Morel, E., Simeone, F. C., Delpivo, C., Clavaguera, S., Seddon, R., & Traas, L. (2025). Environmental science nano the SAbyNA platform : A guidance tool to support industry in the implementation of safe- and processes and nano-enabled products †. *Environmental Science Nano, 12*, 4008–4025. https://doi.org/10.1039/d5en00312a

Chadwick, J. J., Mikulska, I., Radu, A., Chakraborty, S., Zhang, P., Ullah, S., & Lynch, I. (2025). Soil-based zeolite and metal oxide nanomaterial application alters reactive nitrogen losses and lettuce (Lactuca sativa L.) growth. *Environmental Science: Nano, x*, 4966–4979. https://doi.org/10.1039/d5en00526d.

Chen, Q., Riviere, J. E., & Lin, Z. (2023). Toxicokinetics, dose-response and risk assessment of nanomaterials: methodology, challenges, and future perspectives. 1–42. https://doi.org/10.1002/wnan.1808.Toxicokinetics.

Chen, S., Li, S., Qian, S., Xing, J., Liao, J., & Guo, Z. (2024). Stress on the endoplasmic reticulum impairs the photosynthetic efficiency of chlamydomonas. *International Journal of Molecular Sciences, 25*. https://doi.org/10.3390/ijms252413304

Chen, X., Chen, B., Jiang, Y., Zhang, J., Liu, M., Yang, Q., & Liu, H. (2025a). Melatonin improves salt tolerance in tomato seedlings by enhancing photosystem II functionality and calvin cycle activity. *Plants, 14*(12), 1–19. https://doi.org/10.3390/plants14121785

Chen, Y. S., Tian, H. X., Rong, D. C., Wang, L., Chen, S., & Zeng, J. (2025). ROS homeostasis in cell fate, pathophysiology, and therapeutic interventions.

Chen, Z., Guo, Z., Niu, J., Xu, N., Sui, X., Kareem, H. A., Hassan, M. U., Yan, M., Zhang, Q., Wang, Z., Mi, F., Kang, J., Cui, J., & Wang, Q. (2022). Phytotoxic effect and molecular mechanism induced by graphene towards alfalfa (Medicago sativa L.) by integrating transcriptomic and metabolomics analysis. *Chemosphere, 290*(October 2021), 133368. https://doi.org/10.1016/j.chemosphere.2021.133368.

Dai, Y., Guo, Y., Tang, W., Chen, D., Xue, L., Chen, Y., & Guo, Y. (2024). Reactive oxygen species—scavenging nanomaterials for the prevention and treatment of age—related diseases. *Journal of Nanobiotechnology*, 1–42. https://doi.org/10.1186/s12951-024-02501-9

Devi, O., Ojha, N., Laishram, B., Dutta, S., & Kalita, P. (2023). Roles of nano-fertilizers in sustainable agriculture and biosafety. *Environment and Ecology, 41 (1B):*(July), 457–463.

Dhall, S., Nigam, A., Harshavardhan, M., & Mukherjee, A. (2024). Environmental Chemistry and Ecotoxicology A comprehensive overview of methods involved in nanomaterial production and waste disposal from research labs and industries and existing regulatory guidelines for handling engineered nanomaterials. *Environmental Chemistry and Ecotoxicology, 6*(March), 269–282. https://doi.org/10.1016/j.enceco.2024.06.002

Echeverría-Pérez, E. G., Cruz-López, V., Herrera-Rivera, R., Romellón-Cerino, M. J., Rosas-Diaz, J., & Cruz-Martínez, H. (2025). Recent developments of nanomaterials in crop growth and production: the case of the tomato (Solanum lycopersicum). *Agronomy, 15*(7), 1–15. https://doi.org/10.3390/agronomy15071716

El-Saadony, M. T., ALmoshadak, A. S., Shafi, M. E., Albaqami, N. M., Saad, A. M., El-Tahan, A. M., Desoky, E. S. M., Elnahal, A. S. M., Almakas, A., Abd El-Mageed, T. A., Taha, A. E., Elrys, A. S., & Helmy, A. M. (2021). Vital roles of sustainable nano-fertilizers in improving plant quality and quantity-an updated review. In *Saudi Journal of Biological Sciences. 28*, pp. 7349–7359. Elsevier B.V. https://doi.org/10.1016/j.sjbs.2021.08.032.

El-Saadony, M. T., Saad, A. M., Soliman, S. M., Salem, H. M., Desoky, E. S. M., Babalghith, A. O., El-Tahan, A. M., Ibrahim, O. M., Ebrahim, A. A. M., Abd El-Mageed, T. A., Elrys, A. S., Elbadawi, A. A., El-Tarabily, K. A., & AbuQamar, S. F. (2022). Role of nanoparticles in enhancing crop tolerance to abiotic stress: a comprehensive review. *Frontiers in Plant Science, 13*, 946717. https://doi.org/10.3389/FPLS.2022.946717/XML.

Faraz, A., Faizan, M., Hayat, S., & Alam, P. (2022). Foliar application of copper oxide nanoparticles increases the photosynthetic efficiency and antioxidant activity in brassica juncea. *Journal of Food Quality*. https://doi.org/10.1155/2022/5535100.

Fatima, F., Hashim, A., & Anees, S. (2021). Efficacy of nanoparticles as nanofertilizer production: A review. *Environmental Science and Pollution Research, 28*(2), 1292–1303. https://doi.org/10.1007/S11356-020-11218-9/TABLES/4

Gao, M., Chang, J., Wang, Z., Zhang, H., & Wang, T. (2023). Advances in transport and toxicity of nanoparticles in plants. In *Journal of Nanobiotechnology* (Vol. 21). BioMed Central Ltd. https://doi.org/10.1186/s12951-023-01830-5.

Ghorbani, A., Emamverdian, A., Pehlivan, N., Zargar, M., & Razavi, S. M. (2024). Nano-enabled agrochemicals : mitigating heavy metal toxicity and enhancing crop adaptability for sustainable crop production. *1*, 1–37

González-Feijoo, R., Martinez-Castillo, C., Rodríguez-Seijo, A., Pérez-Rodríguez, P., & Arenas-Lago, D. (2024). Nanoagrochemicals versus conventional fertilizers: a field case study with tailor-made nanofertilizers for sustainable crop efficiency of brassica oleracea L. convar. Capitata var. Sabauda. *Agronomy, 14*(9). https://doi.org/10.3390/agronomy14091885.

Gowtham, H. G., Shilpa, N., Singh, S. B., Aiyaz, M., Abhilash, M. R., Nataraj, K., Amruthesh, K. N., Ansari, M. A., Alomary, M. N., & Murali, M. (2024). Toxicological effects of nanoparticles in plants: Mechanisms involved at morphological, physiological, biochemical and molecular levels. In *Plant Physiology and Biochemistry* (Vol. 210). Elsevier Masson s.r.l. https://doi.org/10.1016/j.plaphy.2024.108604.

Guardiola-Márquez, C. E., Martínez-Ballesta, M. del C., García-Sánchez, C. V., Bojorquez-Rodríguez, E. M., Jacobo-Velázquez, D. A., & Martínez-Hernández, G. B. (2024). Effects of foliar and root application of Zn and Fe bio-nanofertilizers on the glucosinolate, Zn and Fe contents of Pak choi (Brassica rapa Subsp. Chinensis) grown under hydroponic and pot cultivation. *Scientia Horticulturae, 338*(September). https://doi.org/10.1016/j.scienta.2024.113654.

Hadrami, D. A. El. (2024, May). *Uptake of nutrients through the stomata and through the cuticle—OMEX*. https://Omexcanada.Com/Blog/Uptake-of-Nutrients-through-the-Stomata-and-through-the-Cuticle/.

Hegde, V., Bhat, M. P., Lee, J. H., Kurkuri, M. D., Kim, C. S., & Lee, K. H. (2024). Carbon-based nanomaterials: Multifaceted role in agrochemical recognition, remediation, and release. In *Nano Today* (Vol. 57). Elsevier B.V. https://doi.org/10.1016/j.nantod.2024.102388

Hopmans, J. W., & Bristow, K. L. (2002). Current capabilities and future needs of root water and nutrient uptake modeling. In *Advances in Agronomy. 77*, pp. 103–183. Academic Press Inc. https://doi.org/10.1016/s0065-2113(02)77014-4.

Hudzovskyi, A., Chemist, A., Demianenko, I., Levtun, I., & Lecturer, S. (2024). The impact of metal oxide-based nanofertilisers on the physicochemical properties of agricultural plants. *BIOLOGICAL SYSTEMS: Theory and Innovation, 15*(3), 28–40. https://doi.org/10.31548/biologiya/3.2024.28.

Isigonis, P., Bouman, E. A., & Varsou, D. (2025). A Nano Risk Governance Portal supporting risk governance of nanomaterials and nano-enabled products. *Computational and Structural Biotechnology Journal, 29*(June), 177–185.

Kah, M. (2015). Nanopesticides and nanofertilizers: Emerging contaminants or opportunities for risk mitigation? *Frontiers in Chemistry, 3*. https://doi.org/10.3389/fchem.2015.00064.

Karacan, H., Talip, K., & Suat, C. (2025). *morphological, enzyme activities, and nutrient uptake of*.

Kekeli, M. A., Wang, Q., & Rui, Y. (2025a). The role of nano-fertilizers in sustainable agriculture: boosting crop yields and enhancing quality. *Plants (Basel, Switzerland), 14*. https://doi.org/10.3390/plants14040554.

Kekeli, M. A., Wang, Q., & Rui, Y. (2025b). The role of nano-fertilizers in sustainable agriculture: boosting crop yields and enhancing quality. In *Plants* (Vol. 14). *Multidisciplinary Digital Publishing Institute (MDPI)*. https://doi.org/10.3390/plants14040554.

Khan, A., Anas, M., Ullah, N., Khattak, W. A., Ali, M., Maaza, M., & Shinwari, Z. K. (2025). Harnessing green nanotechnology: exploring the potential of silver, copper, and their nanocomposites for biomedical and agricultural applications. *Pakistan Journal of Botany, 57*(6), 2197–2220. https://doi.org/10.30848/PJB2025-6(5).

Khepar, V., Ahuja, R., Sidhu, A., & Samota, M. K. (2023). Nano-sulfides of Fe and Mn efficiently augmented the growth, antioxidant defense system, and metal assimilation in rice seedlings. *ACS Omega, 8*(33), 30231–30238. https://doi.org/10.1021/acsomega.3c03012

Khudhur, N. S. (2025). *Passer Journal of Basic and Applied Sciences Detrimental Effects of Metal Nanoparticles on Microorganisms and Biochemical Activities through Soil Quality Index Analysis. 7*(1), 304–317. https://doi.org/10.24271/PSR.2025.449298.1543.

Koley, R., & Mondal, N. K. (2024). Synthesis of calcium-based nanofertilizer and its efficacy towards reduction of oxidative stress and fluoride uptake in rice (Oryza sativa L.). *Plant Nano Biology, 9*(July), 100087. https://doi.org/10.1016/j.plana.2024.100087.

Kumar, N., Kumar, P., & Dubey, R. C. (2025). Nanomaterials in soil science for agricultural productivity and environmental sustainability. In *Discover Environment* (Vol. 3). Springer Nature. https://doi.org/10.1007/s44274-025-00210-0.

Kumar, S., Sharma, Y., Khandelwal, V., Rawat, K., & Patil, A. (2025). Applications of nanotechnology in fertilizers: A review study. In *Sustainable Chemistry for the Environment* (Vol. 10). Elsevier B.V. https://doi.org/10.1016/j.scenv.2025.100247.

Kumar, & Yogendra. (2021). Nanofertilizers and their role in sustainable agriculture. *Annals of Plant and Soil Research, 23*(3), 238–255. https://doi.org/10.47815/apsr.2021.10067.

Lakshmipriya, T., & Gopinath, S. C. B. (2020). Introduction to nanoparticles and analytical devices. In *Nanoparticles in Analytical and Medical Devices* (pp. 1–29). Elsevier. https://doi.org/10.1016/B978-0-12-821163-2.00001-7.

Lee, Z., Lim, J. A., Harikrishna, J. A., Islam, T., Abd Rahim, M. H., & Yaacob, J. S. (2024). Regulation of plant responses to temperature stress: a key factor in food security and for mitigating effects of climate change. *International Journal of Plant Production, 18*(2), 141–159. https://doi.org/10.1007/s42106-024-00282-7

Liu, K., Deng, F., Zeng, F., Chen, Z. H., Qin, Y., & Chen, G. (2025). Plant growth-promoting rhizobacteria improve drought tolerance of crops: a review. In *Plant Growth Regulation*. Springer Science and Business Media B.V. https://doi.org/10.1007/s10725-025-01300-y.

Maaz, T. M., Dobermann, A., Lyons, S. E., & Thomson, A. M. (2025). Review of research and innovation on novel fertilizers for crop nutrition. *Npj Sustainable Agriculture, 3*, 25. https://doi.org/10.1038/s44264-025-00066-0

Malik, S., Muhammad, K., & Waheed, Y. (2023). Nanotechnology: a revolution in modern industry. *Molecules, 28*(2). https://doi.org/10.3390/molecules28020661

Mandal4, S. (2023, August). Improvement in nutritional quality of spices through potential use of titanium dioxide (TiO_2) nanoparticles application.pptx. https://www.Slideshare.Net/Slideshow/Improvement-in-Nutritional-Quality-of-Spices-through-Potential-Use-of-Titanium-Dioxide-Tio-Nanoparticles-Applicationpptx/259570060?Utm_source=chatgpt.Com.

More, S., Bampidis, V., Benford, D., Bragard, C., Halldorsson, T., Bennekou, S. H., Koutsoumanis, K., Machera, K., Naegeli, H., Nielsen, S., Schlatter, J., Schrenk, D., Silano, V., Turck, D., Younes, M., Castenmiller, J., Chaudhry, Q., Cubadda, F., Franz, R., Rincon, A. (2021). Guidance on risk assessment of nanomaterials to be applied in the food and feed chain : human and animal health. *19*(June). https://doi.org/10.2903/j.efsa.2021.6768

Moura, B. C., Rosero-romo, J. J., Monteiro, H., Alberto, A. R., Laranjeira, J., Martin-iglesias, S., Silvan, U., Lanceros-mendez, S., Salazar, D., & Martins, C. F. (2024). Addressing safety and sustainability issues in the development of nano-enabled MULTI-FUNctional materials for metal additive manufacturing. *Sustainable Materials and Technologies, 41*(July), Article e01085. https://doi.org/10.1016/j.susmat.2024.e01085

Movahedi, A., Taghimolaee, M., & Ghayebpanah, S. (2024). *A Review of the Effect of Nanoparticles on Reactive Oxygen Species Production in Different Plants., 1*(1), 131–163.

Murali, M., Gowtham, H. G., Singh, S. B., Shilpa, N., Aiyaz, M., Alomary, M. N., Alshamrani, M., Salawi, A., Almoshari, Y., Ansari, M. A., & Amruthesh, K. N. (2022). Fate, bioaccumulation and toxicity of engineered nanomaterials in plants: Current challenges and future prospects. In *Science of the total environment* (Vol. 811). Elsevier B.V. https://doi.org/10.1016/j.scitotenv.2021.152249

Nabi, F., Sarfaraz, A., Kama, R., Kanwal, R., & Li, H. (2025). Structure-based function of humic acid in abiotic stress alleviation in plants: a review. *Plants, 14*(13), 1–20. https://doi.org/10.3390/plants14131916

Nawaz, T., Gu, L., Fahad, S., Saud, S., Bleakley, B., & Zhou, R. (2024). *Exploring Sustainable Agriculture with Nitrogen-Fixing.*

Pacheco-Sangerman, F., Gómez-Merino, F. C., Peralta-Sánchez, M. G., & Trejo-Téllez, L. I. (2024). Sulfated nutrition modifies nutrient content and photosynthetic pigment concentration in cabbage under salt stress. *Plants, 13*(10). https://doi.org/10.3390/plants13101337.

Panda, S. K., Gupta, D., Patel, M., Vyver, C. Van Der, & Koyama, H. (2024). Functionality of reactive oxygen species (ROS) in plants: toxicity and control in poaceae crops exposed to abiotic stress. *Plants, 13*(15). https://doi.org/10.3390/plants13152071.

Pathak, G., Mangla, S., Kumar, G., Veer, G., Rajeev, B., & Kapoor, K. (2025). Toxicological assessment and risk management of nanoparticles mediated composite materials—critical review : State of the art. *Discover Polymers, 2*(12), 1–27.

Patyka, M. V., Khablak, S. H., Patyka, T. I., Bondareva, L. M., Dolia, M. M., Spychak, V. M., & Lykholat, Y. V. (2025). Evolution of immune mechanisms in monocots and dicots in response to microbial pathogens and abiotic stressors. *Biosystems Diversity, 33*(2), 1–8. https://doi.org/10.15421/012531

Pudhuvai, B., Koul, B., Das, R., & Shah, M. P. (2025). Nano-Fertilizers (NFs) for resurgence in nutrient use efficiency (NUE): a sustainable agricultural strategy. *Current Pollution Reports, 11*(1), 1–29. https://doi.org/10.1007/s40726-024-00331-9

Pulikova, E. P., Ivanov, F. D., Alliluev, I. A., Chernikova, N. P., Krepakova, M. R., Gorovtsov, A. V., Minkina, T. M., Bauer, T. V., Tsitsuashvili, V. S., Garg, M. C., Kumar, S., & Rajput, V. D. (2024). Effects of bulk forms and nanoparticles of zinc and copper oxides on the abundance, nitrogen cycling and enzymatic activities of microbial communities, morphometric parameters and antioxidant status of Hordeum vulgare L. *Environmental Geochemistry and Health, 46*(12), 1–17. https://doi.org/10.1007/s10653-024-02258-y

Robinson, S. (2021). Mechanobiology of cell division in plant growth. *New Phytologist, 231*, 559–564. https://doi.org/10.1111/nph.17369

Salama, D. M., Osman, S. A., Mahmoud, S. H., El-Tanahy, A. M. M., & Abd El-Aziz, M. E. (2024). Improving the productivity and physiological characteristics of lettuce plants using spraying calcium as a nanofertilizer. *Horticulturae, 10*(11). https://doi.org/10.3390/horticulturae10111157.

Sami, R., Elhakem, A., Alharbi, M., Benajiba, N., Almatrafi, M., Abdelazez, A., & Helal, M. (2021). Evaluation of antioxidant activities, oxidation enzymes, and quality of nano-coated button mushrooms (Agaricus Bisporus) during storage. *Coatings, 11*(149), 1–12.

Satya, S., Hashmi, K., Gupta, S., Mishra, P., Khan, T., & Joshi, S. (2024). The vital role of nanoparticles in enhancing plant growth and development. *Engineering Proceedings. 67,* Page 48, 67, 48. https://doi.org/10.3390/ENGPROC2024067048.

Semenova, N. A., Burmistrov, D. E., Shumeyko, S. A., & Gudkov, S. V. (2024). Fertilizers based on nanoparticles as sources of macro- and microelements for plant crop growth: a review. *Agronomy, 14*(8). https://doi.org/10.3390/agronomy14081646.

Shahzadi, S., Fatima, S., Shafiq, Z., Ramzan, M., & Ashraf, S. (2025). RSC Advances A review on green synthesis of silver nanoparticles (SNPs) using plant extracts: a multifaceted approach in photocatalysis, environmental. 3858–3903. https://doi.org/10.1039/d4ra07519f.

Shareef, H. J., Al-Yahyai, R. A., Omar, A. E. D. K., & Barus, W. A. (2021). Foliar nano-fertilization enhances fruit growth, maturity, and biochemical responses of date palm. *Canadian Journal of Plant Science, 101*, 299–306. https://doi.org/10.1139/cjps-2020-0081

Siddique, S. N., Deng, J., & Mohamedelhassan, E. (2023). The application of laponite nanoparticle to lessen the risks of liquefaction: An emerging technique for sand improvement (a review). *Soil Mechanics and Foundation Engineering, 60*(2), 149–157. https://doi.org/10.1007/s11204-023-09876-1

Singh, A., Sharma, A., Singh, O., Rajput, V. D., El-ramady, H. R., Zulfiqar, F., Kumar, R., Al-ghamdi, A. A., & Ghazaryan, K. (2024). Plant Stress In-depth exploration of nanoparticles for enhanced nutrient use efficiency and abiotic stresses management : Present insights and future horizons. *14*(August). https://doi.org/10.1016/j.stress.2024.100576.

Sohail, M. I., Waris, A. A., Ayub, M. A., Usman, M., Zia ur Rehman, M., Sabir, M., & Faiz, T. (2019). Environmental application of nanomaterials: A promise to sustainable future. In *Comprehensive Analytical Chemistry. 87*, pp. 1–54. Elsevier B.V. https://doi.org/10.1016/bs.coac.2019.10.002.

Tavallali, V., & Darvishzadeh, M. D. (2025). Synergistic effects of Fe nanocomplex and nitrophenolate-based biostimulant on growth and physiological performance of tomato seedlings. *BMC Plant Biology, 25*(1). https://doi.org/10.1186/s12870-025-06930-y

Thavaseelan, D., & Priyadarshana, G. (2021, January). *Nanofertilizer use for Sustainable Agriculture.* https://www.Researchgate.Net/Publication/348433424_Nanofertilizer_use_for_Sustainable_Agriculture.

Tirkey, A., Babu, P. J., HemaIswarya, S., & Singh, P. K. (2025). An updated view of nanoparticle interaction with plants: a molecular understanding and beyond. *Journal of Plant Growth Regulation.* https://doi.org/10.1007/s00344-024-11609-z

Vasanth, P., Sivakumar, K., Jeyajothi, R., & Chandrasekaran, P. (2025). Foliar spray of urea and nano-urea in yield and economics of maize (Zea mays L.). *Journal of Applied and Natural Science, 17*(1), 383–388. https://doi.org/10.31018/jans.v17i1.6178.

Vuswas;, R., Tanmayi;, C. S. N., Vuchula Krupa Bindu;, & Mallepu s likhitha; (2024). Yields, nutrient-use efficiencies and production economics of wheat (Triticum aestivum) as influenced by foliar nano urea. *Indian Journal of Agronomy, 69*(3), 269–276. https://doi.org/10.59797/ija.v69i3.5521.

Wang, C. J., & Liu, Z. Q. (2007). Foliar uptake of pesticides-Present status and future challenge. In *Pesticide Biochemistry and Physiology. 87*, pp. 1–8. https://doi.org/10.1016/j.pestbp.2006.04.004.

Wang, L., & Erb, M. (2022). Volatile uptake, transport, perception, and signaling shape a plant's nose. *Essays in Biochemistry, 66*, 695–702. https://doi.org/10.1042/EBC20210092

Wang, X., He, M., Wang, X., Liu, S., Luo, L., Zeng, Q., Wu, Y., Zeng, Y., Yang, Z., Sheng, G., Ren, P., Ouyang, H., & Jia, R. (2024). Emerging Nanochitosan for Sustainable Agriculture. In *International Journal of Molecular Sciences* (Vol. 25). Multidisciplinary Digital Publishing Institute (MDPI). https://doi.org/10.3390/ijms252212261.

Wang, X., Xie, H., Wang, P., & Yin, H. (2023). Nanoparticles in plants: Uptake, transport and physiological activity in leaf and root. *Materials, 16*(8), 3097. https://doi.org/10.3390/MA16083097.

Wang, Y., Dimkpa, C., Deng, C., Elmer, W. H., Gardea-Torresdey, J., & White, J. C. (2022). Impact of engineered nanomaterials on rice (Oryza sativa L.): A critical review of current knowledge. *Environmental Pollution, 297*(October 2021), 118738. https://doi.org/10.1016/j.envpol.2021.118738.

Wei, J., Liu, L., Wei, Z., Qin, Q., Bai, Q., Zhao, C., Zhang, S., & Wang, H. (2025). Silicon nano-fertilizer-enhanced soybean resilience and yield under drought stress. *Plants (Basel, Switzerland), 14*(751), 1–18.

Yadav, M. (2025). Plant Nano Biology Nanoparticle-facilitated targeted nutrient delivery in plants : Breakthroughs and mechanistic insights ☆. *Plant Nano Biology, 12*(April), Article 100156. https://doi.org/10.1016/j.plana.2025.100156

Yoshihara, S., Hirata, S., Yamamoto, K., Nakajima, Y., Kurahashi, K., & Tokumoto, H. (2021). ZnO nanoparticles effect on pollen grain germination and pollen tube elongation. *Plant Cell, Tissue and Organ Culture, 145*, 405–415. https://doi.org/10.1007/s11240-021-02017-2

Zaman, W., Ayaz, A., & Park, S. J. (2025). Nanomaterials in agriculture: a pathway to enhanced plant growth and abiotic stress resistance. *Plants. 14, Page 716*, *14*(5), 716. https://doi.org/10.3390/PLANTS14050716.

Bioproducts as a Renewable Source for Nanofertilizer Development

Ahmad Syahmi Faqihin, M. N. A. Uda, Khalisanni Khalid, Jasni Ismail, Meor Ahmad Faris, Z. A. Arsat, M. K. R. Hashim, Uda Hashim, M. N. Afnan Uda, Maisara A. M. Akhir, Yuri Pamungkas, Widyastuti, F. Abdullah, Muhammad Firdaus Abdul Muttalib, and N. H. Ibrahim

Abstract The rising demand for sustainable agricultural practices has intensified interest in the development of nanofertilizers derived from renewable sources. This paper explores the potential of bioproducts organic materials sourced from agricultural residues, microbial biomass, and plant-based polymers as eco-friendly and sustainable precursors for nanofertilizer production. Bioproducts not only serve as nutrient-rich substrates but also act as natural carriers and encapsulating agents, facilitating controlled nutrient release and improving plant nutrient uptake efficiency. Utilizing green synthesis techniques, these bioproducts offer an environmentally responsible alternative to conventional chemical fertilizers, reducing issues such as nutrient leaching, soil degradation, and water pollution. Despite challenges such

A. S. Faqihin · M. N. A. Uda (✉) · M. K. R. Hashim · F. Abdullah · M. F. A. Muttalib
Faculty of Mechanical Engineering & Technology, Universiti Malaysia Perlis, Arau, Perlis, Malaysia
e-mail: nuraiman@unimap.edu.my

M. N. A. Uda · J. Ismail · M. A. Faris · Z. A. Arsat
Institute of Nano Electronic Engineering, Universiti Malaysia Perlis, Kangar, Perlis, Malaysia

M. N. A. Uda
Centre of Excellence for Biomass Utilization, Universiti Malaysia Perlis, Arau, Perlis, Malaysia

K. Khalid
Agri-Nanotechnologi Program, Bioteknologi and Nanotechnology Research Centre, MARDI, Serdang, Selangor, Malaysia

U. Hashim · M. N. Afnan Uda · N. H. Ibrahim
Faculty of Engineering, Universiti Malaysia Sabah, Kota Kinabalu, Sabah, Malaysia

M. A. M. Akhir
Faculty of Chemical Engineering & Technology, Universiti Malaysia Perlis, Arau, Perlis, Malaysia

Y. Pamungkas
Faculty of Medicine and Health, Institut Teknologi Sepuluh Nopember, Surabaya, Indonesia

Widyastuti
Department of Material and Metallurgical Engineering, Institut Teknologi Sepuluh Nopember, Surabaya, Indonesia

A. Z. Yaser et al. (eds.), *Compost and Liquid Organic Fertilizers*,
https://doi.org/10.1007/978-981-95-8561-8_21

as variability in raw materials, scalability, and regulatory uncertainties, the integration of bioproduct-based nanofertilizers with smart farming systems and circular economy principles holds significant promise. This study highlights the role of bioproducts in revolutionizing nanofertilizer development and contributing to the future of sustainable and resilient agriculture.

Keywords Bioproducts · Nanofertilizers · Green synthesis · Soil health · Controlled nutrient release

1 Introduction

1.1 Definition of Bioproducts

Bioproducts are also known as bio-based products which are substances like materials, chemicals and energy that come from renewable biological sources. These sources include a wide range of organic feedstocks such as agricultural residues like crop stalks, rice husks, and wheat straw; forestry by-products such as wood chips, bark, and sawdust; and industrial biomass including food-processing waste, molasses, algal biomass, and fermentation residues. Alternative to that of petroleum-based materials, bioproducts use naturally occurring and regenerable resources to produce goods and therefore do not rely on fossil fuel derived products.

The best studied example is cellulose, the main structural part of the walls of plant cells. Cellulose-based materials can be engineered and modified to have specific chemical, mechanical as well as physical attributes to be applicable not only in biomedical applications (inclusions in wound dressings and drug delivery systems) but also in agricultural developments (nanofertilizers). The fact of its biodegradability, non-toxicity, and abundance make them especially attractive in green technologies. Green electrochemical sensors and biosensors offer benefits such as biodegradability, non-toxicity, sustainability, low-cost, and sensitive surfaces (Kalambate et al., 2020).

In nanofertilizer development, bioproducts are important as precursors or carriers for nutrient delivery. Nano fertilizers involve nutrients encapsulated or coated in nano-sized vehicles, normally derived from natural polymers such as starch, chitosan or alginate. Figure 1 shows cellulose-based materials. These biopolymers can be further processed into nanoparticles, nanogels or nano capsules so that they can be released in the controlled and targeted manner. This facilitates increased efficiency of nutrient uptake by the plants, the inability of nutrient loss to leach and volatize, and decreased frequency and quantity of fertilizer applied, all of which reduce environmental impact and input cost.

Furthermore, the organic nature of bioproducts ensures that they decompose in soil, enriching soil organic matter and supporting beneficial microbial activity. For example, biochar can persist in soils for up to 556 years and has a positive effect on

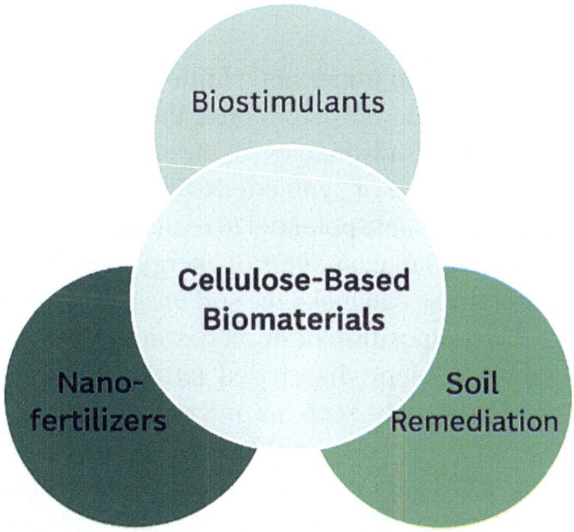

Fig. 1 The multifunctional applications of cellulose-derived biomaterials in sustainable agriculture. These materials could be used as universal bases to produce nanofertilizers, biostimulants and soil remediation agents. Their biodegradability, renewability and tunable surface chemistry allows the control of nutrient release in soils, improving soil health and crop productivity while reducing associated environmental impact

soil organic matter dynamics and C sequestration (Wang et al., 2016). This not only improves soil fertility in the long term but also aligns with sustainable agriculture practices that aim to protect ecosystem health.

The use of nanotechnology in bioproducts is highly promising in agronomic biofortification, where the objective is to boost the nutritional characteristics of crops. For example, the nanoparticles synthesized from plant or microbial based bioproducts can then deliver micronutrients like zinc (Zn) and iron (Fe) into the plants more effectively in case of wheat biofortification. This solves both productivity and nutrition issues and specifically for those with micro-nutrient deficiencies; what is referred to as "hidden hunger" in certain regions. Also, this strategy reduces the rate of environment related issues caused by the excessive use of traditional synthetic fertilizers, which almost always lead to soil deterioration and water contamination.

In summary, bioproducts provide a renewable, functional, and eco-friendly platform for developing next generation nanofertilizers. Nanofertilizers offer a safer, more environmentally friendly alternative to traditional fertilizers, enhancing plant productivity and minimizing environmental pollutants (Fatima et al., 2021). Their adaptability at the nanoscale, combined with their biodegradability and low toxicity, makes them ideal for creating smart, sustainable, and efficient nutrient delivery systems that can boost crop yield, protect soil health, and support global food and environmental security.

1.2 Importance in Sustainable Agriculture

Bioproducts are an important addition to the production of fertilizer for sustainable farming. Synthetic fertilizers raise the amount farmers produce, but these types of fertilizers have disadvantages for the environment and the economy. Reducing overall production and use of synthetic N fertilisers offers large mitigation potential and in many cases realisable potential to reduce emissions (Menegat et al., 2022). Fertilizers are made using large amounts of energy, mostly from non-renewable fossil fuels and their extreme use can make the soil unsafe, harm local waters through chemicals and change the composition of microbes in the earth.

In contrast, bioproduct-based fertilizers, which are developed using renewable biological materials such as microalgae, plant residues, compost, or microbial biomass, offer an eco-friendlier alternative. For example, microalgae as highlighted by Braun and Colla can be harnessed to produce biofertilizers and biostimulants (Braun & Colla, 2022). These algae-derived formulations not only supply essential nutrients but also contain bioactive compounds like amino acids, phytohormones, and polysaccharides that stimulate plant growth, stress resistance, and root development. Moreover, their cultivation can utilize wastewater or carbon dioxide, turning environmental pollutants into valuable resources.

Another key benefit of using bioproducts lies in their biodegradability and low toxicity. Unlike chemical fertilizers, which may accumulate in the soil or run off into aquatic systems and cause eutrophication, bioproduct-based formulations naturally break down into harmless compounds. This significantly reduces the risk of long-term ecotoxicity and helps maintain soil biodiversity, which is essential for nutrient cycling and sustainable land use.

Incorporating bioproducts into nanofertilizer technology further enhances their efficiency. For example, Zinc and iron bio-nanofertilizers formulated with plant growth-promoting rhizobacteria and microalgae significantly increased plant growth (Guardiola-Márquez et al., 2023). At the nanoscale, nutrients can be encapsulated in or bonded to natural polymers like chitosan, starch, or alginate. These bio-based nanocarriers enable targeted and controlled nutrient release, ensuring that plants absorb nutrients more efficiently and only as needed, which minimizes wastage. This controlled delivery mechanism supports the principles of green chemistry by optimizing input use and minimizing environmental impacts.

Employing biomass plays a role in helping develop a circular bioeconomy which centers on recycling, minimizing waste and reviving natural resources. A circular bioeconomy involves safeguarding and regenerating the health of ecosystems, avoiding non-essential products and waste, prioritizing biomass streams for basic human needs, and using renewable energy while minimizing overall energy use (Muscat et al., 2021). Figure 2 shows circular bioeconomy ecosystem. Recycling agricultural and food waste helps farmers and producers to return nutrients to where they came from. Reusing agricultural and food wastes into fertilizers improves soil quality and increases crop quality, with economic benefits from decreased production

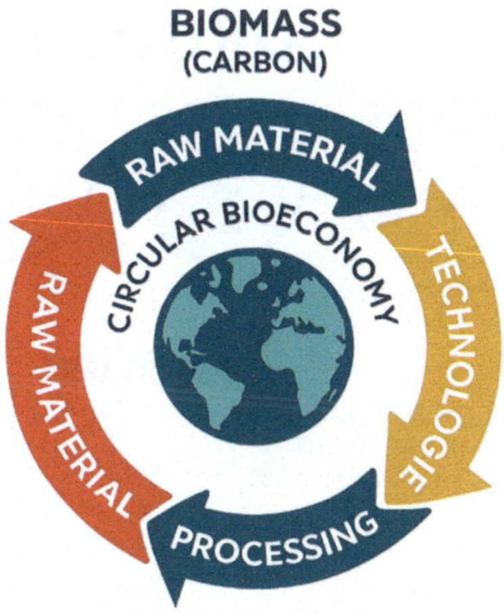

Fig. 2 Concept map of the circular bioeconomy model

and landfilling costs (Muscolo et al., 2021). It helps both protect the environment and boost the economic security of many rural communities.

Beyond environmental advantages, bioproduct-based fertilizers also help to improve soil health over time. For example, algae-based bio-fertilizers, such as microalgae and cyanobacteria, can increase soil fertility and boost crop growth (Ammar et al., 2022). They contribute to the accumulation of organic matter, enhance microbial diversity, and support soil structure and aeration, which in turn boosts water retention and nutrient exchange capacity. These improvements are critical for maintaining long-term soil fertility and productivity, particularly in the face of challenges like climate change, soil erosion, and land degradation.

In summary, the adoption of bioproducts in fertilizer production particularly in nanofertilizer formulations is a forward-thinking strategy that aligns with global goals for sustainable agriculture, climate action, and food security. For example, algal biofertilizer with nanoscale application has huge prospects for further agriculture productivities and fosters suitable development (Mahapatra et al., 2022). It not only reduces reliance on finite resources and chemical inputs but also fosters a regenerative agricultural system that benefits farmers, consumers, and the environment alike.

2 Types of Bioproducts Used in Nanofertilizers

2.1 Organic Waste-Based Bioproducts

Organic wastes such as compost, animal manure, crop residues, and food/agricultural waste have significant potential in the synthesis of nanofertilizers. These wastes are rich in essential nutrients like nitrogen, phosphorus, and potassium, and can serve

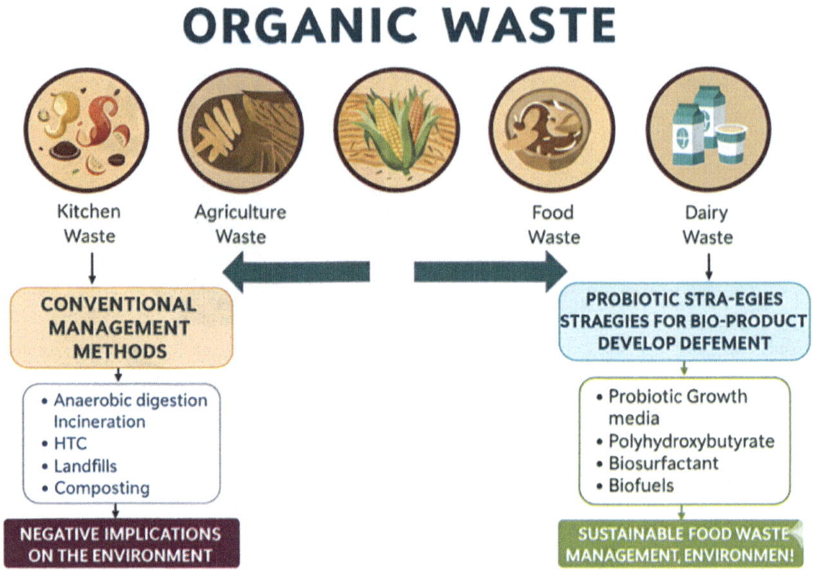

Fig. 3 Organic waste

as natural precursors for nano formulations. Figure 3 shows organic waste. Conversion techniques such as hydrothermal carbonization, pyrolysis, and green synthesis are employed to transform organic waste into functional nanomaterials. Organic waste can be transformed into useful nanomaterials for environmental remediation, bioimaging, sensing, and energy storage (Samaddar et al., 2018).

2.2 *Plant-Based Biopolymers*

Plant-derived biopolymers, including cellulose, starch, alginate, and chitosan, are widely utilized in nanofertilizer formulations. Figure 4 shows biopolymers division. Cellulose, the most abundant biopolymer, and starch, a readily available carbohydrate, offer biodegradable matrices for nutrient encapsulation. Starch-based encapsulation systems can positively influence the controlled release of food ingredients in food and nutritional applications (Zhu, 2017). Alginate, extracted from brown seaweeds, and chitosan, obtained from the shells of crustaceans, are notable for their biocompatibility and gel-forming abilities. These biopolymers serve as carriers for nutrients, enabling sustained and targeted delivery systems that enhance plant uptake and reduce nutrient loss to the environment.

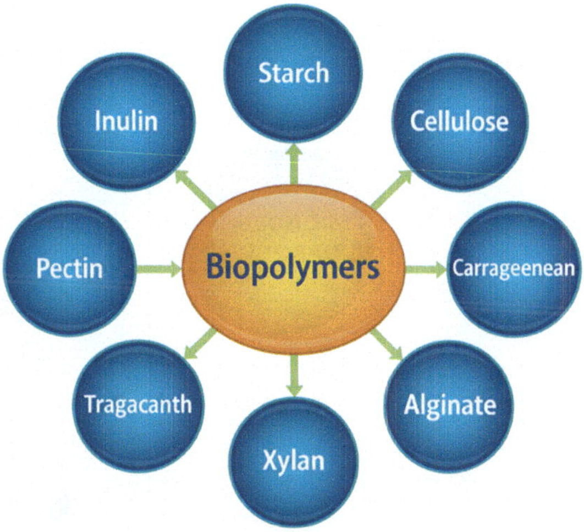

Fig. 4 Biopolymers division

2.3 Microbial Bioproducts

The applications of the microbial bioproducts, such as the use of bacteria, fungi, and its metabolic products are also another important category among nanofertilizers. Some microbes naturally produce nanoparticles via bioreduction processes thus, giving rise to the coefficient production of metallic and non-metallic nanoparticles. Figure 5 shows microbial products. For example, fungi just like Aspergillus and Penicillium species and bacteria such as Bacillus have been used to biosynthesize nanoparticles that have potential to carry essential nutrients for agricultural uses. Biogenic molybdenum nanoparticles combined with Bacillus sp. strain ZH16 significantly improved wheat growth by managing nutrients supply, ionic homeostasis, and arsenic accumulation in soil (Ahmed et al., 2022). The microbial approach is appreciated for its superior sustainability, low toxicity, potential to generate highly functionalized nanoparticles.

2.4 Agro-industrial Byproducts

Agro-industrial byproducts such as sugarcane bagasse, oilseed cakes, rice husks, and fruit peels offer valuable raw materials for nanoparticle synthesis. Figure 6 shows agro industrial waste. These materials are rich in lignocellulosic content and other bioactive compounds, making them suitable for conversion into nano formulations through processes such as sol–gel synthesis and green chemistry approaches

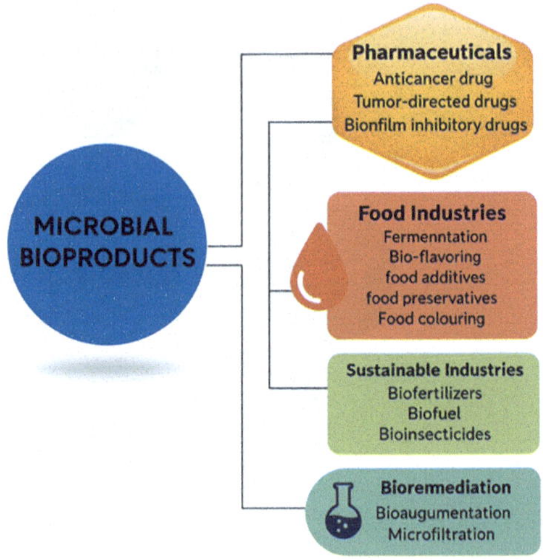

Fig. 5 Use of bio-products in large industries

(Sundararajan et al., 2024). Valorising these byproducts not only reduces industrial waste but also supports a circular economy model, where agricultural residues are transformed into high value nanofertilizer products with enhanced nutrient efficiency and environmental benefits.

3 Applications in Nanofertilizer Development

3.1 Bio-based Nanocarriers

Bio-based nanocarriers obtained from natural polymers including chitosan along with alginate and cellulose and starch have witnessed increasing interest because they successfully trap nutrients while providing controlled and targeted substance release. The material properties of these biopolymers include their destructive capacity while being harmless to the environment and the capability to unite with nutrients and nanoparticles. Chitosan nanoparticles serve as an effective carrier system for encapsulating urea alongside micronutrients which leads to extended availability of nutrients and superior absorption levels by plants. The utilization of chitosan nanoparticles by agricultural science provides advantages for productivity enhancement alongside better absorption of nutrients and more effective pest management methods (Saberi Riseh et al., 2024). The integration of these systems effectively diminishes nutrient waste while supporting the principles of precise agriculture.

Fig. 6 Agro industrial waste

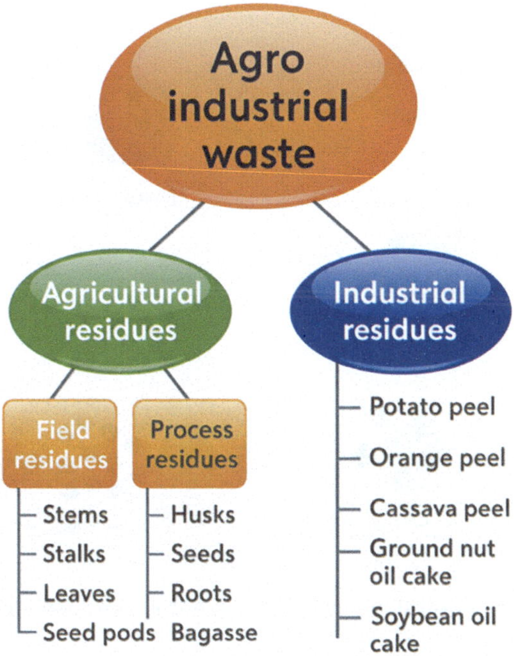

3.2 Green Synthesis of Nanoparticles

Green synthesis refers to eco-friendly techniques for producing nanoparticles. Figure 7 shows green nanoparticles process. Green nanoparticle synthesis using biological agents, such as plants, bacteria, fungi, and algae, offers cost-efficiency, biocompatibility, environmental friendliness, and scalability (Pandit et al., 2022). These methods are preferable to conventional chemical synthesis due to their lower toxicity, reduced energy input, and sustainability. For example, plant extracts rich in polyphenols can reduce metal ions to nanoparticles, acting both as reducing and capping agents. Similarly, microbial activity from species such as *Bacillus subtilis* and *Aspergillus niger* can biosynthesize zinc oxide or iron nanoparticles suitable for soil or foliar application in agriculture.

This diagram represents the consecutive stages of green nanoparticle production using plants. Green plants, after being collected, are dried and crushed into fine green plant powder; bioactive molecules of green plants are extracted by boiling them with water at 60 °C. The obtained filtrate is rich in bioactives having reducing and stabilization properties during nanoparticle synthesis. The formation of nanoparticles in green synthesis is due to the interaction between the plant extract and a precursor solution which eventually gets stabilized as green nanoparticles used for versatile applications (Biomedicine et al., 2021).

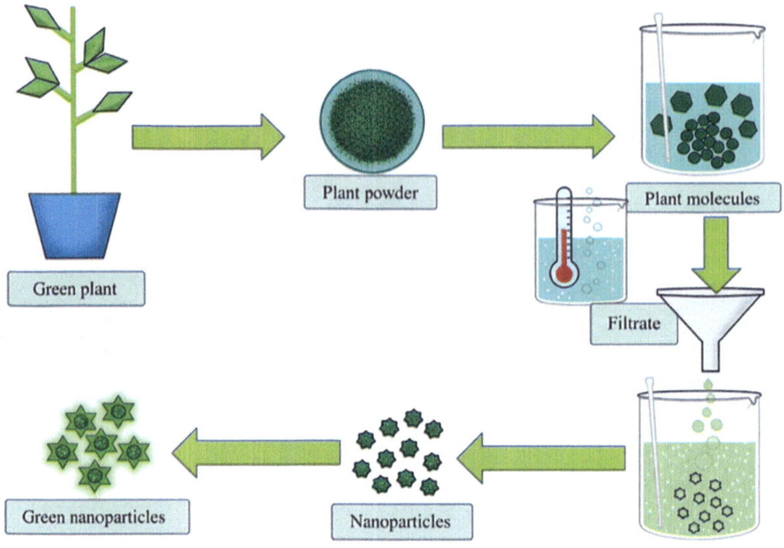

Fig. 7 Schematic diagram of the green synthesis of nanoparticles with plant extract

3.3 *Functionalization of Bioproducts*

The surface properties of bioproducts get modified through functionalization processes to improve nanofertilizer system performance. The application of grafting and cross-linking and coating methods enables biopolymers to establish improved binding mechanisms toward nutrients which creates better stability profiles as well as enhanced uptake and nutrient delivery characteristics. Surface-modified cellulose and chitosan undergo chemical engineering that enables them to produce environmentally triggered nutrient release based on pH changes and temperature variations. Grafting chitosan-cellulose hybrids with acrylic acid creates more robust, flexible, and efficient superabsorbent polymers for controlled release of soil nutrients in agriculture and horticulture (Essawy et al., 2016). The method enhances nutrient utilization efficiency and decreases environmental risks.

3.4 *Dual-Function Bioproducts*

Some bioproducts serve dual functions in nanofertilizer systems as both nutrient carriers and nutrient sources. A prime example is biochar enriched with nitrogen, phosphorus, and potassium (NPK). Biochar, a carbon-rich product derived from pyrolysis of organic matter, has a porous structure that allows it to retain and gradually release nutrients. When doped or infused with mineral nutrients at the nanoscale, biochar becomes a smart fertilizer. Biochar can increase crop productivity by improving soil structure, nutrient use efficiency, aeration, porosity, and water-holding capacity (Alkharabsheh et al., 2021).

4 Benefits of Using Bioproducts

4.1 Renewable and Biodegradable

Biomass is readily available, useful and crucial in making bioproducts, including nanofertilizers for the farming industry. Microalgae biomass can be used to produce various bioproducts, offering high-value nutrients and environmental sustainability, with potential for resource recovery and valorization (Calijuri et al., 2022). Some of other materials for bioproducts from left over from farming, like wheat straw, corn stalks and rice husks, as well as waste from forestry jobs such as bark, sawdust and wood chips. These are considered low-value wastes in the regular process.

By turning biomass into bioproducts, both valuable substances are recovered, and waste is converted into useful materials just like in a circular economy. Because of this process, there will be less organic waste ending up in landfills, emitting methane gas or polluting the environment. Likewise, when microalgae are produced using fertilized water or CO_2 waste, they can be collected and made into fertilizers high in nitrogen, phosphorus and various minerals needed by plants. Microalgal biofertilizers add nutrients to the soil and can be used as a treatment alternative for wastewater treatment (Braun & Colla, 2022).

Furthermore, biomass materials are completely biodegradable which means they don't leave any harmful substances in the soil. Green biomass is biodegradable and can be used to develop high-efficiency urea fertilizer for crops (Majeed et al., 2023). This characteristic helps reduce the negative impact of farming on the environment. Excessive use of synthetic fertilizers has resulted in deterioration of environmental systems, especially water resources, and can cause serious illnesses like cancer (Majeed et al., 2023). Use of bioproducts from biomass helps to avoid problems from harmful chemicals that can damage soil and harm the environment. For example, biochar enhances composting and soil remediation by improving chemical properties, microbial community structure, and environmental effects, while also enhancing plant growth (Wu et al., 2017).

Using renewable biomass allows us to create nanomaterials such as capsules containing nutrients, gel-like carriers or nanoparticles based on natural materials. As the result, plants get the nutrients they require more efficiently and lose less fertilizer. The outcome is more crops being grown while harming the environment less.

It is worth noting that since the production of nanofertilizers from biomass follows green chemistry principles, it utilizes non-toxic, renewed and environmentally friendly materials and chemicals. Offering a non-petroleum option, biomass-based fertilizers lead to using fewer atmosphere-damaging agrochemicals, helping mitigate climate change. Nanofertilizers enhance farm productivity and environmental sustainability, reduce production cost, and mitigate environmental footprint (Babu et al., 2022). Because of their biodegradable nature such inputs naturally break down into simpler substances without creating adverse environmental conditions and help restore soil health.

4.2 Cost-Effective

Utilizing locally available biomass and agricultural waste for nanofertilizer production significantly reduces manufacturing costs while offer a safer, more environmentally friendly alternative to traditional fertilizers, enhancing plant productivity and minimizing environmental pollutants (Fatima et al., 2021). This makes the approach economically viable, especially in developing regions where high input costs can hinder sustainable farming practices. Bioproducts offer a low-cost feedstock option, and the valorization of waste also adds economic value to materials that would otherwise be discarded or underutilized. Biowastes can be used as low-cost raw feedstock to produce bioenergy, value-added products, and other commodities (Mignogna et al., 2024).

4.3 Enhances Soil Health

Bioproduct-derived nanofertilizers also play positive role in improving soil structure and fertility. Nano-biofertilizers, using nanotechnology, increase crop yields and tolerance to abiotic stresses and are environmentally friendly and sustainable substitutes for traditional chemical fertilizers (Zafar et al., 2024). Organic carriers include biochar, compost, and cellulose derivatives that stimulate microbial activities, increase soil organic matters, and promote water-holding capacity. These effects create a biologically active rhizosphere thereby enhancing healthier plant –soil interaction and long-term productivity. Besides, the slow release of nutrients prevents nutrient overload and hence disturb soil microbial balance.

4.4 Reduced Environmental Impact

Compared to conventional chemical fertilizers, bioproduct-based formulations release fewer harmful by-products and reduce risks of leaching and runoff. This helps mitigate water pollution, eutrophication, and toxicity to non-target organisms. Bioproducts from biowaste, such as proteins and biofertilizers, can improve environmental impacts by reducing energy use and improving ecotoxicity and global warming (Fernández-Gutiérrez et al., 2022). Green synthesis techniques further limit the use of hazardous chemicals, resulting in safer nanofertilizer products for both farmers and ecosystems. Green synthesis of nanofertilizers is preferable due to less chemical contamination and safer production (Ndaba et al., 2022).

4.5 Integrated Nutrient Management

Applying nanofertilizers made from bioproducts is crucial for achieving INM because it gives plants the necessary nutrients right away and improves the soil's condition in the future. They adjust the release of nutrients to suit the plant as it grows and goes through different stages. Plants optimize uptake of nitrogen and phosphorus through modifications to growth and development, and engagement with microorganisms that facilitate their capture (Oldroyd & Leyser, 2020). Because nutrients are slowly supplied over time, there is no need to fertilize often, less waste happens, and synthetic chemical products are required less often too.

One good thing about bioproduct-made nanofertilizers is that they can deliver macronutrients and micronutrients in forms that crops can absorb easily. Nanofertilizers increase nutrient uptake efficiency and are promising solutions for replacing conventional fertilizers, promoting sustainable agriculture (Nongbet et al., 2022). Because nanoparticles have a high surface area and are reactive, they can move easily through plants and enhance how plants absorb nutrients. Consequently, farmers harvest more crops, the harvested crops become healthier to eat, and the environment is less affected than when using conventional fertilizers.

In addition, using natural polymers and biodegradable carriers for nanofertilizers helps restore the health of the soil. Nanobiofertilizers improve soil health and plant growth, reducing the use of chemical fertilizers (Nongbet et al., 2022). As these carriers disintegrate, they add valuable matter to the soil and encourage the presence of helpful bacteria. Thanks to this approach, the health of the soil keeps improving which ensures sustainable farming for years.

Furthermore, using bioproduct-based nanofertilizers reduces the amount of nutrients that flow into streams and chemicals that build up in the soil, in turn saving nature, cleaning groundwater and reducing agricultural pollutants. Nanofertilizers offer a safer, more environmentally friendly alternative to traditional fertilizers, enhancing plant productivity and minimizing environmental pollutants (Fatima et al., 2021). The way they help provide nutrients, support healthy soil and protect the environment is what makes them significant in improving agriculture. Nanofertilizers, derived from bioproducts, offer benefits in crop nutrition management by enhancing abiotic stress tolerance and improving agricultural productivity (Al-Mamun et al., 2021a).

5 Challenges and Limitations

5.1 Variability of Biomass

The composition of biomass used in making nanofertilizers can vary which makes it difficult when developing these products. Materials collected from farms and organics are often not the same, meaning their chemical and nutrient content can change greatly depending on many different situations. This means that crop selection can change the biochemical composition of the rest material, for example by making it richer in nitrogen but less in silica. Moreover, some elements in the soil, including how fertile it is and whether it is acidic or alkaline, can shape the qualities of the biomass. For example, root biomass with high cellulose and hemicellulose contents enhances soil microbial diversity and enzyme activities, while low cellulose and hemicellulose contents reduce enzyme activity (Sharma et al., 2020). Variation can also come from the harvest stage, the way the biomass is kept and the techniques for processing it. This results in some bioproducts not being consistent in their nutrients and form.

Because of this wide range, the effectiveness, dependability and quality of nanofertilizers may not be stable. If the nutrients are not always present in the same amount, the release of nutrients can be uneven, plants may not take them up as planned and the product might have shorter shelf-life. Plants adapt to spatial and temporal variability in soil nutrient concentrations by modulating root system architecture and regulating nutrient transporter expression (Aibara & Miwa, 2014). Therefore, it is difficult to ensure products are formulated the same each time, provide the same nutrients and remain reliable which are all critical for anyone using these products. Furthermore, if the nutrient release from biomass is either very slow or very fast, it may cause negative effects on the environment and the soil. Tackling this issue involves applying standard techniques, using strong control methods and mixing various biomass raw materials to obtain a better balance. Consequently, even though bioproducts have great importance in shaping sustainable nanofertilizers, their varying nature is a crucial detail that should be carefully handled. Variability in biomass characteristics, such as moisture content, carbohydrate and ash compositions, bulk density, and particle size/shape distributions, can impact the economics of transforming biomass into value-added products (Al-Mamun et al., 2021a).

5.2 Technical Limitations

There are many technical difficulties in nanoparticle production from bioproducts, such as difficulty in controlling the process, lack of abundant raw materials, environmental friendliness, and raw material expenses (Mohammadinejad & Mansoori, 2020). Obtaining particles with uniform size, a solid structure and proper wrapping of nutrients is quite a challenge when bioproducts are used for manufacturing. Many materials from plants, crop remains or microorganisms tend to have structure that is not uniform. Instead of these materials, synthetic polymers are easier to control because their molecular structures are well defined. For this reason, the size and shape of bio-based nanoparticles often change which may immediately impact how well they deliver nutrients. Nanoparticle size, shape, and charge significantly impact drug delivery efficiency, with cylindrical or needle-like shapes facilitating better cellular uptake compared to spherical NPs (Öztürk et al., 2024).

Particles in nanofertilizers should be of the same size, as this determine their surface area, rate of nutrient release, movement in the soil and interaction with roots. Nanofertilizers use nanoparticles to control the release of nutrients, reducing the need for frequent application and the amount of fertilizer, improving environmental sustainability and crop yields (Yadav et al., 2023a). Floating particles of various sizes may cause the nutrients to move differently, lower the plants' ability to get nutrients and make it difficult to multiply production. Also, because these nanoparticles are composed mainly of biodegradable components, their structure can be easily destroyed or altered either naturally or through chemical reactions in the environment. Because of the instability, nutrients may be released at an early stage, or the formula may not function well during use.

Another important issue is making sure nutrients are efficiently sealed and remain the same quality. Various natural biopolymers such as starch, chitosan or alginate, are commonly used in nanofertilizers because they are biocompatible and can easily decompose. Still, because their molecules differ, the formation of uniformly coated particles becomes difficult. Consequently, the protective power of the nutrients, their distribution inside the fertilizer and the timing of their release may change. Proper control over all these factors is needed for the nanofertilizer to perform and function at their best.

Synthetic chemistry is more consistent and accurate. Scientists can develop synthetic materials that ensure the nanoparticles are made consistently, are securely held and release nutrients as needed. They may cause problems to the environment since they do not break down and can be toxic. Sustainable and eco-friendly methods do work well, yet they lack the precise controls necessary for making them perform well. So, today, extra research is being conducted to improve these bio-based methods such as green chemistry and enzymatic processes, to ensure evenly shaped, steadier and more effective nanoparticles. For example, biopolymer-based polymeric nanoparticles can be synthesized by direct polymerization of monomers or by dispersion of preformed polymers, depending on their desired characteristics and target applications (Pulingam et al., 2022).

5.3 Stability and Shelf-Life

Biologically derived nanofertilizers, those made from natural materials such as starch, chitosan, cellulose, or other plant- and microbial-based polymers are inherently biodegradable. Chitosan nanomaterials are potential next-generation fertilizers with improved nutrient use efficiency and a lower environmental footprint

(Prajapati et al., 2022). While this biodegradability is an advantage in terms of environmental friendliness and sustainability, it also makes these materials more vulnerable to microbial degradation, especially under typical storage and field conditions. Microbial degradation is a key factor in the lifecycle of biodegradable plastics, with composting and biogas plants being the only managed end-of-life options today (García-Depraect et al., 2021). Microorganisms present in the environment, such as bacteria and fungi, are naturally capable of breaking down organic materials. Microorganisms, such as bacteria, fungi, and actinomycetes, play an important role in the decomposition of organic matter, which involves various biochemical mechanisms like hydrolysis, fermentation, and oxidation (Ziliwu & Lase, 2025). This is beneficial after the fertilizer is applied to the soil, as it ensures that no harmful residues are left behind. However, premature microbial degradation can pose a serious problem during storage and transport.

Starch, chitosan and cellulose naturally allow nutrients to be carried well since they are safe, decomposable and can provide protective cover until nutrients are released. For example, nanocellulose and chitosan nanomaterials increase glucose absorption by 1.2, 1.5, and 1.6 fold, respectively, in an in vitro small intestinal epithelium model (Guo et al., 2021). Still, these carriers can fade away quickly if they are exposed to changes in humidity, air and temperature. For example, starch and cellulose are strong at absorbing moisture due to being hydrophilic. Both the structure of the nanocarriers and their ability to promote microbial multiplication are affected by this added water. Despite being moderately antimicrobial, chitosan can experience changes that weaken its structure and cause nutrients to deteriorate. Chitosan's antimicrobial properties are affected by factors such as intrinsic, environmental, microorganism, and physical state (Kong et al., 2010).

The consequences of this happening are major. It means that the nanofertilizer may release its nutrients early, resulting in less effectiveness when it is applied to the field. If the materials inside the fertilizer change, it may not be able to ensure controlled nutrient delivery, a key purpose of nanotechnology in the field of agriculture. For this reason, the time farmers can store their fertilizer is shortened and it becomes less dependable.

One of the main problems with bio-based nanofertilizers is keeping them stable during their storage and transportation. Although synthetic nanofertilizers can be protected from environmental stress with various coatings, bio-base fertilizers rarely have such protection. Also, using a lot of chemical preservatives goes against the main purpose of using environmentally friendly materials.

5.4 Regulatory Barriers

The rules for nanofertilizers made from bioproducts are still weak and scattered, preventing their broader adoption and sale. Nanofertilizers, unlike their traditional counterparts, rely on unique nanoscale substances and methods, raising various risks to health, the environment and safety. Even so, there are no worldwide rules for assessing how these materials affect people, their usefulness and the environment. Most fertilizer standards in different countries do not consider things such as the size of particles, their reactivity or how they may interact with both soil microorganisms and plant systems. As a result, researchers and manufacturers do not have clear instructions for designing, testing and certifying their items for market sale.

For biologically derived nanofertilizers, the challenge is even more complex. While they are often perceived as safer due to their natural and biodegradable components, their nanoparticulate nature can still pose risks such as bioaccumulation or unintended toxicity that are not yet fully understood. Regulatory bodies are cautious about approving these products without comprehensive risk assessments, which require standardized testing protocols for evaluating toxicity, persistence in the environment, and impacts on non-target organisms. However, the tools and methodologies for such assessments are still evolving, and they vary widely between regions and countries.

This lack of clarity slows down innovation. Goal clarity has both a positive direct effect on innovation and a negative moderating effect on the mediation of knowledge inflows between inter-team trust and innovation (Lui et al., 2023). Companies are hesitant to invest in large-scale production or distribution of bioproduct-based nanofertilizers without knowing whether they will receive regulatory approval or face restrictions. Nanotechnology-based agri-products have potential for crop improvement and pest/disease prevention, but require safe use and regulatory approval to ensure consumer and environmental safety (Kumari et al., 2023). Similarly, academic researchers developing new formulations face difficulty in translating their work into commercial applications due to uncertain legal pathways. In some cases, bio-based nanofertilizers may fall into regulatory grey areas treated either as conventional fertilizers, biostimulants, or even as novel substances requiring entirely new risk assessments.

Since TCLP is a challenge, more scientists, regulators and industry representatives are needed to create clear guidelines for nanofertilizers. Governments must set regulations encouraging businesses to engage in nanotechnology research and development while safeguarding public health and safety (Yadav et al., 2023c). It requires designing definitions, following the same routine in testing, setting label requirements and developing special checking systems for nano-agriculture goods. Harmonized regulations between countries are necessary for the smooth spread and trade of these products around the world. Regulatory cooperation in international trade is crucial for resolving issues in the digital economy, avoiding fragmentation and maintaining stability, privacy, and security (Ahmed, 2019). Until such systems are in place, the lack of regulatory clarity will remain a major obstacle to the sustainable and responsible advancement of bioproduct-based nanofertilizer technologies. Nanofertilizers can increase crop yields and minimize environmental impact, but must overcome regulatory obstacles and production economics to gain widespread adoption in the agricultural industry (Yadav et al., 2023b).

5.5 Public Perception and Adoption

While bio-based nanofertilizers can make growing crops more effective and good for the environment, many people do not know about them or aren't willing to use them. Many people in agricultural fields such as farmers, extension agents and certain policymakers, do not understand what nanotechnology is or how it works in farming. Nanotechnology has potential for sustainable agriculture, but barriers like efficient delivery, regulatory concerns, and consumer acceptance must be addressed for its adoption (Hofmann et al., 2020). Because we are not used to such scientific ideas, people may reject or doubt new technologies. It is possible that farmers in short-resource zones prefer to use conventional fertilizers instead of investing in bio-based nanofertilizers, as they are uncertain about their price, use in practice, health effects and future availability.

Besides, the worry about possible health and environmental problems usually caused by the public's uncertainty about anything "nano" can hold back the adoption of innovations. It is feared that, despite coming from nature, nanoparticles may eventually build up in our food and damage the health of the soil and its living things. For example, nanoparticles negatively impact soil health by negatively impacting beneficial bacteria and fungi populations, affecting soil fertility and plant growth (Ameen et al., 2021). If information and rules around these products aren't accessible and evidence-based, these issues are unlikely to go away.

To solve this issue, it is necessary to use education programs and outreach. Conservation education and outreach programs are crucial for educating children and adults, promoting sustainable futures for all (Marsh, 2017). All stakeholders should join efforts to promote training, plan organized events and reveal the important and safe use cases for bio-based nanofertilizers in actual agricultural work. Farmers may become convinced by observing yields rise, input needs drop and soil quality improve in the field and in pilot programs. Using nanotechnology in agricultural and vocational programs creates trust and gives individuals practical skills on a village scale. Nanotechnology can improve sustainable agriculture by reducing chemical use, enhancing food quality, and reducing agricultural inputs (Prasad et al., 2017).

Connecting community leaders, cooperatives of farmers and influencers in agriculture with the technology can greatly improve how people learn about it. For example, cooperative supports and improved agricultural technology adoption significantly boost agricultural income in Pakistan, with less-revenue farmers benefiting more (Khan et al., 2022). Informing the public about what makes nanofertilizers safe, strong and environmentally friendly, in a way all people can understand, will be crucial for winning confidence. Even with all their benefits, widespread use of bio-based nanofertilizers will be difficult without users' trust and increased awareness. Nano-biofertilizers have promising potential for transforming traditional farming techniques into smart agriculture (Sharma et al., 2023).

6 Future Prospects

6.1 Advancements in Green Nanotechnology

The progress in green nanotechnology is important to the future of bio-based nanofertilizers. Green nanotechnology advances in agriculture by enhancing crop quality, combating plant pathogens, and managing diseases and stress, leading to sustainable crop production systems (Sundararajan et al., 2024). Enhancement in green nanotechnology is key to lead the way for sustainable, safe and ecological advances in farm fertilizers. Green nanotechnology is used in fertilizers, insecticides, pesticides, and fungicides for environmental friendliness, cost-effectiveness, and high purity (Sundararajan et al., 2024). Developing green nanotechnology means using renewable sources and safe, low-energy methods to make nanomaterials and nanosystems. In agriculture such progress means better methods for growing crops that are safer for the environment, healthier for the plants, more productive and use nutrients more carefully.

One of the most transformative aspects of green nanotechnology is the biosynthesis of nanoparticles using biological agents such as plant extracts, microbial enzymes, fungi, bacteria, and even algae. Bio-mediated nanoparticle synthesis from plant extracts offers low-cost, non-toxic, and environmentally friendly advantages for therapeutic applications and biomedical research (Karunakaran et al., 2023). These biological entities act as natural reducing and stabilizing agents, replacing toxic solvents or precursors used in conventional chemical synthesis. For example, the use of plant extracts which contain bioactive compounds like flavonoids, phenols, and alkaloids can facilitate the formation of nanoparticles under mild conditions without the need for high temperatures, pressure, or hazardous chemicals. Plant extracts like leaves, stems, barks, fruit, and flowers are rich in flavonoids, phenols, steroids, terpenoids, enzymes, and alkaloids, which play an essential role in reducing metal ions and generating metallic nanoparticles (Soni et al., 2021). This approach not only reduces energy consumption and chemical waste but also makes the nanoparticle production process cost-effective and scalable for agricultural applications.

Enzymes from bacteria and fungi can be part of an alternative system for generating nanoparticles (Qamar & Ahmad, 2021). Those organisms can use metal ions and make them into nanoparticles that fit required qualities such as shape, size and surface charge for nanofertilizer efficiency. Various biometric systems are being researched to make nanoparticles that last longer, reach plant parts and improve their role in the soil–plant system.

Moreover, the ongoing study of bio-based nanomaterials is improving the results obtained from nanofertilizers. Chitosan/alginate nanoparticles show promise for encapsulation, protection, and controlled release of curcumin diethyl diglutarate, improving its stability, digestibility, bioaccessibility, and cellular uptake (Qamar & Ahmad, 2021). A good example is using lignin, a widely available polymer in plants, to mold nanoparticles that function as both carriers and release systems, minimizing the loss of nutrients through leaching or escaping into the air. The antimicrobial nanochitosan is made from chitin, a substance in shellfish exoskeletons and can also boost plants' defense and nutrient delivery. Alginate nanogels, made from seaweed, offer a way to protect and slowly release micronutrients in answer to changes in the environment.

As a consequence, these enhancements guarantee that the production of these nanofertilizers is long-lasting and consistent with green chemistry principles. Green nanotechnology can take advantage of bioproducts in agriculture because it cuts back on polluting chemicals and minimizes damage to the environment during synthesis. By closely cooperating and never stopping, people in robotics can boost these machines, help them be used globally in farming and guarantee their safety and ability to scale.

6.2 Integration with Circular Economy

Bio-based nanofertilizers offer great potential to implement circular economic principles in the agricultural and agro-industrial waste in them to produce high value inputs. Nano biofertilizers are a more economically and environmentally sustainable, highly versatile, and long-lasting agriculture tool, with promising potential for transforming traditional farming techniques into smart agriculture (Sharma et al., 2023). Using agricultural and agro-industrial waste including crop residues, fruit peels, sugarcane bagasse and oilseed cakes to create nano fertilizer is a keyway to develop eco-friendly agricultural methods. Valorization is the making of valuable goods from waste or unvalued materials. Detoxification of such waste would prepare it as a useful resource for creating nanoparticle carriers or to simply provide nutrients for fertilizers.

Rice husks, wheat straw and corn stalks, for example, are full of lignocellulosic material (cellulose, hemicellulose and lignin) which can be fractured into nanoparticles. Researchers are working on designing these particles so they can deliver fertilizers and nutrients in a managed way in the soil. Just so, waste products from peeling bananas, oranges or pomegranates can biodegrade and be used as green nanoparticle making agents because they contain helpful phytochemicals and organic acids. Fruit peel wastes like mango, banana, pomegranate, and apple can successfully synthesize transition metal oxide nanoparticles with various applications like antimicrobial agents, sensors, and photocatalysts (Singh & Chaudhary, 2023).

Agro-industrial byproducts such as sugarcane bagasse (left after extracting juice from sugarcane) and oilseed cakes (the leftovers after processing oil from seeds like soybean, sunflower or canola) are considered especially advantageous. Because it is rich in carbon and fiber, sugarcane bagasse makes an ideal source for biochar and carbon-based nanomaterials which help carry nutrients and improve soil. Unlike oil, oilseed cakes are loaded with proteins, micronutrients and organic matter, allowing them to benefit plants when used in nanoformulations. Similarly, these wastes can be processed so that protein or polysaccharide biopolymers are created to make biodegradable nanocarriers for delivering fertilizer.

Making use of waste to create value, also known as a "waste-to-resource" approach, solves a range of challenges together. This means there's less agricultural waste being burned, thrown into a landfill or rotting (which could result in air pollution, contaminated water or pests). As a result, fewer synthetic inputs are required which are not only costly to create and can be harmful to soil but are also costly. What's more, this kind of system maximizes the use of resources by recycling items from the local environment. This reduces what it costs to produce fertilizers and increases the stability of how they are delivered. It is important because it gives local farmers and industries a chance to manufacture advanced products. Basically, turning biowaste into components for nanofertilizers resolves a problem with waste and provides nourishment, safer surroundings and more sustainable farming. Nano fertilizers offer a promising alternative to chemical fertilizers, improving crop nutrient efficiency, environmental impact, and productivity while enhancing abiotic stress tolerance and promoting sustainable agriculture (Al-Mamun et al., 2021b).

6.3 *Tailor-Made Bioproducts*

The development of tailor-made bioproducts represents a cutting-edge advancement in sustainable agriculture, particularly in the field of bio-based nanofertilizers. For example, algal biorefineries offer potential for sustainable agriculture by producing biofuels, biochemicals, and providing sustainable alternatives to chemical fertilizers, improving soil health and plant growth (Samoraj et al., 2024). This approach focuses on engineering customized fertilizer delivery systems that are specifically designed to meet the unique needs of particular crops, soil types, or agro-ecological zones. By leveraging the versatility of natural biopolymers, researchers aim to fine-tune the physical and chemical properties of these carriers to optimize their performance under specific conditions.

In this area, chitosan is a promising material because it comes from fungi or regular crustaceans, and it breaks down when used. Chitosan has found global attention due to its biocompatibility, biodegradability, and ease of availability in various biomedical, tissue engineering, and biosensor applications (Baranwal et al., 2018). Chitosan is valued because it is biocompatible, forms film, has antimicrobial properties and is very modifiable below the molecular level. Knowing which parts of a plant absorb nutrients and how fast they require them helps researchers modify chitosan using processes such as cross-linking, grafting or ionic gelation to add nutrients, control how fast the chitosan breaks down and guide nutrient release based on a crop's needs. A useful example is with maize or wheat, where nutrition like nitrogen and phosphorus is required early and micronutrients are necessary in the reproductive phase. This situation can be best addressed with a chitosan-based nanocarrier that slowly releases nitrogen and phosphorus in the early stages and micronutrients when the plant is entering reproduction.

Understanding how soil and fertilizer affect each other plays an important role in creating crop and soil-targeted nanocarriers. Certain nutrients might be unavailable in high-pH soils and sandy soils often cause nutrients to move quickly downstream. When researchers pick suitable biopolymer matrices like alginate, starch or cellulose derivatives and modify their surface, they can increase how many nutrients stay in the biopolymer and reduce those that are lost via volatility or through water runoff. For example, chitosan-coated alginate matrices with protein-based biostimulants show potential as sustainable fertilizers, with reduced water uptake and slowed initial nutrient release (Szopa et al., 2025). In addition, these bio-based carriers may be added to bioactive compounds that can advance plant growth, boost the number of microorganisms in the soil around roots or strengthen the soil.

Besides, tailor-made bioproducts often give their plants both nutrient needs and the ability to resist pests or cope with drought. Chitosan nanoparticles transport key nutrients to plants and also help increase their defense against fungal and insect threats. Chitosan-based nanocomposites enhance plant defense, antimicrobial activity, and nutrient uptake (Yu et al., 2021). For tomatoes and cucumbers, the use of nanoparticles has improved crop sizes, fought off illnesses and made fruit better.

Because of this level of accuracy, the farm gets more value from each nutrient which means less frequent applications and a smaller chance of harm to the environment. Long-term such innovation may assist in increasing the output with the same degree of input from the farmer, in a sustainable manner. Research and innovation are key drivers in accelerating the transition to sustainable, healthy, and inclusive food systems, from primary production to consumption, in the agrifood sector (Riccaboni et al., 2021).

In short, combining materials science, agronomy and environmental protection led to the development of tailor-made bio-based nanofertilizers. It could help move fertilizer use away from a single approach and toward useful and targeted solutions that support both productivity and protect the environment. Nanotechnology offers significant prospects for tailoring nanofertilizer production, which can improve plant productivity and minimize environmental pollutants (Fatima et al., 2021).

References

Ahmed, U. (2019). The importance of cross-border regulatory cooperation in an era of digital trade. *World Trade Review, 18*(S1), 99–120. https://doi.org/10.1017/S1474745618000514

Ahmed, T., Noman, M., Rizwan, M., Ali, S., Ijaz, U., Nazir, M. M., ALHaithloul, H. A. S., Alghanem, S. M., Abdulmajeed, A. M., & Bin Li. (2022). Green molybdenum nanoparticles-mediated bio-stimulation of Bacillus sp. strain ZH16 improved the wheat growth by managing in planta nutrients supply, ionic homeostasis and arsenic accumulation. *Journal of Hazardous Materials, 423*, 127024. https://doi.org/10.1016/J.JHAZMAT.2021.127024

Aibara, I., & Miwa, K. (2014). Strategies for optimization of mineral nutrient transport in plants: Multilevel regulation of nutrient-dependent dynamics of root architecture and transporter activity. *Plant & Cell Physiology, 55*(12), 2027–2036. https://doi.org/10.1093/PCP/PCU156

Alkharabsheh, H. M., Seleiman, M. F., Battaglia, M. L., Shami, A., Jalal, R. S., Alhammad, B. A., Almutairi, K. F., & Al-Saif, A. M. (2021). Biochar and its broad impacts in soil quality and fertility, nutrient leaching and crop productivity: A review. *Agronomy, 11*(5), 993. https://doi.org/10.3390/AGRONOMY11050993

Al-Mamun, M. R., Hasan, M. R., Ahommed, M. S., Bacchu, M. S., Ali, M. R., & Khan, M. Z. H. (2021a). Nanofertilizers towards sustainable agriculture and environment. *Environmental Technology & Innovation, 23*, 101658. https://doi.org/10.1016/J.ETI.2021.101658

Al-Mamun, M. R., Hasan, M. R., Ahommed, M. S., Bacchu, M. S., Ali, M. R., & Khan, M. Z. H. (2021b). Nanofertilizers towards sustainable agriculture and environment. *Environmental Technology and Innovation, 23*. https://doi.org/10.1016/J.ETI.2021.101658

Ameen, F., Alsamhary, K., Alabdullatif, J. A., & ALNadhari, S. (2021). A review on metal-based nanoparticles and their toxicity to beneficial soil bacteria and fungi. *Ecotoxicology and Environmental Safety, 213*. https://doi.org/10.1016/J.ECOENV.2021.112027

Ammar, E. E., Aioub, A. A. A., Elesawy, A. E., Karkour, A. M., Mouhamed, M. S., Amer, A. A., & EL-Shershaby, N. A. (2022). Algae as bio-fertilizers: Between current situation and future prospective. *Saudi Journal of Biological Sciences, 29*(5), 3083–3096. https://doi.org/10.1016/J.SJBS.2022.03.020

Babu, S., Singh, R., Yadav, D., Rathore, S. S., Raj, R., Avasthe, R., Yadav, S. K., Das, A., Yadav, V., Yadav, B., Shekhawat, K., Upadhyay, P. K., Yadav, D. K., & Singh, V. K. (2022). Nanofertilizers for agricultural and environmental sustainability. *Chemosphere, 292*(December 2021), 133451. https://doi.org/10.1016/j.chemosphere.2021.133451

Baranwal, A., Kumar, A., Priyadharshini, A., Oggu, G. S., Bhatnagar, I., Srivastava, A., & Chandra, P. (2018). Chitosan: An undisputed bio-fabrication material for tissue engineering and bio-sensing applications. *International Journal of Biological Macromolecules, 110*, 110–123. https://doi.org/10.1016/J.IJBIOMAC.2018.01.006

Biomedicine, N., Shafiq, T., Uzair, M., Iqbal, M. J., & Zafar, M. (2021, May). *Nano biomed eng green synthesis of metallic nanoparticles and their potential in bio-medical applications.* https://doi.org/10.5101/nbe.v13i2.p191-206

Braun, J. C. A., & Colla, L. M. (2022). Use of microalgae for the development of biofertilizers and biostimulants. *BioEnergy Research, 16*(1), 289–310. https://doi.org/10.1007/S12155-022-10456-8

Calijuri, M. L., Silva, T. A., Magalhães, I. B., Pereira, A. S. A. d. P., Marangon, B. B., de Assis, L. R., & Lorentz, J. F. (2022). Bioproducts from microalgae biomass: Technology, sustainability, challenges and opportunities. *Chemosphere, 305*, 135508. https://doi.org/10.1016/J.CHEMOSPHERE.2022.135508

Essawy, H. A., Ghazy, M. B. M., El-Hai, F. A., & Mohamed, M. F. (2016). Superabsorbent hydrogels via graft polymerization of acrylic acid from chitosan-cellulose hybrid and their potential in controlled release of soil nutrients. *International Journal of Biological Macromolecules, 89*, 144–151. https://doi.org/10.1016/J.IJBIOMAC.2016.04.071

Fatima, F., Hashim, A., & Anees, S. (2021). Efficacy of nanoparticles as nanofertilizer production: A review. *Environmental Science and Pollution Research, 28*(2), 1292–1303. https://doi.org/10.1007/S11356-020-11218-9/TABLES/4

Fernández-Gutiérrez, D., Argüelles, A., Castejón Martínez, G., Soriano Disla, J. M., & Lara-Guillén, A. J. (2022). Unlocking new value from urban biowaste: LCA of the VALUEWASTE biobased products. *Sustainability, 14*(22), 14962. https://doi.org/10.3390/SU142214962

García-Depraect, O., Bordel, S., Lebrero, R., Santos-Beneit, F., Börner, R. A., Börner, T., & Muñoz, R. (2021). Inspired by nature: Microbial production, degradation and valorization of biodegradable bioplastics for life-cycle-engineered products. *Biotechnology Advances, 53*. https://doi.org/10.1016/J.BIOTECHADV.2021.107772

Guardiola-Márquez, C. E., López-Mena, E. R., Segura-Jiménez, M. E., Gutierrez-Marmolejo, I., Flores-Matzumiya, M. A., Mora-Godínez, S., Hernández-Brenes, C., & Jacobo-Velázquez, D. A. (2023). Development and evaluation of zinc and iron nanoparticles functionalized with plant growth-promoting rhizobacteria (PGPR) and microalgae for their application as bio-nanofertilizers. *Plants, 12*(20), 3657. https://doi.org/10.3390/PLANTS12203657

Guo, Z., DeLoid, G. M., Cao, X., Bitounis, D., Sampathkumar, K., Ng, K. W., Loo, S. C. J., & Demokritou, P. (2021). Effects of ingested nanocellulose and nanochitosan materials on carbohydrate digestion and absorption in an in vitro small intestinal epithelium model. *Environmental Science: Nano*, *8*(9), 2554–2568. https://doi.org/10.1039/D1EN00233C

Hofmann, T., Lowry, G. V., Ghoshal, S., Tufenkji, N., Brambilla, D., Dutcher, J. R., Gilbertson, L. M., Giraldo, J. P., Kinsella, J. M., Landry, M. P., Lovell, W., Naccache, R., Paret, M., Pedersen, J. A., Unrine, J. M., White, J. C., & Wilkinson, K. J. (2020). Technology readiness and overcoming barriers to sustainably implement nanotechnology-enabled plant agriculture. *Nature Food, 1*(7), 416–425. https://doi.org/10.1038/S43016-020-0110-1

Kalambate, P. K., Rao, Z., Dhanjai, Wu, J., Shen, Y., Boddula, R., & Huang, Y. (2020). Electrochemical (bio) sensors go green. *Biosensors & Bioelectronics*, *163*. https://doi.org/10.1016/J.BIOS.2020.112270

Karunakaran, G., Sudha, K. G., Ali, S., & Cho, E. B. (2023). Biosynthesis of nanoparticles from various biological sources and its biomedical applications. *Molecules*, *28*(11). https://doi.org/10.3390/MOLECULES28114527

Khan, N., Ray, R. L., Kassem, H. S., Ihtisham, M., Siddiqui, B. N., & Zhang, S. (2022). Can cooperative supports and adoption of improved technologies help increase agricultural income? Evidence from a recent study. *Land*, *11*(3). https://doi.org/10.3390/LAND11030361

Kong, M., Chen, X. G., Xing, K., & Park, H. J. (2010). Antimicrobial properties of chitosan and mode of action: A state of the art review. *International Journal of Food Microbiology*, *144*(1), 51–63. https://doi.org/10.1016/J.IJFOODMICRO.2010.09.012

Kumari, R., Suman, K., Karmakar, S., Mishra, V., Lakra, S. G., Saurav, G. K., & Mahto, B. K. (2023). Regulation and safety measures for nanotechnology-based agri-products. *Frontiers in Genome Editing*, *5*. https://doi.org/10.3389/FGEED.2023.1200987

Lui, S., Lai, J., Luo, B. N., & Moran, P. (2023). Will goal clarity lower team innovation? A moderated mediation model of inter-team trust. *Journal of Knowledge Management, 27*(4), 975–992. https://doi.org/10.1108/JKM-10-2021-0787

Mahapatra, D. M., Satapathy, K. C., & Panda, B. (2022). Biofertilizers and nanofertilizers for sustainable agriculture: Phycoprospects and challenges. *Science of the Total Environment, 803*, 149990. https://doi.org/10.1016/J.SCITOTENV.2021.149990

Majeed, Z., Nawazish, S., Baig, A., Akhtar, W., Iqbal, A., Khan, W. M., Bukhari, S. M., Zaidi, A., Show, P. L., & Mansoor, N. (2023). Effect of varying thickness properties of the slow release fertilizer films on morphology, biodegradability, urea release, soil health, and plant growth. *PLoS ONE*, *18*(2 February). https://doi.org/10.1371/JOURNAL.PONE.0278568

Marsh, K. (2017). Conservation education and outreach techniques. *Environmental Education Research, 23*(4), 591–594. https://doi.org/10.1080/13504622.2016.1199662

Menegat, S., Ledo, A., & Tirado, R. (2022). Greenhouse gas emissions from global production and use of nitrogen synthetic fertilisers in agriculture. *Scientific Reports, 12*(1), 1–13. https://doi.org/10.1038/S41598-022-18773-W

Mignogna, D., Szabó, M., Ceci, P., & Avino, P. (2024). Biomass energy and biofuels: Perspective, potentials, and challenges in the energy transition. *Sustainability*, *16*(16), 7036. https://doi.org/10.3390/SU16167036

Mohammadinejad, R., & Mansoori, G. A. (2020). Large-scale production/biosynthesis of biogenic nanoparticles. In *Biogenic nano-particles and their use in agro-ecosystems* (pp. 67–83). https://doi.org/10.1007/978-981-15-2985-6_5

Muscat, A., de Olde, E. M., Ripoll-Bosch, R., Van Zanten, H. H. E., Metze, T. A. P., Termeer, C. J. A. M., van Ittersum, M. K., & de Boer, I. J. M. (2021). Principles, drivers and opportunities of a circular bioeconomy. *Nature Food, 2*(8), 561–566. https://doi.org/10.1038/S43016-021-00340-7

Muscolo, A., Romeo, F., Marra, F., & Mallamaci, C. (2021). Recycling agricultural, municipal and industrial pollutant wastes into fertilizers for a sustainable healthy food production. *Journal of Environmental Management, 300*, 113771. https://doi.org/10.1016/J.JENVMAN.2021.113771

Ndaba, B., Roopnarain, A., Rama, H., & Maaza, M. (2022). Biosynthesized metallic nanoparticles as fertilizers: An emerging precision agriculture strategy. *Journal of Integrative Agriculture, 21*(5), 1225–1242. https://doi.org/10.1016/S2095-3119(21)63751-6

Nongbet, A., Mishra, A. K., Mohanta, Y. K., Mahanta, S., Ray, M. K., Khan, M., Baek, K. H., & Chakrabartty, I. (2022). Nanofertilizers: A smart and sustainable attribute to modern agriculture. *Plants, 11*(19). https://doi.org/10.3390/PLANTS11192587

Oldroyd, G. E. D., & Leyser, O. (2020). A plant's diet, surviving in a variable nutrient environment. *Science, 368*(6486). https://doi.org/10.1126/SCIENCE.ABA0196

Öztürk, K., Kaplan, M., & Çalış, S. (2024). Effects of nanoparticle size, shape, and zeta potential on drug delivery. *International Journal of Pharmaceutics, 666*. https://doi.org/10.1016/J.IJPHARM.2024.124799

Pandit, C., Roy, A., Ghotekar, S., Khusro, A., Islam, M. N., Bin Emran, T., Lam, S. E., Khandaker, M. U., & Bradley, D. A. (2022). Biological agents for synthesis of nanoparticles and their applications. *Journal of King Saud University - Science, 34*(3), 101869. https://doi.org/10.1016/j.jksus.2022.101869

Prajapati, D., Pal, A., Dimkpa, C., Harish, Singh, U., Devi, K. A., Choudhary, J. L., & Saharan, V. (2022). Chitosan nanomaterials: A prelim of next-generation fertilizers; existing and future prospects. *Carbohydrate Polymers, 288*. https://doi.org/10.1016/J.CARBPOL.2022.119356

Prasad, R., Bhattacharyya, A., & Nguyen, Q. D. (2017). Nanotechnology in sustainable agriculture: Recent developments, challenges, and perspectives. *Frontiers in Microbiology, 8*(JUN). https://doi.org/10.3389/FMICB.2017.01014/PDF

Pulingam, T., Foroozandeh, P., Chuah, J. A., & Sudesh, K. (2022). Exploring various techniques for the chemical and biological synthesis of polymeric nanoparticles. *Nanomaterials, 12*(3). https://doi.org/10.3390/NANO12030576

Qamar, S. U. R., & Ahmad, J. N. (2021). Nanoparticles: Mechanism of biosynthesis using plant extracts, bacteria, fungi, and their applications. *Journal of Molecular Liquids, 334*. https://doi.org/10.1016/J.MOLLIQ.2021.116040

Riccaboni, A., Neri, E., Trovarelli, F., & Pulselli, R. M. (2021). Sustainability-oriented research and innovation in "farm to fork" value chains. *Current Opinion in Food Science, 42*, 102–112. https://doi.org/10.1016/J.COFS.2021.04.006

Saberi Riseh, R., Vatankhah, M., Hassanisaadi, M., & Varma, R. S. (2024). A review of chitosan nanoparticles: Nature's gift for transforming agriculture through smart and effective delivery mechanisms. *International Journal of Biological Macromolecules, 260*, 129522. https://doi.org/10.1016/J.IJBIOMAC.2024.129522

Samaddar, P., Ok, Y. S., Kim, K. H., Kwon, E. E., & Tsang, D. C. W. (2018). Synthesis of nanomaterials from various wastes and their new age applications. *Journal of Cleaner Production, 197*, 1190–1209. https://doi.org/10.1016/J.JCLEPRO.2018.06.262

Samoraj, M., Çalış, D., Trzaska, K., Mironiuk, M., & Chojnacka, K. (2024). Advancements in algal biorefineries for sustainable agriculture: Biofuels, high-value products, and environmental solutions. *Biocatalysis and Agricultural Biotechnology, 58*. https://doi.org/10.1016/J.BCAB.2024.103224

Sharma, S., Sharma, N., & Neemisha. (2020). Assessing extracellular enzymatic activity in the soil on addition of root biomass with different biochemical composition. *Current Science, 119*(11), 1807–1814. https://doi.org/10.18520/CS/V119/I11/1807-1814

Sharma, B., Tiwari, S., Kumawat, K. C., & Cardinale, M. (2023). Nano-biofertilizers as bio-emerging strategies for sustainable agriculture development: Potentiality and their limitations. *Science of the Total Environment, 860*, 160476. https://doi.org/10.1016/J.SCITOTENV.2022.160476

Singh, A., & Chaudhary, M. (2023). An overview of physical properties of fruit wastes mediated greensynthesized transition metal oxide nanoparticles and its applications. *Asian Journal of Chemistry, 35*(3), 529–536. https://doi.org/10.14233/AJCHEM.2023.27003

Soni, V., Raizada, P., Singh, P., Cuong, H. N., Saini, A., Saini, R. V., Van Le, Q., Nadda, A. K., Le, T. T., & Nguyen, V. H. (2021). Sustainable and green trends in using plant extracts for the synthesis of biogenic metal nanoparticles toward environmental and pharmaceutical advances: A review. *Environmental Research, 202*. https://doi.org/10.1016/J.ENVRES.2021.111622

Sundararajan, N., Habeebsheriff, H. S., Dhanabalan, K., Cong, V. H., Wong, L. S., Rajamani, R., & Dhar, B. K. (2024). Mitigating global challenges: Harnessing green synthesized nanomaterials for sustainable crop production systems. *Global Challenges, 8*(1), 1–15. https://doi.org/10.1002/gch2.202300187

Szopa, D., Pstrowska, K., & Witek-Krowiak, A. (2025). Chitosan-coated alginate matrices with protein-based biostimulants: A controlled-release system for sustainable agriculture. *Materials, 18*(3). https://doi.org/10.3390/MA18030591

Wang, J., Xiong, Z., & Kuzyakov, Y. (2016). Biochar stability in soil: Meta-analysis of decomposition and priming effects. *GCB Bioenergy, 8*(3), 512–523. https://doi.org/10.1111/GCBB.12266

Wu, S., He, H., Inthapanya, X., Yang, C., Lu, L., Zeng, G., & Han, Z. (2017). Role of biochar on composting of organic wastes and remediation of contaminated soils—A review. *Environmental Science and Pollution Research, 24*(20), 16560–16577. https://doi.org/10.1007/S11356-017-9168-1

Yadav, A., Yadav, K., & Abd-Elsalam, K. (2023a). Nanofertilizers: Types, delivery and advantages in agricultural sustainability. *Agrochemicals, 2*(2), 296–336. https://doi.org/10.3390/AGROCHEMICALS2020019

Yadav, A., Yadav, K., & Abd-Elsalam, K. A. (2023b). Exploring the potential of nanofertilizers for a sustainable agriculture. *Plant Nano Biology, 5*(August), 100044. https://doi.org/10.1016/j.plana.2023.100044

Yadav, A., Yadav, K., & Abd-Elsalam, K. A. (2023c). Exploring the potential of nanofertilizers for a sustainable agriculture. *Plant Nano Biology, 5*. https://doi.org/10.1016/J.PLANA.2023.100044

Yu, J., Wang, D., Geetha, N., Khawar, K. M., Jogaiah, S., & Mujtaba, M. (2021). Current trends and challenges in the synthesis and applications of chitosan-based nanocomposites for plants: A review. *Carbohydrate Polymers, 261*. https://doi.org/10.1016/J.CARBPOL.2021.117904

Zafar, S., Bilal, M., Ali, M. F., Mahmood, A., Kijsomporn, J., Wong, L. S., Kumar, V., & Alotaibi, S. S. (2024). Nano-biofertilizer an eco-friendly and sustainable approach for the improvement of crops under abiotic stresses. *Environmental and Sustainability Indicators, 24*. https://doi.org/10.1016/J.INDIC.2024.100470

Zhu, F. (2017). Encapsulation and delivery of food ingredients using starch based systems. *Food Chemistry, 229*, 542–552. https://doi.org/10.1016/J.FOODCHEM.2017.02.101

Ziliwu, Y. M., & Lase, N. K. (2025). Peran Mikroorganisme dalam Proses Degradasi Bahan Organik. *Hidroponik: Jurnal Ilmu Pertanian Dan Teknologi Dalam Ilmu Tanaman, 2*(1), 132–141. https://doi.org/10.62951/HIDROPONIK.V2I1.235

Fertilizer Runoff Management—Strategies and Practices to Minimise Excess Nutrients in Water Reservoirs

Qaisara Izzaty Mohd Azni, Fatin Shafiqa Shazili, and Noor Aziah Serri

Abstract This chapter explores the practical use of compost and liquid organic fertilizers, emphasizing their contribution to soil health, crop productivity, and water quality preservation. It begins with an in-depth review of nutrient dynamics, examining causes of nutrient runoff together with the intricacies of the nutrient cycle, which includes the water, nitrogen, phosphorus, and carbon cycles. Effective strategies of best management practices (BMPs) for reducing nutrient pollution from agricultural, urban, and residential sources are outlined, offering evidence-based guidance supported by current research and case studies. A review of policy and regulatory frameworks highlights key environmental regulations, national standards, and incentives that promotes sustainable fertilizer use. These insights help readers understand how governance influences fertilizer management and environmental compliance. The chapter also underscores the role of monitoring and assessment in evaluating environmental impacts and shaping policy. It covers water quality monitoring systems, such as Malaysia's National Water Quality Standards (NWQS) and Water Quality Index (WQI), as well as the responsibilities of the Department of Environment. Predictive models and decision-support tools are presented as essential resources for improving nutrient management strategies. In conclusion, collaborating among researchers, policymakers, environmental managers, and practitioners is emphasized as crucial to developing and applying effective, sustainable nutrient management solutions to address fertilizer runoff.

Keywords Fertilizer runoff · Nutrient dynamics · Best management practices (BMPs) · Environmental policy · Water quality monitoring

Q. I. M. Azni · F. S. Shazili · N. A. Serri (✉)
Bioprocess Technology Division, School of Industrial Technology, Universiti Sains Malaysia, Pulau Pinang, Malaysia
e-mail: aziah_serri@usm.my

A. Z. Yaser et al. (eds.), *Compost and Liquid Organic Fertilizers*,
https://doi.org/10.1007/978-981-95-8561-8_22

Abbreviations

AES	Agricultural Environmental Schemes
AN	Ammonium Nitrogen
ANFIS	Adaptive Neuro-Fuzzy Inference System
BB-SRF	Bio-Based Slow-Release Fertilizer
BMPs	Best Management Practices
BOD	Biochemical Oxygen Demand
C	Carbon
CAP	Common Agricultural Policy
CCF	Clay-Coated Fertilizer
CEA	Controlled Environmental Agriculture
COD	Chemical Oxygen Demand
CSOs	Combined Sewer Overflows
DO	Dissolved Oxygen
DS	Dissolved Solid
DOE	Department of Environment
ERS	Environmental Remote Sensing
EQA	Environmental Quality Act
Fe	Iron
FNNN	Feed-forward Neural Network
GAPs	Good Agricultural Practices
GHG	Greenhouse Gases
GIS	Geographic Information Systems
GPS	Global Positioning System
IM	Injection-Molded
K	Potassium
LID	Low Impact Development
Mn	Manganese
Mg	Magnesium
N	Nitrogen
NWQS	National Water Quality Standards
NH_4^+-N	Ammoniacal Nitrogen
NO_3^-N	Nitrate Nitrogen
PO_4^{3-}P	Ortho-phosphate Phosphorus
P	Phosphorus
PATs	Precision Agriculture Technologies
PCA	Principal Component Analysis
PCF	Polymer-Coated Fertilizer
SCF	Sulphur-Coater Fertilizer
SDG	Sustainable Development Goals
SRF	Slow-release Fertilizer
TSS	Total Suspended Solid
UAVs	Unmanned Aerial Vehicles

VRT Variable Rate Technologies
VRI Variable Rate Irrigation
WCF Wax-Coated Fertilizer
WQI Water Quality Index

List of symbols

< Less than
> Greater than
% Percent

1 Understanding Nutrient Dynamics

1.1 Sources of Nutrient Runoff

Nutrient runoff is essentially the flow of excess nutrients, primarily nitrogen (N) and phosphorus (P), from land into water bodies of water like rivers, lakes and oceans. Nutrients come from both point and diffuse sources (nonpoint). Point sources like industry and wastewater treatment plants were under control due to environmental regulations while nonpoint sources such as urban stormwater runoff and agricultural runoff continue to contribute nutrients to receiving waters (Zahoor & Mushtaq, 2023). Understanding the sources and implications of nutrient runoff is essential for developing effective strategies to safeguard our water resources and promote sustainable practices.

Agriculture runoff remains a major environmental concern, despite the role of agrochemicals in enhancing soil productivity. This issue primarily stems from the over application of synthetic fertilizers, exceeding the soil's retention capacity, leading to nutrient leaching into the surrounding environment through rainfall, infiltration, and irrigation (Albou et al., 2024). These chemicals often remain in the aqueous phase for extended periods and are cable of migrating over long distances, thereby impacting aquatic ecosystems and threaten source of drinking water (Singh et al., 2023). Some of the chemicals consists of heavy metals such as cadmium, lead, and zinc that can accumulate within the water and sediment and poses threats to surface water and groundwater (Albou et al., 2024). This problem is further exacerbated by the persistent nature of heavy metals in the environment (Singh et al., 2023; Albou et al., 2024).

On the other hand, urbanization, deforestation, and agriculture contributed significantly to nutrient discharges into the water bodies (McDowell & Hamilton, 2023; Albou et al., 2024). The increasing nutrient loads resulted in the disturbance of biogeochemical systems and alter the dominance of certain algae species over other

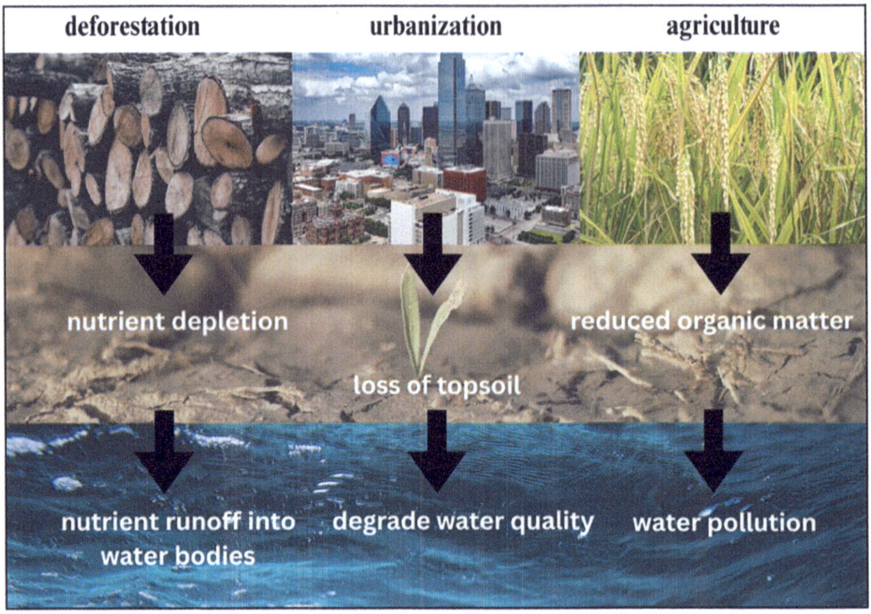

Fig. 1 The effects of deforestation, urbanization, and high-demand agriculture on the soil and water bodies

benthic plants, promote hypoxia, and lead to shifts in aquatic food web. (McDowell & Hamilton, 2023). These activities have causes nutrient depletion, loss of topsoil (soil erosion), and reduce organic matter in the soil, thereby leads to water pollution and degradation of water quality as illustrated in Fig. 1 (Zahoor & Mushtaq, 2023).

Deforestation or removal of trees without reforestation has resulted in significant damage to soil functionality leading to imbalance of the ecosystem (Zaman et al., 2025). When land is cleared for agricultural, urbanization and industrial purposes, it affects soil organic matter content and the structural stability of soils, causing soil erosion and soil degradation problems (McDowell & Hamilton, 2023; Zahoor & Mushtaq, 2023; Albou et al., 2024). Forest management and operations, on the other hand, have the potential to degrade water quality through diffuse (non-point source) pollution from nutrients runoff followed by fertiliser application, as well as increased sedimentation and carbon transport after planting and harvest (Shah et al., 2022). This nutrient transport can result in eutrophication of receiving waters, where algal growth depletes oxygen levels eventually leading to adverse impacts on marine ecosystems (McDowell & Hamilton, 2023).

In conclusion, addressing nutrient runoff is essential for protecting water quality and preserving aquatic ecosystems. By implementing sustainable practices across agricultural and urban landscapes, and fostering community awareness, we can significantly reduce nutrient pollution. There is also a need to obtain better understanding of the dynamic interactions among multiple factors (e.g., sources, land use, characteristic of catchment and climate, management strategies) that control fate and transport of nutrients into the water streams.

1.2 Nutrient Cycle

By definition, nutrient cycling occurs when nutrients move between various parts of a cell, community, or ecosystem and can be reused again by some of these parts. Nutrients cycling in the soil includes water from rain, dead leaves, wood, fungus, bacteria, and other ground living organisms (Fahad et al., 2024). This process involves trees and other plants taking up mineral and non-mineral nutrients from the soil via roots and these nutrients are stored in parts of the plants such as flowers and leaves. The nutrients are transferred back to the soil when the animal consume the plants or when they die. The dead animals will be break down by arthropods, earthworms, fungi, and bacteria in the soil (Fahad et al., 2024). The most important nutrients that are imperative for sustaining life on Earth are include carbon (C), nitrogen (N), and phosphorus (P) as these cycles ensure the availability of nutrients that needed for growth, reproduction, and ecosystem functioning (Zaman et al., 2025). As for plant-soil systems, these cycles are important in nutrient exchange between plants and soil while regulating ecosystem productivity (Zaman et al., 2025). Meanwhile, nitrogen is also essential in plant-role systems by converting inert atmospheric nitrogen into biologically available form that plants can absorb and utilize (Zaman et al., 2025). The nitrogen cycle is comprised of nitrogen fixation, a process that provides a utilizable nitrogen source to plants and higher organisms, and the effective reversal of this process by the nitrification and denitrification, responsible for returning fixed N_2 to the atmosphere (Gopalasingam & Hasnain, 2022). It is also worthy to note that nitrogen cycle is one of the vital cycles on our planet Earth after the carbon cycle (Necula et al., 2022).

Unlike the carbon and nitrogen cycles, phosphorus does not have a significant atmospheric component because it is primarily stored in rocks and minerals as phosphate compounds (Roy et al., 2022). Phosphorus cycle has four major components which is tectonic uplift and exposure of phosphorus-bearing rocks to the forces of weathering; physical erosion and chemical weathering of rocks producing soils and providing dissolved and particulate phosphorus to rivers; riverine transport of phosphorus to lakes and the ocean; and sedimentation of phosphorus associated with organic and mineral matter and burial in sediments (Ruttenberg, 2019).

The carbon, nitrogen, and phosphorus cycles are closely interconnected, as these elements are essential for life and interact within ecosystems. For example, biological nitrogen fixation by both symbiotic and non-symbiotic organisms is an important energy-demanding process that is mainly fuelled by available carbon sources but is also coupled to phosphate availability (Celi et al., 2022). Furthermore, carbon is an essential building block for all life forms on Earth, and its natural recycling from the atmosphere is a constant process from one living organism to another (Limeri &

Reid, 2023). These natural processes, however, are increasingly disrupted by anthropogenic activities such as deforestation, intensive farming, and industrialization, all of which leads to excessive nutrient runoff, hence altering the nutrient balance and degrading water quality (Necula et al., 2022). One major impact is eutrophication, where excessive nitrogen and phosphorus from agricultural runoff enter water bodies, causing harmful algal blooms. The primary driver of eutrophication in ecosystem is runoff of water with phosphorus and nitrogen residue from point sources (Devlin & Brodie, 2023).

Moreover, the disruption of the nitrogen cycle significantly contributes to the release of greenhouse gases, thereby intensifying the effects of climate change (Grace, 2024). This disruption is largely driven due to the rapid increasing of burning fossil fuels, intensive livestock farming, and over application of synthetic fertilizers shown in Fig. 2 (Necula et al., 2022).

In general, the disruption of this biogeochemical cycle can lead to several adverse effects such as nutrient imbalances, reduced crop productivity, and soil infertility (Zaman et al., 2025). Therefore, effective solutions to mitigate nutrient runoff are urgently needed on a global scale. For example, nutrient runoff can be minimized by reduce the use of synthetic fertilizer and adopting an organic alternative such as compost and biofertilizer, which also can improve the soil health and maintaining the nutrient cycle component (Kumar et al., 2022). By maintaining a balanced carbon, nitrogen, and phosphorus cycle, the biodiversity can be protected, enhance food security, and ensure the long-term stability of Earth's ecosystems for future generations.

A recent study by Said et al. (2025) studied on a slow-release fertilizer (SRF) which have emerged as a promising solution in replacing synthetic fertilizers, which

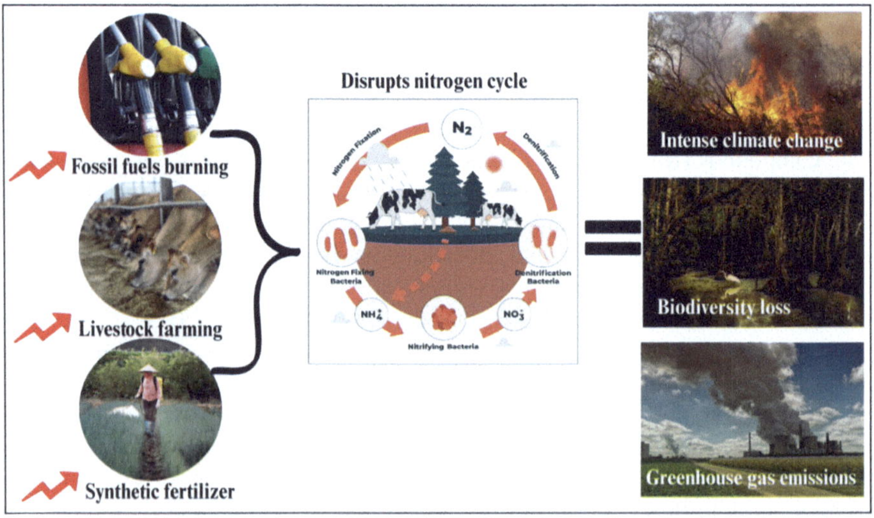

Fig. 2 The impact of nitrogen cycle disruptions

significantly reduce the nutrient runoff. This research study aims to provide an in-depth discussion on environmental and ecological consequences associated with the application of synthetic fertilizers, particularly on their contribution to the increasing greenhouse gases (GHG). While these synthetic fertilizers known to increase the crop yield, the nutrients applied however are not fully taken up by plants. This study highlights that only 10–20% of applied phosphorus are absorbed by crops, whereas about 40–70% of nitrogen applied are lost to the environment via rain event, irrigation, and volatilization. Eventually, these nutrients loss are the result of the eutrophication and water pollution, poses serious threats to aquatic ecosystems and water quality. Therefore, SRF offers a solution by gradually releasing these nutrients over time, hence reducing nutrient loss and improving the efficiency of fertilizer applications compared to the conventional fertilizer by releasing the nutrient based on the plant needs and environmental settings. Moreover, this approaches also comprises of several SRF technologies namely, polymer-coated fertilizers (PCFs), sulphur-coated fertilizers (SCFs), bio-based slow release fertilizers (BB-SRFs), wax-coated fertilizers (WCFs), and clay-coated fertilizers (CCFs). These technologies were designed to modify the materials of the chemical through physical, chemical, or biologically. SRF offer various advantages including reducing the environment impact, enhancing soil health, controlling nutrient release, and improving crop quality. However, further research is necessary to address the complexities associated with the production which results in high production costs. Notable, there is a growing need to replace non-biodegradable coatings into biodegradable and cost-effective materials with the aim to ensure environmental sustainability.

2 Best Management Practices (BPMs)

With a growing population, urbanization is expanding, increasing pressure on agriculture to meet food demands (Zahoor & Mushtaq, 2023). This leads to water pollution from runoff and leaching, harming ecosystems and drinking water sources. Best management practices (BMPs), also known as low impact development (LID), are widely used to mitigate pollution, protect water quality, and enhance sustainability.

2.1 Agricultural BMPs

The agricultural sector's revolution to enhance crop productivity must be driven through the best management practices, which ensure the sustainability of the process by minimizing waste, and optimizing resource use while reducing environmental impact. Precision agriculture technologies (PATs) and cover cropping are the two most used approaches in farming. Both methods have been increasingly adopted worldwide in some countries such as Ethiopia, Africa, and the United States of America (USA) (Getahun et al., 2024; Quintarelli et al., 2022). In this chapter, we covered PAT's technology and its key idea on how the technology is being applied in agricultural sectors worldwide because of the versatilities of the technology and the efficiency of the strategies is better than the cover cropping method in fertilizers runoff management.

Precision agriculture offers innovative tools to optimize crop production while minimizing environmental impacts. Key technologies include remote sensing and drones, which assess crop stress status by providing high-resolution imagery for monitoring crop health, soil conditions, and pests. Secondly, Global Positioning System or GPS-guided machinery ensures accurate planting, fertilization, and harvesting to reduce waste. Lastly, variable rate technologies (VRT), like targeted sprayers and precision planters, help reduce input waste (Getahun et al., 2024). Additionally, data analytics and decision support systems integrate multi-source data streams to guide site-specific interventions, enhancing productivity and sustainability of crop production (Naresh et al., 2024).

Remote sensing and drones are revolutionizing modern agriculture, improving culture monitoring while promoting sustainable practices. The importance of Unmanned Aerial Vehicles or UAVs in sustainability and agricultural monitoring has been increasingly recognized (Kannan et al., 2025; Guebsi et al., 2024). These technologies support precision agriculture by optimizing resource management while reducing pesticide and fertilizer usage (Yadav & Sidana, 2023). Effective weed management through drones has been observed (Roslim et al., 2021), and UAVs facilitate monitoring in small farms (Gokool et al., 2023; Hafeez et al., 2023). Precision techniques, including machine learning and sensor technologies, have been shown to improve sustainable agricultural production (Getahun et al., 2024; Rane & Choudhary, 2023; Sabir et al., 2024). Recent advances in remote sensing allow extensive data analysis, contributing to agricultural practices efficient at varying scales (Omia et al., 2023). The integration of hyperspectral remote sensing increases precision agriculture in climatic conditions (Pande & Moharir, 2023). Through these innovations, the future of agriculture seems to align with sustainability goals by addressing environmental concerns (Singh et al., 2023; Nemade et al., 2023).

GPS technology is a cornerstone of precision agriculture, providing accurate geographic positioning and enabling precise management of farmland. It facilitates site-specific crop management by allowing for precise planting, fertilizing, and harvesting, which reduces waste and enhances efficiency (Getahun et al., 2024; Hu 2023). GPS-guided machinery ensures that inputs are applied accurately, minimizing over-application and reducing environmental impacts such as nutrient runoff and greenhouse gas emissions (Getahun et al., 2024; Sharma, 2023). GPS contributes significantly to sustainable crop production by optimizing resource use and enhancing productivity. These technologies enable data-driven approaches to manage crops, soil, and resources more efficiently, promoting practices that conserve water, reduce chemical usage, and enhance soil health (Getahun et al., 2024; Kayastha, 2024). GPS allows for precise application of fertilizers, pesticides, and water, reducing waste and ensuring that inputs are applied only where needed. This leads to improved nutrient use efficiency and minimizes environmental impacts such as nutrient runoff and pollution (Getahun et al., 2024). GPS-based auto-guidance systems in agricultural vehicles help reduce the overlapping of equipment passes, saving fuel, labour, and time. This also decreases soil compaction, contributing to better soil health (Pandey et al., 2021). By providing accurate positional information, GPS enables better monitoring and management of crop health and growth conditions. This supports informed decision-making and can lead to higher yields and increased profitability (Nazarov et al., 2024; Hu 2023).

Variable rate technology (VRT) is a key advancement in precision agriculture, enabling variable input application using data from soil sensors, satellite images, and weather forecasts to optimize yields (O'Shaughnessy et al., 2019). A crucial application is variable rate irrigation (VRI), which adjusts water distribution based on crop needs and soil conditions, unlike traditional uniform irrigation (Lo et al., 2016). VRI enhances efficiency, conserves water, and promotes deeper root growth, leading to healthier plants and improved yields. It minimizes over- and under-irrigation, ensuring better resource use and higher agricultural productivity.

The implications of improved irrigation efficiency extend beyond agricultural productivity. Enhanced practices increase water conservation efforts, significantly reducing the overall water footprint of agriculture. This is particularly protruding in regions facing water scarcity or the competing demands of water from urban areas. In addition, the application of VRI has been associated with a reduction in groundwater exhaustion, as evidenced by studies that highlight the role of technology in optimizing water use, thus supporting vital aquifers for future generations (Neupane & Guo, 2019). In addition to its contributions to irrigation efficiency, VRT plays a crucial role in minimizing fertilizer and pesticides. Through specific site application methods, VRT ensures that agrochemicals are applied only where they are needed, based on precise health assessments of soil crop and nutrient levels (Fabiani et al., 2020) This targeted application reduces the likelihood of excess chemicals that leach into nearby water courses and mitigates the risk of harmful algae and other forms of aquatic pollution. Consequently, the reduction in chemical flow is a critical component to promoting an agricultural paradigm that prioritizes environmental health, aligning itself with sustainable agricultural practices.

Variable rate technology (VRT) is transforming agriculture by reducing the amount of fertilizer applied, while also optimizing the timing and placement of applications. As a result, issues such as nutrient leaching and runoff into the environment and surrounding water bodies could be significantly reduced (Mcfadden et al., 2023). By optimizing resource use, VRT boosts productivity while promoting long-term environmental conservation. This precision approach ensures targeted input management, lowering contamination risks and supporting a more resilient agricultural future.

In addition, the adoption of VRT offers substantial economic implications for farmers. As highlighted by Irmak et al. (2021), economic benefits come from reductions in entry costs and improvements in crops attributable to more efficient resource management. By applying only, the necessary number of inputs, farmers can significantly reduce their expenses on fertilizers and pesticides. This not only results in cost savings but also helps improve profitability, since yields can be maximized without incurring excessive input costs. In addition, the efficient use of water facilitated by VRT improves irrigation practices, which finally leads to better resource and sustainability management. Empirical studies have corroborated the effectiveness of variable rate technology in several agricultural sectors. Chen et al. (2024) emphasizes the advantages of the VRT in the cotton culture, which indicates that the localized application of water and nutrients correlates with an increase in capsules per plant and better fibre quality. These studies illustrate that VRT's personalized approach not only serves to improve agricultural results but also promotes agricultural practices of the environment.

2.2 *Urban and Residential BMPs*

Sustainable urban and residential agriculture refers to agricultural practices integrated into urban areas, designed not only to produce food but also to improve environmental sustainability, economic resilience and social cohesion (Pradhan et al., 2024). Increasing urbanization resulted to the disturbance of the urban ecosystem such as flooding and drought that induced from the global climate change. In response, growing attention is being directed toward the development of sustainable food systems, as highlighted in a study by (Evans et al., 2022). This study highlights the multiple benefits of urban agriculture including reduction of greenhouse gas emissions, increase biodiversity, control microclimates regulation, increased pollination, and improve health and well-being of the community. Urban agriculture includes integrating food production within or on existing grey infrastructure such as walls, rooftop cultivation, and balcony through innovative systems such as hydroponic, aquaponic, and aeroponic.

Residential agriculture, increasingly popular in urban and suburban areas, enhances sustainability, economic growth, and community cohesion. It boosts biodiversity, reduces carbon footprints, and supports local wildlife by creating green spaces and mitigates climate change by capturing carbon, filtering air pollutants, and managing rainwater (Raihan & Tuspekova, 2022).

From an economic point of view, residential agriculture acts as a catalyst for the creation of jobs and the strengthening of local economies. Reported by Iida et al. (2023), the initiatives that promote urban agriculture can lead to a wide range of job opportunities, ranging from agricultural production and to transformation to retail and marketing. These works can revitalize economically anguished neighbourhoods, improve access to food and stimulate local markets, thus generating a multiplicative effect on regional economic health.

Many studies have illustrated the effectiveness of vertical agriculture and green roofs as sustainable urban agricultural strategies. Green roofs contribute not only to food production, but also improve urban biodiversity, reduce the effect of urban heat island and improve air quality by acting as natural filters for pollution (Fig. 3). Meanwhile, vertical agriculture offers an efficient use of space and resources, increase

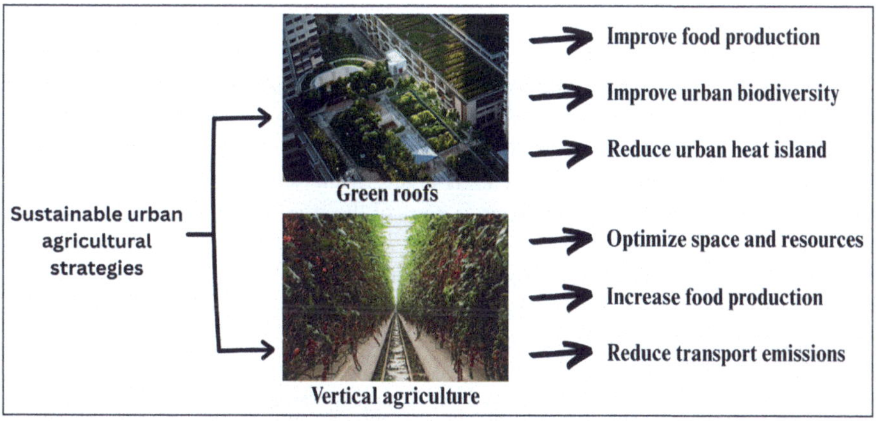

Fig. 3 Advantages of sustainable urban agricultural strategies which are green roofs and vertical agriculture (Daneshyar, 2024)

food productivity, and reduce transportation-related emissions (Fig. 3). The inclusion of various plant species in urban green spaces contributes to the resilience of ecosystems, highlighting interconnectivity between urban agricultural practices and broader environmental health.

The COVID-19 pandemic underlined the fragility of food supply chains and caused a reassessment of urban food security. Marusak et al. (2021) present how disturbances in world supply chains have propelled urban agriculture under spotlights as a means of improving food resilience. While consumers become more aware of the vulnerabilities in food supply, urban farms have positioned themselves as vital resources for communities, facilitating localized food production and consumption that strengthen food security in times of crisis. Noteworthy is the fact that BMPs such as green roofs, raingarden, porous pavement, rainwater harvesting, bioretention ponds, and bioretention swales has been proven to be effective from in minimizing flooding occurrence by lowering peak discharge and flooding volume (Bhusal et al., 2024).

3 Policy and Regulatory Framework

3.1 Existing Regulations

To reduce water pollution from the agricultural activities, various policies and regulations should be implemented and at the same time gives awareness and education to the farmers in managing their agricultural in sustainable way (Zahoor & Mushtaq, 2023). Farmers play a vital role in mitigating the nutrients pollution in water bodies and therefore policies such as reducing nitrogen (N) pollution that can only be successful when farmers adopting sustainable agricultural practices such as intercropping and use moderate amount of fertilizer (Martínez-Dalmau et al., 2021). Policies can be categorized as mandatory measures (quotas, technical requirements, calendars, technologies), voluntary and incentives-based measures, and awareness-raising and private initiatives (Martínez-Dalmau et al., 2021).

A study conducted by Afroz et al. (2014) has discussed the timeline of the water resource policies in Malaysia. In order to control the river pollution that occurred in Malaysia, the pollution-related legislation namely Waters Act 1920 was introduced in 1920s before Environmental Quality Act (EQA) was established in 1974. EQA, 1974 was established to prevent, mitigate, control the pollution, and improve the environmental resources in the country. With the aim to serve as a framework for other laws and regulations, the Act was revised on numerous occasions due to the rising levels of pollution from 1977 (Hezri & Hasan, 2006). It is worth noting that EQA 1974 has act as a pivotal piece of legislation in the country by providing a robust legal framework to combat rising pollution, to enhance sustainable development, and most importantly, protect public health by safeguarding the drinking water sources.

According to Afroz et al. (2014), in order to achieve water pollution control goals, various laws or order and regulations were revised over time to tackle the agro-based water pollution issues. Street, Drainage, and Building Act of 1974, Section 70A of the Act was enacted to protect water resources pollution that caused by soil erosion and siltation where practices for earthwork are required. By means of this law, authorities were given the power to use their discretion to reject any activities that may potentially generate soil erosion and siltation. To further strengthen local authorities' powers, the Local Government Act, 1976 was enacted. The Act under Section 7A, has given authority to prevent inland pollution from public nuisance and to protect the drinking water resources from being polluted.

Additionally, an act was established in 1976 to include provisions for preventing environmental pollution. Town and Country Planning Act of 1976 does not specifically target water pollution, but it outlines regulation for land development, usage, and strategies to enhance physical environment. Moreover, another act to address the primary contributors to inland water pollution namely siltation and soil erosion was established under Act of 1970, the Land Conservation Act (Afroz et al., 2014). These pollutants had mainly come from the non-point sources; therefore, a regulatory policy was implemented at both state and federal levels in Malaysia, with the aim to mitigate the pollutants impact on water quality (Afroz et al., 2014).

3.2 *Incentives for Sustainable Practices*

Agriculture plays a crucial role in food security, economic development, and environmental sustainability. However, conventional farming systems including the use of synthetic chemical, fertilizer, pesticides and so on are often not compatible with natural environmental systems (Epule et al., 2023). In a long-term effect, the fertilized soil becomes overloaded with chemical toxins, and low in nutrients which is destroying soil health and environment (Panhwar et al., 2019).

Therefore, shifting to sustainable agriculture is now recognized as an important incentive for any development programs. Environmentally sustainable systems are constantly exploring for ways to increase the efficiency of resource utilisation, such as water, land, and nutrients (Andretta et al., 2021). Align with sustainable practices, Good Agricultural Practices (GAPs) play a critical role in promoting sustainability by ensuring efficient resource use, environmental conservation, and food safety, which directly influences the availability and effectiveness of sustainable incentives.

GAPs is a set of agricultural management practices used at the farm and post-production levels for producing safe and quality artificial products and food. However, the term GAP is used differently depending on the farmer's demands, the type of agriculture, and the producing country. For example, Palm oil GAP in Malaysia includes nine management techniques: land preparation, soil conservation, weed control, fertiliser application, pruning, pest and disease control, harvesting, and record keeping (Mohd Suib et al., 2023). Government institutions have the holistic vision required to determine the needs and pathways to implement sustainable agriculture practices.

To further encourage farmers to adopt GAPs, financial incentives such as subsidies, tax rebates, and grants are often introduced by governments and agricultural agencies. For instance, the farmers more likely to prefer subsidies or cost shared programs to promote the efficient use of nutrients and water, which helps minimize runoff and soil degradation (Palm-Forster et al., 2017). In a previous review highlighted a higher subsidy not only looking at the feasibility of applying strategies in a conventional technique but looks at the indirect factors that enhance cost-mitigation, improve nutrient removal efficiency and minimise sediment transport to environment (Luna Juncal et al., 2023).

On the other hand, policy and regulatory incentives are essential tools used by governments to encourage sustainable agricultural practices and maintain long-term environmental protection. Environmental regulations have been implemented in various countries to ensure sustainable agricultural practices, reduce environmental pollution and promote biodiversity conservation. Some regulations, such as those aimed at reducing pollution and greenhouse gas emissions, have been effective in improving air and water quality and reducing the negative impact of agricultural practices on the environment (Pande et al., 2023). For instance, Agri-Environmental Schemes (AES) implemented in the European Union under the Common Agricultural Policy (CAP) require farmers to meet specific environmental standards in exchange for financial support, such as maintaining soil cover, reducing pesticide use, or preserving biodiversity (Zindler et al., 2024).

To summarise, incorporating sustainable farming techniques is not only an option, but a initiative for ensuring the future of food supply, environmental health, and rural economies. Conventional farming practices, which depend significantly on synthetic fertilisers and pesticides, may result in short-term production but represent major long-term risks to soil fertility, water quality, and general ecosystem health. This focusses on the significance of transitioning to sustainable agriculture systems that prioritise efficient resource utilisation, environmental conservation, and the well-being of future generations.

4 Monitoring and Assessment Techniques

4.1 Significance of Evaluation and Monitoring

The urge for crop production increase with the rapid population that is expected to reach 9 billion people in 2050–2060 (Zahoor & Mushtaq, 2023) with expected increasing of food production by 50–70% as stated by Food and Agriculture Organization of the United Nations. Fertilizers and manure are vital to enhance nutrients in the soil in order to promote growth of crop production. Nutrients such Nitrogen (N), Phosphorus (P), and Potassium (K) are the crucial nutrients for growth of crop and development (Kleinman et al., 2011; Naeem et al., 2020). In addition, high concentration of Phosphate (P) or animal manure has long been known to increase crop production. However, the improper use and application of these have deteriorate the environment, affected the living beings. Excessive usage of fertilizers causes nutrient stock buildup in the soil which then resulted to soil leaching and runoff during raining event or irrigation (Cai et al., 2025) Moreover, high intensity of rainfall resulted in high amount of P accumulate in the water bodies thereby exacerbate water pollution issue (Kleinman et al., 2011). On top of that, P is significantly less soluble than nitrate, yet nitrate is more easily mobilized during surface runoff, further intensifying water quality issues (Change et al., 2022).

Leaching and surface runoff will release excess nutrients, heavy metals from fertilizer into the waterway and polluted the water and contribute to problem such as eutrophication (Zahoor & Mushtaq, 2023). It is evident that agriculture-related water pollution has become a significant issues world widely but worsen in the developing country due to conventional agricultural practices, lack of education among the farmers (Zahoor & Mushtaq, 2023; Cai et al., 2025). In light of this, water quality monitoring is vital to safeguard public health and ecosystems by providing critical information in order to improve water quality. There is also a need to determine and pinpoint 'hotspot' regions that are prone to nutrient loss continuously with the aim to mitigate the adverse effects of the excessive nutrient runoff (Masud et al., 2025). By assessing these 'hotspot' regions, policymakers and farmers could established and implement several options of more effective management strategies to not only minimize ecological harm but also enhances agricultural sustainability while protecting water quality (Masud et al., 2025). Water quality monitoring methods ranges from traditional approaches to advanced methods with advanced technologies such as smart monitoring systems and real-time data monitoring, Internet of Things (IoT), and artificial intelligence (Masud et al., 2025; Miller et al., 2023).

The integration of advanced technologies has revolutionized data collection and thus providing real-time insights that significantly enhance accuracy, promote effective management strategies, mitigate environmental risks, improve rapid detection and provide prediction that able to help researchers and policymakers to foresee any changes that could prevent any unwanted crisis. These improvements have become essential in this era where water pollution issues have emerged as a global challenge that threaten public health risk and destructing the environment.

4.2 Water Quality Monitoring

Due to endless problem of water pollution resulted from agriculture activities, environmental monitoring has been implemented and should play a vital role in mitigating this problem (Masud et al., 2025). Eutrophication is a process where high input of nutrients depletes oxygen levels in the water and kills off aquatic life (McDowell & Hamilton, 2023) has become a global issue as it effects drinking water source, recreation, reduce the aesthetics value of tourist attraction, and may pose risks for public health (McDowell & Hamilton, 2023). Therefore, daily water quality monitoring is crucial to prevent further pollution and control accurate water quality assessment allows specific authorities to plan for sustainable water resources management (Loi et al., 2022). Every state holds the responsibility for monitoring water quality, with each country and state applying their own quality standards to evaluate the water quality (Frincu, 2025). This monitoring typically comprises of hydrological, physical, chemical, and biological parameters which may be measured on site or analysed through laboratory testing.

There are two main methods of water monitoring that are widely used and recognized in Malaysia which are Water Quality Index (WQI) and National Water Quality Standards (NWQS) (Norlida et al., 2021; Zainudin, 2010). To evaluate the contamination level of the water, Water Quality Index (WQI) was proposed by Malaysia Department of Environment (DOE) and has been a practised for over 25 years (Ibrahim et al., 2021) and classification of NWQS is based on WQI which helps to determine which category falls into under the NWQS systems. NWQS comprises of six classes namely I, IIA, IIB, III, IV, and V with Class I representing the highest quality and Class IV representing the poorest quality as per shown in Tables 1 and 2 (Norlida et al., 2021).

4.3 Modelling and Predictive Tools

Several monitoring technologies and systems have been established, integrating real-time data communication technologies to provide ongoing water quality assessments (Masud et al., 2025). These modern approaches utilize water quality indices, multivariate statistical techniques, environmental remote sensing (ERS), and geographic

Table 1 Water classes and explanation (Norlida et al., 2021; Zainudin, 2010)

Class	Explanation
Class I	Conservation of natural environment Water Supply I—Practically no treatment necessary Fishery I—Very sensitive aquatic species
Class IIA Class IIB	Water Supply II—Conventional treatment Fishery II—Sensitive aquatic species Recreational use body contact
Class III	Water Supply III—Extensive treatment required Fishery III—Common, of economic value, and tolerant species; livestock drinking
Class IV	Irrigation
Class V	None of the above

Table 2 Water quality index classification by DOE (Norlida et al., 2021; Zainudin, 2010)

Parameter	Unit	Class				
		I	II	III	IV	V
Dissolved oxygen (DO)	mg/L	>7	5–7	3–5	1–3	<1
pH	–	>7	6–7	5–6	<5	>5
Ammoniacal nitrogen (AN)	mg/L	<0.1	–0.3	0.3–0.9	0.9–2.7	>2.7
Biochemical oxygen demand (BOD)	mg/L	<1	1–3	3–6	6–12	>12
Chemical oxygen demand (COD)	mg/L	<10	10–25	25–50	50–100	>100
Total suspended solid (TSS)	mg/L	<25	25–50	50–150	150–300	>300
Water Quality Index (WQI)	–	<92.7	76.5–92.7	51.9–76.5	31.0–51.9	>31.0

information systems (GIS) that play a crucial role in water quality assessment as these methods facilitate extensive data collection and analysis, enabling the application of indices that quantify water cleanliness and determine overall water quality status (Arman et al., 2025).

A study by Masud et al. (2025) proposed a water runoff quality monitoring system that encapsulates screen-printed electrochemical sensor electrodes that use an injection-molded (IM) process. The innovative method creates a Lego-like structures, allowing the customization and assembling of sensor arrays designed to measure specific environmental parameters. It is worth to note that this system able to reduce the labour-intensive water sampling work while providing a real-time data that could determine the high-risk nutrient loss region and to optimize the fertilizer application.

Several studies have incorporated statistical and mathematical methods such as chemometrics approach to enhance the interpretation of environmental data, which can often be multifaceted. Loi et al. (2022) has done a comprehensive water quality monitoring in Malaysia (Selangor River basin, Langat River basin, and Klang River basin) by applying chemometrics approaches such as agglomerative hierarchical cluster analysis, principal component analysis, factor analysis, and Man-Kendall trend analysis. Fourteen parameters such as pH, colour, hardness, dissolved solid (DS), total suspended solid (TSS), turbidity, Biochemical Oxygen Demand (BOD), Chemical Oxygen Demand (COD), ammoniacal nitrogen (NH_4^+–N), nitrate nitrogen (NO_3^-–N), ortho-phosphate phosphorus (PO_4^{3-}–P), iron (Fe), manganese (Mn) and magnesium (Mg) were collected to provide information for overall water quality. From the study, the water pollution from the three locations were causes by the abundance of ammonium and nitrate. Hence, the study aimed to provide future framework to reduce the impact from the excessive nutrients and believed that it could be done by adopting sustainable agricultural practices.

Daw et al. (2018) utilized geographic information systems (GIS) to calculate water quality index (WQI) with NWQS as its base in his research. GIS is used to classify water quality and to determine principal component analysis (PCA) pollution sources of the Nerus River basin. The study showed the correlation between land use and water quality and concluded that land use was negatively linked to nutrients and organic factors. The river was classified as Class II which showed that the river was slightly polluted and Class III which showed moderately polluted during the studies from year 2000 and 2013.

A study by Al-Adhaileh and Alsaade (2021) has used artificial intelligence approaches to predict water quality with the aim to control water pollution, ensuring the quality of drinking and wastewater treatment and promote sustainable and green environment. To predict the water quality index (WQI), the adaptive neuro-fuzzy inference system (ANFIS) algorithm, while feed-forward neural network (FNNN) and K-nearest neighbours were applied for water quality classification. The study showed that ANFIS model performed better than other methods in predicting Water Quality Index (WQI). ANFIS also provide more precise WQI predictions.

Emerging technologies in this modern era have proven to be imperative for achieving more efficient and accurate data collection while reducing the burden on farmers, and supporting the pursuit sustainable nitrogen fertilizer management in agriculture. To date, a study conducted by Cai et al. (2025) has highlighted how new technologies and innovations such as Internet of Things (IoT), integration and deep learning models have become pivotal in managing N fertilizer rate by providing real-time data on environmental and soil conditions in response to emerging agricultural challenges. IoT tools include soil moisture sensors, weather monitoring systems, and crop health detectors have appeared imperative for the farmers to be able to respond fast to the unexpected weather shift in which this helps to reduce the nitrogen loss and improve crop yields. This review explores the evolving statergies for optimizing nitrogen development fertilizer rate optimization strategies by observing their objectives, benefits, and limitations. The study also highlights the policies and emerging technologies that is required to accelerate the adoption of these approaches. Ultimately, this study goal is to strike a balance ensuring food security and safeguarding environmental sustainability. This paper has outlined several inevitable challenges though having modern technologies and modelling tools where the adoption of nitrogen fertilizer rate proposition strategies is low in number globally. These challenges can be tackled by combining technical, economic, educational, social, and policy-related factors. Hence, overcoming these obstacles is essential to enhance nitrogen use efficiency and promoting sustainability agriculture practices and supporting progress toward the Sustainable Development Goals (SDGs).

5 Conclusion

Rapid population growth has led to high demand of crops, resulting in extensive use of synthetic fertilizers that rich with nutrients such as nitrogen, ammonia, and nitrogen oxide. The increasing usage of this fertilizer has adversely impacted water quality where it affected health of human and aquatic organisms, disrupt soil nutrient cycle, contribute to the intensification of global warming and biodiversity loss. Furthermore, the misapplication of the fertilizer such as over application of the fertilizer, lack of knowledge of the farmers has further exacerbated these issues. Therefore, the needs for comprehensive data collection, increased awareness regarding sustainable agriculture and best management practices (BMPs), the effective use of modelling tools, and implementation of policies is vital and should be urgently addressed and adopted globally. By managing the fertilizer runoff and adopting sustainable agriculture, we are able to preserve the quality of the soil and water for the next future generations. Cutting-edge technology and artificial intelligence-based modelling tools such as ANFIS, FNNN, IoT, ERS, and GIS have demonstrated significant potential in optimizing nutrient management of urban farms. To ensure the successful implementation of these collaborative efforts of the farmers, researchers, industry, and policymakers should be done in order to make certain of the application of the nutrient management approach while ensuring environmental protection. By combining this collaboration while integrating advanced technologies, and sustainable practices, the impacts of the fertilizer runoff will be significantly reducing. Hence, research on this topic should be continuously studied in order to adapt the strategies to local context across the world.

Acknowledgements The work was supported by the Ministry of Higher Education (MOHE), via Fundamental Research Grant Scheme (FRGS) [FRGS/1/2022/STG02/USM/02/6].

References

Afroz, R., et al. (2014). Water pollution: Challenges and future direction for water resource management policies in Malaysia. *Environment and Urbanization ASIA, 5*(1), 63–81. https://doi.org/10.1177/0975425314521544

Al-Adhaileh, M. H., Alsaade, F. W. (2021). Modelling and prediction of water quality by using artificial intelligence. *Sustainability (Switzerland), 13*(8), 1–18. https://doi.org/10.3390/su13084259

Albou, E. M. et al. (2024). Agricultural practices and their impact on aquatic ecosystems—a mini-review. *Ecological Engineering and Environmental Technology, 25*(1), 321–331. https://doi.org/10.12912/27197050/175652

Andretta, I. et al. (2021). Environmental impacts of feeding crops to poultry and pigs. In *Environmental impact of agro-food industry and food consumption* (pp. 59–79). https://doi.org/10.1016/B978-0-12-821363-6.00001-1

Arman, N. Z., et al. (2025). Water quality assessment of Johor River Basin, Malaysia, using multivariate analysis and spatial interpolation method. *Environmental Science and Pollution Research*, 1766–1782. https://doi.org/10.1007/s11356-024-35692-7

Bhusal, A., et al. (2024). Evaluating the effectiveness of best management practices in adapting the impacts of climate change-induced urban flooding. *Atmosphere, 15*(3), 1–18. https://doi.org/10.3390/atmos15030281

Cai, S., Zhao, X., & Yan, X. (2025). Towards precise nitrogen fertilizer management for sustainable agriculture. *Earth Critical Zone, 2*(February), 100026. https://doi.org/10.1016/j.ecz.2025.100026

Celi, L. et al. (2022). Interconnectig soil organic matter with nitrogen and phosphorus cycling. In *In multi-scale biogeochemical processes in soil ecosystems*. https://doi.org/10.1002/9781119480419.ch3

Change, C., et al. (2022). *Impact of climate change on catchment nutrient dynamics: insights from around the world*.

Chen, C., Hussain, Z., Liu, J., Zaman, M., & Akhlaq, M. (2024). Innovative technologies in sprinkler irrigation: An overview. *Agrociencia*, 1–18.

Daneshyar, E. (2024). Residential rooftop urban agriculture: Architectural design recommendations. *Sustainability (Switzerland), 16*(5). https://doi.org/10.3390/su16051881

Daw, M. M., Omaer, L., & Alabyad, M. (2018). Impacts of land-use changes on water quality by an application of GIS analysis: A case study of Nerus River. *Int J Eng Technol, 7*, 155–164.

Devlin, M., & Brodie, J. (2023). Nutrients and eutrophication. In *Marine pollution-monitoring, management and mitigation* (pp. 75–100). https://doi.org/10.1007/978-3-031-10127-4_4

Epule, T. E. et al. (2023). Nutrient management in agriculture: the contributions of organic and conventional nutrients to food security and the environment in Africa. In *Organic farming (second edition) global perspectives and methods* (pp. 181–201). https://doi.org/10.1016/B978-0-323-99145-2.00018-5.

Evans, D. L., et al. (2022). Ecosystem service delivery by urban agriculture and green infrastructure—a systematic review. *Ecosystem Services, 54*(January). https://doi.org/10.1016/j.ecoser.2022.101405

Fabiani, S., Vanino, S., Napoli, R., Zajíček, A., Duffková, R., Evangelou, E., & Nino, P. (2020). Assessment of the economic and environmental sustainability of Variable Rate Technology (VRT) application in different wheat-intensive European agricultural areas: A Water-Energy-Food nexus approach. *Environmental Science & Policy, 114*, 366–376.

Fahad, S., et al. (2024). *Environment, climate, plant and vegetation growth, environment, climate, plant and vegetation growth*. https://doi.org/10.1007/978-3-030-49732-3

Frincu, R. M. (2025). Artificial intelligence in water quality monitoring : A review of water quality assessment applications. *Water Quality Research Journal, 60*(1). https://doi.org/10.2166/wqrj.2024.049

Getahun, S., Kefale, H., & Gelaye, Y. (2024). Application of precision agriculture technologies for sustainable crop production and environmental sustainability: A systematic review. *The Scientific World Journal, 2024*(1), 2126734.

Gokool, S., Mahomed, M., Kunz, R., Clulow, A., Sibanda, M., Naiken, V., Chetty, K., & Mabhaudhi, T. (2023). Crop monitoring in smallholder farms using unmanned aerial vehicles to facilitate precision agriculture practices: A scoping review and bibliometric analysis. *Sustainability, 15*(4), 3557.

Gopalasingam, C. C., & Hasnain, S. S. (2022). Frontiers in metalloprotein crystallography and cryogenic electron microscopy. *Current Opinion in Structural Biology, 75*, 1–10. https://doi.org/10.1016/j.sbi.2022.102420

Grace, J. (2024). The carbon cycle. In *Encyclopedia of biodiversity* (3rd edn, pp. 380–392).

Guebsi, R., Mami, S., & Chokmani, K. (2024). Drones in precision agriculture: A comprehensive review of applications, technologies, and challenges. *Drones, 8*(11), 686.

Hafeez, A., Husain, M. A., Singh, S. P., Chauhan, A., Khan, M. T., Kumar, N., Chauhan, A., & Soni, S. K. (2023). Implementation of drone technology for farm monitoring & pesticide spraying: A review. *Information Processing in Agriculture, 10*(2), 192–203.

Hezri, A. A., & Hasan, M. N. (2006). Towards sustainable development? *The Evolution of Environmental Policy in Malaysia, 30*, 37–50.

Hu, Y. (2023). Agricultural management system based on GPS and GIS. In *World Conference on Communication & Computing (WCONF)* (pp. 1–6). https://doi.org/10.1109/WCONF58270.2023.10234990

Ibrahim, T. N. B. T., et al. (2021). Seasonal effects on spatial variations of surface water quality in a tropical river receiving anthropogenic influences. *Sains Malaysiana, 50*(3), 571–593. https://doi.org/10.17576/jsm-2021-5003-02

Iida, A., Yamazaki, T., Hino, K., & Yokohari, M. (2023). Urban agriculture in walkable neighborhoods bore fruit for health and food system resilience during the COVID-19 pandemic. *Npj Urban Sustainability, 3*(1), 4.

Irmak, S., Sharma, V., Haghverdi, A., Jhala, A., Payero, J. O., & Drudik, M. (2021). Maize crop coefficients under variable and fixed (uniform) rate irrigation and conventional and variable rate fertilizer management in three soil types. *Agricultural Water Management, 243*, 106489. https://doi.org/10.1016/j.agwat.2020.106489

Kannan, S., Desouza, V. A., Nughal, J., Das, S., Bhattacharyya, U. K., & Yadav, A. (2025). Artificial intelligence and drone-assisted plant disease diagnosis: A review of advancements and challenges. *International Journal of Advanced Biochemistry Research, 9*, 358–366. https://doi.org/10.33545/26174693.2025.v9.i3e.3959

Kayastha, S. (2024). *Growing green: Sustainable agriculture meets precision farming: A review*. Bhartiya Krishi Anusandhan Patrika [Preprint], (of). https://doi.org/10.18805/bkap697

Kleinman, P. J. A., et al. (2011). Managing agricultural phosphorus for water quality protection: Principles for progress. *Plant and Soil, 349*(1–2), 169–182. https://doi.org/10.1007/s11104-011-0832-9

Kumar, S., et al. (2022). Biofertilizers: An ecofriendly technology for nutrient recycling and environmental sustainability. *Current Research in Microbial Sciences, 3*(100094), 1–26. https://doi.org/10.1016/j.crmicr.2021.100094

Limeri, L., & Reid, J. (2023). Plant nutrition and nutrient cycling. In *Intoductory biology* (Vol. 2).

Lo, T. H., Heeren, D. M., Martin, D. L., Mateos, L., Luck, J. D., & Eisenhauer, D. E. (2016). Pumpage reduction by using variable-rate irrigation to mine undepleted soil water. *Transactions of the ASABE, 59*(5), 1285–1298.

Loi, J. X., et al. (2022). Water quality assessment and pollution threat to safe water supply for three river basins in Malaysia. *Science of the Total Environment, 832*(March), 155067. https://doi.org/10.1016/j.scitotenv.2022.155067

Luna Juncal, M. J., et al. (2023). Towards nutrient neutrality: A review of agricultural runoff mitigation strategies and the development of a decision-making framework. *Science of the Total Environment, 874*(162408), 1–28. https://doi.org/10.1016/j.scitotenv.2023.162408

Martínez-Dalmau, J., Berbel, J., & Ordóñez-Fernández, R. (2021). Nitrogen fertilization. A review of the risks associated with the inefficiency of its use and policy responses. *Sustainability (Switzerland), 13*(10), 1–15. https://doi.org/10.3390/su13105625

Marusak, A., Sadeghiamirshahidi, N., Krejci, C. C., Mittal, A., Beckwith, S., Cantu, J., Morris, M., & Grimm, J. (2021). Resilient regional food supply chains and rethinking the way forward: Key takeaways from the COVID-19 pandemic. *Agricultural Systems, 190*, Article 103101.

Masud, M., et al. (2025). Real-time monitoring of fertilizer runoff at the watershed scale using a low-cost solar-powered Lego-like electrochemical water quality monitoring system. *Computers and Electronics in Agriculture, 232*(August 2024), 110064. https://doi.org/10.1016/j.compag.2025.110064

McDowell, R. W., & Hamilton, D. P. (2023). *Nutrients and eutrophication: Introduction, marine and freshwater research*. Springer Nature Switzerland. https://doi.org/10.1071/MF13059

Mcfadden, J., Njuki, E., & Griffin, T. (2023). Precision agriculture in the digital era: Recent adoption on U.S. farms.

Miller, M., et al. (2023). IoT in water quality monitoring—are we really here? *Sensors, 23*(2), 1–12. https://doi.org/10.3390/s23020960

Mohd Suib, N. A. B., et al. (2023). The Influence of Good Agricultural Practice (GAP) on the productivity and well-being of Malaysian Sustainable Palm Oil (MSPO)-certified independent smallholders in Malaysia. *Agriculture (Switzerland), 13*(990), 1–19. https://doi.org/10.3390/agriculture13050990

Moura, N. C. B., Pellegrino, P. R. M., & Martins, J. R. S. (2016). Best management practices as an alternative for flood and urban storm water control in a changing climate. *Journal of Flood Risk Management, 9*(3), 243–254. https://doi.org/10.1111/jfr3.12194

Naeem, M., Ansari, A. A., & Gill, S. S. (2020). *Contaminants in agriculture: Sources, impacts and management*. Cham: Springer International Publishing.

Naresh, R., Jadav, S. K., Singh, M., Patel, A., Singh, B., Beese, S., & Pandey, S. K. (2024). Role of hydroponics in improving water-use efficiency and food security. *International Journal of Environment and Climate Change, 14*(2), 608–633. https://doi.org/10.9734/ijecc/2024/v14i23976

Nazarov, A., Sulimin, V., & Shvedov, V. (2024). Harnessing precision agriculture technologies to enhance efficiency and yield in the agro-industrial complex. In *BIO web of conferences* (Vol. 121, p. 02010). https://doi.org/10.1051/bioconf/202412102010

Necula, D. C., et al. (2022). Nitrogen emissions from agriculture and livestock sector, among the causes of climate change. *Scientific Papers: Animal Science & Biotechnologies/Lucrari Stiintifice: Zootehnie si Biotehnologii, 55*(2), 18–23. https://research.ebsco.com/linkprocessor/plink?id=a93c1906-3614-3e37-9d1e-ea7bfcdad29e

Nemade, S., Ninama, J., Kumar, S., Pandarinathan, S., Azam, K., Singh, B., & Ratnam, K. M. (2023). Advancements in agronomic practices for sustainable crop production: A. *International Journal of Plant & Soil Science, 35*(22), 679–689.

Neupane, J., & Guo, W. (2019). Agronomic basis and strategies for precision water management: A review. *Agronomy, 9*(2), 87. https://doi.org/10.3390/agronomy9020087

Norlida, M. H., Aufa, M. M., Fadzli, A. M. N., Shah, M. M. S., & Czahari, M. (2021). Effect of different fertilizer management on water quality in the paddy field (pp. 15–28). https://doi.org/10.32789/agrofood.2021.1002

O'Shaughnessy, S. A., Evett, S. R., Colaizzi, P. D., Andrade, M. A., Marek, T. H., Heeren, D. M., Lamm, F. R., & LaRue, J. L. (2019). Identifying advantages and disadvantages of variable rate irrigation: An updated review. *Applied Engineering in Agriculture, 35*(6), 837–852.

Omia, E., Bae, H., Park, E., Kim, M. S., Baek, I., Kabenge, I., & Cho, B. K. (2023). Remote sensing in field crop monitoring: A comprehensive review of sensor systems, data analyses and recent advances. *Remote Sensing, 15*(2), 354.

Palm-Forster, L. H., Swinton, S. M., Shupp, R. S. (2017). Farmer preferences for conservation incentives that promote voluntary phosphorus abatement in agricultural watersheds. *Journal of Soil and Water Conservation, 72*(5), 493–505. https://doi.org/10.2489/jswc.72.5.493

Pande, C. B., & Moharir, K. N. (2023). Application of hyperspectral remote sensing in precision farming and sustainable agriculture under climate change: A review. *Climate Change Impacts on Natural Resources, Ecosystems and Agricultural Systems*, 503–520.

Pande, S., et al. (2023). The impact of environmental regulations on agriculture. *International Journal of Advanced Research in Science, Communication and Technology (IJARSCT), 3*(2), 326–331. https://doi.org/10.48175/568

Pandey, P. C., Tripathi, A. K., & Sharma, J. K. (2021). An evaluation of GPS opportunity in market for precision agriculture. In *Elsevier eBooks* (pp. 337–349). https://doi.org/10.1016/b978-0-12-818617-6.00016-0

Panhwar, Q. A., et al. (2019). Fertilizer management strategies for enhancing nutrient use efficiency and sustainable wheat production. In *Organic farming: Global perspectives and methods* (pp. 17–39). https://doi.org/10.1016/B978-0-12-813272-2.00002-1

Pradhan, P., Subedi, D. R., Dahal, K., Hu, Y., Gurung, P., Pokharel, S., Kafle, S., Khatri, B., Basyal, S., Gurung, M., & Joshi, A. (2024). Urban agriculture matters for sustainable development. *Cell Reports Sustainability, 1*(9).

Quintarelli, V., et al. (2022). Cover crops for sustainable cropping systems: A review. *Agriculture, 12*(12), 2076. https://doi.org/10.3390/agriculture12122076

Raihan, A., & Tuspekova, A. (2022). Role of economic growth, renewable energy, and technological innovation to achieve environmental sustainability in Kazakhstan. *Current Research in Environmental Sustainability, 4*, Article 100165.

Rane, N. L., & Choudhary, S. P. (2023). Remote sensing (RS), UAV/drones, and machine learning (ML) as powerful techniques for precision agriculture: Effective applications in agriculture. *International Research Journal of Modernization in Engineering Technology and Science, 5*(4), 4375–4392.

Roslim, M. H. M., Juraimi, A. S., Che'Ya, N. N., Sulaiman, N., Manaf, M. N. H. A., Ramli, Z., Motmainna, M. (2021). Using remote sensing and an unmanned aerial system for weed management in agricultural crops: A review. *Agronomy, 11*(9), 1809. https://doi.org/10.3390/agronomy11091809

Roy, S., et al. (2022). Plant growth-promoting rhizobacteria: an alternative for NPK fertilizers. In *Microbes and microbial biotechnology for green remediation* (pp. 149–167). https://doi.org/10.1016/B978-0-323-90452-0.00007-4

Ruttenberg, K. (2019). Phosphorus cycle. In: Encyclopedia of Ocean sciences (3rd edn, pp. 447–460). https://doi.org/10.1016/B978-0-12-409548-9.10807-3.

Sabir, R. M., Mehmood, K., Sarwar, A., Safdar, M., Muhammad, N. E., Gul, N., Athar, F., Majeed, M. D., Sattar, J., Khan, Z., & Fatima, M. (2024). Remote sensing and precision agriculture: A sustainable future. In *Transforming agricultural management for a sustainable future: Climate change and machine learning perspectives* (pp.75–103). Springer Nature.

Said, N. S. M., et al. (2025). Bridging the gap in nutrient management: Transitioning from conventional to sustainable slow-release fertilizers in modern agriculture. *Journal of Cleaner Production, 513*(May), 145731. https://doi.org/10.1016/j.jclepro.2025.145731

Shah, N. W., et al. (2022). The effects of forest management on water quality. *Forest Ecology and Management, 522*(120397), 1–23. https://doi.org/10.1016/j.foreco.2022.120397

Sharma, S. (2023). Precision agriculture: Reviewing the advancements, technologies, and applications in precision agriculture for improved crop productivity and resource management. *Reviews in Food and Agriculture, 4*(2), 45–49. https://doi.org/10.26480/rfna.02.2023.45.49

Singh, N. K., et al. (2023). Fate of pesticides in agricultural runoff treatment systems: Occurrence, impacts and technological progress. *Environmental Research, 237*(P2), 117100. https://doi.org/10.1016/j.envres.2023.117100

Yadav, N., & Sidana, N. (2023). Precision agriculture technologies: Analysing the use of advanced technologies, such as drones, sensors, and GPS, in precision agriculture for optimizing resource management, crop monitoring, and yield prediction. *Journal of Advanced Zoology, 44*.

Zahoor, I., & Mushtaq, A. (2023). Water pollution from agricultural activities: A critical global review. *International Journal of Chemical and Biochemical Sciences, 23*(1), 164–176.

Zainudin, Z. (2010). Benchmarking river water quality in Malaysia. *Jurutera*, (February), 12–15.

Zaman, W., Ayaz, A., & Puppe, D. (2025). Biogeochemical cycles in plant–soil systems: Significance for agriculture, interconnections, and anthropogenic disruptions. *Biology, 14*(4), 1–29. https://doi.org/10.3390/biology14040433

Zindler, M., et al. (2024). Improving agri-environmental schemes: Suggestions from farmers and nature managers in a Central European region. *Environmental Management, 73*, 826–840. https://doi.org/10.1007/s00267-023-01922-w

Metagenomic Assessment of Microbial Diversity at Five Sewage Treatment Plants (STP) Across Sabah, Malaysia and Impact on Water Quality Parameters

M. R. Abd Rahman, A. Raidin, P. C. Lee, K. N. Balakrishnan, H. Roslie, R. Razali, I. M. Maliki, and K. F. Rodrigues

Abstract Bacterial metagenomics is an established technology that can be applied to assess the microbial diversity across sewage samples and its contribution to the effective treatment of sewage water in effluent treatment plants. The primary objective of this study was to establish a baseline for the microbial population diversity across five Sewage Treatment Plants (STP) in Sabah, Malaysia. The secondary objective was to elucidate the microbial diversity and correlate this with the quality of the treated effluent. Water samples were collected from the intake, processing and release points within each of the five facilities and 16S rRNA gene amplicon sequencing was performed on Oxford Nanopore MinION using the 16S PCR barcoding kit. Sequences were pre-processed, and data were analysed using EPI2ME platform to classify them based on taxonomy. Water quality parameters of the samples were provided by an accredited laboratory. The microbial profiles from the five treatment plants indicated a diverse range of human gut microflora which included *Megamonas funiformis, Anaerobutyricum hallii, Agathobacter rectalis, Ligilactobacillus ruminis, Phascolarctobacterium faecium, Megasphaera elsdenii.* Interestingly, no pathogenic bacteria were detected in the influent. Treatment of the effluent using aerobic and anaerobic processes resulted in a significant decrease in the total microbial populations and a shift in the microbial diversity as the nutrient load declined during treatment. The water quality parameters at all five discharge points of all STP were well within the range of the Environmental Quality (Sewage) Regulation 2009 which implies that the current microbial populations have attained an ecological equilibrium and that the STPs are operating at a high efficiency. The metagenomic data will

M. R. A. Rahman · A. Raidin
Jabatan Perkhidmatan Pembetungan Sabah, Kota Kinabalu, Sabah, Malaysia

P. C. Lee · K. N. Balakrishnan · H. Roslie · R. Razali · I. M. Maliki · K. F. Rodrigues (✉)
Biotechnology Research Institute, Universiti Malaysia Sabah, Jalan UMS, Kota Kinabalu, Sabah, Malaysia
e-mail: kennethr@ums.edu.my

P. C. Lee
Faculty of Science and Technology, Universiti Malaysia Sabah Jalan UMS, Kota Kinabalu, Sabah, Malaysia

A. Z. Yaser et al. (eds.), *Compost and Liquid Organic Fertilizers*,
https://doi.org/10.1007/978-981-95-8561-8_23

establish a baseline that can be applied to monitor the operation of the STPs and carry out necessary interventions in the event of changes in the quality of effluent.

Keywords Sewage · Effluent · Metagenomics · Microbial populations · MinION

1 Introduction

Sewage treatment plants (STP) represent man-made ecological niches that function as biological systems for the processing of effluent that is composed of carbohydrates, lipids, protein and compounds that can be biodegraded or bio transformed via microbial processes. The elucidation of microbial populations with the objective of establishing a base line population across the different processes at a STP constitutes the first objective of this study, followed by correlation of the microbial diversity with the quality of treated effluent. Microbial metagenomics is a powerful analytical method that can be applied to detect both culturable and unculturable microbes from environmental samples. This method has wide ranging applications in situations in which microbial meta data is required to make informed decisions in the design and development of effluent treatment processes. The method facilitates the rapid identification of microbial populations using 16S rRNA sequencing using Next Generation Sequencing platforms (Bukin et al., 2019). Multiple studies conducted globally have revealed insights into the microbial communities involved in the bioremediation of effluent in wastewater treatment facilities (Motteran et al., 2025).

Metagenomics within the scope of effluent treatment has implications for both public health and safety, as well as bioprocess engineering for more efficient treatment of wastewater. The early detection of microbial pathogens, including bacteria and viruses prior to them evolving into a threat to public health has formed the basis for several studies (Mkilima, 2025), one of which identified the presence of viral agents linked to gastroenteritis at a US Air Force Academy (Mechikoff et al., 2024). Bacteria that have evolved antibiotic resistance are another major cause for concern and the presence of Antibiotic Resistance Bacteria (ARB) in wastewater can provide a preliminary insight into the status of these genes with a community. A global study conducted by Tarek et al. (2024) revealed a common set of ARG that could be detected in sewage across multiple countries and tracked them as the wastewater was processed at facilities, and subsequent meta-analyses of the prevalence of ARBs before and after the COVID-19 pandemic highlighted regional variations in ARB (Alhazmi et al., 2024). These global findings were also supported by investigations conducted in Brazil, which revealed ARB implicated in resistance to multiple drugs in untreated sewage (Ramos et al., 2024).

The dynamic nature of the composition of most sewage treatment plants (STP) presents a challenge to bioprocess engineers due to the flux in the baseline microbial nutrients and its impact on the beneficial microbial communities within specific treatment tanks. The general physical design of a STP comprises an intake with physical gratings for the removal of large particulates as well as non-biodegradable content such as plastic waste that can result in clogging of the piping. The filtered effluent is then subjected to anaerobic and aerobic treatment which facilitate the key biochemical processes which include denitrification, general bioremediation, biotransformation, biodegradation, and flocculation.

The bacterial consortia, that inhabit STP are the primary enablers of these processes. The genera that are responsible for the reduction of Nitrate and Nitrite to Nitrogen gas are *Pseudomonas*, *Bacillus* and *Paracoccus*. The genera *Thiobacillus*, *Sulfurimonas* and *Thioalkalivibrio* are involved in Sulfur-oxidizing autotrophic denitrification, *Acidobacteria* and *Bacteroidetes* contribute to denitrification via their involvement in anammox. *Nitrospira* is actively involved in the oxidation of Nitrite to Nitrate which leads to denitrification (Huang et al., 2023). The process of flocculation of suspended particulates has been reported to be promoted by *Bacillus subtilis*, *Bacillus velezensis* and *Phanerochaete chrysosporium* all of which are dominant in activated sludge (Vimala, 2019).

In addition to bacteria, algae and fungi have been reported to contribute to the process of bioremediation and biotransformation of effluents. In sewage treatment plants, several bacterial genera play essential roles in the biodegradation of organic matter during the secondary treatment phase. *Pseudomonas* species are highly versatile and capable of degrading a wide range of organic pollutants, while *Bacillus* species contribute to the breakdown of proteins and other complex compounds. Nitrosomonas and Nitrobacter are key players in the nitrification process, converting ammonia into nitrate, which is crucial for nitrogen removal. Additionally, some strains of *Escherichia*, such as *E. coli*, also participate in organic matter degradation, although they are more commonly used as indicators of fecal contamination. Together, these bacteria significantly reduce biochemical oxygen demand (BOD) and help purify wastewater before it is released into the environment.

The general design of sewage treatment plants currently operation in Sabah, Malaysia comprises three basic elements, an intake which facilitates buffering of the volume of influent and serves a filter for large particulates, an anaerobic and aerobic treatment system, and finally a settlement tank where the residual sludge is retrieved as sediment, prior to release of the treated effluent into water bodies in compliance with the Environmental Quality (Sewage) regulations. A schematic representation of the STP is depicted in Fig. 1.

Raw sewage enters the treatment plant through the influent intake, where it first undergoes pre-treatment to remove large solids and grith through physical screening. The wastewater is then sent to a treatment tank for biological treatment, where microorganisms break down organic pollutants under controlled conditions. After

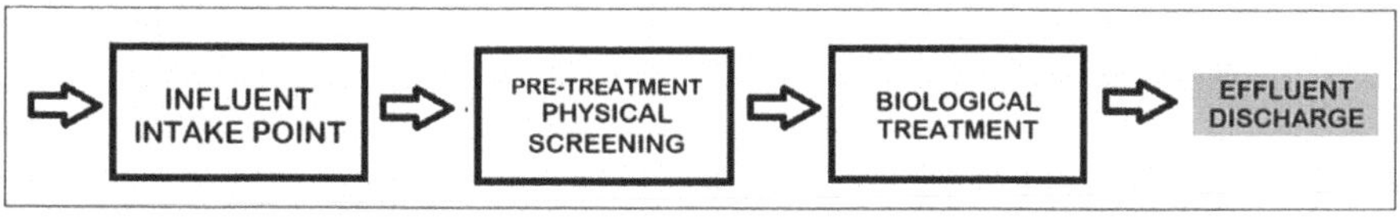

Fig. 1 Schematic representation of the general layout of an STP in Sabah, Malaysia

biological treatment, the water is allowed to settle, separating the sludge from the clear supernatant. The treated water is then tested to ensure that it meets regulatory standards before being discharged into a nearby water body.

The primary objective of this investigation was to determine the baseline microbial population of each of the five STP using a metagenomic approach that focused on bacterial diversity, using current methods in high throughput DNA sequencing. This data was then examined within the context of the water quality parameters to elucidate the relationship between the bacterial population and the quality of the discharged water.

2 Material and Methods

2.1 Study Site and Sample Site Selection

The wastewater samples were collected in triplicates at each sampling point from a wastewater treatment plant located in STP Kuala Penyu (5.562575562210634, 115.59049988711246), STP Kundasang (5.986543905165384, 116.57777701879384), STP Tenom (4.911998350249564, 115.9207615306945), STP Kudat (6.909009953323166, 116.84433295087561), and the regional sewage treatment plant Kota Kinabalu, Sabah (5.916661725965121, 116.07655835457037), in September 2024. The sampling points were: (1) the point where raw sewage enters the plant (influent), (2) the point where the sewage is treated, and (3) the point where the treated sewage (effluent) is released into the environment through the discharge system. Samples were collected randomly at each sampling point with the assistance of the Jabatan Perkhidmatan Pembetungan Sabah (JPP, Sabah).

2.2 Wastewater Sampling

Necessary precautions and personal protective equipment were taken during sample collection. A total of 500 mL of wastewater samples were collected in sterile polypropylene bottles (Nalgene™) and transported to the laboratory in a cold chain environment with information indicating sampling points. The wastewater samples were transported directly to biosafety level-2 (BSL-2) safety cabinet, located at the Biotechnology Research Institute at UMS for immediate processing to minimize the degradation of bacterial DNA. The water quality parameters: pH, Temperature, Biochemical Oxygen Demand in 5 days at 20 oC (BOD), Chemical Oxygen Demand (COD), Suspended solids and Ammoniacal Nitrogen (AMN) were measured and reported by an independent accredited laboratory.

2.3 Wastewater Sample Processing: Filtration, DNA Extraction and PCR Amplification

The collected samples were filtered with 0.22-μm MCE filter membranes (Membrane Solutions Corp), which were connected to disposable funnels and post-filter paper was stored in −20 °C prior to DNA extraction. DNA extraction was carried out using the DNeasy PowerWaterR kit (Qiagen, Germany) following to the manufacturer's instructions. The DNA concentration was evaluated using Qubit 3.0 fluorometer with Qubit dsDNA HS (high sensitivity, 0.2–100 ng) Assay Kit according to the manufacturer's protocols. The concentration of all DNA samples was made sure to be within the range of 1–100 ng/10 μl which is acceptable for metagenomic sequencing. PCR was then carried out with QuantiNova® SYBR Green PCR kit (Qiagen, Germany) to confirm the presence of bacterial DNA by targeting the 16S rRNA gene region with the following primers: forward: (5'—GGGTTGCGCTCGTTGC—3') and reverse: (5'—ATGGYTGTCGTCAGCTCGTG—3') (Cao et al., 2018).

2.4 MinION Library Preparation and Sequencing

The DNA extracted from the triplicate samples at each sampling point was pooled to obtain a total of 14 individual samples, representing different sampling points. Then, the pooled samples were subjected for amplicon sequencing targeting the V1–V9 regions of the 16S rRNA gene and the subsequent library were prepared using 16S Barcoding Kit 24 V14 (SQK-16S114.24) following the manufacturer's instructions. Barcoded 16S rRNA gene amplicons were purified using AMPure XP (Beckman Coulter, Brea, USA) and the final libraries of 50 fmol was loaded onto Spot-ON of the flow cell (R10.4.1) and sequenced using the MinION nanopore sequencer (Oxford Nanopore Technologies, United Kingdom) for approximately 20 h.

2.5 MinION Data Analysis

MinKNOW software version 6.0.11 (Oxford Nanopore Technologies, UK) was used for base calling in a high-accuracy mode and the reads were filtered with minimum Q score of 9 to remove low-quality reads. The base called reads were uploaded in 16 s metagenomic workflow (https://epi2me.nanoporetech.com), a cloud-based algorithm including analytical workflow for 16S rRNA classification, was used to analyze the processed reads to identify the bacterial species. Each file was classified in real-time using the NCBI 16S rRNA gene blast database, which was made available by Oxford Nanopore. Sequences have been assigned for taxonomy classification using Silva database to ensure the highest degree of accuracy, and the default settings have been used to categorize every read. The EPI2ME output was evaluated based on the presence of bacterial species; the species that were detected were those with the highest read count.

3 Results and Discussion

Sample collection: The samples collected from at the influent point comprised raw sewage with a variable consistency and a pH which ranged from 6.5 to 7.6 across the five STP. The pH of the final effluent ranged from 4.40 to 7.50. The concentration and purity of the DNA ranged from 2.3 ng to 46 ng and the average A260:A280 ratio was 2.0.

3.1 Sewage Treatment Plant Kuala Penyu (STP KP)

The STP KP is designed to accommodate one intake point into which effluent is received from the sewage system in Kuala Penyu, and retained in a sludge holding tank to facilitate sedimentations, following which the influent is transferred in batches to one of two reactor tanks. These tanks are closed, and sunlight does not enter the reaction tanks. The influent is subjected to intermittent aeration for two hours aeration followed by one hour of settlement and 1 h of decanting prior to release into the Kuala Penyu bay (Table 1).

The influent contained a high count of pathogenic bacterium *Arcobacter* sp. implying that there may be some ingress of human or animal waste into the system, or alternatively an ingress of saline water from the adjacent marshes. The presence of *Desulfuromonas thiophila*, is indicative of the anoxic nature of the influent as this bacterium has been reported to be a Sulfur-reducing bacterium from anoxic freshwater sediment (Finster et al., 1997). The sludge holding tank was observed to have a lower count of the bacteria in the water column, implying that the microbes have settled to the bottom of the tank. The dominant species was *Sulfovorum* sp., a chemolithoautotroph that has been reported to be a sulfur oxidizing bacterium. The remaining bacteria are common environmental bacteria. One pathogen, *Legionella*

Table 1 Microbial profile at the intake, biological tanks (process treatment) and effluent at STP Kuala Penyu indicating the species with the highest count

Intake	Biological tank	Effluent
Acetobacter sp. *Desulforomoas thiophilia* *Acetoanerobium sticklandii* *Sulfuricurvum kujinese*	*Sulfovorum* sp. XGS-02 *Arcobacter* sp. FWKO B *Desulforomoas thiophilia*	*Sulfovorum* sp. XGS-02 *Cornebacterium canis* *Casimicrobium huifangae* *Megamonas funiformis* *Paludibaculum fermentans*

oakridgenesis, that has been previously reported as present in groundwater (Costa et al., 2005). A cellulose degrading bacterium *Maribellus comscasis* was also detected is anaerobic in nature and marine in origin as documented by Zeng et al. (2021) and its presence in the influent may have been the consequence of the location of the STP in the vicinity of a mangrove swamp connected to the Kuala Penyu bay. The treated effluent was dominated by the well documented bacterium *Casimicrobium huifangae*, that is associated with treated sewage (Song et al., 2020). The presence of a high count of *Megamonas funiformis*, which was not present in the influent, is an indication that this fecal bacterium (Sakon et al., 2008) has established residence in the effluent, and this will require interventions such as increasing the aerobic capacity of the system. The count of *Paludibaculum fermentans* is indicative of conditions that are similar to that associated with littoral regions of lakes (Kulichevskya et al., 2014). The STP in Kuala Penyu is the only STP which hosts covered reactor tanks, and the lack of sunlight is probably one of the factors which contributes to its distinct microbial profile.

3.2 Centralized Sewage Treatment Plant Kundasang

The sewage treatment facility is in the highlands at an elevation of 1.900 m AMSL and the average temperature of 19.7 °C is likely to be one of the factors which contributes to variations in the bacterial population. The influent is derived from the town sewage systems. The influent is subjected to sedimentation prior to the transfer using pumps to two biological treatment treatment tanks in batches. The influent is aerated for cycles of 2 h, 1 h of settling (sedimentation) and followed by decanting and release (Table 2).

The influent comprised of bacteria associated with both animal and human gut microbiome, indicating that the raw sewage is a composite of animal and human waste. The biochemical analysis indicated that the pH was neutral, and the BOD and COD were both high due to the presence of organic matter. The presence of Ammoniacal Nitrogen is an indication that the raw influent has remained under anaerobic conditions for an extended period resulting in the release of Ammonia. Dissolved

Table 2 The dominant bacteria detected at the intake, biological tanks and effluent at STP Kundasang

Intake	Biological tank	Effluent
Megamonas funiformis *Anaerobutyricum hallii* *Agathobacter rectalis* *Ligilactobacillus ruminis*	*Megamonas funiformis* *Candidatus Megaera* *Chromulinavorax destructans* *Casimicrobium huifangae* *Succinivibrio* *Dextrinosolvens* *Candidatus bodocaedibacter*	*Casimicrobium huifangae* *Sulfurovum sp.*

Ammonia increases the pH of raw sewage. The reactor tanks 1 and 2 comprise different bacterial populations even though they are fed from the same source of influent. This is indicative of each tank functioning as a unique ecosystem. The count of faecal bacteria was reduced in tank 1, implying that the aerobic treatment is effective. Interestingly, tank 2 was more effective at reducing the faecal microbial load and this may be attributed to difference in the methodology used for treatment, with tank 1 comprising intermittent cycles of aeration and sedimentation and tank 2 subjected to continuous aeration for six hours, followed by decantation and release into the settling chamber. The treated effluent from both Tank 1 and Tank 2 was distinct regarding its microbial composition. This indicates that these two tanks support different microbial communities which may be the direct result of adopting two different aerobic procedures for the two tanks. The final effluent had a pH of 4.40 and this is likely the result of the biological activity of Sulfurovum sp. which has been reported to oxidize Sulfur and reduce nitrate. Highest number of *Aliarcobacter* recorded in KUN2 (7%) followed by KUN3 (5%) and KUN5 (2%). *Aliarcobacter* is commonly found in wastewater and is recognized as an emerging zoonotic and waterborne pathogen. This genus thrives in various aquatic environments, with wastewater being significant reservoir due to its high organic nutrient load. Studies indicate a high detection rate of *Aliarcobacter* in wastewater treatment plants (WWTPs), suggesting these facilities serve as hotspots for the bacteria. Both influent and effluent wastewater samples often show the presence of *Aliarcobacter*, indicating itstolerance to treatment processes and potential for environmental dissemination. Interestingly, 17% of Sulfurovum was noticed in KUN5, a species that can oxidize sulfur compounds in wastewater, contributing to sulfur removal and the reduction of hydrogen sulfide, a toxic and malodorous gas common in sewage systems. Their presence in wastewater indicates active sulfur and nitrogen cycling and highlights their importance in maintaining microbial ecosystem balance. The presence of high number of Lysobacter in KUN2 (5%), KUN3 (11%) and KUN4 (6%) is a good indication as they are primarily recognized for their ability to produce bioactive compounds, including enzymes and antibiotics, which contribute to organic matter breakdown and pathogen control. They are commonly associated with activated sludge systems due to their adaptive strategies for surviving in competitive microbial communities. Similarly, presence of Dyella species in KUN2 (3%), KUN3 (5%) and KUN4 (5%) can break down other organic compounds, including biphenyls, and could be useful in treating wastewater containing such substances.

Among all the samples, a high number of bacterial diversities was observed in KUN2 followed by KUN4 and KUN1 values above 3 are common communities with many species that are relatively evenly distributed. A higher H suggests both high richness (many species) and good evenness (species are relatively equally abundant). A value close to 1 (like 0.905) under Simpson index indicates a highly diverse community where no single species dominates.

Table 3 Microbial profile at the intake, processing tanks and effluent at STP Tenom

Intake	Biological tank	Effluent
Megamonas funiformis *Acetoanaerobium sticklandii* *Phascolarctobacterium faecium* *Megasphaera elsdenii*	*Megamonas funiformis* *Ligilactobacillus ruminis* *Sellimonas intestinalis* *Aquirufa nivalisilvae* *Megamonas funiformis* *Roseomonas fluvialis* *Enhydrobacter sp.* *Dermatobacter hominis*	*Aquirufa nivalisilvae* *Succinivibrio dextrinosolvens* *Megamonas funiformis*

3.3 Centralized Sewage Treatment Plant Tenom

The influent comprised human faecal bacteria with a lower count of animal ruminant bacteria implying that the sewage was derived primarily from human settlements. The sludge holding tank served a sedimentation chamber as was evinced by the lower count of microbes in the water column. The reactor tanks 1 and 2 exhibited different profiles of microbes, implying that different protocols or methods have been implemented at these two tanks. The treated effluent was found to have a marked reduction in the faecal bacterial count implying that the treatment in both the reactor tanks was effective in reducing their load. The increase in pH from 6.5 to 7.5 and the decrease in Ammoniacal Nitrogen from 25.9 to 20.7 is likely to be the result of the activity of the anaerobe *Roseomonas fluvialis* (Table 3).

3.4 Centralized Sewage Treatment Plant Kudat

The layout of the facility consists of an intake point, followed by retention tanks and two treatment tanks which operate under the following conditions: two hours of aeration and one hour of settlement, followed by decanting and release (Table 4).

The influent contained a very low count of microbes associated with human faeces a near neutral pH of 7.5, and Ammoniacal Nitrogen (13 mg/L), the COD and BOD were 185 and 86 (mg/L) respectively. The count reduced significantly in reactor

Table 4 Microbial profile at the intake, processing tanks and effluent at STP4 indicating the dominant Genera present in each of the samples

Intake	Biological Tank	Effluent
Megamonas funiformis *Succinivibrio dextrinosolvens*	*Candidatus* *Bodocaedibacter* *Vickermanii* *Megamonas funiformis* *Algoriphagus sanaruensis* *Cornybacterium canis*	*Algoriphagus sanaruensis* *Leptolyngbya boryana*

tank 2, indicating that it is functioning well below its stated capacity and effectively treating the influent. Reactor tanks 1 and 2 exhibited a difference in bacterial composition, implying differences in protocols or methods for treatment. The final effluent did not exhibit a significant change in pH, however the BOD and COD were reduced to 14 and 34 (mg/L) respectively. The Ammoniacal nitrogen increased to 19 mg/L. This can be attributed to the presence of untreated protein wastes in the effluent.

3.5 Regional Sewage Treatment Plant, Kota Kinabalu

The influent was dominated by human gut microbiota, implying that the major component consisted of sewage discharged from residential and commercial housing. The first phase of sewage treatment involved Demand Aeration (DA) in four individual tanks. Demand aeration led to a significant reduction in the human gut microflora and the tanks were dominated by aerobic bacteria associated with activated sludge. This finding implies that the sludge accumulated at the bottom of the demand aeration tanks is being actively recycled during the process of aeration. DAT 2 and DAT 3 exhibited a higher microbial count in terms of copy number as compared to DAT 1 and DAT 4. The tanks contained bacteria that are involved in the degradation of insoluble organic compounds as well as bacteria that are symbiotic with microalgae, thus implying that microalgae may be thriving in the DAT due to exposure to the sun. The second phase involved intermittent aeration, and this was associated with a decrease in the count of the bacterial population which can be attributed to settlement of the bacteria in the sludge at the bottom of the tank. The third phase involved decantation, and the microbial population was dominated by Fluvibacter sp., an aerobic bacterium associated with treated sewage. The low count indicates that almost all the nutrients have been depleted. The fourth stage involved release of the effluent into a flowing river. The final effluent was dominated by four species of microbes which have been reported to be associated with treated sewage.

According to the biochemical report provided by JPP, several key water quality parameters were analyzed for both the influent and the final discharge. The pH of the influent was recorded at 7.6, indicating a slightly alkaline condition, while the final discharge exhibited a lower pH of 6.80, suggesting a shift toward acidity post-treatment. The Biochemical Oxygen Demand (BOD) showed a notable reduction from 26.0 mg/L in the influent to 15.0 mg/L in the final discharge, reflecting a decrease in biodegradable organic matter. Similarly, the Chemical Oxygen Demand (COD) decreased significantly from 270.0 mg/L to 49.0 mg/L, indicating effective removal of both biodegradable and non-biodegradable organic compounds. Total Suspended Solids (TSS) were also reduced from 72.0 mg/L in the influent to 26.0 mg/L in the final discharge, demonstrating the system's efficiency in removing particulate matter. However, an increase in Ammonia Nitrogen was observed, rising from 12.2 mg/L in the influent to 22.4 mg/L in the final discharge, which may suggest limitations in the nitrification process or potential release from organic nitrogen sources during treatment (Table 5).

Table 5 Microbial profile at the intake, processing tanks and effluent at RSTP, Kota Kinabalu

Influent	Biological tank	Effluent
Megamonas funiformis	*Megamonas funiformis*	*Enhydrobacter sp.,*
Phascolarctobacterium faecium	*Runella sp. SP2*	*Casimicrobium huifangae*
Acetoanaerobium sticklandii	*Candidatus megaera*	*Haliscomenobacter hydrossis*
Lachnospira eligens	*Casimicrobium huifangae*	*Urbifossiella limbaea*
Agathobacter rectalis	*Roseomonas fluvialis*	
Anaerobutyricum hallii	*Candidatus Bodocaedibacter*	

4 Conclusions

The study compared bacterial diversity at five centralized sewage treatment plants (STPs) in Kuala Penyu, Kundasang, Tenom, and Kudat, highlighting the effects of effluent treatment on microbial diversity. In Kuala Penyu, the influent was dominated by *Acetobacter* sp. and *Desulforomoas thiophilia*, indicating anoxic conditions of the incoming wastewater, while the effluent showed a reduction in *D. thiophilia* and high counts of *Casimicrobium huifangae* and *Megamonas funiformis*, suggesting that the major constituent of the influent was of human fecal origin. Kundasang's influent contained gut-associated bacteria from both humans and animals, with the effluent showing effective reduction of faecal bacteria but maintaining sulfur oxidizers like *Sulfurovum* sp. Tenom's influent was primarily composed of human faecal bacteria, with the effluent demonstrating a marked reduction in these bacteria and improved pH and ammoniacal nitrogen levels due to the activity of *Roseomonas fluvialis*. Kudat's influent had low human-associated microbes, and the effluent showed effective reduction of organic matter. There was a significant improvement in the quality of the treated effluent at the five STPs indicating that the overall microbiological ecosystems have been well established and functioning within the scope of the conditions under which this study was conducted. To improve effluent treatment, the study suggests enhancing aeration, optimizing treatment protocols, regular microbial and biochemical monitoring, and incorporating advanced treatment technologies. These measures aim to improve the microbial diversity and overall treatment efficiency, leading to better environmental and public health outcomes.

Acknowledgements The authors wish to acknowledge the assistance provided by the technical staff at each of the Centralized Sewage Treatment Plants located at Kudat, Kuala Penyu, Kundasang and Tenom.

References

Alhazmi, S. M., BaniMustafa, A. A., Alindonosi, A. R., & Almutairi, A. F. (2024). Metagenomic meta-analysis of antibiotic-resistance genes in wastewater: A perspective from the covid-19 pandemic. *Water, 16*(24), 3571.

Bukin, Y. S., Galachyants, Y. P., Morozov, I. V., Bukin, S. V., Zakharenko, A. S., & Zemskaya, T. I. (2019). The effect of 16S rRNA region choice on bacterial community metabarcoding results. *Scientific Data, 6*(1), 1–14.

Costa, J., Tiago, I., da Costa, M. S., & Veríssimo, A. (2005). Presence and persistence of Legionella spp. in groundwater. *Applied and Environmental Microbiology, 71*(2), 663–671. https://doi.org/10.1128/AEM.71.2.663-671.2005

Finster, K., Coates, J. D., Liesack, W., & Pfennig, N. (1997). *Desulfuromonas thiophila* sp. nov., a new obligately sulfur-reducing bacterium from anoxic freshwater sediment. *International Journal of Systematic Bacteriology, 47*(3), 754–758. https://doi.org/10.1099/00207713-47-3-754

Huang, S., Fu, Y., Zhang, H., Wang, C., Zou, C., & Lu, X. (2023). Research progress of novel bio-denitrification technology in deep wastewater treatment. *Frontiers in Microbiology, 14*, 1284369.

Kulichevskaya, I. S., Suzina, N. E., Rijpstra, W. I. C., Damsté, J. S. S., & Dedysh, S. N. (2014). *Paludibaculum fermentans* gen. nov., sp. nov., a facultative anaerobe capable of dissimilatory iron reduction from subdivision 3 of the Acidobacteria. *International Journal of Systematic and Evolutionary Microbiology, 64*(Pt 8), 2857–2864. https://doi.org/10.1099/ijs.0.066175-0

Li, J., Cao, J., Zhu, Y. G., Chen, Q. L., Shen, F., Wu, Y., Xu, S., Fan, H., Da, G., Huang, R. J., & Wang, J. (2018). Global survey of antibiotic resistance genes in air. *Environmental Science & Technology, 52*(19), 10975–10984.

Mechikoff, M. A., Collins, J. P., Golder, P., Ingersoll, C. M., McGarry, R. E., Lu, X. J., & Balboni, A. L. et al. (2024). Application of pan-viral metagenomic sequencing on United States air force academy wastewater to uncover potential causes of acute gastroenteritis. *Military Medicine, usae518*.

Mkilima, T. (2025). Wastewater as a sentinel for emerging antimicrobial resistance threats: A review. *Environmental Quality Management, 35*(1), Article e70115.

Motteran, F., de Oliveira Silva, B. K., de Lima Costa, J., da Silva Júnior, L. P., Campos, H. C. G., Tatsumi, C., & Kato, M. T. et al. (2025). Microbial diversity, dynamics, and metabolic analysis of microorganisms present in wastewater treatment plants in urban areas. *Brazilian Journal of Chemical Engineering*, 1–21.

Ramos, S., Júnior, E., Alegria, O., Vieira, E., Patroca, S., Cecília, A., & Nunes, A., et al. (2024). Metagenomics insights into bacterial diversity and antibiotic resistome of the sewage in the city of Belém, Pará, Brazil. *Frontiers in Microbiology, 15*, 1466353.

Sakon, H., Nagai, F., Morotomi, M., & Tanaka, R. (2008). Sutterella parvirubra sp. nov. and Megamonas funiformis sp. nov., isolated from human faeces. *International Journal of Systematic and Evolutionary Microbiology, 58*(Pt 4), 970–975. https://doi.org/10.1099/ijs.0.65456-0

Song, Y., Jiang, C. Y., Liang, Z. L., Wang, B. J., Jiang, Y., Yin, Y., Zhu, H. Z., Qin, Y. L., Cheng, R. X., Liu, Z. P., Liu, Y., Jin, T., Corvini, P. F., Rabaey, K., Wang, A. J., & Liu, S. J. (2020). *Casimicrobium huifangae* gen. nov., sp. nov., a ubiquitous "most-wanted" core bacterial taxon from municipal wastewater treatment plants. *Applied and Environmental Microbiology, 86*(4), e02209–e02219. https://doi.org/10.1128/AEM.02209-19

Tarek, M. H., Hubbart, J. A., & Garner, E. (2024). Tracking sources and dissemination of indicator antibiotic resistance genes at a watershed scale. *ACS ES&T Water, 4*(2), 399–412.

Vimala, R. T. V. (2019). Production and flocculating performance of bioflocculant by bacterial strain and its application for municipal wastewater treatment. *Journal of Pure & Applied Microbiology, 13*(3).

Zheng, R., Cai, R., Liu, R., Liu, G., & Sun, C. (2021). Maribellus comscasis sp. nov., a novel deep-sea Bacteroidetes bacterium, possessing a prominent capability of degrading cellulose. *Environmental Microbiology, 23*(8), 4561–4575. https://doi.org/10.1111/1462-2920.15650

Conclusion and Future Perspectives

Abu Zahrim Yaser

Abstract This chapter synthesizes the core findings and methodologies explored throughout this volume, providing a definitive summary of current best practices in compost and liquid fertilizer.

This chapter synthesizes the core findings and methodologies explored throughout this volume, providing a definitive summary of current best practices in compost and liquid fertilizer production. We look forward, charting the trajectory of innovation by examining cutting-edge research, emerging applications, and their potential for fostering sustainable systems globally. The future of waste management and nutrient cycling is poised for significant transformation, driven by technological integration and a renewed focus on ecological harmony. The immediate future of composting is being shaped by advanced technologies that promise to increase efficiency, reduce processing times, and maximize nutrient recovery. These innovations range from nanoscale applications to digital intelligence, fundamentally altering how we manage organic waste streams. The following paragraphs detail these exciting new frontiers in an industry previously dominated by traditional methods.

Rapid composting technologies focus on drastically shortening the time required to produce mature compost, typically through hyperthermophilic composting, enhanced aeration, mechanical agitation, and optimized microbial activity. These methods are crucial for handling large volumes of organic waste efficiently in urban settings where space and time are limiting factors. Research in this area seeks to reduce bulking agent as well as to balance speed with product quality, ensuring that the rapid process still yields a safe and beneficial end-product. The application of cutting-edge enzyme technology in composting can significantly accelerate the degradation of complex organic materials, such as cellulose and lignin. This targeted enzymatic breakdown improves overall process efficiency and potentially reduces

A. Z. Yaser (✉)
Universiti Malaysia Sabah, Jalan UMS, Sabah, Malaysia
e-mail: zahrim@ums.edu.my

A. Z. Yaser et al. (eds.), *Compost and Liquid Organic Fertilizers*,
https://doi.org/10.1007/978-981-95-8561-8_24

composting time frames. Advanced research is focused on identifying and engineering novel enzymes such as fruit-jaggery-effective microorganism solution; that function optimally under specific, industrial composting conditions.

Artificial Intelligence (AI) and Machine Learning (ML) are being integrated into composting facilities to optimize operational parameters such as temperature, moisture content, and aeration in real-time. These smart systems enable predictive modelling for process control, ensuring consistent product quality and maximizing efficiency. The application of AI reduces human error and energy consumption, marking a significant leap toward "smart" waste management.

Integrating anaerobic digestion (AD) with composting allows for a two-stage process that maximizes resource recovery: biogas for energy and nutrient-rich digestate for composting. This synergistic approach optimizes energy generation while producing a stable organic fertilizer, closing nutrient and energy loops simultaneously. The synergistic relationship between microbes (composting) and earthworms (vermicomposting) can enhance the efficiency and nutrient quality of the final product. This integrated approach, often referred to as vermicomposting, improves soil structure, microbial biodiversity, and nutrient availability. Research explores optimizing the balance of these organisms to treat different waste types more effectively than either method alone.

Composting sewage and industrial sludge effectively manages a significant urban waste problem, stabilizing potentially hazardous materials into a usable soil amendment. Key challenges involve managing heavy metal concentrations and ensuring the complete eradication of pathogens during the thermophilic phase. Research is focused on developing safe and compliant protocols that allow for the beneficial reuse of this abundant resource. Novel composting techniques are being developed to safely treat and degrade certain organic industrial byproducts using specialized microbial communities under controlled conditions. This innovative approach offers an environmentally friendly alternative to traditional disposal methods for specific waste types. Strict monitoring and regulation are essential to ensure the complete transformation of these materials and the safety of the final product. Innovations in the synthesis of biodegradable polymers derived from organic waste streams present promising solutions for sustainable packaging and material use. The focus then shifts to understanding the optimal conditions for their degradation within composting environments, ensuring a closed-loop lifecycle.

The impact of composting extends far beyond simply producing agricultural nutrients; its potential intersects with various scientific disciplines and industrial needs. Beyond its traditional role as a soil amendment, compost exhibits potential for use in a variety of non-agricultural applications, such as in biofiltration systems and erosion control. Its rich organic matter and microbial content can contribute to natural solutions for various environmental challenges. This section explores these alternative uses, highlighting the broad utility of compost in sustainable infrastructure development.

Compost leachate (CL) contains essential macro and micronutrients, making it a valuable, cost-effective organic liquid fertilizer alternative to commercial fertilizers. However, raw leachate often contains high salt content and potential phytotoxic substances, requiring careful characterization and appropriate dilution rates before application. Developing standardized dilution guidelines based on the biowaste source is crucial to optimize plant nutrient uptake and prevent soil pollution. Nanotechnology offers the potential to revolutionize agriculture by enabling the controlled and targeted release of nutrients from organic sources like compost leachate. Nano-fertilizers, with their small size and high surface area, significantly enhance nutrient absorption efficiency by plants while minimizing environmental loss through leaching. Research into synthesis methods is ongoing to ensure these innovative fertilizers are cost-effective, environmentally friendly, and safe for long-term use.

While technology is crucial, the success of composting fundamentally relies on human engagement and societal integration. Understanding the social dynamics, cultural acceptance, and community benefits is key to replicating these successes worldwide.

The manufacturer's authorised representative in the EU is Springer Nature Customer Service Centre GmbH, Europaplatz 3, 69115 Heidelberg, Germany. If you have any concerns regarding our products, please contact ProductSafety@springernature.com

Printed and bound by CPI Group (UK) Ltd, Croydon, CR0 4YY
21/09/2026
02219986-0016